Springer Series in Statistics

Advisors:
P. Bickel, P. Diggle, S. Fienberg, K. Krickeberg,
I. Olkin, N. Wermuth, S. Zeger

Springer
New York
Berlin
Heidelberg
Barcelona
Hong Kong
London
Milan
Paris
Singapore
Tokyo

Springer Series in Statistics

P.P.B. Eggermont
V.N. LaRiccia

Maximum Penalized Likelihood Estimation

Volume I: Density Estimation

With 30 Figures

 Springer

P.P.B. Eggermont
Department of Food and Resource
 Economics
University of Delaware
Newark, DE 19716
USA
eggermon@udel.edu

V.N. LaRiccia
Department of Food and Resource
 Economics
University of Delaware
Newark, DE 19716
USA
lariccia@udel.edu

Library of Congress Cataloging-in-Publication Data
Eggermont, P.P.B. (Paulus Petrus Bernardus)
    Maximum penalized likelihood estimation / P.P.B. Eggermont, V.N. LaRiccia.
       p. cm. — (Springer series in statistics)
    Includes bibliographical references and index.
    Contents: v. 1. Density estimation
    ISBN 0-387-95268-3 (alk. paper)
      1. Estimation theory.   I. LaRiccia, V.N. (Vincent N.)   II. Title.   III. Series.
QA276.8 .E377 2001
519.5'44—dc21   2001020450

Printed on acid-free paper.

Production managed by Allan Abrams; manufacturing supervised by Jerome Basma.
Photocomposed copy prepared from the authors' $\mathcal{A}_{\mathcal{M}}\mathcal{S}$-TEX files.
Printed and bound by Maple-Vail Book Manufacturing Group, York, PA.
Printed in the United States of America.

9 8 7 6 5 4 3 2 1

ISBN 0-387-95268-3   SPIN 10798867

Springer-Verlag   New York Berlin Heidelberg
A member of BertelsmannSpringer Science+Business Media GmbH

To Jeanne, Theresa and Dave, and Tyler

To Cindy

Preface

This is Volume I of a proposed two volume text on the theory and practice of maximum likelihood estimation. It is intended for graduate students in statistics; applied, industrial, and engineering mathematics; and operations research, as well as for researchers and practitioners in the field. The present volume treats mostly nonparametric density estimation, whereas Volume II is concerned with indirect (or inverse) estimation problems. The material is divided into an introductory chapter and six parts:

> Parametric density estimation,
> Nonparametric density estimation,
> Convexity and optimization,
> Nonparametric least squares (well-posed and ill-posed),
> Generalized deconvolution problems with random sampling,
> Expectation-Maximization algorithms.

The first three parts constitute Volume I, and the others form Volume II. Each part consists of a number of theoretical chapters, in which the maximum (penalized) likelihood estimators for the problems under discussion are introduced and their asymptotic behavior thoroughly analyzed, as well as an "in action" chapter, in which computational issues are discussed and the small sample behavior of the estimators is demonstrated, using simulated and real data. The material is liberally sprinkled with exercises that cover modifications of the results presented and fill in missing details of the mathematical development, but sometimes deal with open problems. Computational projects are also indicated. We briefly discuss each part.

Parametric density estimation

Here, we treat the standard asymptotic theory (consistency, best asymptotic normality) of parametric maximum likelihood estimators in the regular case. Computational issues, such as Newton's method, the method of scores, and Expectation-Maximization (EM) algorithms are discussed. Additional topics include robustness, ridge regression, skewed heavy-tailed densities (log-normal, Gamma, Weibull), and mixtures of normals. Simulation studies illustrating the small sample behavior of maximum likelihood estimators and the effects of misspecification of the parametric model are given in a separate chapter.

Nonparametric density estimation

Under the usual nonparametric assumptions, we prove the almost sure bound of $\mathcal{O}(n^{-2/5})$ on the L^1 error of kernel density estimators, as well as the (best) asymptotic normality of an estimator of the entropy based on kernel estimators. The principal tools, discrete parameter submartingales and Devroye's exponential inequalities, are discussed in detail, but at an elementary level. The optimality of kernels is discussed (and later in the convexity part, optimal kernels of high order are computed).

In a chapter on maximum penalized likelihood estimation, we prove the a.s. bound of $\mathcal{O}(n^{-2/5})$ for the mple using Good's first roughness penalization functional (the Good estimator). We also consider the roughness penalization of the log-density. Maximum smoothed likelihood estimation for log-concave densities is discussed briefly, as is minimum (smoothed) distance estimation.

In the chapter on monotone and unimodal densities, we analyze in detail the pool-adjacent-violators algorithm for computing the Grenander estimator, the solution to the monotone maximum likelihood estimation problem. We show that the kernel-estimator-made-monotone always has smaller error than the kernel estimator itself, for any reasonable measure of the global error (all L^p norms, Kullback-Leibler, Hellinger), and that the Grenander operator is a contraction in this sense. We also consider unimodal density estimation, with just about the same conclusion.

We discuss smoothing parameter selection (mostly) from the L^1 perspective: Devroye's double kernel method and variations, the various plug-in methods, as well as a pilotless version of the Hall-Wand estimator. For some of these methods, we prove that the selected smoothing parameter H satisfies $H \asymp n^{-1/5}$ almost surely. A discrepancy principle is discussed for kernel estimation and for the Good estimator.

Simulation studies comparing the various estimators are given in a separate chapter. We also compare various kernels with themselves and with the Good estimator. (It turns out that the GOOD estimator is remarkably GOOD!)

Convexity and optimization

All of the estimators discussed, except the kernel estimator, are defined as solutions to convex minimization problems. In this part, we give a self-contained treatment of such problems, in particular the existence of the estimators is established. We also lay the groundwork for ill-posed indirect estimation and for the convergence of EM and EM-like algorithms for such problems. Various inequalities used in other parts of the text, that derive from convexity, are either proved or stated as exercises.

As the above description makes clear, no attempt was made to be encyclopedic. The problems and estimators considered, and the general approach to the questions involved, are mostly determined by the interests and backgrounds of the authors and their friends. Some alternatives (with references) are briefly mentioned when appropriate, and the interested reader should follow these up, despite our editorial comments.

Why a new text ?

Why a new text on statistical estimation, and why did it take on the present form ? Our interest in the field started in 1992 with maximum (smoothed) likelihood estimation for indirect estimation problems, more specifically, for nonparametric deconvolution based on independent, identically distributed data. At that time, the literature on this topic was limited, and did not appeal to us. Thus, the idea of writing a monograph on indirect estimation was born. It soon became clear that maximum penalized likelihood estimation (mple) for indirect problems is quite hard and that it was perhaps prudent to start with plain nonparametric density estimation. The standard mple method involved the roughness penalization proposed by GOOD (1971), but it turned out to have been in disuse. The method of choice was that of kernel estimation, so *that* had to be included. Here, there was already a huge and growing literature, of which we liked the often unquoted work of DEVROYE and coauthors, in particular, the seminal text, DEVROYE and GYÖRFI (1985). While attending the 1997 Symposium on Nonparametric Function Estimation in Montreal, the idea to include a chapter on maximum likelihood estimation for monotone and unimodal densities was born. With general nonparametric density estimation included, it seemed reasonable to include the general parametric case also. As far as mathematical background was concerned, a good deal of indirect estimation is naturally explained in the context of convex minimization problems. Somewhat surprisingly, even in the context of parametrics and (sub)martingales, convexity arguments are frequently and freely used. So it was decided to include a relatively elementary treatment of convexity and convex minimization problems.

During the preparation of this text it became clear that the main prerequisite for the text is an introductory course on finite-dimensional vector spaces (introductory functional analysis would be helpful, but is not necessary) and an acquaintance with probability theory, including the law of the iterated logarithm (no measure theory is required). Parts of the text have been used in various classes and seminars. Professors David Mason (University of Delaware) and Uwe Einmahl (Vrije Universiteit Brussel) have used

Chapter 2 in their courses. We have used the chapters on nonparametric density estimation and convexity at the Univeristy of Delaware. Early on, much of the material was tried out on students during informal seminars. We thank Chris Venaccio, Tim Loomer, Joe Collins, André Acusta, and Carmelita Perlitz for their patience. Paul Deheuvels, Luc Devroye, Alexander Goldenshluger, David Mason, and Andrei Zaitsev have read all or parts of the manuscript, and we thank them for their comments.

Acknowledgments

We started this project as members of the DEPARTMENT OF MATHEMATICAL SCIENCES and finished it as members of FOOD AND RESOURCE ECONOMICS. It is our pleasant duty to thank John Nye, Dean of the College of Agriculture, and Bobby Gempesaw, then Chair of Food and Resource Economics (now Associate Provost) for their vision and (repeated and ultimately successful) efforts to rescue the statistics program at the University of Delaware, and to thank the members of Food and Resource Economics for the welcome they extended to us.

As with any intellectual endeavor, we were influenced by many people, from our teachers to anonymous referees, and we thank them all. However, four individuals must be explicitly mentioned. First of all, we wish to thank Zuhair Nashed of our old department for listening and helping us with both mathematical and departmental issues: In our new department, we can only say we miss him. Next, we wish to thank Paul Deheuvels for his interest and encouragement in our project, especially very early on, and for his painstaking review of the next-to-last version of the manuscript. A special thanks goes to Luc Devroye for his many suggestions and interest in the text and for urging us on to finish, but more importantly, for the many discussions at the Crab Trap, which helped shape our basic view of the field. Finally, it is our privilege to thank David Mason. His nitpicking critique is (now) greatly appreciated, as are some suggestions that helped shape the text. We must also thank him for single-handedly keeping the Statistics and Probability Seminar alive and for giving it such colorful names as the *Industrial Applied Statistics Seminar* or the *Probability, Informatics and Statistics Seminar*. Without his efforts, it is doubtful whether there would have been anything left to move to our new home.

Newark, Delaware   *Paul Eggermont, Vince LaRiccia*
April 13, 2001

Contents

Notations, Acronyms, and Conventions

The numbering and referencing conventions are as follows. Items of importance, such as formulas, theorems, and exercises, are labeled as $(Y.X)$, with Y the current section number and X the current (consecutive) item number. The exceptions are tables and figures, which are independently labeled following the same system. A reference to Item $(Y.X)$ is to the item with number X in section Y of the current chapter. References to items outside the current chapter take the form $(Z.Y.X)$, with Z the chapter number and X and Y as before. References to the literature take the standard form AUTHOR (YEAR) or AUTHOR#1 and AUTHOR#2 (YEAR), and so on. The references are arranged in alphabetical order by the first author, and by year.

We tried to limit our use of acronyms to some very standard ones, as in the following list.

iid	independent, identically distributed.
rv	random variable.
m(p)le	maximum (penalized) likelihood estimation (or estimator).
pdf	probability density function.
cdf	(cumulative) distribution function.
EM	Expectation–Maximization.
a.s.	almost surely or, equivalently, with probability 1.

Some of the standard notations throughout the text are as follows.

$\mathbb{1}(x \in A)$	The indicator function of the set A.
$\mathbb{1}_A(x)$	Also the indicator function of the set A.
$\mathbb{1}(X \leqslant x)$	The indicator function of the event $\{X \leqslant x\}$.
$(x)_+$	The maximum of 0 and x (x a real-valued expression).
$\asymp,\ \asymp_{as}$	Asymptotic equivalence, and the almost sure version. See Definition (1.3.6).
$=_{as}$	Almost sure equality.
$\leqslant_{as}$	Almost surely less than or equal. Likewise for $\geqslant_{as}$.

$\preccurlyeq$
For symmetric matrices $A, B \in \mathbb{R}^{n \times n}$, we write $A \preccurlyeq B$ if $C = B - A$ is semi-positive definite, i.e., $x^\top C x \geqslant 0$ for all $x \in \mathbb{R}^n$.

$\mathcal{O}_P, o_P$
See Definition (1.3.4), and (1.3.5).

Opie-One
An Opie-One estimator θ_n for θ_o is such that
$$\sqrt{n}\,(\theta_n - \theta_o) = \mathcal{O}_P(1)\,.$$

$\int_{\mathbb{R}} f \, dG$
This is shorthand for $\int_{\mathbb{R}} f(x)\, dG(x)$. Similarly for $\int_{\mathbb{R}} f\, g$.

$\varphi * \psi$
The convolution of two functions, defined as
$$\varphi * \psi(x) = \int_{\mathbb{R}} \varphi(x - y)\, \psi(y)\, dy\,, \quad x \in \mathbb{R}\,.$$

$\varphi * d\Psi$
The convolution of a function φ and a distribution Ψ, defined as
$$\varphi * d\Psi(x) = \int_{\mathbb{R}} \varphi(x - y)\, d\Psi(y)\,, \quad x \in \mathbb{R}\,.$$

F_n
The empirical distribution function of the data $X_1, \cdots, X_n$.

A_h
Typically, $A_h(x) = h^{-1} A(h^{-1}x)$, and similarly for other capital symbols. For lowercase symbols, this does not usually apply.

$(A^m)_h$
Following the previous convention,
$$(A^m)_h(x) = h^{-1}\big\{\, A(h^{-1}x)\,\big\}^m\,.$$

$A_h * dF_n$
The kernel estimator with kernel A, written as the convolution of A_h and the empirical distribution function.

$\mathcal{A}_h dF_n$
A boundary kernel estimator; so it is not quite a convolution of a kernel A and the empirical distribution function.

f_h
This is typically the large sample asymptotic estimator under consideration. For kernel estimation, it is $A_h * f$ or $A_h * f_o$, where f or f_o is the "true" density.

f^{nh}
The estimator based on $X_1, X_2, \cdots, X_n$. In kernel estimation, it equals $A_h * dF_n$. But it also denotes the maximum penalized likelihood estimator, the GOOD (1971) mple, the log-penalization, and the monotone and unimodal estimator. The context *should* make it clear.

g_κ
The kernel obtained with fractional integration by parts. See § 4.3.

T
The one-sided exponential kernel or pdf, which is given by $T(x) = \exp(-|x|)\, \mathbb{1}(x \geqslant 0)$. But also the histogram operator, see § 8.2, in particular, (8.2.3).

$\mathfrak{B}$
The two-sided exponential kernel, $\mathfrak{B}(x) = \frac{1}{2}\exp(-|x|)$.

ϕ
The normal kernel or density $\phi(x) = (2\pi)^{-1/2}\exp(-\frac{1}{2}x^2)$.

ϕ_σ
The scaled normal $\phi_\sigma(x) = \sigma^{-1}\phi(\sigma^{-1}x)$. Sometimes we also use :

$\phi(\,\cdot\,;\mu,\sigma)$	$\phi(\,x\,;\mu,\sigma) = \phi_\sigma(\,x - \mu\,)$.
$\|\cdot\|_p$	The $L^p(\Omega)$ norm, $1 \leqslant p \leqslant \infty$. See (1.3.8).
$L^p(\Omega)$	Space of (equivalence classes of measurable) functions f on $\Omega \subset \mathbb{R}$ with $\|f\|_p < \infty$.
$W^{m,p}(\Omega)$	Sobolev space of all functions in $L^p(\Omega)$) with m-th derivative in $L^p(\Omega)$ also.
$\mathrm{H}(\varphi, \psi)$	The Hellinger distance. See (1.3.18).
$\mathrm{KL}(\varphi, \psi)$	The Kullback-Leibler distance. See (1.3.19).
$\mathrm{PHI}(\varphi, \psi)$	The Pearson's φ^2 distance. See (1.3.20).
$a \vee b$	The function with values $[\,a \vee b\,](x) = \max(\,a(x)\,,\,b(x)\,)$.
$a \wedge b$	The function with values $[\,a \wedge b\,](x) = \min(\,a(x)\,,\,b(x)\,)$.
$\mathrm{LCM}(F)$	The least concave majorant of the distribution F.
$\mathrm{lcm}(f)$	The function, continuous from the left, which is equal almost everywhere to the derivative of $\mathrm{LCM}(F)$, where F is the distribution corresponding to the density f.
$\mathbb{R}_+, \mathbb{R}_{++}$	$\mathbb{R}_{++} = (\,0\,,\,\infty)$ (0 not included), $\mathbb{R}_+ = \mathbb{R}_{++} \cup \{\,0\,\}$ (somewhat pedantic, but $\ldots$).
$\delta f(x; h)$	The Gateaux variation of f at x in the direction h. For differentiable functions, it is just $h\,f'(x)$. See § 10.1.
$\mathrm{var}_n(A; h)$	See (7.3.27).
$\mathrm{var}_o(A; h)$	See (7.4.3).
$\mathrm{var}_o(A; h; x)$	See (7.4.11).

1

Parametric and Nonparametric Estimation

1. Introduction

This text is concerned with statistical estimation, in particular, with the estimation of probability densities of random variables. Typically, one considers a sample $X_1, X_2, \cdots, X_n$ of independent, identically distributed random variables with common density f, and one seeks to estimate f from the data. Both the parametric and the nonparametric settings are treated, with the emphasis on the latter. It is perhaps useful to briefly discuss how statistical estimation is interpreted in this text. In rough outline, statistical estimation deals with the following issues.

(a) The construction of estimators.

(b) Quantifying the notion of accuracy of the estimator.

(c) Deriving bounds on the accuracy for a fixed "amount" n of data, and asymptotically for $n \to \infty$. The ultimate goal is to obtain the distribution of the accuracy, viewed as a random variable, again for fixed n or asymptotically.

(d) Quantifying notions of optimality of the estimator. One would like to know the limits on the accuracy of any estimator, and whether the estimators under consideration reach these limits. For parametric estimation, this is well understood, but for nonparametric problems, this is perhaps a bit ambitious.

(e) Settling the relevance or adequacy of the estimators for finite (small) n, given the model.

(f) Certifying the adequacy of the indicated model. The latter goes by the name of "goodness-of-fit testing", which falls outside the scope of this work.

(g) Finally, the actual computation of the estimators is of concern. Typically, statistical estimators are defined implicitly as solutions to nonlinear equations or to minimization problems. Effective ways to compute and/or approximate them must be determined.

One would classify items (a) through (d) as belonging to estimation *theory*. However, in (d), the "application", that is, the connection with the "real world", must be kept in mind. Parts (e) through (g) definitely belong to the realm of estimation *practice*, the application of estimation to actual problems. In (g), it is more or less assumed that the implicitly defined estimators exist and are unique, and that one has a way of resolving the nonuniqueness if the need arises. So estimation *theory* creeps back in.

To illustrate what we have in mind, let f_o be a univariate probability density function (pdf), with corresponding distribution function (cdf) F_o. The simplest *density estimation* problem is where the data

(1.1) $X_1, X_2, \cdots, X_n$ are independent, identically distributed (iid),
univariate random variables, with common density $f_o(x)$,

and we wish to estimate f_o . Here, n is commonly referred to as the sample size. It is customary to encode the data $X_1, X_2, \cdots, X_n$ in the *empirical distribution function* F_n, defined as

(1.2) $$F_n(x) = \frac{\#\{X_i : X_i \leqslant x\}}{n} \quad , \quad -\infty < x < \infty ,$$

the fraction of the observations not exceeding x. What is lost by doing so is the order in which the observations occurred, but assuming iid observations, this is irrelevant, since F_n is a *sufficient statistic* for F_o.

How one goes about estimating f_o is influenced by the availability (or lack thereof) of information on the "model" for f_o. In the *parametric* model, the pdf f_o is assumed to belong to a parametric family

(1.3) $$\mathfrak{F} \stackrel{\text{def}}{=} \{f(\cdot;\theta) : \theta \in \Theta\} ,$$

described by a (low-dimensional) parameter θ belonging to the set of all possible parameters Θ. Thus, there exists a $\theta_o \in \Theta$ such that

(1.4) $$f_o(x) = f(x;\theta_o) , \quad -\infty < x < \infty .$$

It is obligatory to mention the standard example for density estimation, viz. the family of normal densities, parametrized by $\theta = (\mu, \sigma)$,

(1.5) $$f(x;\theta) = \frac{1}{\sqrt{2\pi}\,\sigma} \exp\left(-\frac{(x-\mu)^2}{2\,\sigma^2}\right) , \quad -\infty < x < \infty .$$

In all parametric problems, the tacit assumption is that the parameter θ represents a single or perhaps a few (very few) real numbers. Thus, although the problem reduces to that of estimating these few real numbers, it should be kept in mind that the goal is still to estimate the density $f(\cdot;\theta_o)$. For now, we assume that θ represents a single real number.

The standard method for estimating θ_o, dating back all the way to FISHER (1922), is by *maximum likelihood estimation*. In this method, the estimator (assuming it exists) is any value of θ for which the likelihood is maximal. Under the model (1.1)–(1.3), this amounts to solving

$$(1.6) \qquad \begin{array}{ll} \text{maximize} & \prod_{i=1}^{n} f(X_i\,;\theta) \\[2mm] \text{subject to} & \theta \in \Theta\,, \end{array}$$

but it is more natural to consider the averaged logarithm of the likelihood, written as

$$(1.7) \qquad \tfrac{1}{n} \sum_{i=1}^{n} \log f(X_i\,;\theta) = \int_{\mathbb{R}} \log f(x;\theta)\, dF_n(x)\,.$$

The reason for this is that under reasonable conditions, the log-likelihood converges as $n \to \infty$

$$\int_{\mathbb{R}} \log f(x;\theta)\, dF_n(x) \longrightarrow \int_{\mathbb{R}} \log f(x;\theta)\, dF_o(x)\,, \qquad (\theta \text{ fixed})$$

almost surely, say, by the strong law of large numbers. Thus, the maximum likelihood estimation problem is equivalent to

$$(1.8) \qquad \begin{array}{ll} \text{minimize} & -\int_{\mathbb{R}} \log f(x;\theta)\, dF_n(x) \\[2mm] \text{subject to} & \theta \in \Theta\,, \end{array}$$

and we refer to (1.8) as such. The questions of interest are whether (1.8) has a unique solution θ_n, how it may be computed, and what can be said about the error $\theta_n - \theta_o$ or $f(\cdot\,;\theta_n) - f(\cdot\,;\theta_o)$. One reason for the popularity of maximum likelihood estimation is that under reasonable conditions,

$$(1.9) \qquad \sqrt{n}\,(\theta_n - \theta_o) \longrightarrow_{\mathrm{d}} Y \sim N(0,\sigma^2)\,,$$

i.e., $\sqrt{n}\,(\theta_n - \theta_o)$ converges in distribution to a normally distributed random variable Y, with mean 0 and variance σ^2 given via

$$(1.10) \qquad \frac{1}{\sigma^2} = \int_{\mathbb{R}} \frac{|g(x;\theta_o)|^2}{f(x;\theta_o)}\, dx\,,$$

where $g(x;\theta)$ denotes the partial derivative of $f(x;\theta)$ with respect to θ. (For a precise definition of convergence in distribution, see § 3.) Moreover, with some technical caveats, the variance σ^2 in (1.9) is a lower bound for the variance of all *unbiased* estimators, which usually is expressed by saying that (1.10) is the *Cramér-Rao lower bound*.

There are other methods for estimating θ_o, such as the method of moments, the method of maximal spacings, and quantile regression. Mention must be made of least-squares estimation, especially in connection with nonparametric estimation later on (even though the authors have mixed feelings about it). The idea of least-squares estimation is that assuming f_o

were known, an ideal choice of θ would minimize

$$(1.11) \qquad \int_{\mathbb{R}} |f(x;\theta) - f_o(x)|^2 \, dx \; ,$$

the latter being equal to 0 if the model is correct. Because f_o is unknown, this method cannot be used. However, by rewriting (1.11) as

$$(1.12) \qquad \int_{\mathbb{R}} |f(x;\theta)|^2 \, dx - 2 \int_{\mathbb{R}} f(x;\theta) \, dF_o(x) + \int_{\mathbb{R}} |f_o(x)|^2 \, dx \; ,$$

we see that the first term can be computed exactly, and that the last term is independent of θ. The second term may be estimated by

$$-2 \int_{\mathbb{R}} f(x;\theta) \, dF_n(x) \; ,$$

and so the *least-squares* estimator of the parameter θ_o is the solution to

$$(1.13) \qquad \begin{aligned} \text{minimize} \quad & -2 \int_{\mathbb{R}} f(x;\theta) \, dF_n(x) + \int_{\mathbb{R}} |f(x;\theta)|^2 \, dx \\ \text{subject to} \quad & \theta \in \Theta \; . \end{aligned}$$

Of course, the questions that concerned us for maximum likelihood estimation apply here as well. In particular, under reasonable conditions, the analogue of (1.9) holds, with σ^2 given by (hold on to your hat)

$$(1.14) \qquad \sigma^2 = \frac{\displaystyle\int_{\mathbb{R}} |g(x;\theta_o) - E|^2 \, f(x;\theta_o) \, dx}{\left\{ \displaystyle\int_{\mathbb{R}} |g(x;\theta_o)|^2 \, dx \right\}^2} \; ,$$

where $g(x;\theta)$ is still the partial derivative of $f(x;\theta)$ with respect to θ, and

$$E = \int_{\mathbb{R}} g(x;\theta_o) \, f(x;\theta_o) \, dx \; .$$

Apart from the fact that this is *ugly* compared with the maximum likelihood case, it is also worse, in that the last σ^2 is larger than the one in (1.10). (It has to be if the Cramér–Rao lower bound deserves its name.)

This concludes the introductory description of parametric estimation. It is treated in much greater detail in Part I.

If for whatever reason a parametric model for f_o is not forthcoming, we are dealing with what is customarily referred to as a *nonparametric* estimation problem. Another way of saying this is that one wishes to make as few assumptions as possible about the density f_o in (1.1). What does this actually mean ? Without any assumptions, the density f_o is just a nonnegative, integrable function with integral equal to 1. Thus, it is an infinite-dimensional object, and infinitely many parameters are required to describe f_o. By way of example, an approximate way to describe f_o is by means of the probabilities $p_i = \mathbb{P}[\, X_1 \in (\, i\,h\, , (i+1)h\,]\,]$,

(1.15) $$p_i = \int_{i\,h}^{(i+1)\,h} f_o(x)\, dx \ , \quad i = 0, \pm1, \pm2, \cdots \ ,$$

where h is a small, positive number. It is clear that a finite amount of data, as in (1.1), will not suffice to determine the infinitely many p_i, let alone f_o. Thus, even in the nonparametric setting, some assumptions are needed, with the purpose of reducing the infinite number of parameters to (almost) finitely many. The necessity of some assumptions also surfaces when considering maximum likelihood or least-squares estimation, that is, (1.8) or (1.13). The maximum likelihood problem

(1.16)
$$\text{minimize} \quad -\tfrac{1}{n} \sum_{i=1}^{n} \log f(X_i)$$
$$\text{subject to} \quad f \text{ is a continuous pdf}$$

has no solution. Loosely speaking, the "solution" of (1.16) would be a sum of point masses at the observations

(1.17) $$f^n(x) = \tfrac{1}{n} \sum_{i=1}^{n} \delta(\, x - X_i\,) \ , \quad -\infty < x < \infty \ ,$$

where $\delta(x)$ is the unit mass at 0, but this is not a density. Note that the distribution corresponding to f^n is F_n, the empirical distribution function.

What can be done about (1.16) not having a solution? One approach is to replace the continuity assumption on f in (1.16) by the requirement that f be constant on the intervals $(\, ih\, , (i+1)\,h\,]$, $i = 0, \pm1, \pm2, \cdots$. The maximum likelihood solution for f is then given by a histogram as in (1.15), see THOMPSON and TAPIA (1990), although the histogram estimator has the inconvenience of being discontinuous. An alternative is to "spread out" the point masses a bit to obtain

(1.18) $$f^{nh}(x) = \tfrac{1}{n} \sum_{i=1}^{n} A_h(\, x - X_i\,) \ , \quad -\infty < x < \infty \ ,$$

where $A_h(x) = h^{-1}A(h^{-1}x)$ for a nice bell-shaped density A, say, with zero mean and finite variance. By choosing A as the uniform density on $[-\tfrac{1}{2}, \tfrac{1}{2}]$, the kernel estimator comes close to a histogram. Note that f^{nh} may also be written as

(1.19) $$f^{nh}(x) = A_h * dF_n(x) \overset{\text{def}}{=} \int_{\mathbb{R}} A_h(x - y)\, dF_n(y) \ .$$

We use either notation throughout. The estimator (1.18) is referred to as the kernel estimator with "kernel" A and smoothing parameter h. (The standard symbol for the kernel is K, which we wish to reserve for something else, see § 2 and Volume II.) Kernel estimators date back to AKAIKE (1954), ROSENBLATT (1956), WHITTLE (1958), PARZEN (1962), and WATSON and LEADBETTER (1963).

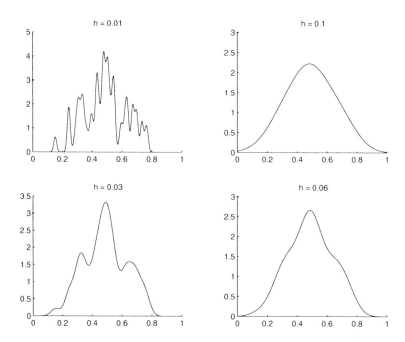

Figure 1.1. Kernel density estimators for the Buffalo snow fall data (divided by 1.3 times the largest observed annual snowfall), with the Gaussian kernel, and various values of the smoothing parameter h.

Based on the above motivation, one would not guess that kernel estimators can be any good. In fact, they are, *provided* the smoothing parameter is chosen properly, as we show in Part II. A preview is given in Figure 1.1, in which estimators for the density governing the annual snow fall in Buffalo, New York, are given, for various values of h, and with A the standard normal density. (For the data, see Appendix 1.) It is clear that the top two estimators in Figure 1.1 cannot be "right". For $h = 0.01$, the density seems much too rough (the estimator is undersmoothed), and for $h = 0.1$, the density is much too smooth and all detail is washed out (the estimator is oversmoothed). The two estimators at the bottom of Figure 1.1 seem reasonable, but therein lies the problem. Which one of these two reasonable estimators is closest to the truth ? There is obviously a need for *rational* procedures for selecting the smoothing parameter h, that is, the h selected should depend only on the data. We devote two chapters in Part II to this, one on the theory, and the other on the practical performance of smoothing parameter selection procedures.

Returning to the maximum likelihood problem (1.16), we ask the question of how one can incorporate the information that the solution of (1.16) should be a pdf, or a smooth pdf. One way to do this is by the method of sieves, about which more is said in §5. The oldest and the authors' favorite method is to consider maximum penalized likelihood estimation, that is, to add a roughness penalization term to the negative log-likelihood. Thus, the problem (1.16) is replaced by

(1.20)
$$\text{minimize} \quad - \int_{\mathbb{R}} \log f(x) \, dF_n(x) + h^2 \, R(f)$$

$$\text{subject to} \quad f \text{ is a continuous density} ,$$

where R is the roughness penalization term and h is the smoothing parameter, analogous to its use in kernel density estimation. The authors have a soft spot for the choice of I.J. GOOD (1971)

(1.21)
$$R(f) = \int_{-\infty}^{\infty} \left| \frac{d}{dx} \sqrt{f(x)} \right|^2 dx ,$$

with $R(f) = +\infty$ when the derivative of $\sqrt{f}$ is not square integrable on the line. This leads to the estimator f, implicitly defined by $\varphi = \sqrt{f}$ with

(1.22)
$$\varphi(x) = \frac{1}{n} \sum_{i=1}^{n} \frac{\mathcal{B}_h(x - X_i)}{\varphi(X_i)} , \quad x \in \mathbb{R} ,$$

where $\mathcal{B}_h(x) = (2h)^{-1} \exp(h^{-1}|x|)$ is the scaled, two-sided exponential kernel. This estimator turns out to be a remarkably GOOD estimator, both in theory and practice (provided h is chosen properly).

In view of (1.16)–(1.20), one may likewise define a penalized version of the least-squares problem (1.13) by seeking to

(1.23)
$$\text{minimize} \quad - 2 \int_{\mathbb{R}} f(x) \, dF_n(x) + \int_{\mathbb{R}} |f(x)|^2 \, dx + h^2 \, R(f)$$

$$\text{subject to} \quad f \text{ is a continuous pdf} ,$$

with $R(f)$ the roughness penalty and h the smoothing parameter, as before. Now, the choice

(1.24)
$$R(f) = \int_{\mathbb{R}} |f'(x)|^2 \, dx ,$$

with $R(f) = +\infty$ when f' is not square integrable, is intriguing. With this penalization, the solution f of (1.23) satisfies the boundary value problem

(1.25)
$$- h^2 f'' + f = dF_n(x) , \quad -\infty < x < \infty ,$$
$$f(x) \longrightarrow 0 \quad \text{for } |x| \to \infty ,$$

and is given by the pdf

(1.26)
$$f(x) = \mathcal{B}_h * dF_n(x) , \quad -\infty < x < \infty .$$

See Exercise (11.3.25). In effect, $\mathcal{B}_h$ is the *Green's function* for the boundary value problem (1.25), see, e.g., COURANT and HILBERT (1953). Thus, in this rather natural way, we get a particular kernel estimator, and it seems but a small leap of faith to assume that one can get other kernel estimators by appropriate penalization in (1.23). So, it seems natural to view kernel estimators as least-squares estimators. There is a way to get kernel estimators via maximum likelihood, viz. as the solution to the maximum *smoothed* likelihood problem

$$\text{minimize} \quad -\int_{\mathbb{R}} [\, A_h * \log f \,](x)\, dF_n(x)$$

(1.27)

$$\text{subject to} \quad f \text{ is a pdf},$$

but at first glance, this seems less natural. That it is in fact *very* natural is explained in detail in the next section.

Why are we so concerned with interpreting kernel density estimators as solutions to minimization problems? The question is simply this. While kernel estimators are without equal for the standard density estimation problem (1.1), how does one generalize them to more complicated settings? One important example is that of indirect estimation problems, discussed in § 2 and in Volume II. Another example concerns the estimation of densities with an *a priori* known shape. Some standard shape restrictions are monotonicity and/or convexity of a density on $(0, \infty)$, and unimodality or log-concavity of densities on $(-\infty, \infty)$, or that the tails of f_o may be monotone or convex. By way of example, to estimate a smooth, unimodal density, one may consider the solution of the problem

$$\text{minimize} \quad -\int_{\mathbb{R}} [\, A_h * \log f \,](x)\, dF_n(x)$$

(1.28)

$$\text{subject to} \quad f \text{ is a unimodal density}.$$

In Chapter 6, we study this problem in great detail.

In the remainder of this introductory chapter, we continue to set the tone of the text by elaborating on the topics mentioned above and discussing to what extent they are to be or not to be covered.

2. Indirect problems, EM algorithms, kernel density estimation, and roughness penalization

Here, we discuss the simplest of indirect estimation problems. It is a bit of an advertisement for Volume II, but it also sheds an unexpected light on kernel density estimation and GOOD's roughness penalization.

Let $Y_1, Y_2, \cdots, Y_n$ be iid univariate random variables with common pdf f_o. The goal is still to estimate f_o, but now we are interested in the situation in which the Y_i are contaminated by iid noise $Z_1, Z_2, \cdots, Z_n$, independent of the Y_i. Thus, the (iid) observations are

(2.1)
$$X_i = Y_i + Z_i, \quad i = 1, 2, \cdots, n.$$

Assuming that the common distribution K of the Z_i is known, the common distribution of the X_i has density $g = \mathcal{K}f_o$, where for any pdf φ,

$$(2.2) \qquad \mathcal{K}\varphi(x) = \varphi * dK(x) = \int_{\mathbb{R}} \varphi(x-z)\, dK(z) , \qquad -\infty < x < \infty .$$

If K has a density k, then this may be rewritten as $\mathcal{K}\varphi = k * \varphi$, the convolution of k and φ,

$$(2.3) \qquad k * \varphi(x) = \int_{\mathbb{R}} k(x-y)\,\varphi(y)\, dy , \qquad -\infty < x < \infty .$$

Thus, the *nonparametric deconvolution problem* is to estimate f_o based on the iid observations $X_1, X_2, \cdots, X_n$ with common density $\mathcal{K}f_o$.

We are interested in maximum penalized likelihood estimation for the deconvolution problem, but the approach is driven by algorithmic considerations. With G_n denoting the empirical distribution function of $X_1, \cdots, X_n$ in (2.1), the maximum likelihood estimator of f_o would be any solution of

$$(2.4) \qquad \begin{aligned} &\text{minimize} \quad -\int_{\mathbb{R}} \log \mathcal{K}f(x)\, dG_n(x) \\ &\text{subject to} \quad f \text{ is a pdf} , \end{aligned}$$

but, of course, it is not so obvious that solutions exist. Be that as it may, one may compute reasonable estimators by the following algorithm, essentially due to SHEPP and VARDI (1982). Let f_1 be an initial guess for the pdf f_o, and define for $q \geqslant 1$,

$$(2.5) \qquad f_{q+1}(y) = f_q(y) \cdot \frac{1}{n} \sum_{i=1}^{n} \frac{k(X_i - y)}{k * f_q(X_i)} , \qquad y \in \mathbb{R} .$$

This is an example of an EM algorithm, see (2.4.26) and Volume II, and as such has many wonderful properties, e.g., the sequence $\{ f_q \}_{q \geqslant 1}$ converges to a solution of (2.4) provided one exists. Unfortunately, in the present case, the solution does *not* exist, and indeed, practical computations based on this algorithm show that the f_q at first get better, and then get less and less smooth as the iteration progresses. A cure was provided by SILVERMAN, JONES, WILSON, and NYCHKA (1990), who figured that if smoothness is the problem, one could solve it by adding a smoothing step to the EM algorithm. Thus, for A a smooth symmetric pdf, let $A_h(x) = h^{-1}A(h^{-1}x)$, and consider the EMS algorithm

$$(2.6) \qquad \begin{aligned} &\varphi_q(y) = f_q(y) \cdot \frac{1}{n} \sum_{i=1}^{n} \frac{k(X_i - y)}{k * f_q(X_i)} , \qquad y \in \mathbb{R} , \\ &f_{q+1} = A_h * \varphi_q , \end{aligned}$$

that is, one step of the EM algorithm, followed by a smoothing step [S]. This works quite well in practice, but the question is what is being computed. Also, the convergence of this algorithm is not clear. Order is restored if we implement the smoothing somewhat differently. Define a nonlinear smoothing operator associated with A_h by

(2.7) $\mathcal{N} f(x) = \exp\big([A_h * \{\log f\}](x) \big) , \quad x \in \mathbb{R} ,$

and consider the NEMS algorithm

$$\psi_q = \mathcal{N} f_q ,$$

(2.8) $\varphi_q(y) = \psi_q(y) \cdot \dfrac{1}{n} \displaystyle\sum_{i=1}^{n} \dfrac{k(X_i - y)}{k * \psi_q(X_i)} , \quad y \in \mathbb{R} ,$

$$f_{q+1} = A_h * \varphi_q ,$$

that is, one nonlinear smoothing step $[\,N\,]$, followed by one EM step, followed by another linear smoothing step $[\,S\,]$. The crucial feature is still that the f_q are smooth, due to the last smoothing step. In fact, from a practical point of view, there does not appear to be much of a difference between the NEMS and EMS algorithms. The theoretical advantage of the NEMS algorithm is that it is an EM algorithm for the maximum smoothed likelihood problem

(2.9)

$$\text{minimize} \quad L_{nh}(f) \overset{\text{def}}{=} -\int_{\mathbb{R}} \log \mathcal{K} \mathcal{N} f(x) \, dG_n(x)$$

$$\text{subject to} \quad f \text{ is a pdf} ,$$

which has a unique, continuous solution f^{nh}, and the NEMS iterates converge to it in a suitable sense.

What does all of this have to do with kernel density estimation ? Let us backtrack, and assume that the $Z_i = 0$ for all i. Then, G_n has the same meaning as F_n in (1.2), with $X_1, X_2, \cdots, X_n$ iid random variables with distribution function F_o. Moreover, the operator $\mathcal{K}$ in (2.2) is just the identity operator, i.e., $\mathcal{K}\varphi = \varphi$ for all φ, and the objective function in (2.9) reads as

(2.10) $-\displaystyle\int_{\mathbb{R}} \log \mathcal{N} f(x) \, dG_n(x) = -\int_{\mathbb{R}} [\, A_h * \log f \,](x) \, dG_n(x) .$

This is the plain density estimation problem, of course. Thus, the maximum smoothed likelihood density estimation problem is

(2.11)

$$\text{minimize} \quad \Lambda_{nh}(f) \overset{\text{def}}{=} -\int_{\mathbb{R}} [\, A_h * \log f \,](x) \, dG_n(x)$$

$$\text{subject to} \quad f \text{ is a pdf} .$$

We now show that the solution of (2.11), respectively (1.27), is given by the kernel density estimator $f = A_h * dG_n$, provided A is symmetric. First, rewrite $\Lambda_{nh}(f)$ as a repeated integral, and interchange the order of integration so

$$
\begin{aligned}
\Lambda_{nh}(f) &= -\int_{\mathbb{R}}\int_{\mathbb{R}} A_h(x - y) \log f(y)\, dy\, dG_n(x) \\
&= -\int_{\mathbb{R}} \left(\int_{\mathbb{R}} A_h(x - y)\, dG_n(x) \right) \log f(y)\, dy \\
&= -\int_{\mathbb{R}} [\, A_h * dG_n](y) \log f(y)\, dy\;, \\
&= -\int_{\mathbb{R}} f^{nh}(y) \log f(y)\, dy\;,
\end{aligned}
$$

(2.12)

where in the next to last line we used the fact that A is symmetric, so that $A_h(x - y) = A_h(y - x)$ for all x, y. For reasons that may not (yet) be entirely clear, it is useful to introduce the Kullback-Leibler distance (or divergence) between two nonnegative integrable functions φ and ψ

$$
\text{(2.13)} \qquad \text{KL}(\varphi, \psi) = \int_{\mathbb{R}} \left\{ \varphi(y) \log \frac{\varphi(y)}{\psi(y)} + \psi(y) - \varphi(y) \right\} dy\;.
$$

Note that for any $x \geqslant 0$, $y > 0$, and $t = x/y$,

$$
\text{(2.14)} \qquad x \log(x/y) + y - x = y\left(t \log t + 1 - t \right) \geqslant 0\;,
$$

with equality if and only if $x = y$. Thus, the integrand in (2.13) is nonnegative, and $\text{KL}(\varphi, \psi)$ is well defined if we admit the value $+\infty$.

Let $f^{nh} = A_h * dG_n$, and consider $\Lambda_{nh}(f) - \Lambda_{nh}(f^{nh})$. Then, from (2.12),

$$
\Lambda_{nh}(f) - \Lambda_{nh}(f^{nh}) = \int_{\mathbb{R}} f^{nh}(x) \log \frac{f^{nh}(x)}{f(x)}\, dx\;.
$$

Since f and f^{nh} are pdfs, we may add $f(x) - f^{nh}(x)$ to the integrand without changing the value of the integral, so that

$$
\text{(2.15)} \qquad \Lambda_{nh}(f) - \Lambda_{nh}(f^{nh}) = \text{KL}(f^{nh}, f) \geqslant 0\;,
$$

with equality if and only if $f = f^{nh}$. Thus, f^{nh} is the minimum of $\Lambda_{nh}(f)$ over all pdfs f.

This is a long road to get kernel estimators out of maximum smoothed likelihood estimation, but the authors think it is rather telling. Somewhat tongue-in-cheek one might say that kernel estimators are so good precisely because they are maximum (smoothed) likelihood estimators!

We return to the general deconvolution problem. Just as the GOOD maximum penalized likelihood estimator, see (1.22), performs quite well for plain density estimation, so does the maximum smoothed likelihood estimator for the deconvolution problem. As a matter of fact, from the point of view of L^1 error in the estimators, the NEMS estimator significantly outperforms the Fourier deconvolution kernel estimators. These kernel estimators may be thought of as penalized least-squares estimators, i.e., the solutions to problems of the form

$$
\text{(2.16)} \qquad \text{minimize} \quad \int_{\mathbb{R}} |\mathcal{K} f(x)|^2 \, dx - 2 \int_{\mathbb{R}} \mathcal{K} f(x) \, dG_n(x) + h^2 \, R(f)
$$

$$
\text{subject to} \quad f \text{ is a pdf} ,
$$

with suitable roughness penalization functionals R and smoothing parameter h. This is analogous to the situation for plain density estimation
(1.23). All of this is the main topic of Volume II. For those who cannot
wait, see, e.g., STEFANSKI and CARROLL (1990), DEY, RUYMGAART and
MAIR (1996) and GOLDENSHLUGER (2000) for the Fourier (L^2) estimators,
and EGGERMONT and LARICCIA (1997) for more references and for some
experimental comparisons.

We finish this section by relating the NEMS algorithm and its associated maximum *smoothed* likelihood problem (2.9) for the nonparametric
deconvolution problem to maximum *penalized* likelihood estimation, using
GOOD's roughness penalization. Suppose for the sake of argument that we
replaced the negative log-likelihood $L_{nh}(f)$ for the deconvolution problem
by a smoothed version

$$
\text{(2.17)} \qquad \widetilde{L_{nh}}(f) \overset{\text{def}}{=} - \int_{\mathbb{R}} A_h * dG_n(x) \, \log \mathcal{K} \mathcal{N} f(x) \, dx .
$$

Surely this would not make much of a difference. Since f and $A_h * dG_n$ are
pdfs, this new negative log-likelihood may be written as

$$
\text{(2.18)} \qquad \widetilde{L_{nh}}(f) = \text{KL}(\, A_h * dG_n \, , \, \mathcal{K} \mathcal{N} f \,) + R_h(f) ,
$$

with

$$
\text{(2.19)} \qquad R_h(f) = \int_{\mathbb{R}} f(x) - \mathcal{N} f(x) \, dx .
$$

Thus, we may view the smoothed maximum likelihood problem (2.9) as a
penalized version of the original problem (2.4), with penalization $R_h(f)$.
Now what kind of penalization is involved here ? We determine the behavior
of $R_h(f)$ for $A_h = \mathfrak{B}_h$, see (1.22). Formally, using the Green's function
property of $\mathfrak{B}$,

$$
\mathfrak{B}_h * \log f - \log f = h^2 \, (\mathfrak{B}_h * \log f)'' = h^2 \, \mathfrak{B}_h * (\log f)'' \approx h^2 \, (\log f)'' .
$$

It follows that for strictly positive f (dropping the argument y everywhere),

$$
\begin{aligned}
f - \mathcal{N} f &= f \left(1 - \exp(\mathfrak{B}_h * \log f - \log f) \right) \\
&= f \left(1 - \exp(h^2 \, \mathfrak{B}_h * (\log f)'') \right) \\
&\approx -h^2 \, f \left(\mathfrak{B}_h * (\log f)'' \right) \approx -h^2 \, f \, (\log f)'' .
\end{aligned}
$$

Then, after integration by parts,

$$
R_h(f) \approx -h^2 \int_{\mathbb{R}} f (\log f)'' = h^2 \int_{\mathbb{R}} f' \, (\log f)' ,
$$

and thus,

$$(2.20) \qquad R_h(f) \approx h^2 \int_{\mathbb{R}} \frac{|f'(y)|^2}{f(y)} \, dy \ .$$

Admittedly, this derivation is a bit suspect, but it is surprising that the GOOD penalization (1.21) pops up this way. A similar derivation leading to GOOD's roughness penalization may be done for any other reasonable smoother A.

3. Consistency of nonparametric estimators

One of the main themes of this text is the concern with consistency of nonparametric estimators and with convergence rates. Let us consider the kernel estimator f^{nh} of (1.11) for the nonparametric density estimation problem. The estimator f^{nh} is consistent if $f^{nh} \longrightarrow f_o$ in a suitable sense, for $n \to \infty$, and $h = h(n)$ properly chosen. A more precise way of saying that $f^{nh} \longrightarrow f_o$ is that

$$(3.1) \qquad \text{dist}(f^{nh}, f_o) \longrightarrow 0$$

for a suitable "distance" of f^{nh} to f_o. Of course, consistency being established, one would like to know how fast $\text{dist}(f^{nh}, f_o)$ tends to 0. Similar observations apply to parametric estimation. Regardless, one aspect of consistency is the choice of the distance function. Another one is the mode of convergence in (3.1). We elaborate this point first.

Let $\{a_n\}_n$ be a sequence of real-valued functions of the random variables $X_1, X_2, \cdots, X_n$, denoted as

$$a_n = a_n(X_1, X_2, \cdots, X_n) \ .$$

There are many ways to define the convergence of a_n, of which we need the following four.

(3.2) DEFINITION. *We say that* $\{a_n\}_n$ *converges*
(a) *in expectation to 0 if* $\lim_{n \to \infty} \mathbb{E}[\,|a_n|\,] = 0$;
(b) *in probability to 0 if* $\lim_{n \to \infty} \mathbb{P}[\,|a_n| > \varepsilon\,] = 0$ *for all* $\varepsilon > 0$;
(c) *almost surely to 0 if* $\mathbb{P}[\lim_{n \to \infty} |a_n| = 0] = 1$;
(d) *in distribution to a random variable* Y *at rate* b_n *if*
$$\lim_{n \to \infty} \mathbb{P}[\,a_n/b_n \leqslant y\,] = \mathbb{P}[\,Y \leqslant y\,]$$
at all continuity points y *of the distribution of* Y.

The notations for these four notions of convergence are

$$(3.3) \qquad a_n \longrightarrow_{\mathbb{E}} 0 \ , \quad a_n \longrightarrow_p 0 \ , \quad a_n \longrightarrow_{as} 0 \ , \quad a_n/b_n \longrightarrow_d X \ .$$

It is also useful to have the big Oh and little oh notation available.

(3.4) DEFINITION. *We say that*

(a) $a_n =_{as} \mathcal{O}(b_n)$ *if* $\displaystyle\limsup_{n\to\infty} \left| \frac{a_n}{b_n} \right| < \infty$ *almost surely;*

(b) $a_n =_{as} o(b_n)$ *if* $\displaystyle\limsup_{n\to\infty} \left| \frac{a_n}{b_n} \right| = 0$ *almost surely.*

There are in probability versions of these, denoted as

$$(3.5) \qquad\qquad a_n = \mathcal{O}_P(b_n) \quad \text{and} \quad a_n = o_P(b_n) \ .$$

At times, it is useful to describe the exact rate of convergence.

(3.6) DEFINITION. *We say that $a_n \asymp_{as} b_n$ if*

$$0 < \liminf_{n\to\infty} \frac{a_n}{b_n} \leqslant \limsup_{n\to\infty} \frac{a_n}{b_n} < \infty \ , \qquad \text{almost surely} \ .$$

If both a_n and b_n are determinstic, we write $a_n \asymp b_n$.

We now consider the choice of the distance function in (3.1), beginning with the parametric case. Let $d \geqslant 1$ be a fixed integer. We assume that $\theta \in \mathbb{R}^d$, i.e., $\theta = (\theta_1, \theta_2, \cdots, \theta_d)$, with the θ_i denoting scalar variables. Of the various ways to measure the distance between an estimator θ^n and the true θ_o, the Euclidean distance $\| \theta^n - \theta_o \|_2$ seems to be preferable. Here, $\| \cdot \|_2$ is the case $p = 2$ of

$$(3.7) \qquad\qquad \| \theta \|_p = \left\{ \sum_{i=1}^{d} | \theta_j |^p \right\}^{1/p} \ , \quad 1 \leqslant p < \infty \ .$$

These distance measures are referred to as the ℓ^p norms on $\mathbb{R}^d$. However, convergence in the sense of $\| \theta^n - \theta_o \|_2$ converging in one of the previously discussed interpretations is equivalent to the same type of convergence in the other *norms*. This is a finite-dimensional phenomenon.

In the nonparametric case, things are decidely more complicated, due to the infinite-dimensional setting. The question is what is the best way to measure distances between two pdfs φ and ψ. The primary choice is any one of the L^p norms $\| \varphi - \psi \|_p$, where for measurable functions φ,

$$(3.8) \qquad\qquad \| \varphi \|_p = \left\{ \int_{\mathbb{R}} | \varphi(x) |^p \, dx \right\}^{1/p} \ , \quad 1 \leqslant p < \infty \ ,$$

and for $p = \infty$,

$$(3.9) \qquad \| \varphi \|_\infty = \inf \left\{ K \ : \ | \varphi(x) | \leqslant K \text{ almost everywhere} \right\} \ .$$

If φ is continuous, then $\| \varphi \|_\infty = \sup \left\{ | \varphi(x) | \ : \ x \in \mathbb{R} \right\}$. The set of all functions with $\| f \|_p < \infty$ is denoted by $L^p(\mathbb{R})$. The cases of (most) interest are $p = 1, 2$ and ∞. The L^2 norm enjoys the most popularity in the literature, but the authors believe that its use is a conceptual mistake and, following DEVROYE and GYÖRFI (1985), prefer the L^1 norm for the reasons outlined below. See also RACHEV (1991). In this text, the L^∞ norm is used in one significant instance when discussing unimodal density estimation (§ 6.6), but only to obtain L^1 error bounds.

Scaling invariance. One of the features of the L^1 norm is that it is the only L^p norm which is *scaling invariant*. Let $\sigma > 0$ and consider the scaled densities

$$(3.10) \qquad \varphi_\sigma(x) = \sigma^{-1} \varphi(\sigma^{-1} x) , \quad -\infty < x < \infty ,$$

and likewise for ψ_σ. Then, it is an easy exercise to show that for $1 \leqslant p \leqslant \infty$,

$$(3.11) \qquad \| \varphi_\sigma - \psi_\sigma \|_p = \sigma^{1-1/p} \| \varphi - \psi \|_p ,$$

provided we take $1/p = 0$ for $p = \infty$. Thus, $p = 1$ is the only invariant case.

(3.12) EXERCISE. Verify (3.11).

Thus, if for the Buffalo snow fall data set we decide to measure snowfall in centimeters rather than inches, or in fractions of 1.3 times the largest observed annual snowfall as in Figure 1.1, and go on to estimate the pdf of the snowfall accordingly, then the L^1 distance to the true pdf does not change. The pdfs themselves do change, of course. The above scaling invariance extends to invariance under any monotone (differentiable) transformation of the random variable, see Exercise (4.1.44).

Estimating probabilities. Another feature of the L^1 distance between two pdfs is its close relationship to the *total variation* distance between two distributions. If Φ and Ψ are distributions, then the total variation distance between them is

$$(3.13) \qquad \mathrm{TV}(\Phi, \Psi) = \sup_B | \Phi(B) - \Psi(B) | ,$$

where the supremum is over all (measurable) events B and for any distribution F,

$$F(B) = \int_B dF(x)$$

is the probability assigned by F to the event Ω. Thus, the total variation distance measures the largest possible difference in the probabilities assigned to events. The relevance of the total variation distance to statistics is obvious.

It is a standard exercise to show that if Φ and Ψ have densities φ and ψ, then

$$(3.14) \qquad \mathrm{TV}(\Phi, \Psi) = 2 \| \varphi - \psi \|_1 .$$

Thus, the L^1 errors give us rather sharp information on differences in probabilities.

(3.15) EXERCISE. Verify (3.14).

Cauchy sequences. A final technical observation is that when we have a sequence of densities $\{\varphi_n\}_{n \geqslant 1}$ which is a Cauchy sequence in L^1, i.e.,

$$(3.16) \qquad \|\varphi_n - \varphi_m\|_1 \longrightarrow 0 \quad \text{for } n, m \to \infty ,$$

then the sequence has a limit, which is again a pdf. This is a useful feature when considering the existence of solutions to the various maximum penalized or smoothed likelihood estimation problems, because often one can construct Cauchy sequences of what one hopes are approximations to a solution. In contrast, if the sequence of densities $\{\varphi_n\}_n$ is a Cauchy sequence in L^2, then the limit need not be a pdf.

(3.17) EXERCISE. Let $\varphi_n(x) = n^{-1}\psi(n^{-1}x)$ for some bounded pdf ψ. Show that the φ_n are pdfs and that $\{\varphi_n\}_n$ is a Cauchy sequence in L^2, but that its limit is the zero function.

Of course, the L^1 norm has *some* drawbacks.

Measuring tail behavior. One of the drawbacks of the L^1 distance is that differences in the tails of the pdfs are largely ignored. If tail estimation is (more) important, then one could use the Kullback-Leibler, Hellinger, or Pearson's φ^2 distances, defined, respectively, as

$$(3.18) \qquad \text{KL}(\varphi, \psi) = \int_{\mathbb{R}} \left\{ \varphi(x) \log \frac{\varphi(x)}{\psi(x)} + \psi(x) - \varphi(x) \right\} dx ,$$

$$(3.19) \qquad \text{H}(\varphi, \psi) = \int_{\mathbb{R}} \left| \sqrt{\varphi}(x) - \sqrt{\psi}(x) \right|^2 dx ,$$

$$(3.20) \qquad \text{PHI}(\varphi, \psi) = \int_{\mathbb{R}} \frac{|\varphi(x) - \psi(x)|^2}{\psi(x)} dx .$$

The following inequalities between these distances hold, for all (sub)pdfs φ, and ψ (i.e., they are nonnegative, with integral $\leqslant 1$),

$$(3.21) \qquad \tfrac{1}{4} \|\varphi - \psi\|_1^2 \leqslant \text{H}(\varphi, \psi) \leqslant \text{KL}(\varphi, \psi) \leqslant \text{PHI}(\varphi, \psi) ,$$

the proofs of which are delayed until Chapter 10. (Some of the individial inequalities may be improved on.) Note that $\text{H}(\varphi, \psi) \leqslant \|\varphi - \psi\|_1$, so that the Hellinger distance between pdfs is always bounded. This is not true for the Kullback-Leibler and Pearson's φ^2 distances.

Some comments on these distances are in order. First, only the Hellinger distance is symmetric. We already encountered the Kullback-Leibler distance in § 2. The Hellinger distance is quite reasonable since square-root

densities are square integrable. Apart from this, it is also useful for theoretical reasons, such as for establishing minimal conditions for the consistency
of estimators, see LE CAM (1970), or IBRAGIMOV and HAS'MINSKII (1981),
but we shall not address this. We extensively use Hellinger distances when
discussing the GOOD estimator (1.22) in § 5.2.

It is a nice exercise to show that these new distances are also scaling
invariant.

(3.22) EXERCISE. Show that KL, H, and PHI are scaling invariant.

Ease of mathematical handling. Another drawback of the L^1 error
is that it is not so easy to work with. By way of example, an analogue for
the L^1 error of (1.11)–(1.12) does not exist, and that will cost us dearly.
In fact, this ease of mathematical handling of the integrated squared error
is an important factor in its popularity.

At this point, the question arises whether the choice of distance really
matters. The answer turns out to be yes and shows a major difference
between parametric and nonparametric problems: For parametric problems, all reasonable ways of measuring the difference between an estimator
and the "true" parameter are equivalent (at least asymptotically), but for
nonparametric problems, this is not the case. The following example and
exercises give some details regarding the difference between convergence in
L^∞-norm and L^1-norm.

(3.23) EXAMPLE. Let $\{U_j\}_j$ be a sequence of iid random variables with
uniform $(0,1)$ distribution. Define a sequence of random functions $\{Y_j\}_j$
on $(-1, 2)$ by

$$(3.24) \qquad Y_j(x) = \begin{cases} j, & |x - U_j| \leqslant j^{-2}, \\ 0, & \text{otherwise}. \end{cases}$$

Let $x \in (-1, 2)$ be fixed. We first show that $Y_j(x) \longrightarrow_{\text{as}} 0$. If $x < 0$
or $x > 1$, this is obvious. For $0 < x < 1$, we use that the almost sure
convergence is equivalent to

$$(3.25) \qquad \lim_{m \to \infty} \mathbb{P}\big[\, Y_j(x) < \varepsilon \text{ for all } j \geqslant m \,\big] = 1 \quad \text{for all } \varepsilon > 0 \,.$$

Since the events $Y_j(x) < \varepsilon$ are independent, we have for all $\varepsilon > 0$,

$$\mathbb{P}\big[\, Y_j(x) < \varepsilon \text{ for all } j \geqslant m \,\big] = \prod_{j=m}^{\infty} \mathbb{P}\big[\, Y_j(x) < \varepsilon \,\big] \,.$$

Now, for all j large enough to avoid edge effects, $\mathbb{P}\big[\, Y_j(x) < \varepsilon \,\big] = 1 - (2/j^2)$,
and $\prod_{j=1}^{\infty} \{\, 1 - (2/j^2) \,\}$ is a convergent infinite product, so

$$\lim_{m \to \infty} \mathbb{P}\big[\, Y_j(x) < \varepsilon \text{ for all } i \geqslant m \,\big] = \lim_{m \to \infty} \prod_{j=m}^{\infty} \{\, 1 - (2/j^2) \,\} = 1.$$

So (3.25) holds, and $Y_j(x) \longrightarrow_{as} 0$. The same holds for $x = 0$ and $x = 1$. On the other hand,

$$\| Y_j \|_1 = 2\,j^{-1} \longrightarrow_{as} 0 \,, \qquad \| Y_j \|_\infty = j \longrightarrow_{as} \infty \,.$$

Summarizing, the sequence $\{Y_j\}_j$ converges a.s. to 0 in L^1–norm, converges a.s. to ∞ in L^∞–norm, and converges a.s. pointwise to 0.

(3.26) EXERCISE. (a) Replace (3.24) by

(3.24′) $$Y_j(x) = \sqrt{j}\,, \qquad |x - U_j| \leqslant j^{-1}\,,$$

and repeat the above example. (In particular, for fixed x, the sequence $\{Y_j(x)\}_j$ does not converge a.s. to 0.)

(b) For any (random) function Y on $[a, b]$, show that $\| Y \|_1 \leqslant (b-a)\, \| Y \|_\infty$.

(c) Construct a sequence of random functions $\{Y_j\}_j$ on $\mathbb{R}$ such that

$$\| Y_j \|_1 \longrightarrow_{as} \infty \,, \qquad \| Y_j \|_\infty \longrightarrow_{as} 0 \,.$$

(3.27) EXERCISE. Construct similar examples to illustrate the difference between L^1 and L^2.

Finally, we note that in Example (3.23), we actually proved a special case of the Borel-Cantelli lemma. Because later on we have occasion to refer to it, we state a version of it here.

(3.28) BOREL-CANTELLI LEMMA. Let $\{Y_n\}_{n \geqslant 1}$ be a sequence of random variables, and let $c > 0$. Then, the following statements hold.

(a) $\sum_{n=1}^{\infty} \mathbb{P}[\,|Y_n| \geqslant c\,] < \infty$ implies $\limsup_{n \to \infty} |Y_n| \leqslant_{as} c$;

(b) If the Y_n are mutually independent, then $\sum_{n=1}^{\infty} \mathbb{P}[\,|Y_n| \geqslant c\,] = \infty$ implies $\limsup_{n \to \infty} |Y_n| \geqslant_{as} c$.

We list the exercises as they have occurred throughout this section.

EXERCISES : (3.12), (3.15), (3.17), (3.21), (3.26), (3.27).

4. The usual nonparametric assumptions

In this section, we briefly discuss the assumptions needed to establish consistency and convergence rates in (nonparametric) kernel density estimation. Here, we only consider the kernel estimator (1.18) with $A = \mathfrak{B}$, the two-sided exponential kernel, but the conclusions apply to general kernels. The goal is to establish bounds on the L^1 error $\| \mathfrak{B}_h * dF_n - f_o \|_1$.

Since the kernel estimator $\mathcal{B}_h * dF_n(x)$ is the mean of the iid random variables $\mathcal{B}_h(x - X_i)$, with expected values $\mathcal{B}_h * dF_o(x) = \mathcal{B}_h * f_o(x)$ and variances $(\mathcal{B}_h)^2 * f_o - (\mathcal{B}_h * f_o)^2$, its mean is $\mathcal{B}_h * f_o$ and its variance is

$$(4.1) \quad \mathbb{E}[\,|\,[\,\mathcal{B}_h * (dF_n - dF_o)\,](x)\,|^2\,] =$$
$$(nh)^{-1} \left\{ (\mathcal{B}^2)_h * dF_o(x) - (\mathcal{B}_h * dF_o(x))^2 \right\} \leqslant$$
$$(nh)^{-1} \left\{ (\mathcal{B}^2)_h * dF_o(x) \right\} .$$

From the triangle inequality

$$|\,\mathcal{B}_h * dF_n(x) - f_o(x)\,| \leqslant$$
$$|\,\mathcal{B}_h * dF_o(x) - f_o(x)\,| + |\,[\,\mathcal{B}_h * (dF_n - dF_o)\,](x)\,| ,$$

it then follows that

$$(4.2) \quad \mathbb{E}[\,|\,\mathcal{B}_h * dF_n(x) - f_o(x)\,|\,] \leqslant$$
$$|\,\mathcal{B}_h * dF_o(x) - f_o(x)\,| + (nh)^{-1/2} \sqrt{(\mathcal{B}^2)_h * dF_o(x)} .$$

and so, upon integration,

$$(4.3) \quad \mathbb{E}[\,\|\,\mathcal{B}_h * dF_n - f_o\,\|_1\,] \leqslant$$
$$\|\,\mathcal{B}_h * f_o(x) - f_o(x)\,\|_1 + (nh)^{-1/2} \|\,\sqrt{(\mathcal{B}^2)_h * f_o}\,\|_1 .$$

At this point, it seems clear that some assumptions are needed to get convergence rates. The *usual nonparametric assumptions* are designed to deal with each term in (4.3) separately. Thus, an opaque way of phrasing the usual nonparametric assumptions is

$$(4.4) \qquad \limsup_{h \to 0} h^{-2} \|\,\mathcal{B}_h * f_o(x) - f_o(x)\,\|_1 = C ,$$

$$(4.5) \qquad \limsup_{h \to 0} \|\,\sqrt{(\mathcal{B}^2)_h * f_o}\,\|_1 = C' ,$$

for finite constants C and C'. A somewhat more meaningful way is

$$(4.6) \qquad \|\,(f_o)''\,\|_1 < \infty ,$$

$$(4.7) \qquad \mathbb{E}[\,|X|^\kappa\,] < \infty \quad \text{for some } \kappa > 1 .$$

These last two conditions are referred to as the *usual nonparametric assumptions*. The conditions (4.4) and (4.6) are just about equivalent, but condition (4.7) is slightly stronger than (4.5), see § 4.2.

With these nonparametric assumptions, it follows from (4.3) that

$$(4.8) \qquad \mathbb{E}[\,\|\,\mathcal{B}_h * dF_n - f_o\,\|_1\,] \leqslant c \left\{ h^2 + (nh)^{-1/2} \right\} ,$$

for a suitable constant c. Asymptotically, $c = \min(C, C')$. The conclusion is that

$$(4.9) \qquad \|\,f_o - \mathcal{B}_h * dF_n\,\|_1 \longrightarrow_{\mathbb{E}} 0 , \quad \text{provided } h \longrightarrow 0 ,\ nh \longrightarrow \infty .$$

Moreover, the right-hand side of (4.8) is minimized (asymptotically) for $h \asymp n^{-1/5}$, which gives

$$(4.10) \qquad \mathbb{E}[\, \| \, \mathfrak{B}_h * dF_n - f_o \, \|_1 \,] = \mathcal{O}(\, n^{-2/5} \,) \, .$$

There is of course no indication that this is a sharp bound, but in fact it is, see DEVROYE (1987). What is surprising is that the same rate applies to the almost sure convergence of $\| \, \mathfrak{B}_h * dF_n - f_o \, \|_1$, even for data-driven choices of h. We come back to all of this in Chapters 4 and 7.

5. Parametric vs nonparametric rates

We have seen that the difference between parametric and nonparametric estimation lies in the dimensionality of the parameter to be estimated. In this section, we show that this results in differences between the convergence rates of some natural estimators. The assumption is that these natural estimators achieve the optimal convergence rate, or at least cannot be drastically improved on.

To elaborate on this, and to make it more precise, suppose that the density f_o to be estimated belongs to some class $\mathcal{F}$ of densities, e.g., the class of all densities on $(-\infty, \infty)$, the class of all decreasing densities on $(0, \infty)$, or the class $PDF(C, C')$ of all densities f for which

$$(5.1) \qquad \limsup_{h \to 0} h^{-2} \, \| \, f - \mathfrak{B}_h * f \, \|_1 \leqslant C \, ,$$

$$(5.2) \qquad \limsup_{h \to 0} \, \| \, \sqrt{(\mathfrak{B}^2)_h * f} \, \|_1 \leqslant C' \, ,$$

for fixed, known constants C, C'.

Let f^n be an estimator of f_o, that is, f^n is a function of the data

$$(5.3) \qquad f^n(x) = f^n(x \, ; X_1, X_2, \cdots, X_n) \, , \qquad -\infty < x < \infty \, .$$

Thus, as in kernel estimation, f^n may incorporate a smoothing parameter h, as long as the choice of h is data driven, $h = h_n(X_1, X_2, \cdots, X_n)$. The best possible (expected) estimation error is then

$$(5.4) \qquad \mathfrak{R}_{\mathcal{F}}(n) = \inf_{f^n} \, \sup_{f_o \in \mathcal{F}} \, \mathbb{E}[\, \| \, f^n - f_o \, \|_1 \,] \, ,$$

where we picked the L^1 error for reasons discussed in §3. The above quantity is known as the (expected) minimax error for the class $\mathcal{F}$. In this text, we are not overly concerned with determining the rate at which the minimax error tends to 0, although we assume that we know what it is. For the generic parametric estimation problem, the maximum likelihood estimator has expected minimax error $\mathcal{O}(n^{-1/2})$, so that $\mathfrak{R}_{\mathcal{F}}(n) = \mathcal{O}(n^{-1/2})$. This rate goes by the name of the parametric rate, even though there are exceptional parametric problems in which one achieves better rates. We come back to this in Part I. For nonparametric problems, one usually achieves a lower convergence rate. By way of example, for the class $PDF(C, C')$, one has

$$(5.5) \quad 0 < \liminf_{n \to \infty} n^{2/5} \, \mathfrak{R}_{PDF(C,C')}(n) \leqslant$$

$$\limsup_{n \to \infty} n^{2/5} \, \mathfrak{R}_{PDF(C,C')}(n) < \infty \, .$$

In § 4, we showed that this was the upper bound. Thus, it would be nice if the parametric convergence rate characterized parametric problems, as in the following "theorem".

(5.6) "THEOREM". If $\mathfrak{R}_{\mathcal{F}}(n) = \mathcal{O}(n^{-1/2})$, then $\mathcal{F}$ is a parametric family with a finite- dimensional parameter.

Unfortunately, the theorem fails, a counter example being provided by the class $\mathcal{F}$ of pdfs whose Fourier transforms (characteristic functions) are known to vanish outside of the interval $(-1, 1)$, see § 4.7. Thus, the di- mensionality of the parameter to be estimated is not an indicator of the achievable convergence rate.

The following distinction appears to come closer. Although for many parametric problems there exists unbiased estimators with finite variance, this does not happen for problems with a nonparametric convergence rate, as shown by DEVROYE and LUGOSI (2000).

(5.7) THEOREM. [DEVROYE and LUGOSI (2000)] The minimax error sat- isfies $\limsup_{n \to \infty} \sqrt{n} \, \mathfrak{R}_{\mathcal{F}}(n) = \infty$ if and only if for every $n \geqslant 1$ and for every estimator f^n at least one of the following two statements holds.

(a) f^n is biased

$$\sup_{f_o \in \mathcal{F}} \| f_o - \mathbb{E}[f^n] \|_1 > 0 \, ;$$

(b) f^n has infinite variance

$$\sup_{f_o \in \mathcal{F}} \| \sqrt{\mathbb{E}[(f^n)^2]} \, \|_1 = +\infty \, .$$

It should be noted that if (a) fails, i.e., $\mathbb{E}[f^n] = f_o$, and (b) holds, then

$$\| \sqrt{\mathbb{E}[(f^n)^2] - (f_o)^2} \, \|_1 = +\infty \, ,$$

and this resembles the variance of the estimator being ∞. The proof of the theorem is simple and ingenious.

PROOF. $\Longrightarrow$: Let $f_o \in \mathcal{F}$, and suppose that f^n is an unbiased estimator of f_o. Let n and p be natural numbers. Later, we take p fixed and let $n \to \infty$. Consider the following estimator of f_o:

$$(5.8) \qquad f_{pn}(x) = \frac{1}{n} \sum_{i=1}^{n} f^p\left(x \, ; X_{pi+1}, X_{pi+2}, \cdots, X_{pi+p} \right) \, ,$$

$-\infty < x < \infty$. Obviously, f_{pn} is unbiased. Since f_{pn} is the sum of n independent, identically distributed random variables, then for each x, the variance of $f_{pn}(x)$ satisfies

$$\mathrm{Var}[\, f_{pn}(x)\,] = \tfrac{1}{n}\,\mathrm{Var}[\, f^p(x)\,] \; .$$

It follows that

(5.9) $\mathbb{E}[\,\| f_{pn} - f_o \|_1 \,] = \| \mathbb{E}[\, | f_{pn} - \mathbb{E}[\, f_{pn}\,]\, | \,] \|_1 \leqslant n^{-1/2}\,\| \sqrt{\mathrm{Var}[\, f^p\,]} \,\|_1 \; .$

This holds for any $f_o \in \mathcal{F}$. Now, choose f_o, depending possibly on the sample size pn, so as to be within a factor 2 of the minimax error, i.e,

$$\mathfrak{R}_{\mathcal{F}}(pn) \leqslant 2\,\mathbb{E}[\,\| f_{pn} - f_o \| \,] \; ,$$

so from (5.9),

$$(pn)^{1/2}\,\mathfrak{R}_{\mathcal{F}}(pn) \leqslant 2\,p^{1/2}\,\| \sqrt{\mathrm{Var}[\, f^p\,]} \,\|_1 \; .$$

Now, keep p fixed, and let $n \to \infty$. Then, the left-hand side tends to ∞, by assumption. Since p is fixed, this implies that

$$\| \sqrt{\mathrm{Var}[\, f^p\,]} \,\|_1 \longrightarrow \infty \quad \text{as} \quad n \to \infty \; .$$

In other words, the variance *equals* ∞.   Q.e.d.

(5.10) EXERCISE. Prove the only if part of the theorem.

EXERCISES : (5.10).

6. Sieves and convexity

Roughness penalization in its various forms is a standard way of making sense out of the nonparametric maximum likelihood problems (1.16) and (2.4). A second method for obtaining smooth estimators was proposed by GRENANDER (1981), and its idea is to restrict the minimization in the maximum likelihood problem to classes of smooth densities, given the colorful name of *sieves*. In this section, we discuss sieves and make the point that the method of sieves is similar to roughness penalization.

The identifying feature of a sieve is that it is a *small* set of densities, as opposed to the *big* class of all densities. There are essentially two kinds of sieves. One kind is obtained by considering finite-dimensional subspaces of $L^1(\mathbb{R})$, and then a sieve consists of all pdfs in a particular subspace. Since each subspace is finite-dimensional, the pdfs it contains cannot be arbitrarily rough. The other kind of sieve is obtained by considering compact subsets of $L^1(\mathbb{R})$. For the present purpose, compactness of a set means that the set can be approximated well by finite-dimensional subspaces, in a sense to be made precise below. However, things get more interesting for particular kinds of compact sets. We discuss the two kinds of sieves in some detail.

The simplest example of sieves is when we have a nested sequence of finite-dimensional subspaces of continuous functions

$$(6.1) \qquad V_1 \subset V_2 \subset \cdots \subset V_m \subset \cdots \subset L^1(\mathbb{R}) \,,$$

which is dense in $L^1(\mathbb{R})$, i.e., for all $\varphi \in L^1(\mathbb{R})$,

$$(6.2) \qquad \lim_{m \to \infty} \inf \{\, \| v - \varphi \|_1 \, : \, v \in V_m \,\} = 0 \,.$$

(6.3) EXAMPLE. A useful and simple example of such subspaces is when V_m consists of all step functions that vanish outside of the interval $[-m, m]$, and with jumps at the points j/m, $j = 0, \pm 1 \pm 2, \cdots, \pm m^2$. One verifies that (6.2) does indeed hold, and V_{2m}, $m = 1, 2, \cdots$, is a nested sequence as in (6.1). Moreover, $\dim(V_m) = 2 \, m^2 + 1$. The pdfs in V_m are histograms.

With these subspaces in hand, the method of sieves determines estimators in the deconvolution problem (2.4) as solutions to

$$(6.4) \qquad \begin{aligned} \text{minimize} \quad & -\int_{\mathbb{R}} \log \mathcal{K} f(x) \, dG_n(x) \\ \text{subject to} \quad & f \in V_m \,, \ f \text{ is a pdf} \,, \end{aligned}$$

for a suitably chosen value of m, similar to choosing the smoothing parameter in penalization problems. Variations on the above theme abound, of which we mention two.

For the first one, recall that if f is a pdf, then $\sqrt{f} \in L^2(\mathbb{R})$. Alternatively, if $\varphi \in L^2(\mathbb{R})$, then $\varphi^2 \in L^1(\mathbb{R})$, and φ^2 is nonnegative. So it makes sense to choose a sequence of finite-dimensional subspaces of $L^2(\mathbb{R})$

$$(6.5) \qquad W_1 \subset W_2 \subset \cdots \subset W_m \subset \cdots \subset L^2(\mathbb{R}) \,,$$

which is dense in $L^2(\mathbb{R})$, analogous to (6.2), and to consider the problem

$$(6.6) \qquad \begin{aligned} \text{minimize} \quad & -\int_{\mathbb{R}} \log[\, \mathcal{K}(\varphi^2)\,](x) \, dG_n(x) \\ \text{subject to} \quad & \varphi \in W_m \,, \ \| \varphi \|_2 = 1 \,. \end{aligned}$$

If $\varphi^{n,m}$ denotes the solution to (6.6), then $f = (\varphi^{n,m})^2$ is the estimator of the pdf. Subspaces of $L^2(\mathbb{R})$ satisfying (6.5) may be constructed using any convenient orthonormal basis for $L^2(\mathbb{R})$, see, e.g., SANSONE (1991).

In the second variation, one defines exponential families

$$(6.7) \qquad f(x) = \varphi_o(x) \exp\big(w(x) \big) \,, \quad -\infty < x < \infty \,,$$

where φ_o is some fixed pdf and $w \in W_m$, say. The minimization problem may then be formulated accordingly. We note the special case of density estimation, where $\mathcal{K}f = f$ for all f,

$$\text{minimize} \quad -\int_{\mathbb{R}} w(x)\, dG_n(x)$$

(6.8) $\qquad$ subject to $\quad w \in W_m$, and

$$\int_{\mathbb{R}} \varphi_o(x)\, \exp\big(w(x)\big)\, dx = 1 \ .$$

Here, the objective function is simple when compared with (6.6), say, but the pdf constraint is more complicated. A related problem is considered in § 5.3.

In the above, it should be observed that the sieves serve the second purpose of *discretizing* the estimation problem. The original problems being nonparametric, that is, infinite-dimensional, a crucial step in any computation is the approximation of the infinite-dimensional problem by a finite-dimensional one. It is the task of numerical analysis to investigate discretization errors, but in this text, we assume that the discretization effects are negligible compared with the effects of noisy data. Of course, one had better make sure, experimentally say, that this is indeed the case.

A cursory inspection of estimation by the method of sieves reveals that the dimension of the subspaces V_m plays the role of the smoothing parameter, and the question arises of how it should be chosen. On the one hand, we want the subspaces V_m or W_m to be big, so that our unknown density f_o may be approximated well by elements from V_m or W_m. On the other hand, bigger subspaces contain rougher pdfs, and that means that the estimators will be rougher. This is the analogue of the bias-variance trade off for kernel density estimation.

On to the second kind of sieves, obtained by considering nested sequences of compact subsets of continuous functions

(6.9) $\qquad C_1 \subset C_2 \subset \cdots \subset C_m \subset \cdots \subset L^1(\mathbb{R})$,

which are dense in $L^1(\mathbb{R})$. We do not go into the details, except to note that compactness here may be interpreted to mean *almost finite dimensional* in the sense that if $C \subset L^1(\mathbb{R})$ is compact, then for any nested sequence of finite-dimensional subspaces V_m which is dense in $L^1(\mathbb{R})$, as in (6.1)–(6.2), one has

(6.10) $\qquad \displaystyle\sup_{\varphi \in C} \inf_{v \in V_m} \| v - \varphi \|_1 \longrightarrow 0 \quad \text{as } m \to \infty$.

For related material, see Appendix 3.

So in the above interpretation, admitting compact sets as sieves does not drastically alter the state of affairs, especially keeping in mind the need for discretization. We arrive at a different interpretation when considering nested compact subsets of the form

(6.11) $\qquad C_M = \big\{ f \text{ is a pdf} : R(f) \leqslant M \big\}$, for a suitable roughness functional $R(f)$. In the actual application of these sets, the constant M must be chosen appropriately. We restrict attention to convex roughness functionals, that is, functionals satisfying

$$(6.12) \qquad R(\lambda f + (1 - \lambda) g) \leqslant \lambda R(f) + (1 - \lambda) R(g) ,$$

for all f, g, and all $0 \leqslant \lambda \leqslant 1$. Now, the sieved version of (6.4) reads as

$$(6.13) \qquad \begin{aligned} &\text{minimize} \quad -\int_{\mathbb{R}} \log \mathcal{K}f(x)\, dG_n(x) \\ &\text{subject to} \quad R(f) \leqslant M , \ f \text{ is a pdf} . \end{aligned}$$

It is important to note that the objective function in (6.13) is likewise convex. It follows from the theory of convex minimization problems, studied in detail in Part III, that (6.13) is equivalent to

$$(6.14) \qquad \begin{aligned} &\text{minimize} \quad -\int_{\mathbb{R}} \log \mathcal{K}f(x)\, dG_n(x) + h^2\, R(f) \\ &\text{subject to} \quad f \text{ is a pdf} , \end{aligned}$$

where the *Lagrange multiplier* h^2 is chosen such that the solution f^{nh} of (6.14) satisfies

$$(6.15) \qquad R(f^{nh}) = M .$$

(In exceptional cases, the solution f of (6.13) satisfies $R(f) < M$. This possibility is ignored here.) It is now apparent that this version of the method of sieves and maximum penalized likelihood estimation are just about equivalent, especially upon noting that typically the constant M is unknown. In view of (6.15), this means that one may omit all references to M, and consider h^2 as the primary unknown. However, one sees that choosing M in the sieve context is the same as choosing h in the penalization framework.

Another interesting kind of compact sieves are the *convolution sieves*, although the name *kernel sieves* comes to mind also. These sieves are indexed by a smoothing parameter h and take the form

$$(6.16) \qquad C_h = \big\{ A_h * dP : P \text{ is a probability distribution} \big\} .$$

So in this case, the solutions to the (sieved) maximum likelihood problem are restricted to being kernel estimators. The resulting problem may be phrased as a problem to estimate a distribution function

$$(6.17) \qquad \begin{aligned} &\text{minimize} \quad -\int_{\mathbb{R}} \log\big(A_h * dP(x) \big)\, dF_n(x) \\ &\text{subject to} \quad P \text{ is a probability distribution} . \end{aligned}$$

Denote the solution by $P^{n,h}$. Then, the estimator for f_o is $f^{nh} = A_h * dP^{n,h}$. For the plain density estimation problem ($\mathcal{K}f = f$ for all f), WALTER and BLUM (1984) give explicit solutions for the case in which A is the two-sided exponential kernel. The optimal P consists of discrete point masses at the X_i, but the weights are not necessarily equal.

This description of the method of sieves must suffice, and serves as the authors' justification for studying estimation from the penalization point of view only. The above also hints at the important role played in this text by convexity.

7. Additional notes and comments

Ad § 1 : There are numerous papers on the failure and optimality of parametric maximum likelihood estimation. We stay optimistic, and only quote the optimality paper, YATRACOS (1998). The derivation of the least-squares method (1.13), and (1.23) leading to kernel density estimators is very well known in nonparametric density estimation, see STONE (1984). For the method of maximum spacings, see PYKE (1965), CHENG and AMIN (1983), RANNEBY (1984), and the recent GOSH and RAO JAMMALA-MADAKA (2001). See also § 2.8.

Ad § 2 : GOOD (1971) uses *Bayesian* arguments to justify his roughness penalization functional (1.21). That it pops up in a rather unexpected way in § 2, via nonlinear smoothing of EM algorithms, lends further support to his insights.

Ad § 3 : The L^1 setting for density estimation was originally advocated by DEVROYE and GYÖRFI (1985), and the authors wholeheartedly embrace their point of view. However, there is a huge literature on the L^2 setting, to which the reader is welcome.

Ad § 5 : This section is largely based on the last section of DEVROYE and LUGOSI (2000).

Ad § 6 : The model (6.8) comes back in disguised form in §§ 5.3 and 4, under the name of log splines. See the references there.

Part I:

Parametric Density Estimation

2

Parametric
Maximum Likelihood Estimation

1. Introduction

In this chapter, we study maximum likelihood estimation for parametric families of probability density functions, typically involving only a small number of scalar parameters. The questions of interest include: does the estimator exist, is it unique, what are its properties (small sample and asymptotic), how can the estimator be computed, and, most urgently, why should parameters (and pdfs) be estimated this way. There is also the worry of whether the true pdf actually belongs to the parametric family specified. In this chapter, we discuss these questions at length, sometimes in general, sometimes ·for specific estimation problems, to illustrate the issues and to contrast the details with the nonparametric case studied in Part II. In Chapter 3, some "experimental" illustrations and implications are discussed.

The specific problem is as follows. Let $X = (X_1, X_2, \cdots, X_n)^T \in \mathbb{R}^{n \times d}$ denote a sample of size n of an $\mathbb{R}^d$-valued random variable with joint pdf (with respect to Lebesgue measure) $f(x; \theta_o)$, $x \in \mathbb{R}^{n \times d}$. The pdf of X is known to belong to the parametric family $\mathfrak{F} = \{ f(x; \theta) : \theta \in \Theta \}$, where $\Theta \subset \mathbb{R}^p$ is the set of all possible parameters. We wish to estimate θ_o on the basis of X. The (averaged) negative log-likelihood of θ given X is defined as

$$(1.1) \qquad L_n(\theta) = -\tfrac{1}{n} \log f(X; \theta) .$$

A maximum likelihood estimator (mle) of θ, denoted by θ_n, is any solution of the problem

$$(1.2) \qquad \begin{aligned} &\text{minimize} \quad L_n(\theta) \\ &\text{subject to} \quad \theta \in \Theta . \end{aligned}$$

An important special case arises when $X_1, X_2, \cdots, X_n$ are iid random variables with common pdf $\varphi(x; \theta_o)$, with $x \in \mathbb{R}^d$. The negative log-likelihood of θ is then

(1.3) $$L_n(\theta) = -\frac{1}{n} \sum_{i=1}^{n} \log \varphi(X_i\, ; \theta)\, ,$$

in view of the fact that the joint pdf of the X_i is given by

(1.4) $$f(x_1, x_2, \cdots, x_n\, ; \theta) = \prod_{i=1}^{n} \varphi(x_i\, ; \theta)\, .$$

The obvious questions that now need answering include: How is θ_n to be computed. For that matter, does θ_n exist and is it unique (and what do we do when the answer is negative)? And most urgently, why do we even want to estimate θ_o by way of θ_n? In this section, we present five examples of maximum likelihood estimation to illustrate some of these questions. This will give an indication of the strengths and weaknesses of the maximum likelihood methodology. In the remainder of this chapter, we discuss these questions in more detail. The examples are the multivariate normal family of pdfs, the exponential model, the linear (multivariate normal) model, mixture problems, and the Weibull distribution, the last one as an example of heavy-tailed distributions.

The multivariate normal model. A random variable $X \in \mathbb{R}^d$ is said to be a (nondegenerate) multivariate normal with mean μ and covariance matrix Σ if its pdf is given by

(1.5) $$f(x; \mu, \Sigma) = \frac{\exp\left(-\frac{1}{2}(x - \mu)^{\top}\Sigma^{-1}(x - \mu)\right)}{\left((2\pi)^d \det(\Sigma)\right)^{1/2}}\, , \quad x \in \mathbb{R}^d\, ,$$

with $\mu = \mathbb{E}[X] \in \mathbb{R}^d$ and $\Sigma \in \mathbb{R}^{d \times d}$ a positive-definite matrix and $\det(\Sigma)$ the determinant of Σ. The above is denoted succinctly and to the point as

(1.6) $$X \sim N_d(\mu, \Sigma)\, ,$$

or sometimes also as $X \sim N(\mu, \Sigma)$ when the dimension is clear from the context. The multivariate normal is by far the most often postulated model in "real" applications.

Suppose now that $X_1, X_2, \cdots, X_n$ are iid with $X_1 \sim N(\mu_o, I)$, where I is the identity matrix, and $\mu_o \in \mathbb{R}^d$ is the unknown parameter we wish to estimate. The parametric family is then

$$f(x; \mu) = (2\pi)^{-d/2} \exp\left(-\frac{1}{2}\|x - \mu\|^2\right)\, , \quad x \in \mathbb{R}^d\, ,$$

with $\mu \in \mathbb{R}^d$ the parameter. Here, $\|\cdot\|$ denotes the Euclidean norm on $\mathbb{R}^n$. The negative log-likelihood of μ given $X_1, X_2, \cdots, X_n$ is, in accordance with (1.3),

(1.7) $$L_n(\mu) = \frac{1}{2n} \sum_{i=1}^{n} \|X_i - \mu\|^2 + \text{ other terms}\, ,$$

where the "other terms" are independent of μ. It is easy to verify that, apart from the "other terms",

$$L_n(\mu) = \tfrac{1}{2} \left\| \mu - \tfrac{1}{n} \sum_{i=1}^{n} X_i \right\|^2 + \tfrac{1}{n} \sum_{i=1}^{n} \left\| X_i \right\|^2 - \tfrac{1}{n^2} \left\| \sum_{i=1}^{n} X_i \right\|^2 .$$

It follows that the mle of μ is

(1.8)
$$\mu_n = \tfrac{1}{n} \sum_{i=1}^{n} X_i .$$

So, in this case, the mle exists and is unique. What are its properties? It is a standard exercise in elementary probability to show that $\mu_n \sim N(\mu_o, \tfrac{1}{n} I)$ or, equivalently, in a more standard representation,

(1.9)
$$\sqrt{n}\,(\mu_n - \mu_o) \sim N(0, I) .$$

There is even a result that says that μ_n is the "best" estimator available. We will come back to this in § 2.

The exponential model. Let $\lambda_o \in \mathbb{R}^d$ have positive components, and let the random variable $X \in \mathbb{R}^d$ with nonnegative components have a multivariate exponential distribution with mean $1/\lambda_o \in \mathbb{R}^d$ (component wise division). We wish to estimate λ_o on the basis of the iid sample $X_1, X_2, \cdots, X_n$ of X. The parametric family of pdfs is given by

(1.10)
$$f(X\,;\lambda) = \{ \prod_{j=1}^{d} \lambda_j \} \exp(-\lambda^{\scriptscriptstyle T} X) , \quad X \in \mathbb{R}_+^d ,$$

where $\lambda = (\lambda_1, \lambda_2, \cdots, \lambda_d)^{\scriptscriptstyle T}$. The negative log-likelihood is then

(1.11)
$$L_n(\lambda) = \tfrac{1}{n} \sum_{i=1}^{n} \lambda^{\scriptscriptstyle T} X_i - \sum_{j=1}^{d} \log \lambda_j , \quad \lambda \in \mathbb{R}_+^d .$$

Now, we may write $L_n(\lambda) = \tfrac{1}{n} \sum_{j=1}^{d} \{ \lambda_j \sum_{i=1}^{n} X_{ij} - n \log \lambda_j \}$, where X_{ij} is the j-th component of X_i, so that minimizing $L_n(\lambda)$ is the "sum" of d unrelated problems. We now resort to differential calculus to see that the mle λ_n of λ_o exists and is unique and is given via

(1.12)
$$\frac{1}{\lambda_n} = \tfrac{1}{n} \sum_{i=1}^{n} X_i .$$

What are the properties of λ_n? For simplicity, let us assume $d = 1$. From (1.12), it is not hard to figure out that $\mathbb{E}[\lambda_n^{-1}] = \lambda_o^{-1}$, and that λ_n^{-1} has a Gamma distribution. Also, it is easy to see that by the central limit theorem,

(1.13)
$$\sqrt{n}\,(\lambda_n^{-1} - \lambda_o^{-1}) \longrightarrow_{\mathrm{d}} \Lambda \sim N(0, \lambda_o^{-2}) .$$

It turns out that asymptotically this is "best" possible. But what if the interest is in λ_o and not in λ_o^{-1}? One can show that $\mathbb{E}[\lambda_n] = n\lambda_o/(n-1)$ for $n \geqslant 2$, so λ_n is not a good estimator of λ_o: It is biased, but the obvious modification $\widetilde{\lambda}_n = (n-1)\lambda_n/n$ is an unbiased estimator. There is an analogue of (1.13), viz.

$$(1.14) \qquad \sqrt{n}\,(\,\widetilde{\lambda}_n - \lambda_o\,) \longrightarrow_d \Lambda \sim N(0, \lambda_o^{-o})\,,$$

and this too turns out to be asymptotically "best". When things go well with maximum likelihood estimation, this is the type of result one is shooting for. All of this will be made precise in § 2.

(1.15) EXERCISE. Show that λ_n as given by (1.12) indeed minimizes $L_n(\lambda)$ of (1.11).

The linear model. The linear multivariate normal model for a random variable X models the distribution as (assumes that the distribution is)

$$(1.16) \qquad X \sim N_n(Z\beta_o, \sigma_o^2 I),$$

where $Z \in \mathbb{R}^{n \times p}$ is a known matrix, $\beta_o \in \mathbb{R}^p$ is a vector of unknown parameters, and $\sigma_o^2 \in \mathbb{R}$ is also an unknown parameter. This is the basic model underlying both regression analysis and the design of experiments. The parametric family of distributions is given by

$$f(X\,;\beta,\sigma) = \frac{1}{(2\pi\sigma^2)^{n/2}} \exp\left(-\frac{\|X - Z\beta\|^2}{2\sigma^2}\right)\,, \qquad X \in \mathbb{R}^d\,.$$

In accordance with (1.1), the maximum likelihood problem for estimating β_o and σ_o is

$$(1.17) \qquad \begin{aligned} &\text{minimize} \quad (2n\sigma^2)^{-1}\|X - Z\beta\|^2 + \log\sigma \\ &\text{subject to} \quad \beta \in \mathbb{R}^d\,,\ \sigma > 0\,. \end{aligned}$$

It is clear that the optimal β is not influenced by the choice of σ, so that β_n solves the *linear least-squares* problem

$$(1.18) \qquad \begin{aligned} &\text{minimize} \quad \|Z\beta - X\|^2 \\ &\text{subject to} \quad \beta \in \mathbb{R}^d\,. \end{aligned}$$

It follows that the solution β_n is unique if and only if Z has full column rank. In that case,

$$(1.19) \qquad \beta_n = (Z^{\top} Z)^{-1} Z^{\top} X.$$

If Z does not have full column rank, then there are infinitely many solutions. However, although β_n need not be unique, the quantity $Z\beta_n$ is uniquely determined : It is the orthogonal projection of X onto the range of Z. Then, going back to (1.17), taking $\beta = \beta_n$ gives a nice minimization problem for determining σ_n, with a unique solution given by

$$(1.20) \qquad \sigma_n^2 = \tfrac{1}{n}\|Z\beta_n - X\|^2\,.$$

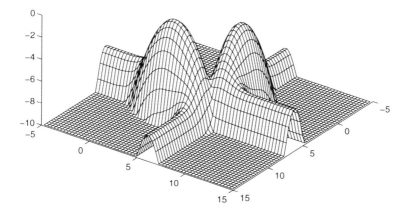

Figure 1.1. Graph of $\max(-L_n(\theta), -10)$ as function of μ_1 and μ_2, for a mixture of two normals, with $w = (0.8, 0.2)$, and $\sigma_1 = \sigma_2 = 1$. The sample size was $n = 100$, from a mixture of two normals, with w and σ as above and $\mu = (7, 3)$.

What can we say about β_n, $Z\beta_n$, and σ_n? If Z has full column rank, we have again

$$(1.21) \qquad \sqrt{n}\,(Z\beta_n - Z\beta_o) \sim N_n(0, \sigma_o^2 I) \;,$$

but we will save the properties of σ_n for later.

(1.22) EXERCISE. Show that if (β_n, σ_n) is the solution of (1.17), then β_n solves (1.18) and σ_n is given by (1.20).

Mixture of normals. The following situation is not uncommon. One observes a random variable with pdf belonging to a known parametric family $\{\, f(x\,;\theta) \,:\, \theta \in \Theta \,\}$, except that once in a while (assumed to be a random event independent of the observations) something goes wrong and a random variable with a pdf from a different family $\{\, g(x\,;\xi) \,:\, \xi \in \Xi \,\}$ is observed. Actually, it could be that the parametric family stays the same, but the parameter has changed. This may be modeled by a mixture of the two families. This setting has many applications, see, e.g., TITTERINGTON, SMITH, and MAKOV (1985), or MCLACHLAN and BASFORD (1988). Here, we restrict our attention to mixtures of univariate normals, as follows. For $\theta_j = (\mu_j, \sigma_j)$, let $\phi(x\,;\theta_j) = \sigma_j^{-1}\,\phi\big(\sigma_j^{-1}(x - \mu_j)\big)$, with $\phi(x)$ the standard

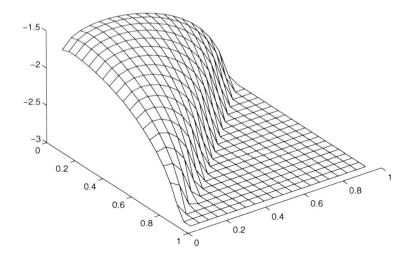

Figure 1.2. Graph of $\max(-L_n(\theta), -3)$ as function of w_1 and w_2, for a sample of size 100, for a mixture of three normals, with $\mu = (3, 7, 5)$ and $\sigma_1 = \sigma_2 = \sigma_3 = 0.5$. The sample size was $n = 100$, from a mixture of three normals, with μ and σ as above and $w = (0.2, 0.2, 0.6)$.

normal distribution. Let $\theta = (\theta_1, \theta_2, \cdots, \theta_m)$. The parametric family is

$$(1.23) \qquad f(x\,;w,\theta) = \sum_{j=1}^{m} w_j\, \phi(x\,;\theta_j) \ , x \in \mathbb{R}\ ,$$

where $w = (w_1, w_2, \cdots, w_m)^{\top}$, in which the w_j are nonnegative and add up to 1. In the above, one may let the number of components m be an unknown parameter as well, but it is definitely a parameter of a different nature compared with w and θ. The maximum likelihood estimator of (w, θ) is any solution of

$$\text{minimize} \quad L_n(w, \theta) = -\tfrac{1}{n} \sum_{i=1}^{n} \log\Big\{ \sum_{j=1}^{m} w_j\, \phi(X_i\,;\theta_j) \Big\}$$

(1.24)

$$\text{subject to} \quad \sum_{j=1}^{m} w_j = 1 \ , \ w \geqslant 0 \quad \text{component wise}\ .$$

This is certainly a messy problem. For simplicity, assume that the weights w_j are known and strictly positive. Then, $L_n(w, \theta)$ does not admit a global minimum, because for $(m \geqslant 2)$ we may take θ_i, $i \geqslant 2$, to be anything with $\sigma_i > 0$, set $\mu_1 = X_i$ for some i, and let $\sigma_1 \to 0^+$. Then, $L_n(w, \theta) \to -\infty$. This corresponds to a point mass at $x = X_i$. However, lots of local minima exist, usually. The difficulty is in finding the "correct" local minimum.

In Figure 1.1, we show the graph of $-L_n(w, \theta)$ for $m = 2$, as function of μ_1 and μ_2, with $w = (0.8, 0.2)$ and $\sigma_1 = \sigma_2 = 1$. Two local maxima, as well as a saddle point, are present, and who knows what else. This means that blindly applying some minimization technique may create unpredictable results. If, however, we fix $\theta_1, \cdots, \theta_m$, then $L_n(w, \theta)$ is a nice (convex) function of w that admits a unique minimum. This means that most minimization techniques will do a good job in this setting. The graph of $-L_n(w, \theta)$ for $m = 3$ as function of (w_1, w_2) is shown in Figure 1.2. Moreover, the constraints on w are easy to handle, as the exercise below makes clear.

(1.25) EXERCISE. Assume θ is fixed. Show that (1.24) has the same solutions as

$$\text{minimize} \quad -\frac{1}{n} \sum_{i=1}^{n} \log \Big\{ \sum_{j=1}^{m} w_j \, \phi(X_i \, ; \theta_j) \Big\} + \sum_{j=1}^{m} w_j \, ,$$

$$\text{subject to} \quad w \geqslant 0 \text{ component wise .}$$

Right-skewed distributions with heavy tails. Many data sets encountered by statisticians appear to be well modeled by right-skewed heavy-tailed distributions. Some parametric families that have these characteristics are the Weibull, log-normal, and Gamma distributions. Some areas of applications are time-till-failure models, see, e.g., SOBCZYK and SPENCER (1992), and the examples discussed in § 3.5.

The pdf of a three-parameter Weibull random variable is given by

$$(1.26) \qquad f(x \, ; \theta) = \frac{\alpha}{\lambda} \left(\frac{x - \gamma}{\lambda} \right)^{\alpha - 1} \exp \left[-\left(\frac{x - \gamma}{\lambda} \right)^{\alpha} \right] \, , \qquad x \geqslant \gamma \, ,$$

where $\theta = (\gamma, \lambda, \alpha) \in \mathbb{R} \times \mathbb{R}_{++} \times \mathbb{R}_{++}$. Here, $\mathbb{R}_{++} = (0, \infty)$, to distinguish it from $\mathbb{R}_+ = [0, \infty)$. The Weibull family of pdfs includes the exponential distribution, J-shaped distributions, and unimodal right-skewed distributions. The pdf of the three-parameter log-normal distribution is given by

$$(1.27) \quad f(x \, ; \theta) = \frac{1}{\sqrt{2\pi} \, \sigma \, (x - \gamma)} \exp \left[-\frac{\big(\log(x - \gamma) - \mu\big)^2}{2\sigma^2} \right] \, , \qquad x \geqslant \gamma \, ,$$

and $= 0$ otherwise, with $\theta = (\gamma, \mu, \sigma) \in \mathbb{R} \times \mathbb{R} \times \mathbb{R}_{++}$. Although for $\alpha > 1$ their graphs look similar, the log-normal has heavier tails than the Weibull distribution.

We briefly discuss parameter estimation for the Weibull distribution. Substantially, the same observations apply for the log-normal distribution. Given an iid sample $X_1, X_2, \cdots, X_n$ of the random variable X with pdf

$f(x;\theta_o)$, we wish to estimate θ_o. The negative log-likelihood of θ is

$$(1.28) \quad L_n(\theta) = -\log\alpha + \alpha\log\lambda +$$

$$-\frac{\alpha-1}{n}\sum_{i=1}^{n}\log(X_i - \gamma) + \frac{1}{n}\sum_{i=1}^{n}\left(\frac{X_i - \gamma}{\sigma}\right)^\alpha .$$

The associated maximum likelihood problem is not as clean as for the normal model, say. As a matter of fact, the mle does not exist, since for $\alpha < 1$, and for all λ, the choice $\gamma_n = X_{1,n} = \min\{X_1, X_2, \cdots, X_n\}$ gives $L_n(\gamma_n, \lambda, \alpha) = -\infty$. So $L_n(\theta)$ does not admit a global minimum. However, if the true α_o satisfies $\alpha_o > 1$, then one can expect a local minimum of $L_n(\theta)$ with $\alpha_n > 1$. This solution θ_n is still referred to as the mle. If this local minimum does not exist, then it seems all one can do is fudge the estimator by taking $\gamma_n = X_{1,n} - (1/n)$, say, and determining λ_n and α_n by minimizing $L_n(\theta)$. The performance of the mle for all three families on simulated data is discussed in Chapter 3.

Much more interesting is how one decides which parametric family of pdfs one chooses to model the right-skewed, heavy-tailed data. The consequences of this choice are illustrated on some real data sets (in which one does *not* know the correct model) in the next chapter.

(1.29) EXERCISE (Shifted Exponentials). Let $X_1, X_2, \cdots, X_n$ be iid random variables with common pdf

$$f(x;\gamma_o, \lambda_o) = \begin{cases} \lambda_o\exp\left(-\lambda_o(x - \gamma_o)\right) , & x \geq \gamma_o , \\ 0 & , \quad\text{otherwise} . \end{cases}$$

Let $X_{1,n} \leqslant X_{2,n} \leqslant \cdots \leqslant X_{n,n}$ denote the order statistics. Show that the maximum likelihood estimators of γ and λ are given by

$$\gamma_n = X_{1,n} , \qquad \lambda_n = \left\{\frac{1}{n}\sum_{i=2}^{n} X_{i,n} - \gamma_n\right\}^{-1} .$$

(1.30) EXERCISE. Let $X_1, X_2, \cdots, X_n$ be iid random variables with the shifted double exponential as common pdf

$$f(x;\theta) = (2\sigma)^{-1}\exp\left(-\sigma^{-1}|x - \mu|\right) , \quad x \in \mathbb{R} .$$

Show that the mle for μ and σ are given by

$$\mu_n = \text{median}\left(X_1, X_2, \cdots, X_n\right) , \qquad \sigma_n = \frac{1}{n}\sum_{i=1}^{n} |X_i - \mu_n| .$$

This concludes our discussion of some typical examples of parametric maximum likelihood estimation

The remainder of the chapter is organized as follows. In §2, we characterize the "best" estimators for fixed n and consider the convergence of maximum likelihood estimators for fairly general parametric families. In

§§ 3 and 4, we consider some basic computational procedures for solving the maximum likelihood problem (1.2), viz. the Newton-Raphson method (sequential quadratic programming) and the EM algorithm. In §§ 5 and 6, we briefly consider robust estimation and ridge regression. In these sections, the emphasis is somewhat different. For the ideal model, the mle's exist and have the standard (asymptotic) optimality properties, but these properties are irrelevant when the data have a few "gross" errors, or many small errors. Now, one would like the estimators to be more or less insensitive (robust) with respect to such failures. This is accomplished by suitably modifying the maximum likelihood problem. Finally, in § 7, maximum likelihood estimation for some right-skewed heavy-tailed distributions is examined.

EXERCISES : (1.15), (1.22), (1.25), (1.29), (1.30).

2. Optimality of maximum likelihood estimators

The main reason for the popularity of maximum likelihood estimation is that it is based on a well-founded probabilistic principle, which implies that mle's often are optimal in a statistically meaningful sense. This optimality may be for the finite sample case, or only asymptotically as the sample size tends to ∞. In this section, we concentrate on the asymptotic distribution of mle's, mainly because in general it is impossible to determine the finite sample distribution (other than through simulations). It is only in simple examples that the mle can be determined explicitly in closed form, and only in these cases that the distribution of the mle can be determined analytically. However, the touchstone for asymptotic properties is still what one would call "best" for finite sample problems.

Finite sample optimality. Let $X = (X_1, X_2, \cdots, X_n) \in \mathbb{R}^{n \times d}$ be a random variable with pdf $f(x; \theta_o)$ as in (1.1). For a function g defined on $\mathbb{R}^{n \times d}$, let

$$(2.1) \qquad \mathbb{E}_\theta[\, g(X)\,] = \int_{\mathbb{R}^{n \times d}} g(x) f(x; \theta) \, dx \ .$$

Here, the function g can be real-, vector-, or matrix-valued. Because $\mathbb{E}_{\theta_o}$ is rather awkward looking, we let $\mathbb{E}_o = \mathbb{E}_{\theta_o}$. Now, let θ_n be an estimator (not necessarily an mle) of θ_o. What would it mean for θ_n to be a "best" estimator ? It certainly seems desirable that θ_n be an unbiased estimator, irrespective of the value of θ_o, i.e.,

$$(2.2) \qquad \mathbb{E}_o\big[\, \theta_n \,\big] = \theta_o \ , \quad \text{for all } \theta_o \in \Theta \ .$$

If (2.2) holds, we say that θ_n is a *globally* unbiased estimator of θ_o. In the scalar case, $\theta \in \mathbb{R}$, another desirable feature of an estimator is that

$\mathbb{E}\left[\,|\,\theta_n - \theta_o\,|^2\,\right]$ be as small as possible. In the multiparameter case $\theta \in \mathbb{R}^p$, we would like that for all $\theta_o \in \Theta$,

$$(2.3) \qquad \mathbb{E}_o\left[\,(\theta_n - \theta_o)(\theta_n - \theta_o)^{\top}\,\right] \preccurlyeq \mathbb{E}_o\left[\,(\widetilde{\theta}_n - \theta_o)(\widetilde{\theta}_n - \theta_o)^{\top}\,\right],$$

for any other globally unbiased estimator $\widetilde{\theta}_n$ of θ_o. Here, we write $A \preccurlyeq B$ for symmetric matrices A, $B \in \mathbb{R}^{p \times p}$ if $B - A$ is semipositive-definite. If an estimator θ_n satisfies (2.2) and (2.3), it is called a globally minimum variance unbiased estimator. LEHMANN (1983) calls these uniformly minimum variance unbiased estimators (UMVU), but later on we use "uniformly" in its Advanced Calculus sense. However, we do use the acronym UMVU, but sometimes we just call them "best" estimators. (Actually, one could argue that the "best" estimator θ_n would be one for which $\mathbb{E}_o\left[\,(\theta_n - \theta_o)(\theta_n - \theta_o)^{\top}\,\right]$ is as small as possible, in the sense of (2.3), over all possible estimators, be they unbiased or not. But that is another story.) If $\mathbb{E}_o[\,\theta_n\,] = \theta_o$, then

$$(2.4) \qquad \mathbb{V}\mathrm{ar}_o\left[\,\theta_n\,\right] = \mathbb{E}_o\left[\,(\theta_n - \theta_o)(\theta_n - \theta_o)^{\top}\,\right]$$

is the variance-covariance matrix, or covariance matrix for short, of θ_n.

Do UMVU's in fact exist, and do mle's have this distinction? In § 1, we saw some examples of globally unbiased mle's, but that is far from being UMVU's. We also saw examples of biased mle's that were easily transformed into globally unbiased estimators. We should also keep in mind the situation in which we only want to estimate some function $h(\theta_o)$ of θ_o, say, the third moment or the 95th percentile of the pdf $f(x;\theta_o)$, or the probability of some event. Under what circumstances are mle's "best" estimators? The following two theorems spell it out. The first one gives a lower bound on the variance of a globally unbiased estimator. In some instances in which the mle is unbiased, it actually achieves this bound and, hence, is UMVU. The second theorem covers some other instances where the mle is UMVU. The bound is expressed in terms of the *Fisher information matrix* $\Phi(\theta) \in \mathbb{R}^{p \times p}$, defined as

$$(2.5) \qquad \Phi(\theta) = \tfrac{1}{n}\,\mathbb{E}_\theta\left[\,\nabla^2_\theta\{\,-\log f(X;\theta)\,\}\,\right],$$

assuming its existence. Here, and below, ∇^s_θ denotes the derivative of order s with respect to θ, so

$$\nabla_\theta \log f(x;\theta) \in \mathbb{R}^p\,, \quad \nabla^2_\theta \log f(x;\theta) \in \mathbb{R}^{p \times p}\,, \quad \nabla^3_\theta \log f(x;\theta) \in \mathbb{R}^{p \times p \times p}$$

are, respectively, the gradient of $\log f(x;\theta)$ with respect to θ, the matrix of second-order derivatives, and the three-way array of third-order derivatives with respect to θ. Loosely speaking, $\Phi(\theta)$ is the Hessian (the "second derivative") of the expected value of the negative log-likelihood function. The basic assumption is that

$$(2.\mathrm{A}1) \qquad \begin{array}{c} \Phi(\theta) \text{ exists and is a positive-definite} \\ \text{matrix for all } \theta \text{ close enough to } \theta_o\,. \end{array}$$

We consider the case of estimating $h(\theta_o)$, where $h : \Theta \longrightarrow \mathbb{R}^q$ is continuously differentiable, with the assumption that

(2.A2) $\nabla h(\theta_o)$ has full row rank .

Note that ∇ (without subscripts) denotes the gradient with respect to *all* of the variables.

(2.6) THEOREM. *If $\widetilde{\gamma}_n$ is any globally unbiased estimator of $h(\theta_o)$, then*

$$\nabla h(\theta_o) \left[n\Phi(\theta_o) \right]^{-1} \left[\nabla h(\theta_o) \right]^{T} \prec \mathbb{V}\mathrm{ar}_o \left[\widetilde{\gamma}_n \right] .$$

For a proof, see RAO (1973). For $p = 1$, $\nabla h(\theta_o) \left[n\Phi(\theta_o) \right]^{-1} \left[\nabla h(\theta_o) \right]^{T}$ is the Cramér-Rao lower bound for the variance of any globally unbiased estimator of $h(\theta_o)$. In fact, the above theorem shows that the multivariate case is "the same".

(2.7) THEOREM. *Let θ_n be an mle of θ_o, and assume that $h(\theta_n)$ is a globally unbiased estimator of $g(\theta_o)$. If $g(\theta_o)$ has a unique globally minimum variance unbiased estimator, then it is $h(\theta_n)$.*

For a proof, see LEHMANN (1983). So the situation is somewhat disappointing. However, the mle behaves about as well as one can expect from any general estimation procedure. Note that there are cases in which no globally unbiased estimators exist. The following example/exercise is from LEHMANN (1983).

(2.8) EXERCISE. Consider a discrete random variable with binomial distribution B(p,m), i.e., for some $0 \leqslant p \leqslant 1$,

$$\mathbb{P}[\, X = k \,] = \binom{m}{k} \, p^k \, (1 - p)^{m-k} , \quad 0 \leqslant k \leqslant m .$$

Suppose we wish to estimate $h(p) = 1/p$. Show that no unbiased estimator of $h(p)$ exists, in the sense of (2.2). Likewise, when p is restricted *a priori* to an interval $[a, b]$, with $0 < a < b \leqslant 1$.

This concludes our description of finite sample optimality of estimators. There is a beautiful theory providing definitive answers to the existence of UMVU's and related questions, involving such concepts as minimal sufficient statistics and exponential families. It takes LEHMANN (1983) several hundred pages to properly treat this, and the interested reader is referred there.

Almost sure convergence (consistency). If one cannot find a function g such that $g(\theta_n)$ is an unbiased estimator of θ_o, then in general very little can be said about the optimality of the mle for a fixed sample size n. However, for the setting of iid observations $X_1, X_2, \cdots, X_n$ as in (1.3), the mle does have *asymptotic* optimality properties under fairly mild regularity conditions. So, for the remainder of this section, we assume that

$$X_1, X_2, \cdots, X_n \text{ are iid, with common pdf } f(x, \theta_o) \ .$$

In this case, the negative log-likelihood is given by

$$L_n(\theta) = -\tfrac{1}{n} \sum_{i=1}^{n} f(X_i ; \theta) \ .$$

Now, consider the maximum likelihood estimation problem (1.2)–(1.3). The following amazing result of WALD (1949) shows that if problem (1.2) has any solution, then $\theta_n \longrightarrow_{as} \theta_o$ (the true value). We present a slightly stronger version of WALD's theorem, but this has no practical consequences. Here, we should also mention the weak formulation of WOLFOWITZ (1949).

First, we state the assumptions required. For any $\theta \in \Theta$, and $\varrho > 0$ and $r > 0$, define

(2.9)
$$M(x ; \theta, \varrho) = \sup \{ f(x ; \xi) \ : \ \| \xi - \theta \| < \varrho \} \ ,$$
$$m(x ; r) = \sup \{ f(x ; \theta) \ : \ \| \theta \| > r \} \ .$$

We assume that for all sufficiently small ϱ and for all sufficiently large r,

(2.A3)
$$\mathbb{E}_o \big[\, | \log M(X; \theta, \varrho) | \, \big] < \infty \quad \text{and}$$
$$\mathbb{E}_o \big[\, | \log m(X; r) | \, \big] < \infty \ ,$$

and we make the following blanket assumptions, recalling that KL denotes the Kullback-Leibler distance (1.3.18),

(2.A4)   For all $x \in \mathbb{R}^d$, $f(x ; \theta)$ is a continuous function of θ .

(2.A5) $\text{KL} \big(f(\cdot ; \theta_o), f(\cdot ; \theta) \big) > 0$ for $\theta \neq \theta_o$.

(2.A6)   For all $x \in \mathbb{R}^d$, $f(x ; \theta) \longrightarrow 0$ for $\| \theta \| \longrightarrow \infty$.

(2.A7) $\mathbb{E}_o[\, | \log f(X; \theta_o) | \,] < \infty$.

(2.A8) $\Theta \subset \mathbb{R}^p$ is closed .

(2.A8') Equivalently, $\Theta \cap \{ \| \theta \| \leqslant r \}$ is compact for every $r > 0$.

Here, we use $A \subset B$ for sets A and B to mean that A is a subset of B. It need not be a *proper* subset.

(2.10) WALD's THEOREM. *Suppose* (2.A3) *through* (2.A8) *hold, and let* $\{ \theta_n \}_n$ *be a sequence of estimators (not necessarily mle's) that satisfies* $\limsup_{n \longrightarrow \infty} L_n(\theta_n) - L_n(\theta_o) \leqslant_{as} 0$. *Then,* $\theta_n \longrightarrow_{as} \theta_o$.

(2.11) COROLLARY. *Under the assumptions* (2.A3) *through* (2.A8), *if the mle's* θ_n *exist for all* n *large enough, then* $\theta_n \longrightarrow_{as} \theta_o$.

For the proof of the theorem, we need the following three lemmas and theorem.

(2.12) LEMMA. If $\theta \neq \theta_o$, then $\mathbb{E}_o\big[-\log f(X; \theta)\big] > \mathbb{E}_o\big[-\log f(X; \theta_o)\big]$.

PROOF. It suffices to show that $\mathbb{E}_o\big[\log\{f(X; \theta_o)/f(X; \theta)\}\big] > 0$, but this is just $\mathrm{KL}\big(f(\,\cdot\,; \theta_o), f(\,\cdot\,; \theta)\big)$, the Kullback-Leibler divergence. This is positive, by assumption (2.A5).   Q.e.d.

(2.13) LEMMA.
$$\lim_{\varrho \to 0} \mathbb{E}_o[-\log M(X; \theta, \varrho)] = \mathbb{E}_o[-\log f(X; \theta_o)] + \mathrm{KL}\big(f(\,\cdot\,; \theta_o), f(\,\cdot\,; \theta)\big) .$$

PROOF. By (2.9) and (2.A4), we have that $\log M(X; \theta, \varrho) \searrow \log f(X; \theta)$ for $\varrho \longrightarrow 0$, whence by monotone convergence and (2.A3)–(2.A7),
$$\mathbb{E}_o\big[\log M(X; \theta, \varrho)\big] \searrow \mathbb{E}_o\big[\log f(X; \theta)\big] ,$$
and the result follows.   Q.e.d.

(2.14) LEMMA. For $r \to \infty$, we have $\mathbb{E}_o\big[-\log m(X; r)\big] \longrightarrow \infty$.

PROOF. Assumption (2.A6) implies that $m(x; r) \nearrow \infty$ for all $x \in \mathbb{R}^d$. We cannot immediately draw the conclusion of the lemma from the monotone convergence theorem, because there is a negative part, possibly. So consider the negative and positive parts separately. From (2.A6), it does indeed follow by monotone convergence that $\mathbb{E}_o\big[-\log(m(X; r) \wedge 1)\big] \longrightarrow \infty$. For the negative part, we note that for all $r_o > 0$, if $r > r_o$, then
$$0 \geqslant \mathbb{E}_o\big[-\log(m(X; r) \vee 1)\big] \geqslant \mathbb{E}_o\big[-\log(m(X; r_o) \vee 1)\big],$$
and so, by (2.A3), if r_o is sufficiently large, and for all $r > r_o$,
$$\mathbb{E}_o\big[\,\big|\log(m(X; r) \vee 1)\big|\,\big] \leqslant \mathbb{E}_o\big[\,\big|\log(m(X; r_o) \vee 1)\big|\,\big] < \infty .$$
Since
$$\mathbb{E}_o\big[-\log m(X; r)\big] = \mathbb{E}_o\big[-\log(m(X; r) \wedge 1)\big] + \mathbb{E}_o\big[-\log(m(X; r) \vee 1)\big] ,$$
the conclusion of the lemma follows. Note that in the above equality on the right, the event $\{\, m(X; r) = 1\,\}$ has been accounted for twice, but in this case, $\log(\, m(X; r) \wedge 1) = 0$, so no harm is done.   Q.e.d.

The following notation seems useful. For $\Xi \subset \Theta$, let

(2.15) $$L_n(\Xi) = \inf\{\, L_n(\theta) : \theta \in \Xi\,\} .$$

We also have occasion to use open balls $B(\theta, \varrho)$ defined by

(2.16) $$B(\theta, \varrho) = \{\, \xi : \|\,\xi - \theta\,\| < \varrho\,\} .$$

(2.17) THEOREM. *Assume* (2.A3) *through* (2.A8). *Let* Ξ *be a closed subset of* Θ, *with* $\theta_o \notin \Xi$. *Then*,

$$\liminf_{n\to\infty} L_n(\Xi) - L_n(\theta_o) > 0 \quad almost\ surely .$$

PROOF. We begin by noting that $\mathbb{E}_o\big[-\log f(X; \theta)\big]$ is well defined, but could equal $+\infty$, cf. the discussion at the start of the proof of Lemma (2.14). Define

$$\Xi(\infty) \stackrel{\text{def}}{=} \big\{ \theta \in \Xi : \mathbb{E}_o\big[-\log f(X; \theta)\big] = +\infty \big\} .$$

By the strong law of large numbers for all $\theta \in \Xi(\infty)$, we then have that $L_n(\theta) \longrightarrow_{as} +\infty$, and so the infimum of $L_n(\theta)$ over Ξ does not occur on $\Xi(\infty)$. We may thus assume without loss of generality that

$$\mathbb{E}_o\big[|\log f(X; \theta)|\big] < \infty \quad \text{for all} \quad \theta \in \Xi .$$

Now to the actual proof. To motivate this, note that by the strong law of large numbers for fixed θ, we have $L_n(\theta) \longrightarrow_{as} \mathbb{E}_o\big[-\log f(X; \theta)\big]$. Then,

$$L_n(\theta) - L_n(\theta_o) \longrightarrow_{as} \mathbb{E}_o\big[\log\{f(X; \theta_o)/f(X; \theta)\}\big] ,$$

which is positive for $\theta \neq \theta_o$ by Lemma (2.12). The difficulty is that we must do this uniformly in $\theta \in \Xi$. So here it goes.

By Lemma (2.14) and (2.A7), there exists positive numbers r and δ such that

(2.18) $$\mathbb{E}_o\big[-\log m(X; r)\big] = \mathbb{E}_o\big[-\log f(X; \theta_o)\big] + \delta_0 .$$

Let $\Xi_r = \{ \theta \in \Xi : \|\theta\| \leqslant r \}$. Then,

$$L_n(\Xi) = \min\big(L_n(\Xi_r), L_n(\Xi \backslash \Xi_r) \big) ,$$

where $\Xi \backslash \Xi_r = \{ x \in \Xi : x \notin \Xi_r \}$. By the definition (2.9) of m,

(2.19) $$L_n(\Xi \backslash \Xi_r) \geqslant -\frac{1}{n} \sum_{i=1}^{n} \log m(X_i; r) .$$

Now, consider $L_n(\Xi_r)$. Let

$$\delta(\theta) = \tfrac{1}{2} \text{KL}\big(f(\,\cdot\,; \theta_o), f(\,\cdot\,; \theta) \big) .$$

By Lemma (2.13) for each $\theta \in \Xi_r$, there exists a $\varrho(\theta) > 0$ such that

(2.20) $$\mathbb{E}_o\big[-\log M\big(X; \theta, \varrho(\theta)\big)\big] = \mathbb{E}_o\big[-\log f(X; \theta_o)\big] + \delta(\theta) .$$

Since Ξ_r is compact and $\{ B\big(\theta, \varrho(\theta)\big) : \theta \in \Xi_r \}$ is an open covering of Ξ_r, there exists a finite subcover, i.e., there exist $\theta_1, \theta_2, \cdots, \theta_k \in \Xi_r$ such that

$$\Xi_r \subset \bigcup_{j=1}^{k} B(\theta_j, \varrho_j) ,$$

where $\varrho_j = \varrho(\theta_j)$ for all j. Likewise, let $\delta_j = \delta(\theta_j)$. Then,

$$(2.21) \quad L_n(\Xi_r) \geqslant L_n(\overset{k}{\underset{j=1}{\cup}} B(\theta_j, \varrho_j)) = \min_{1 \leqslant j \leqslant k} -\frac{1}{n} \sum_{i=1}^{n} \log M(X_i; \theta_j, \varrho_j) \ .$$

Now, by the strong law of large numbers,

$$-\frac{1}{n} \sum_{i=1}^{n} \log M(X_i, \theta_j, \varrho_j) \longrightarrow_{\text{as}} \mathbb{E}_o\left[-\log M(X; \theta_j, \varrho_j) \right] \ ,$$

$$-\frac{1}{n} \sum_{i=1}^{n} \log m(X_i, r) \longrightarrow_{\text{as}} \mathbb{E}_o\left[-\log m(X; r) \right] \ .$$

Since $L_n(\theta_o) \longrightarrow_{\text{as}} \mathbb{E}_o[-\log f(X; \theta_o)]$, it follows that

$$\liminf_{n \to \infty} L_n(\Xi) - L_n(\theta_o) \geqslant_{\text{as}} \min\{ \delta_j : 0 \leqslant j \leqslant k \} > 0 \ ,$$

with δ_0 defined in (2.18) and $\delta_j = \delta(\theta_j)$. This concludes the proof.   Q.e.d.

PROOF OF WALD'S THEOREM (2.10). The proof goes by way of contra-diction. Let $\varepsilon > 0$ be arbitrary, and suppose that infinitely many θ_n satisfy

$$\theta_n \in \Xi \overset{\text{def}}{=} \{ \xi \in \Theta : \| \xi - \theta_o \| \geqslant \tfrac{1}{2}\varepsilon \} \ .$$

So, a subsequence $\{ \theta_{n_k} \}_k$ lies in Ξ. Obviously, $\Xi \subset \Theta$ is closed, and $\theta_o \notin \Xi$, so that by Theorem (2.17),

$$\liminf_{k \to \infty} L_{n_k}(\theta_{n_k}) - L_{n_k}(\theta_o) \geqslant \liminf_{k \to \infty} L_{n_k}(\Xi) - L_{n_k}(\theta_o)$$

$$\geqslant \liminf_{n \to \infty} L_n(\Xi) - L_n(\theta_o) > 0 \quad \text{almost surely} \ ,$$

but this contradicts the premise of the theorem. The conclusion is that almost surely only finitely many θ_n lie in Ξ. Since $\varepsilon > 0$ was arbitrary, the theorem follows.   Q.e.d.

(2.22) REMARK. In the above proof, establishing that the set Ξ is closed and bounded is the crucial step, which is why assumption (2.A8) was made.

We have just shown that if maximum likelihood estimators exists, then they converge a.s. to the true value. What can be said about the existence of maximum likelihood estimators? Even for specific parametric families $f(\,\cdot\,; \theta)$, this is typically a messy question, and as the Weibull model of §1 demonstrates, nonexistence is a serious possibility. Be that as it may, asymptotically, there is no difficulty, because local minimizers exist and have the requisite properties.

(2.23) THEOREM. Let $A \subset \Theta$ be an open neighborhood of θ_o. Under the assumptions (2.A1) through (2.A8), $L_n(\theta)$ has a local minimum θ_n in A for all n large enough, and $\theta_n \longrightarrow_{\text{as}} \theta_o$.

PROOF. The proof is the multivariate analogue of the proof in RAO (1973), p. 364. Consider the set ∂A, the *boundary* of the set A. The set ∂A is obviously closed and $\theta_o \notin \partial A$. By Theorem (2.17), there exists a $\delta > 0$ such that for all n large enough and for all $\theta \in \partial A$,

$$L_n(\theta) \geqslant_{as} L_n(\theta_o) + \delta .$$

This implies that the minimum of $L_n(\theta)$ on the set $A \cup \partial A$ occurs in A. But then this is a local minimum. Since obviously $L_n(\theta_n) \leqslant L_n(\theta_o)$, WALD's Theorem (2.10) shows that $\theta_n \longrightarrow_{as} \theta_o$.   Q.e.d.

Asymptotic optimality. Besides the almost sure convergence of the (local) minimum θ_n of $L_n(\theta)$, we would also like to know its asymptotic distribution. To say anything about this at all, we need some added assumptions.

(2.A9)   Let $A \subset \Theta$ be an open neighborhood of θ_o .

(2.A10)   For a.e. $x \in \mathbb{R}^d$ and for $s = 1, 2, 3$,

$$\nabla_\theta^s \{ \log f(x\,;\,\theta) \} \text{ is a continuous function of } \theta \in A .$$

(2.A11)   There exists an integrable function N on $\mathbb{R}^d$

such that for all $\theta \in A$ and $s = 1, 2$,

$$\left\| \nabla_\theta^s f(x\,;\,\theta) \right\| \leqslant N(x) , \quad x \in \mathbb{R}^d .$$

(2.A12)   There exists a function M on $\mathbb{R}^d$

with $\mathbb{E}_o\left[M(X) \right] < \infty$, such that for all $\theta \in A$,

$$\left\| \nabla_\theta^3 \{ \log f(x\,;\,\theta) \} \right\| \leqslant M(x) , \quad x \in \mathbb{R}^d .$$

Since $\mathbb{R}^p$, $\mathbb{R}^{p \times p}$ and $\mathbb{R}^{p \times p \times p}$ are finite dimensional, the choice of norms on these spaces has no effect on the final results, so we use the Frobenius norm (the square root of the sum of squares of the components).

(2.24) ASYMPTOTIC NORMALITY THEOREM. *Let $f(x; \theta)$ satisfy conditions (2.A1) through (2.A12), and let θ_n be local minima of $L_n(\theta)$ in the set A. Then,*

$$\sqrt{n}\,(\theta_n - \theta_o) \longrightarrow_d Y \sim N\left(0, \left[\Phi(\theta_o)\right]^{-1}\right) .$$

PROOF. Let $g(\theta) = \nabla L_n(\theta) \in \mathbb{R}^p$, and denote its components by $g_j(\theta)$, $j = 1, 2, \cdots, p$. By (2.A10), we may use Taylor's theorem with remainder on the functions $g_j\left(\theta_o + t\,(\theta_n - \theta_o)\right)$, $0 \leqslant t \leqslant 1$, to conclude that

$$g_j(\theta_n) = g_j(\theta_o) + (\theta_n - \theta_o)^T \left[\nabla\{g_j\}\right](\theta_o) + \mathcal{R}_j\,(\theta_n - \theta_o) ,$$

where $\mathcal{R}_j \in \mathbb{R}^{1 \times p}$ is defined as

(2.25) $\mathcal{R}_j = \frac{1}{2}\,(\theta_n - \theta_o)^T \left[\nabla^2\{g_j\}\right](\xi_j) , \quad j = 1, 2, \cdots, p ,$

in which $\xi_j = \theta_o + t_j(\theta_n - \theta_o)$ for some t_j, $0 < t_j < 1$. We write this compactly as

$$(2.26) \qquad \nabla L_n(\theta_n) = \nabla L_n(\theta_o) + H_n(\theta_o)(\theta_n - \theta_o) + R_n(\theta_n - \theta_o),$$

in which $H_n(\theta) = [\nabla^2 l_n](\theta)$ and

$$(2.27) \qquad R_n = [\, \mathcal{R}_1^{\tau} \,|\, \mathcal{R}_2^{\tau} \,|\, \cdots \,|\, \mathcal{R}_p^{\tau} \,]^{\tau}.$$

Since $\nabla L_n(\theta_n) = 0$ if θ_n is an mle, and assuming that $H_n(\theta_o) + R_n$ is invertible, we obtain that

$$(2.28) \qquad \theta_n - \theta_o = -[\, H_n(\theta_o) + R_n \,]^{-1} \nabla L_n(\theta_o).$$

From the lemmas below we have that $H_n(\theta_o) \longrightarrow_p \Phi(\theta_o)$, the Fisher information matrix, see (2.5), and $R_n \longrightarrow_p 0$. Since $\Phi(\theta_o)$ is assumed to be invertible, then

$$E_n \overset{\text{def}}{=} [\,\Phi(\theta_o)\,]^{-1} (H_n(\theta_o) - \Phi(\theta_o) + R_n) \longrightarrow_p 0$$

as well. Consequently,

$$[\, H_n(\theta_o) + R_n \,]^{-1} = [\, I + E_n \,]^{-1} [\Phi(\theta_o)]^{-1} \longrightarrow_p [\Phi(\theta_o)]^{-1}.$$

We need to borrow one more result, viz. $\nabla L_n(\theta_o) \longrightarrow_d \Lambda \sim N(0, \Phi(\theta_o))$, to complete the proof. \hfill Q.e.d.

Here are the three lemmas we jumped the gun on.

(2.29) LEMMA. $\sqrt{n}\, L_n(\theta_o) \longrightarrow_d \Lambda \sim N(0, \Phi(\theta_o))$.

(2.30) LEMMA. $[\nabla^2 L_n](\theta_o) \longrightarrow_p \Phi(\theta_o)$.

(2.31) LEMMA. $R_n \longrightarrow_p 0$.

PROOF OF LEMMA (2.29). Observe that $\nabla L_n(\theta_o) = -\frac{1}{n} \sum_{i=1}^{n} Z_i$ is a sum of iid random variables $Z_i = \nabla_\theta \log f(X_i; \theta_o)$, with means

$$\mathbb{E}_o[\, Z_i \,] = \int_{\mathbb{R}^d} \nabla_\theta f(x; \theta_o)\, dx = \nabla_\theta \Big\{ \int_{\mathbb{R}^d} f(x; \theta_o)\, dx \Big\} = 0,$$

because the integral is independent of θ (it equals 1). The differentiation under the integral sign is the content of Lebesgue's theorem and is allowed here by assumption (2.A11). The covariance matrix of Z_i is given by

$$\mathbb{C}\text{ov}(Z_i) = \int_{\mathbb{R}^d} f(x; \theta_o)\, \{\, \nabla_\theta \log f(x; \theta_o) \,\} \{\, \nabla_\theta \log f(x; \theta_o) \,\}^{\tau}\, dx$$

$$= \int_{\mathbb{R}^d} \{\, \nabla_\theta f(x; \theta_o) \,\} \{\, \nabla_\theta f(x; \theta_o) \,\}^{\tau}\, \frac{dx}{f(x; \theta_o)}.$$

Now,

$$f(x;\theta)\,\nabla^2_\theta \log f(x;\theta) = \nabla^2_\theta f(x;\theta) - \frac{\{\nabla_\theta f(x;\theta)\}\{\nabla_\theta f(x;\theta)\}^{\,r}}{f(x;\theta)}\,,$$

and the first term on the right integrates to 0, again by Lebesgue's theorem and (2.A11). It follows that $\mathbb{C}\mathrm{ov}(Z_i) = n\,\Phi(\theta_o)$. By (2.A1), the multivariate central limit theorem applies and gives the desired result.   Q.e.d.

PROOF OF LEMMA (2.30). The matrix $H_n(\theta_o)$ is given as a sum of iid (matrix-valued) random variables

$$H_n(\theta_o) = -\frac{1}{n}\sum_{i=1}^n \nabla^2_\theta \log f(X_i\,;\,\theta_o)\,,$$

which have finite means by (2.A1). Then, the weak law of large numbers implies that $H_n(\theta_o) \longrightarrow_p \Phi(\theta_o)$.   Q.e.d.

PROOF OF LEMMA (2.31). Recall the definitions (2.25) and (2.27) of R_n and the ξ_j. As a first step, note that $\theta_n - \theta_o \longrightarrow_{as} 0$ by Corollary (2.11). So, to show that $R_n \longrightarrow_p 0$, it suffices to show that $[\nabla^2\{g_j\}](\xi_j)$ converges in probability to some finite limit.

Since $\theta_n - \theta_o \longrightarrow_{as} 0$, then the ξ_j of (2.25) satisfy $\xi_j \longrightarrow_{as} \theta_o$, for $j = 1, 2, \cdots, p$, and so the ξ_j lie in an open neighborhood A of θ_o. By (2.A12), there exists a function M with $\mathbb{E}[\,M(X)\,] < \infty$, such that

$$\big\| \,[\nabla^2\{g_j\}](\xi_j)\, \big\| \leqslant \frac{1}{n}\sum_{i=1}^n M(X_i) \qquad \text{for all } j\,,$$

at least for all n large enough. It follows that $[\nabla^2\{g_j\}](\xi_j) \longrightarrow_p \Gamma_j$ for finite $\Gamma_j \in \mathbb{R}^{p\times p}$, by the weak law of large numbers.   Q.e.d.

(2.32) REMARK. In view of Theorem (2.6) for the finite sample case, Theorem (2.24) implies that the mle is a *best asymptotically normal* (BAN) estimator, with "best" analogous to the sense of globally minimum variance unbiased estimators in the finite sample case. It should be noted, though, that many other estimators may have this property. See Theorem (3.8).

(2.33) REMARK. It should be observed that the proof of Theorem (2.24) involves little more than Taylor's theorem with remainder, and the standard convergence theorems from probability. Actually, the little more is the a.s. convergence of θ_n, and that was quite a bit harder to show.

(2.34) REMARK. The above development is the classical approach to asymptotics, see, e.g., the proof presented in CRAMÉR (1946). The only blemish on our presentation is condition (2.A10), which with some care can be eliminated. This was the state of affairs 50 years ago. Stronger results under milder regularity conditions can be proven using the modern approach to asymptotic efficiency due to LE CAM (1970), see IBRAGIMOV and HAS'MINSKII (1981).

(2.35) REMARK. Working from (2.25) and using the strong, instead of the weak, law of large numbers in Lemmas (2.30) and (2.31), and applying the law of the iterated logarithm in Lemma (2.29), one can show that

$$(2.36) \qquad \| \theta_n - \theta_o \| =_{\mathrm{as}} \mathcal{O}\big((n^{-1} \log \log n)^{1/2} \big) .$$

Such a rate is called the *parametric rate of convergence*, cf. § 1.4.

(2.37) EXERCISE. Under the conditions of Theorem (2.24) show that (2.36) holds.

Finally, the conditions of Theorem (2.24) are satisfied by many familiar parametric families of pdfs. However, as the following exercise illustrates, there are parametric families for which they are not met.

(2.38) EXERCISE. Verify whether the following parametric families satisfy the assumptions of (2.A1) through (2.A12).
(a) The univariate normal family (or multivariate $N_d(\mu, \sigma^2 I)$–family).
(b) The shifted exponential family

$$f(x\,;\,\lambda, \gamma) = \begin{cases} \lambda \exp\big(-\lambda(x - \gamma)\big) , & x \geqslant \gamma , \\ 0 & , \quad \text{otherwise} , \end{cases}$$

with $(\lambda, \gamma)^{T} \in \mathbb{R}_{++} \times \mathbb{R}$.
(c) The log-normal family $f(x\,;\,\theta)$, defined as

$$f(x\,;\,\theta) = \frac{1}{\sqrt{2\pi}\, \sigma\, (x - \gamma)} \exp\left[-\frac{\big(\log(x - \gamma) - \mu\big)^2}{2\sigma^2} \right] , \quad x \geqslant \gamma ,$$

and $= 0$ otherwise, with $\theta = (\gamma, \mu, \sigma)^{T} \in \mathbb{R} \times \mathbb{R} \times \mathbb{R}_{++}$.
(d) The logistic family

$$f(x; \theta) = \sigma^{-1} \lambda\big(\sigma^{-1} (x - \mu)\,;\, \alpha \big) , \quad x \in \mathbb{R} ,$$

where

$$\lambda(\, t\,;\, \alpha) = \alpha\, e^{-t} (1 + e^{-t})^{-\alpha - 1} , \quad t \geqslant 0 ,$$

with $\theta = (\mu, \sigma, \alpha)^{T} \in \Theta = \mathbb{R} \times \mathbb{R}_{++} \times \mathbb{R}_{++}$. For more about different techniques of applied parametric estimation for the various distributions considered here, see, e.g., JOHNSON, KOTZ and BALAKRISHNAN (1994), and references therein.

(2.39) EXERCISE. (a) Suppose $\{\, f(x\,;\,\theta) \,:\, \theta \in \Theta \,\}$ is a parametric family that satisfies the assumptions (2.A1) through (2.A12). Consider the parametric family

$$\varphi(x \,;\, w, \theta) = \sum_{j=1}^{m} w_j \, f(x \,;\, \theta_j) \,, \quad x \in \mathbb{R} \,,$$

with $(w^{\scriptscriptstyle T}, \theta^{\scriptscriptstyle T})^{\scriptscriptstyle T} \in S_{m-1} \times \Theta^m$, in which $S_{m-1} = \{\, w \in \mathbb{R}^m_+ \,:\, \sum_{j=1}^{m} w_j = 1 \,\}$ is the unit simplex in $\mathbb{R}^m$. Show that this new parametric family satisfies (2.A1) through (2.A12).
(b) Draw your conclusions about mixtures of normals and shifted exponentials.

(2.40) EXERCISE. Let $X_1, \dots, X_n$ be iid random variables with pdf

$$f(x; \gamma, \mu) = \begin{cases} \gamma (x - \mu)^{\gamma - 1} \exp\big(-(x - \mu)^{\gamma}\big) \,, & x \geq \mu \,, \\ 0 & , \quad \text{otherwise} \,, \end{cases}$$

where γ is a *known* positive number and $\mu \in \mathbb{R}$ is the unknown parameter.
(a) Show that if $\gamma = 3$, the conditions (2.A1) through (2.A12) are satisfied, and if $\gamma = 1$, they are not. (The actual cutoff point is $\gamma = 2$.) Conclude that the complete family, i.e., with γ unknown as well, does not satisfy (2.A1) through (2.A12). In other words, the Weibull family does not satisfy the assumptions.
(b) If $\gamma = 1$, show that $X_{1,n} = \min(X_1, X_2, \cdots, X_n)$ is the mle of μ, and that the pdf of $X_{1,n}$ is given by

$$f(x; \lambda, \gamma, n) = \begin{cases} n \exp\big(-n(x - \mu)\big) \,, & x \geq \mu \,, \\ 0 & , \quad \text{otherwise} \,. \end{cases}$$

Also, show that $\sqrt{n}\,(X_{1,n} - \gamma) \longrightarrow_p 0$ and $n\,(X_{1,n} - \gamma) \longrightarrow_d Y$, where Y is an exponential random variable with $\lambda = 1$.

The following two exercises concern the case in which one wants to estimate a *function* of θ_o, cf. Theorems (2.6) and (2.7).

(2.41) EXERCISE. Let $h : \mathbb{R}^p \longrightarrow \mathbb{R}^p$. Give conditions (similar to (2.A9) through (2.A12)?) such that the mle of $h(\theta_o)$ is $h(\theta_n)$.

(2.42) EXERCISE. Let $h : \mathbb{R}^p \longrightarrow \mathbb{R}^p$. Give conditions similar to (2.A9) through (2.A12) such that

$$\sqrt{n}\,\big(h(\theta_n) - h(\theta_o)\big) \longrightarrow_d Y \sim N\big(0, V(\theta_o)\big) \,,$$

where

$$V(\theta_o) = \nabla h(\theta_o) \big[\, \Phi(\theta_o)\,\big]^{-1} \big[\, \nabla h(\theta_o)\,\big]^{\scriptscriptstyle T}$$

(2.43) EXERCISE. (a) Let $A \overset{\text{def}}{=} \{ \theta : \| \theta - \theta_o \| < \varepsilon \}$ for some $\varepsilon > 0$. Assume that the parametric family $f(x; \theta)$ satisfies for all $\theta \in A$,

$$\| \nabla_\theta f(x; \theta) \| \leqslant M(x) , \quad x \in \mathbb{R}^d ,$$

with $\int_{\mathbb{R}^d} M(x)\, dx < \infty$. Show that for all $\theta \in A$,

$$\int_{\mathbb{R}^d} | f(x; \theta) - f(x; \theta_o) |\, dx \leqslant m \, \| \theta - \theta_o \| ,$$

for a suitable constant m.

(b) Suppose that the assumptions (2.A1) through (2.A12) hold. Let θ_n be a minimizer of $L_n(\theta)$ on A. Show that

$$\int_{\mathbb{R}^d} | f(x; \theta_n) - f(x; \theta_o) |\, dx =_{\text{as}} \mathcal{O}\big((n^{-1} \log \log n)^{1/2} \big) .$$

(2.44) EXERCISE. (a) State and prove analogues of Theorems (2.23) and (2.24) for the least-squares estimation problem (1.1.13).

(b) For the scalar case $\Theta \subset \mathbb{R}$, the Cramér-Rao lower bound implies that

$$\left\{ \int_{\mathbb{R}} | f_\theta(x; \theta) |^2\, dx \right\}^2 \leqslant \left\{ \int_{\mathbb{R}} \frac{| f_\theta(x; \theta) |^2}{f(x; \theta)}\, dx \right\} \times$$

$$\left\{ \int_{\mathbb{R}} | f_\theta(x; \theta) - t |^2 f(x; \theta)\, dx \right\} ,$$

where $t = \int_{\mathbb{R}} f_\theta(x; \theta) f(x; \theta)\, dx$ and $f_\theta(x; \theta)$ is the derivative of $f(x; \theta)$ with respect to θ. Find an independent proof for this inequality, under the conditions stated in (a) (specialized to the scalar case).

EXERCISES : (2.37), (2.38), (2.39), (2.40), (2.48), (2.42), (2.43), (2.44).

3. Computing maximum likelihood estimators

In this section and the next, we study the computation of the maximum likelihood estimator θ_n in the problem (1.2), under the assumption that θ_n exists and is unique. Actually, in view of WALD's Theorem (2.10) and the Asymptotic Normality Theorem (2.24), this assumption may be weakened to the extent that $L_n(\theta)$ has a minimum in a neighborhood of θ_o (the true parameter). Assuming the existence of θ_n, how can we get hold of it ? Experience shows that it is uncommon that solutions can be found in closed form. This being the case, we must have recourse to *iterative* optimization methods. In the past, statisticians have shown a predilection toward Newton-Raphson methods for the solution of $\nabla L_n(\theta) = 0$. This is not necessarily the same as finding (local) minima of $L_n(\theta)$, but it is satisfactory nevertheless, as we discuss in the remainder of this section.

The basis for Newton-Raphson procedures is the linear approximation of $\nabla L_n(\theta)$ at a guess ξ for θ_n

(3.1) $$\nabla L_n(\theta) = \nabla L_n(\xi) + H_n(\xi)(\theta - \xi) + r_n ,$$

with $H_n(\xi) = \nabla^2 L_n(\xi)$, as in (2.26), and with r_n the remainder. As a matter of fact, (3.1) is the definition of r_n. The usefulness stems from the fact that for θ close to ξ, the reminder r_n is negligible compared with the linear term, see (2.25)–(2.27), and Lemma (2.31). Instead of solving the equation $\nabla L_n(\theta) = 0$ directly, in the Newton-Raphson method, the right-hand side of (3.1) is set equal to 0, ignoring the term r_n, and the resulting equation is solved for θ. Starting from an initial guess $\xi = \theta^1$, new and improved guesses θ^q are computed by

(3.2) $$\theta^{q+1} = \theta^q - \left[H_n(\theta^q) \right]^{-1} \nabla L_n(\theta^q) , \quad q = 1, 2, \cdots .$$

In general, it is hard to guarantee that this iterative process will converge, but the appeal of the method is partly explained by the fact that if this procedure converges, it converges very rapidly. The following derivation is instructive. Let us assume that $\nabla L_n(\theta_n) = 0$, and that $\theta^1 - \theta_n$ is small. Then,

(3.3) $$\theta^2 - \theta_n = \theta^1 - \theta_n - \left[H_n(\theta^1) \right]^{-1} \left(\nabla L_n(\theta^1) - \nabla L_n(\theta_n) \right) .$$

Taylor series with remainder gives

(3.4) $$\nabla L_n(\theta_n) = \nabla L_n(\theta^1) + H_n(\theta^1)(\theta_n - \theta^1) + \varrho_n ,$$

with $\varrho_n \leqslant c_n \| \theta^1 - \theta_n \|^2$, where c_n is the maximum of $\| \nabla^3 L_n(\theta) \|$ on the set $\mathrm{rect}(\theta^1, \theta_n)$, the rectangle with sides parallel to the coordinate axes, and with opposite vertices θ^1 and θ_n,

(3.5) $$\mathrm{rect}(\xi, \theta) = \left\{ \eta \in \mathbb{R}^p \;\middle|\; \begin{matrix} \forall q \; \exists\, t_q \in [0,1] : \\ \eta_q = t_q\, \xi_q + (1 - t_q)\, \theta_q \end{matrix} \right\} .$$

Now, substituting (3.4) into (3.3) gives $\theta^2 - \theta_n = \left[H_n(\theta^1) \right]^{-1} \varrho_n$ so that

(3.6) $$\| \theta^2 - \theta_n \| \leqslant c_n \| \left[H_n(\theta^1) \right]^{-1} \| \| \theta^1 - \theta_n \|^2 .$$

This shows that the error is *squared* in each iterative step.

This result may be exploited in an unexpected fashion, and provides for further appeal of the Newton-Raphson method. Note that estimators $\widehat{\theta}_n$ satisfying

(3.7) $$\sqrt{n} \,(\widehat{\theta}_n - \theta_o) = \mathcal{O}_P(1) , \quad n \longrightarrow \infty ,$$

are a dime a dozen. (Since the rate $1/\sqrt{n}$ is "standard", we refer to such estimators as Opie-One estimators.) So, asymptotically, these $\widehat{\theta}_n$ are pretty good guesses not only for θ_o, but also for θ_n, the mle or the local minimum in a neighborhood of θ_o, cf. Theorem (2.23). So $\widehat{\theta}_n$ should be a good starting point for an iterative procedure to compute mle's.

(3.8) THEOREM. Let $\theta^1 - \hat{\theta}_n$, with $\hat{\theta}_n$ satisfying (3.7), and let the sequence $\{\theta^q\}_{q \geqslant 1}$ be generated by (3.2). Then, under the assumptions (2.A1) through (2.A12), for all $q \geqslant 2$,

$$\sqrt{n} \, (\theta^q - \theta_o) \longrightarrow_d Y \sim N\big(0, [\, \Phi(\theta_o)\,]^{-1}\big) \ .$$

PROOF. It suffices to prove the theorem for $q = 2$, since then the conclusion of the theorem implies $\sqrt{n} \, (\theta^2 - \theta_o) = \mathcal{O}_P(1)$.

Because the theorem makes asymptotic statements only, we can invoke Theorem (2.23) to claim the existence of local minimizers θ_n of $L_n(\theta)$ and WALD's Theorem (2.24) to show that

(3.9) $$\sqrt{n} \, (\theta_n - \theta_o) \longrightarrow_d Y \sim N(0, [\, \Phi(\theta_o)\,]^{-1}) \ .$$

Since the θ_n are (local) minimizers of $L_n(\theta)$, they satisfy $\nabla L_n(\theta_n) = 0$. Taylor series with remainder gives, cf. (2.25)–(2.27),

(3.10) $$\nabla L_n(\theta_n) = \nabla L_n(\theta^1) + H_n(\theta^1)\,(\theta_n - \theta^1) + \varrho_n \ ,$$

with $\| \varrho_n \| \leqslant c_n \| \theta^1 - \theta_n \|^2$, where

(3.11) $$c_n = \sup\{ \, \| \nabla^2 L_n(\theta) \| \ : \ \theta \in \text{rect}(\theta^1, \theta_n) \} \ ,$$

with $\text{rect}(\xi, \theta)$ as in (3.5). Assumption (2.A12) implies that $c_n = \mathcal{O}_P(1)$. Now, substituting (3.10) into (3.3) gives

(3.12) $$\| \theta^2 - \theta_n \| \leqslant c_n \| \big[H_n(\theta^1) \big]^{-1} \| \, \| \theta^1 - \theta_n \|^2 \ ,$$

whence $n \| \theta^2 - \theta_n \| = \mathcal{O}_P(1)$. This implies that $\sqrt{n} \, (\theta^2 - \theta_n) \longrightarrow_p 0$, and therefore,

$$\sqrt{n} \, (\theta^2 - \theta_o) = \sqrt{n} \, (\theta^2 - \theta_n) + \sqrt{n} \, (\theta_n - \theta_o) \quad \text{and} \quad \sqrt{n} \, (\theta_n - \theta_o)$$

asymptotically have the same distribution, and the conclusion follows from (3.9).   Q.e.d.

The proof of the above theorem actually shows that $n \| \theta^2 - \theta_n \| = \mathcal{O}_P(1)$, so that θ^2 and θ_n are just about the same random variable.

The above theorem says that best asymptotically normal estimators are a dime a dozen as well. The moral is that asymptotically maximum likelihood estimation is easy. The challenge is to figure out how the asymptotically optimal estimator (the mle) may be applied in finite sample problems. Typically, small adjustments are required and suggested by analyzing the results of simulation studies. The adjusted estimator will typically outperform any ad hoc estimator.

There is another way of looking at the Newton-Raphson method, which is particularly useful when the constraint $\theta \in \Theta$ comes into play. The linearization of $\nabla L_n(\theta)$ may also be viewed as the quadratic approximation to $L_n(\theta)$ itself

(3.13) $$L_n(\theta) = L_n(\xi) + \nabla L_n(\xi)\,(\theta - \xi) + \tfrac{1}{2}(\xi - \theta)^\top H_n(\xi)\,(\theta - \xi) + r_n \ ,$$

where r_n is the remainder and, presumably, $r_n = \mathcal{O}(\|\theta - \xi\|^3)$. In the *Sequential Quadratic Programming* approach to minimizing $L_n(\theta)$ over Θ, one starts out with an initial guess θ^1 and successively solves the problems

$$
(3.14) \qquad
\begin{aligned}
\text{minimize} \quad & \nabla L_n(\theta^q)(\theta - \theta^q) + \tfrac{1}{2}(\theta - \theta^q)^{\scriptscriptstyle T} H_n(\theta^q)(\theta - \theta^q) \\
\text{subject to} \quad & \theta \in \Theta \,,
\end{aligned}
$$

with the solution being denoted by θ^{q+1}. If θ^{q+1} lies inside Θ ("the constraints are not active") rather than on the boundary of Θ ("the constraints are active"), then θ^{q+1} is still given by (3.2). Actually, the constraints may be approximated as well, but we will not dwell on this. This procedure is most effective when the set Θ can be described by linear equalities and inequalities. Nonnegativity constraints are especially ubiquitous. If $\theta \in \Theta$ reduces to $\theta \geqslant 0$ (component wise), then (3.14) is the standard quadratic programming problem. A standard algorithm for solving it is algorithm NNLS in Lawson and Hanson (1995). For a review of general sequential quadratic programming, see Boggs and Tolle (1995).

The prime concern with the computational efficacy of the Newton-Raphson method deals with the computation of $H_n(\theta^q)$. If this is deemed too unwieldy, it may be replaced by its expected value $\Psi(\theta) = \mathbb{E}_\theta[H_n(\theta)]$. The resulting method is known as the method of scores,

$$
(3.15) \qquad \theta^{q+1} = \theta^q - \big[\Psi(\theta^q)\big]^{-1} \nabla L_n(\theta^q) \,.
$$

From an asymptotic point of view, this is indistinguishable from the Newton-Raphson method itself, see Exercise (3.17). Yet another alternative is useful when computing $\Psi(\theta)$ is unwieldy as well. This would especially be the case when the number of components of θ is large. In the modified Newton-Raphson method, the matrix $H_n(\theta)$ is not updated at each step, so the resulting algorithm looks like

$$
(3.16) \qquad \theta^{q+1} = \theta^q - \big[H_n(\theta^\ell)\big]^{-1} \nabla L_n(\theta^q) \,,
$$

with $H_n(\theta^\ell)$ updated once in a while. There is an analogue for the method of scores. These modified approaches are useful when the initial guess θ^1 is perhaps not very accurate.

(3.17) EXERCISE. Prove Theorem (3.8) for the method of scores. [Hint : (3.12) must be replaced by

$$
(3.12') \qquad \|\theta^2 - \theta_n\| \leqslant c_n\left(n^{-1/2}\|\theta^1 - \theta_n\| + \|\theta^1 - \theta_n\|^2\right),
$$

which allows the same conclusion $n\|\theta^2 - \theta_n\| = \mathcal{O}_P(1)$ to be drawn.]

(3.18) EXERCISE. The pdf of the two-parameter Weibull distribution is given by

$$
f(x; \sigma, \alpha) = \frac{\alpha}{\sigma}\left(\frac{x}{\sigma}\right)^{\alpha-1} \exp\left[-\left(\frac{x}{\sigma}\right)^\alpha\right] \,, \qquad x \geq 0 \,,
$$

and $= 0$ for $x < 0$, with $\sigma > 0$ and $\alpha > 0$.

(a) Determine $H_n(\theta)$ and $\nabla L_n(\theta)$ for this distribution.

(b) Work out the details of the Newton-Raphson procedure and the method of scores for the estimation of σ and α.

EXERCISES: (3.17), (3.18) .

4. The EM algorithm

We now consider the EM algorithm for the maximum likelihood problem. The EM algorithm is an iterative procedure, which is based on a thorough understanding of the probabilistic model. The crucial feature is the introduction of "missing data", i.e., unobserved random variables that would make the estimation problem much easier had they been observed. This motivates an iterative procedure, in which alternately the missing data are estimated on the basis of the observed data and the current understanding of the estimation problem (i.e., the current guess for the parameter), and then the unknown parameter is estimated on the basis of the observed and the estimated missing data. This last step is done by maximum likelihood, and constitutes the M-step of the EM algorithm. The first step is called the E-step of the algorithm, E for Expectation, not Estimation, unfortunately, as the sequel makes clear.

In its present form, the EM algorithm was formalized by DEMPSTER, LAIRD, and RUBIN (1977), who studied the common thread in earlier works, such as that of HARTLEY (1958) on estimating multinomial probabilities when sampling from finite populations, and work on linear models by HEALY and WESTMACOTT (1956) and HARTLEY and HOCKING (1971). A nice review of the (then) current state of affairs is provided by RUUD (1991).

We recall the original maximum likelihood problem. The observed data are $X = (X_1, X_2, \cdots, X_n)$, which is a random variable with pdf $f(x\,;\theta_o)$, and we wish to estimate θ_o by solving (1.1), repeated here for convenience

(4.1)
$$\text{minimize} \quad L_n(\theta) \stackrel{\text{def}}{=} -\tfrac{1}{n} \log f(X\,;\theta)$$
$$\text{subject to} \quad \theta \in \Theta .$$

The pdf $f(x\,;\theta_o)$ belongs to a known parametric family $\{\, f(x\,;\theta)\,:\,\theta \in \Theta\,\}$ with parameter space Θ. The EM algorithm introduces the missing data Y with pdf $g(y\,;\theta_o)$, which belongs to an also known parametric family $\{\, g(y\,;\theta)\,:\,\theta \in \Theta\,\}$ with the same parameter space Θ as before. We require that the distribution $k(x\,|\,Y)$ of X conditioned on Y is known explicitly. For convenience, we also assume that it does not depend on θ_o.

The reason for introducing Y is that typically θ_o would be easy to estimate if Y was actually observed, and also typically, the data X provide information regarding the distribution of Y. The distribution $h(y \mid X, \theta)$ of Y conditioned on X is given by

$$(4.2) \qquad h(y \mid X, \theta_o) = \frac{k(X \mid y)\, g(y\,; \theta_o)}{f(X\,; \theta_o)} \,,$$

and we assume that different θ give different $h(y \mid X, \theta)$. The maximum likelihood problem for estimating θ_o based on X and Y would be

$$(4.3) \qquad \begin{aligned} &\text{minimize} \quad -\tfrac{1}{n} \log\{\, k(X \mid Y)\, g(Y\,; \theta)\,\} \\ &\text{subject to} \quad \theta \in \Theta \,, \end{aligned}$$

but, unfortunately, Y is not observed. Is it possible to estimate the missing data Y? Or, equivalently, could we estimate $g(Y\,; \theta)$? The only avenue open is to estimate $g(Y\,; \theta)$ via the conditional expectation

$$(4.4) \qquad -\log g(Y^{\text{est}}\,; \theta) = \mathbb{E}\big[\, -\log g(Y\,; \theta) \,\big|\, X\,; \theta_o \,\big] \,,$$

but unfortunately we do not know θ_o either. Although it sure seems that we are spinning our wheels, let us replace θ_o in (4.4) by our best guess so far, denoted by ξ_1. The missing data are then estimated via

$$(4.5) \qquad -\log g(Y^{\text{est}}\,; \theta) = \mathbb{E}\big[\, -\log g(Y\,; \theta) \,\big|\, X\,; \xi_1 \,\big] \,,$$

and consequently, (4.3) is replaced by

$$(4.6) \qquad \begin{aligned} &\text{minimize} \quad Q(\theta \mid \xi_1) \overset{\text{def}}{=} -\frac{1}{n} \int_\Omega h(y \mid X, \xi_1) \log g(y\,; \theta)\, dy \\ &\text{subject to} \quad \theta \in \Theta \,. \end{aligned}$$

This constitutes the E-step of the EM algorithm. The M-step consists of solving (4.6). The solution is denoted by ξ_2. In case the solution is not unique, a solution must be selected. Because it seems to hard to say anything about this in reasonable generality, we leave it at that. An added wrinkle is that it is often sufficient to determine ξ_2 such that $Q(\xi_1 \mid \xi_1) > Q(\xi_2 \mid \xi_1)$. The resulting algorithm is then called a GEM algorithm (G for Generalized). The ξ_2 "thus" obtained is denoted by

$$(4.7) \qquad \xi_2 = M(\xi_1) \,.$$

The (G)EM algorithm now consists of repeatedly applying the above to construct the sequence $\{\xi_k\}_k$ by

$$(4.8) \qquad \xi_{k+1} = M(\xi_k) \,, \quad k = 1, 2, \cdots \,.$$

So far so good. What have we wrought, though? In the above, there is no reference to the original maximum likelihood problem (4.1), so some explanation is required. In fact, the (G)EM algorithm is a minimization algorithm for (4.1), as the following theorem makes clear.

(4.9) THEOREM. *Let* $\{\xi_k\}_k$ *be generated by the (G)EM algorithm. Then, for all* $k \geqslant 1$,

$$L_n(\xi_k) - L_n(\xi_{k+1}) \geqslant Q(\xi_k \mid \xi_k) - Q(\xi_{k+1} \mid \xi_k) > 0 ,$$

unless $\xi_k = \xi_{k+1}$, *i.e.,* ξ_k *is a fixed point of the (G)EM iteration.*

PROOF. For $\theta, \xi \in \Theta$, let

$$H(\theta, \xi) = -\frac{1}{n} \int_\Omega h(y \mid X, \xi) \log h(y \mid X, \theta) \, dy .$$

A simple calculation shows that

$$L_n(\theta) = H(\theta, \xi) + Q(\theta \mid \xi) + \text{other terms} ,$$

with $Q(\theta \mid \xi)$ as in (4.6), and where the "other terms" are independent of θ and ξ. Then,

$$L_n(\xi) - L_n(\theta) = \{ Q(\xi \mid \xi) - Q(\theta \mid \xi) \} + \{ H(\theta, \xi) - H(\xi, \xi) \} .$$

Now, note that

$$H(\theta, \xi) - H(\xi, \xi) = \frac{1}{n} \int_\Omega h(y \mid X, \xi) \log \frac{h(y \mid X, \xi)}{h(y \mid X, \theta)} \, dy ,$$

which is a Kullback-Leibler number and, hence, is nonnegative. So we obtain

$$L_n(\xi) - L_n(\theta) \geqslant Q(\xi \mid \xi) - Q(\theta \mid \xi) ,$$

with equality if and only if $\theta = \xi$. It follows that

$$L_n(\xi_k) - L_n(\xi_{k+1}) \geqslant Q(\xi_k \mid \xi_k) - Q(\xi_{k+1} \mid \xi_k) > 0 ,$$

by the very construction of ξ_{k+1}.   Q.e.d.

(4.10) COROLLARY. (a) *If* $L_n(\theta)$ *has a unique minimum* θ_n, *then* θ_n *is a fixed point of* M, *i.e.,* $\theta_n = M(\theta_n)$.
(b) *If* θ_n *is an mle, then* $M(\theta_n)$ *is also an mle, and* $L_n(\theta_n) = L_n(M(\theta_n))$.

The above shows that the (G)EM algorithm provides better and better estimators for θ_o, irrespective of the initial point ξ_1. However, this does not imply, as is sometimes stated in the literature, that they converge to θ_n. In the full generality implied, all kind of things can and will go wrong, as the case studies of Chapter 3 make clear. So we will not go into proving or even stating theorems that show that the (G)EM algorithm for parametric problems converges. For some results, see WU (1983). For convex (nonparametric) problems, we are in a much better position, as elaborated in Volume II.

Linear models. We illustrate the EM algorithm as it arose originally in linear models, see HARTLEY and HOCKING (1971), although practical significance can no longer be attached to it. Much of the classical theory of the design of experiments dealt with setting up experiments in such a way that (a) the model (1.16) applied, (b) the resulting estimators and tests were optimal in an appropriate sense, and (c) the estimator β_n as given by (1.19) is easy to calculate, in the sense that $(Z^T Z)^{-1} Z^T$ was explicitly known and "simple". However, experiments being what they are, there are always some observations "missing", which means that one or more rows of Z have to be deleted. But then the explicit, simple expression for β_n is no longer valid. A very reasonable approach is then to see if the missing data can be estimated, because then the explicit simple solution for β_n applies. If necessary, this procedure can be iterated. The following discussion concentrates on β only. Once β_n has been obtained, the mle for σ is given by (1.20).

Assume that the designed experiment is described by the linear model

$$(4.11) \qquad Y = \begin{bmatrix} U \\ X \end{bmatrix} \sim N\left(\begin{bmatrix} W \\ Z \end{bmatrix} \beta_o , \sigma^2 I \right) ,$$

with the mle given by

$$(4.12) \qquad \beta_n = CY .$$

For the missing data case, let us suppose that U went unobserved. Then, obviously,

$$X \sim N(Z\beta_o , \sigma_o^2 I) ,$$

and the mle is then given as in (1.19),

$$(4.13) \qquad \beta_n = (Z^T Z)^{-1} Z^T X ,$$

which is supposedly hard to compute. If we had a good estimate for U, then (4.12) could be used again. How would we estimate U? The distribution of U conditioned on X is given by

$$U \sim N(W\beta_o , \sigma^2 I) .$$

If we knew β_o, it would be reasonable to estimate U by $W\beta_o$. The mle would then again be given by (4.12) so that

$$\beta_n = C \begin{bmatrix} W\beta_o \\ X \end{bmatrix} .$$

Now, ignoring the distinction between β_o and the estimator β_n, we can use this as a basis for an iterative procedure to compute what we believe will be β_n, viz. with β^1 some a priori guess for β_o, compute

$$(4.14) \qquad \beta^{k+1} = C \begin{bmatrix} W\beta^k \\ X \end{bmatrix} , \quad k = 1, 2, \cdots .$$

It is not hard to show, irrespective of the initial guess β^1, that $\{\beta^k\}_k$ actually converges to β_n given by (4.13), see Exercise (4.19), but what has it got to do with the EM algorithm?

The point of course is that (4.14) *is* the EM algorithm, as we now show. The distribution $g(Y; \beta, \sigma)$ of Y satisfies

$$- \log g(Y; \beta, \sigma) = \sigma^{-2} \left[\| U - W\beta \|^2 + \| X - Z\beta \|^2 \right] + \text{ other terms },$$

where the "other terms" do not depend on β. The distribution of U conditioned on X is $N(W\beta_o, \sigma^2 I)$. Then, assuming a guess β^1 for β_o, and ignoring terms independent of β, we obtain

$$Q(\beta \mid \beta^1) = \tfrac{1}{n} \mathbb{E}\left[- \log g(Y; \beta, \sigma) \,\middle|\, X, \beta^1 \right],$$

which may be written as

$$
\begin{aligned}
Q(\beta \mid \beta^1) &= (n\sigma^2)^{-1} \, \mathbb{E}\left[\| U - W\beta \|^2 + \| X - Z\beta \|^2 \,\middle|\, X, \beta^1 \right] \\
&= (n\sigma^2)^{-1} \left\{ \mathbb{E}\left[\| U - W\beta \|^2 \,\middle|\, X, \beta^1 \right] + \| X - Z\beta \|^2 \right\}.
\end{aligned}
$$
(4.15)

However,

$$
\begin{aligned}
\mathbb{E}\left[\| U - W\beta \|^2 \,\middle|\, X, \beta^1 \right] &= \mathbb{E}\left[\| U - W\beta^1 \|^2 + \| W\beta - W\beta^1 \|^2 \right. \\
&\qquad \left. - 2 \, (W\beta^1 - W\beta)^T (U - Z\beta^1) \,\middle|\, X, \beta^1 \right] \\
&= n\sigma^2 + \| W\beta - W\beta^1 \|^2,
\end{aligned}
$$

hence

$$Q(\beta \mid \beta^1) = (n\sigma^2)^{-1} \left\{ \| W\beta - W\beta^1 \|^2 + \| X - Z\beta \|^2 + n\sigma^2 \right\}$$

or

$$Q(\beta \mid \beta^1) = \frac{1}{n\sigma^2} \left\| \begin{bmatrix} W\beta^1 \\ X \end{bmatrix} - \begin{bmatrix} W \\ Z \end{bmatrix} \beta \right\|^2 + 1.$$
(4.16)

Minimizing $Q(\beta \mid \beta^1)$ with respect to β is just the original least-squares problem with all of the data observed, so the solution is given by $\beta = \beta^2$ with

$$\beta^2 = C \begin{bmatrix} W\beta^1 \\ X \end{bmatrix}.$$
(4.17)

So the EM algorithm is precisely the heuristically derived algorithm (4.14), and it actually converges. This can be shown directly, see Exercise (4.19), or also by general EM considerations, see Theorem (4.9). As with any iterative procedure, it is important to choose a good initial value β^1. A useful criterion is to select β^1 in such a way that $\mathbb{E}\left[W\beta^1 \right] = \mathbb{E}\left[U \right]$. This is not as impossible as it seems. See Exercise (4.18) below.

(4.18) EXERCISE. Consider the two factor design model with no interaction

$$Y_{ij} = \mu_o + \alpha_i^o + \beta_j^o + e_{ij}, \quad i, j = 1, 2,$$

where the e_{ij} are iid $N(0, \sigma_o^2)$ random variables. Let

$$Y = \left[Y_{11}, Y_{12}, Y_{21}, Y_{22} \right]^T, \quad \theta_o = \left[\mu_o, \alpha_1^o, \alpha_2^o, \beta_1^o, \beta_2^o \right]^T.$$

(a) Determine Z such that $Y \sim N(Z\theta_o, \sigma_o^2 I)$.

(b) Determine the normal equations for this model.

(c) With the added assumption that $\alpha_1^o + \alpha_2^o = 0$ and $\beta_1^o + \beta_2^o = 0$, determine C such that $\theta_n = CY$.

(d) Assuming Y_{11} is missing, derive the EM algorithm.

(e) Show that the EM algorithm converges in one step for the initial guess θ^1 given by

$$\left[\tfrac{1}{2}\left(Y_{12}+Y_{12}\right),\ \tfrac{1}{2}\left(Y_{12}-Y_{22}\right),\ -\tfrac{1}{2}\left(Y_{12}-Y_{22}\right),\ \tfrac{1}{2}\left(Y_{21}-Y_{22}\right),\ -\tfrac{1}{2}\left(Y_{12}+Y_{12}\right)\right]^{\mathsf{T}}.$$

(f) For the initial guess θ^1 of part (e), show that

$$\mathbb{E}\left[W\theta^1\right] = \mathbb{E}\left[Y_{11}\right] = \mu_o + \alpha_1^o + \beta_1^o .$$

(4.19) EXERCISE. (a) Show that β_n, see (4.13), is a fixed point of the iteration (4.14).

(b) Show that

$$\beta^{k+1} - \beta_n = (W^{\mathsf{T}}W + Z^{\mathsf{T}}Z)^{-1}W^{\mathsf{T}}W(\beta^k - \beta_n) , \quad k = 1, 2, \cdots .$$

(c) Assuming that Z has full column rank, show that

$$\left\| (W^{\mathsf{T}}W + Z^{\mathsf{T}}Z)^{-1}W^{\mathsf{T}}W \right\| < 1 .$$

(d) Conclude that $\beta^k \longrightarrow \beta_n$ as $n \to \infty$.

Mixtures of distributions. We briefly discussed mixture models in § 1. Here, we consider various maximum likelihood problems for mixtures. We will see that EM algorithms seem to be well suited for such problems and work quite well in practice. One instance of mixture models is when the distributions are known, but the mix (the mixing distribution) is unknown. In the simplest case, the random variable X has the pdf

$$(4.20) \qquad g(x\,;w) = w\,\phi_1(x) + (1-w)\,\phi_2(x) ,$$

where the ϕ_j are known distributions. Because we may think of X as being an observation drawn from the pdf ϕ_1 with probability w and from ϕ_2 with probability $1 - w$, we may view this as a case of missing data. What is missing is which pdf X is drawn from. To derive the EM algorithm, consider the random variable Y with pdf

$$(4.21) \qquad f(y\,;w) = \begin{cases} w & ,\quad 0 < y < 1 , \\ 1-w & ,\quad 1 < y < 2 , \end{cases}$$

and for which the distribution of X conditioned on Y is given by

$$(4.22) \qquad k(x\,|\,Y) = \begin{cases} \phi_1(x) , & 0 < Y < 1 , \\ \phi_2(x) , & 1 < Y < 2 . \end{cases}$$

The distribution of Y conditioned on X then satisfies

$$(4.23) \qquad h(y \mid X ; w) = \frac{k(x \mid y) \, f(y ; w)}{g(x ; w)} .$$

Given the complete data set, i.e., the iid sample $(X_1, Y_1), \cdots, (X_n, Y_n)$, the negative log-likelihood is given by

$$(4.24) \qquad \Lambda_n(w) = -\frac{1}{n} \sum_{i=1}^{n} \log\{ k(X_i \mid Y_i) \, f(Y_i) \} ,$$

and $Q(w \mid w^{\text{old}}) = \mathbb{E}\left[\Lambda_n(w) \mid X, w^{\text{old}} \right]$ by

$$Q(w \mid w^{\text{old}}) = -\frac{1}{n} \sum_{i=1}^{n} \int_{[0.2]} \log\{ k(X_i \mid y) \, f(y ; w^{\text{old}}) \} \frac{k(X_i \mid y) f(y)}{g(X_i ; w^{\text{old}})} \, dy ,$$

or, written in full,

$$Q(w \mid w^{\text{old}}) = -\frac{1}{n} \sum_{i=1}^{n} \frac{w^{\text{old}} \, \phi_j(X_i) \, \log\{ w \, \phi_1(X_i) \}}{w^{\text{old}} \, \phi_1(X_i) + (1 - w^{\text{old}}) \, \phi_2(X_i)} +$$

$$\frac{(1 - w^{\text{old}}) \, \phi_j(X_i) \, \log\{ (1 - w) \, \phi_2(X_i) \}}{w^{\text{old}} \, \phi_1(X_i) + (1 - w^{\text{old}}) \, \phi_2(X_i)} .$$

So

$$(4.25) \quad Q(w \mid X, w^{\text{old}}) =$$

$$- w^{\text{new}} \log w - (1 - w^{\text{new}}) \log(1 - w) + \text{other terms} ,$$

where the "other terms" are independent of w and

$$(4.26) \qquad w^{\text{new}} = \frac{1}{n} \sum_{i=1}^{n} \frac{w^{\text{old}} \, \phi_1(X_i)}{w^{\text{old}} \, \phi_1(X_i) + (1 - w^{\text{old}}) \, \phi_2(X_i)} .$$

This completes the E-step. Actually, we jumped the gun on calling w^{new} the new w. We do see, though, that the above w^{new} is the minimizer of $Q(w \mid w^{\text{old}})$, so that the above iteration formula is the desired EM algorithm. In Volume II, the convergence of this EM algorithm is studied in detail.

Mixtures of unknown normals. Another, and significantly more complicated, situation arises when the pdfs in the mixture are unknown as well. Here, we consider the mixture of two normals

$$(4.27) \qquad f(x ; \theta) = w \, \phi(x ; \theta_1) + (1 - w) \, \phi(x ; \theta_2) ,$$

where $\theta = (w, \theta_1, \theta_2)^T$, with $\theta_j = (\mu_j, \sigma_j)$. We derive an EM algorithm for the mle of θ given the iid sample $X_1, X_2, \cdots, X_n$ with common pdf belonging to this parametric family. It turns out that the previous setup is still adequate, up to a point. Again, the missing data are which distribution each X_i is drawn from. Introduce the random variable Y with pdf $g(y)$

$$(4.28) \qquad g(y ; w) = \begin{cases} w , & 0 < y < 1 , \\ 1 - w , & 1 < y < 2 , \end{cases}$$

and such that the distribution of X conditioned on Y is

(4.29)
$$k(x \mid Y ; \theta) = \begin{cases} \phi(x ; \theta_1) , & 0 < Y < 1 , \\ \phi(x ; \theta_2) , & 1 < Y < 2 . \end{cases}$$

Then, the joint distribution of (X, Y) satisfies

(4.30)
$$j(x, y ; \theta) = w \, \phi(x ; \theta_1) \, \mathbb{1}_{[0,1]}(y) + (1 - w) \, \phi(x ; \theta_2) \, \mathbb{1}_{[1,2]}(y) ,$$

and the distribution of Y conditioned on X is given by

(4.31)
$$h(y \mid X, \theta) = \frac{w \, \phi(X ; \theta_1) \mathbb{1}_{[0,1]}(y) + (1 - w) \, \phi(X ; \theta_2) \mathbb{1}_{[1,2]}(y)}{w \, \phi(X ; \theta_1) + (1 - w) \, \phi(X ; \theta_2)} .$$

Here, $\mathbb{1}_{[a,b]}$ is the indicator function of the interval $[a, b]$. The negative log-likelihood of θ given the iid sample $(X_1, Y_1), (X_2, Y_2), \cdots , (X_n, Y_n)$ is

(4.32)
$$\Lambda_n(\theta) = -\frac{1}{n} \sum_{i=1}^{n} \log\{ w \, \phi(X_i ; \theta_1) \mathbb{1}_{[0,1]}(Y_i) +$$

$$(1 - w) \, \phi(X_i ; \theta_2) \mathbb{1}_{[1,2]}(Y_i) \} ,$$

and so $Q(\theta \mid \theta^{old}) = \mathbb{E}\big[\Lambda_n(\theta) \mid X, \theta^{old} \big]$ may be written as

(4.33)
$$Q(\theta \mid \theta^{old}) = -\frac{1}{n} \sum_{i=1}^{n} \{ \psi_1(X_i ; \theta^{old}) \log\{ w \, \phi(X_i ; \theta_1) \} +$$

$$\psi_2(X_i ; \theta^{old}) \log\{ (1 - w) \, \phi(X_i ; \theta_2) \} \} ,$$

with $\psi_j(x ; \theta^{old})$ given by

(4.34)
$$\psi_1(x ; \theta^{old}) = \frac{w^{old} \, \phi(x ; \theta_1^{old})}{f(x ; \theta^{old})} ,$$

$$\psi_2(x ; \theta^{old}) = \frac{(1 - w^{old}) \, \phi(x ; \theta_2^{old})}{f(x ; \theta^{old})} .$$

This completes the E-step of the EM algorithm.

For the M-step, we see that minimizing $Q(\theta \mid \theta^{old})$ reduces to three unrelated problems for determining w, θ_1, and θ_2. The new w is exactly as before in (4.26)

(4.35)
$$w^{new} = \frac{1}{n} \sum_{i=1}^{n} \psi_1(X_i ; \theta^{old}) ,$$

and then the μ_j and σ_j $(j = 1, 2)$ are given by

$$(4.36) \qquad \mu_j^{\text{new}} = \frac{\sum\limits_{i=1}^{n} X_i \, \psi_j(X_i \, ; \theta^{\text{old}})}{\sum\limits_{i=1}^{n} \psi_j(X_i \, ; \theta^{\text{old}})}$$

and

$$(4.37) \qquad (\sigma_j^{\text{new}})^2 = \frac{\sum\limits_{i=1}^{n} (X_i - \mu_j^{\text{new}})^2 \, \psi_j(X_i \, ; \theta^{\text{old}})}{\sum\limits_{i=1}^{n} \psi_j(X_i \, ; \theta^{\text{old}})} \; .$$

This completes the M-step of the EM algorithm.

In this case, the EM algorithm does not always converge, and even if it does, there is no guarantee that it converges to an mle. An experimental discussion is given in § 3.3.

(4.38) EXERCISE. Repeat the above for a mixture of m normals. The parametric family is

$$f(x ; \, , \theta) = \sum_{j=1}^{m} w_j \, \phi(x - \mu_j ; \sigma_j) \, , \qquad x \in \mathbb{R} \, ,$$

where $\theta = (w^T, \theta_1, \theta_2, \cdots, \theta_m)^T$ with $\theta_j = (\mu_j, \sigma_j)$ and $w \in S_{m-1}$, cf. Exercise (2.39).

(4.39) EXERCISE. Repeat the above for a mixture of m exponential distributions. The parametric family is

$$f(x ; \theta) = \sum_{j=1}^{m} w_j \, \theta_j \exp(-\theta_j \, x) \, , \qquad x \geqslant 0 \, .$$

(4.40) EXERCISE. The parametric family

$$f(x ; \varepsilon, \mu, \sigma) = (1 - \varepsilon) \, \phi(x ; \mu, \sigma) + \varepsilon \, \phi(x ; \mu, 3\sigma) \, ,$$

with $\theta = (\varepsilon, \mu, \sigma) \in [0, 1] \times \mathbb{R} \times \mathbb{R}_+$ and $\phi(x ; \mu, \sigma)$ the normal (μ, σ^2) distribution, has been proposed by HUBER (1977) as a model for experimental data with "errors". Derive the EM algorithm for this parametric family.

Empirical Bayes estimation and mixtures. In the previous examples of EM algorithms, some "natural" data were missing. In the setting of iid observations $X_1, X_2, \cdots, X_n$ with pdf $k(x ; \theta_o)$, what data are missing? The following is a well known approach, which goes by the name of *empirical Bayes* estimation, see ROBBINS (1956). Let us assume that $\theta = \theta_o$ is itself a random variable, with unknown pdf $f^*(\theta)$. The problem then is to estimate $f^*(\theta)$. One would hope that the estimator $f^n(\theta)$ has most of its mass in a very small (generalized) rectangle around its mean θ_n, so that a good estimator for θ_o would result. Alternatively, if it is discovered that the estimator $f^n(\theta)$ is spread out over a large region, then this casts doubt on the parametric family $k(x;\theta)$ as a model for the random variable X, or on θ_o being a fixed parameter (a degenerate random variable). This also handles mixtures quite naturally. It should be noted that we have eased into nonparametric maximum likelihood estimation: Rather than estimating a parameter $\theta_o \in \mathbb{R}^p$, we now must estimate a probability density function that belongs naturally to an infinite dimensional set.

In the above setting with θ_o being a random variable, the complete data set consists of the iid observations $(X_1, \Theta_1), (X_2, \Theta_2), \cdots, (X_n, \Theta_n)$ of the random variable (X, Θ). (Warning: Θ is now a random variable, not the the parameter space.) The pdf of (X_1, Θ_1) is $k(x,\theta)f^*(\theta)$. So the maximum likelihood estimation problem for the complete data set is

$$\text{(4.41)} \quad \text{minimize} \quad \Lambda_n(f) \overset{\text{def}}{=} -\frac{1}{n}\sum_{i=1}^{n} \log\{\, k(X_i,\Theta_i)f(\Theta_i)\,\}$$

$$\text{subject to} \quad f \text{ is a pdf}.$$

What does the EM algorithm for this problem look like? For the E-step, we need the distribution of the Θ_i conditioned on the X_i. Since the (X_i, Θ_i) are assumed to be iid, it suffices to consider $h(\theta_i|X_i)$, given an initial guess $f_1(\theta)$ for $f^*(\theta)$,

$$\text{(4.42)} \qquad h(\theta\,|\,x\,;f_1) = \frac{k(x,\theta)\,f_1(\theta)}{\mathcal{K}f_1(x)} ,$$

where

$$\text{(4.43)} \qquad \mathcal{K}f(x) = \int_\Theta k(x,\theta)f(\theta)\,d\theta , \quad x \in \mathbb{R}^d .$$

Then,

$$Q(f\,|\,f_1) = \mathbb{E}\big[\,\Lambda(f)\,\big|\,X_1, \cdots, X_n, f_1\,\big]$$
$$= -\frac{1}{n}\sum_{i=1}^{n} \mathbb{E}\big[\,\log\{k(X_i,\Theta_i)f(\Theta_i)\,\}\,\big|\,X_i, f_1\,\big],$$

or

$$\text{(4.44)} \qquad Q(f\,|\,f_1) = -\int_\Theta f_2(\theta)\log\frac{f_1(\theta)}{f(\theta)} + \text{ other terms },$$

where the "other terms" are independent of f, and

$$\text{(4.45)} \qquad f_2(\theta) = f_1(\theta)\cdot\frac{1}{n}\sum_{i=1}^{n}\frac{k(X_i,\theta)}{\mathcal{K}f_1(X_i)} , \quad \theta \in \Theta .$$

This concludes the E-step. The M-step consists of minimizing $Q(f \mid f_1)$ over all pdfs f. Since

$$Q(f \mid f_1) = \mathrm{KL}(f_2, f) - \mathrm{KL}(f_2, f_1) ,$$

with KL the Kullback-Leibler distance, the answer is given by $f = f_2$. Thus, the EM iteration takes the form

$$(4.46) \qquad f_{q+1}(\theta) = f_q(\theta) \cdot \frac{1}{n} \sum_{i=1}^n \frac{k(X_i, \theta)}{K f_q(X_i)} , \qquad \theta \in \Theta .$$

Again, the connection with the original maximum likelihood estimation problem is not so clear, but since $Q(f_q \mid f_q) - Q(f_{q+1} \mid f_q) = \mathrm{KL}(f_{q+1}, f_q)$, Theorem (4.9) implies

$$(4.47) \qquad L_n(f_q) - L_n(f_{q+1}) \geqslant \mathrm{KL}(f_{q+1}, f_q) ,$$

which is positive unless $f_{q+1} = f_q$. In Volume II, we elaborate on this.

A final comment. In the above, we occasionally referred to *the* EM algorithm rather than to *an* EM algorithm for the computation of mle's. Intuitively, it seems clear that in settings with "natural" missing data, there is only one EM algorithm. Indeed, the authors have never come across a situation in which there are two distinct EM algorithms.

EXERCISES: (4.18), (4.19), (4.38), (4.39), (4.40).

5. Sensitivity to errors: M-estimators

Maximum likelihood estimation of parameters in parametric families of pdfs as discussed in §§ 1-4 is quite definitely based on the assumptions (a) $X_1, X_2, \cdots, X_n$ are iid random variables, and (b) the pdf of X_1 belongs to the parametric family under discussion $\mathcal{F} = \{ f(x; \theta) : \theta \in \Theta \}$. How certain of these assumptions can one be? The assumption (a) seems reasonable enough, but the existence of Sampling Theory is an indication of a plethora of pitfalls. Above and beyond this, a not uncommon problem is the occurrence of a few "gross" errors, ranging from data entry problems (1.23 is recorded as 12.3 or 123) to undetected experimental failures that result in improbable, but not *a priori* impossible observations. Another common problem is that the data may contain many small errors (such as roundoff). As to assumption (b), it is almost certainly false in the strict sense but perhaps not unreasonable in that the parametric family fairly accurately describes the random variable. This is often referred to as (various degrees of) misspecification of the model.

Question arise regarding the sensitivity of mle's (and other estimators) with respect to these errors. Here, we are mostly concerned with a theoretical discussion of the way mle's are influenced by errors in the observations.

The theoretical discussion is conducted with nonparametric maximum likelihood estimation in mind. This explains (we hope) the rather personal view of the field. In Chapter 3, we discuss possible violations of assumption (b) from a practical point of view. The study of questions like these falls under the heading of *robust statistics*. In this section, we give a very limited treatment of how errors in the data, both gross and small, effect the mle's. We base our treatment on a special version of the modified likelihood approach, viz. by modifying the parametric family of pdfs. For the authorized version of M-estimators, and robust statistics in general, the reader is referred to HUBER (1981) and ROUSSEEUW and LEROY (1987).

In order to discuss the sensitivity of mle's with respect to errors, we need to quantify such errors in a statistically meaningful way. Before doing so, it is helpful to introduce the notion of statistical functionals. Consider the maximum likelihood estimation problem for a parametric family of pdfs $\{ f(x;\theta) : \theta \in \Theta \}$, based on an iid sample $X_1, X_2, \cdots, X_n$ from the pdf $f(x;\theta_o)$. Assuming that the unique mle θ_n exists, we may view θ_n as a function of $X_1, X_2, \cdots, X_n$, i.e., $\theta_n = \theta(X_1, X_2, \cdots, X_n)$. However, it is not just any function of the X_i: Since the likelihood is independent of the order of the X_i, so is the mle θ_n. For univariate random variables, it is then standard practice to list the X_i in increasing order as $X_{1,n}, X_{2,n}, \cdots, X_{n,n}$, and so $\theta_n = \theta(X_{1,n}, X_{2,n}, \cdots, X_{n,n})$. The $X_{i,n}$ are called the *order statistics*. A convenient way to encode the order statistics is by means of the empirical cdf

$$(5.1) \qquad F_n(x) = \frac{1}{n} \sum_{i=1}^{n} \mathbb{1}_+(x - X_i) , \quad x \in \mathbb{R} ,$$

where $\mathbb{1}_+(x)$ is the indicator function of $\mathbb{R}_+$. Note that in (5.1), we may replace the X_i by the $X_{i,n}$ if we so desire. We may now rewrite θ_n as

$$(5.2) \qquad \theta_n = T(F_n) ,$$

and $T(F_n)$ is referred to as a statistical function or functional. Some familiar examples are the sample mean and variance of a sample $X_1, X_2, \cdots, X_n$,

$$(5.3) \qquad \mu_n = \frac{1}{n} \sum_{i=1}^{n} X_i = \int_{\mathbb{R}^d} x \, dF_n(x) ,$$

$$(5.4) \qquad \sigma_n^2 = \frac{1}{n} \sum_{i=1}^{n} (X_i - \mu_n)^2 = \int_{\mathbb{R}^d} (x - \mu_n)^2 \, dF_n(x) ,$$

and the negative log-likelihood function

$$(5.5) \qquad L_n(\theta) = -\frac{1}{n} \sum_{i=1}^{n} \log f(X_i;\theta) = -\int_{\mathbb{R}^d} \log f(x;\theta) \, dF_n(x) .$$

Here, the integrals are Riemann-Stieltjes (or Riemann-Lebesgue) integrals. The usefulness of the notions of statistical functions and empirical distribution functions is that the appropriate mathematics is set up to deal with functions, not samples. By way of example, it is now possible to compare the iid sample $X_1, X_2, \cdots, X_n$ with the distribution function it is allegedly drawn from, by looking at the "distance"

$$(5.6) \qquad \| F_n - F \|_\infty \overset{\text{def}}{=} \sup_{x \in \mathbb{R}^d} | F_n(x) - F(x) | \; .$$

This is the well-known Kolmogorov-Smirnov statistic for testing whether $X_1, X_2, \cdots, X_n$ is an iid sample from the cdf F. The distance $\| F_n - F \|_\infty$ has been extensively studied, both for finite sample size and asymptotically. We mention the law of the iterated logarithm: If $\int_{\mathbb{R}^d} | x |^2 dF(x) < \infty$, and $X_1, X_2, \cdots, X_n$ are iid with common distribution F, then

$$(5.7) \qquad \limsup_{n \to \infty} \left(n^{-1} \log \log n \right)^{-1/2} \| F_n - F \|_\infty =_{\text{as}} \sqrt{2} \; .$$

For this, and everything else you always wanted know about $F_n(x) - F(x)$ but were afraid to ask, see SHORACK and WELLNER (1986).

How would one specify "gross" errors? Or better yet, how could one tell the absence of gross errors? This is almost equivalent to how one detects whether two iid samples $X_1, X_2, \cdots, X_n$ and $Y_1, Y_2, \cdots, Y_n$ are from the same distribution. It seems reasonable to compare samples in terms of their empirical distribution functions. Denoting the empirical distribution functions of $X_1, X_2, \cdots, X_n$ and $Y_1, Y_2, \cdots, Y_n$ by $F_n(x)$ and $G_n(x)$, respectively, we now specify that two samples $X_1, X_2, \cdots, X_n$ and $Y_1, Y_2, \cdots, Y_n$ are close if the corresponding empirical distribution functions are close in the sense that $\| F_n - G_n \|_\infty$ is small.

(5.8) EXERCISE. (a) Let F_n and G_n be two empirical distribution functions for two samples of size n. Show that $\| F_n - G_n \|_\infty$ can only take on the values k/n, $k = 0, 1, \cdots, 2n$.
(b) Suppose $X_i = Y_i$, $i = 1, 2, \cdots, n-1$, and

$$Y_n = X_n + \text{ big number} \; ,$$

where the "big number" denotes a big number. With F_n and G_n, the empirical distributions functions of the X_i and Y_i, respectively, show that

$$\| F_n - G_n \|_\infty = \tfrac{1}{n} \; .$$

(c) How sensitive is the answer in (b) to the size of the big number? (Consider really small numbers.)
(d) Let F_n be a fixed empirical distribution function based on a random sample of size n. Show that the set

$$\{ G : G \text{ is a cdf} , \; \| G - F_n \|_\infty < \tfrac{1}{n} \}$$

contains no empirical distribution functions for samples of size n besides F_n.

Now, consider the sensitivity of mle's to "gross" errors in the X_i. Note that one "gross" error changes the empirical cdf by $1/n$, whereas for two iid samples from the same distribution, the empirical cdfs will differ by $1/\sqrt{n}$ or thereabouts, cf. (5.7). So let $X_1, X_2, \cdots, X_n$ be iid from the distribution $F(\,\cdot\,; \theta_o)$, except that a few X_i have been grossed out. We assume that the mle θ_n exists and is unique, or that some way of selecting a θ_n exists, such that the estimator θ_n is a function of the empirical distribution function F_n, as in (5.2). In this notation, one would expect or hope that

$$(5.9) \qquad\qquad \theta_o = T(F_o) \,,$$

where $F_o = F(\,\cdot\,; \theta_o)$. In other words, one would hope that for an infinite iid sample $X_1, X_2, \cdots$ from $F(\,\cdot\,; \theta_o)$, the estimator would be θ_o (almost surely).

The absence of oversensitivity to "gross" errors is now guaranteed if there exists a constant r (not too big) such that uniformly in n,

$$(5.10) \qquad\qquad \| T(F_n) - T(G_n) \| \leqslant r \| F_n - G_n \|_\infty \,.$$

The norm on the left is any norm on $\Theta \subset \mathbb{R}^p$. The absence of oversensitivity is usually called robustness, although the technical definition is somewhat different. The estimate (5.10) says that T is Lipschitz continuous. Relaxing this to mere continuity is not satisfactory, see Exercise (5.8)(d). A consequence of (5.10) is the following. Suppose we can partition the data into the collection $\{ X_i : i \in I \}$ of "correct" observations, and the collection $\{ X_i : i \notin I \}$ of "gross" errors. The correct observations are encoded in

$$(5.11) \qquad\qquad F_I(x) = \tfrac{1}{|I|} \sum_{i \in I} \mathbb{1}_+(x - X_i) \,, \qquad x \in \mathbb{R}^d \,.$$

Then, (5.10) implies that

$$(5.12) \qquad\qquad \| T(F_n) - T(F_I) \|_\infty \leqslant \frac{2\, r\, (n - |I|)}{n} \,,$$

irrespective of whether $\{ X_i : i \notin I \}$ are the "gross" errors or not, actually. We will take (5.12) as our practical definition of robustness with respect to gross errors.

(5.13) EXERCISE. If T satisfies (5.9)–(5.10) and $X_1, X_2, \cdots, X_n$ are iid from $F(\,\cdot\,; \theta_o)$, show that $\| \theta_n - \theta_o \| =_{\mathrm{as}} \mathcal{O}\big((n^{-1} \log \log n)^{1/2} \big)$.

Are mle's in fact robust in the sense of (5.10)? The answer is a resounding NO, unfortunately. (If the answer were yes, Exercise (5.13) would just about wipe out §2.) A simple example is the estimation of the mean and standard deviation of a (univariate) normal distribution, based on an iid sample $X_1, X_2, \cdots, X_n$. The mle's are

$$(5.14) \qquad \mu_n = \tfrac{1}{n} \sum_{i=1}^{n} X_i \,, \qquad \sigma_n^2 = \tfrac{1}{n} \sum_{i=1}^{n} (X_i - \mu_n)^2 \,,$$

and these are obviously not robust with respect to gross errors. They are apparently robust with respect to many small errors, but in general, mle's are not even guaranteed to be robust with respect to a *few* small changes. In the following, we set out to devise robust estimators for the mean and standard deviation μ and σ of a univariate normal distribution, but first we make a scenic detour.

Robust parametric families of pdfs. Do parametric families exist for which maximum likelihood estimation is robust with respect to gross errors? In particular, we would like to have robust estimators of location and scale, that is, the parameters $\theta = (\mu, \sigma)$ in the parametric family

$$(5.15) \qquad f(x\,;\theta) = \sigma^{-1}\,\psi\big(\sigma^{-1}(x - \mu)\big)\,, \qquad x \in \mathbb{R}\,,$$

where $\psi(x)$ is a fixed (known) pdf. Recall the mle's for the shifted two-sided exponential distribution of Exercise (1.30). The estimator for μ (the median of $X_1, X_2, \cdots, X_n$) certainly is robust with respect to gross errors, but the mle for the scale parameter is not robust. Note that the median does not change if the largest order statistic is replaced by an arbitrary larger value. It is also worth noting that the median is a continuous function of the values $X_1, X_2, \cdots, X_n$ (in the usual topology on $\mathbb{R}^n$).

Here is an example of a univariate parametric location/scale family for which the mle of the scale is robust with respect to gross errors. The trick is of course to allow gross errors as if they were the usual thing. For the mean, the double exponential was nice. The exponential has fairly heavy tails, but not heavy enough to give a robust mle for the scale. The Cauchy distribution is then a logical next choice, but the following Cauchy-like family seems to be more accessible (and has a very interesting feature). Let $X_1, X_2, \cdots, X_n$ be iid with common pdf

$$(5.16) \qquad f(x\,;\theta) = \frac{\sigma}{2\left(\sigma + |x - \mu|\right)^2}\,, \qquad x \in \mathbb{R}\,,$$

with $\theta = (\mu, \sigma) \in \mathbb{R} \times \mathbb{R}_+$. This is actually a special case of (5.15). The negative log-likelihood is given by

$$(5.17) \qquad L_n(\theta) = -\log(\sigma/2) + \frac{2}{n}\sum_{i=1}^{n}\log\big(\sigma + |\,X_i - \mu\,|\big)\,.$$

Does $L_n(\theta)$ have a global minimum? The affirmative answer seems to be easy to see, but let us have a closer look. For given μ, the function $L_n(\theta)$ attains an absolute minimum as a function of σ: It is certainly a continuous function of σ and $L_n(\theta) \to \infty$ for $\sigma \to \infty$ (because then $\log(\sigma + |\,X_i - \mu\,|) = \log\sigma + \mathcal{O}(\sigma^{-1})$) and also for $\sigma \longrightarrow 0^+$ (actually, this fails if more than half of the X_i are equal to each other, and equal to μ, so we must add "almost surely"). It is also obvious that we may a priori restrict μ to the interval $[X_{1,n}, X_{n,n}]$. If μ is outside this interval, then replacing μ by $X_{1,n}$ or $X_{n,n}$, whichever is closer, reduces every $|\,X_i - \mu\,|$ and so $L_n(\theta)$ decreases.

Since $L_n(\theta)$ is a continuous function of $\theta = (\mu, \sigma) \in [X_{1,n}, X_{n,n}] \times \mathbb{R}_+$, and blows up for $\sigma \longrightarrow 0^+$ and $\sigma \longrightarrow \infty$, an absolute minimum exists.

To find mle's, we need to differentiate $L_n(\theta)$ with respect to μ and σ, but we appear to be in trouble with differentiation with respect to μ. However, $|X_i - \mu|$ is differentiable with respect to μ for $\mu \neq X_i$, and at $\mu = X_i$ it is a nice enough function: There are plenty of lines that can act as a tangent line, viz. all lines with slopes in the interval $[-1, 1]$. So we get the equations

$$(5.18) \qquad \frac{2}{n} \sum_{i=1}^{n} \frac{\text{sign}(\mu - X_i)}{\sigma + |X_i - \mu|} = 0 \ ,$$

$$(5.19) \qquad -\sigma^{-1} + \frac{2}{n} \sum_{i=1}^{n} \frac{1}{\sigma + |X_i - \mu|} = 0 \ .$$

Here, we define $\text{sign}(x) = 1$ for $x > 0$, and $= -1$ for $x < 0$, whereas its value for $x = 0$ is anything in the interval $[-1, 1]$ that makes (5.18) hold true. The last equation may be rewritten as

$$(5.20) \qquad \frac{1}{n} \sum_{i=1}^{n} \frac{|X_i - \mu|}{\sigma + |X_i - \mu|} = \frac{1}{2} \ .$$

Note that for given μ, the equation (5.20) has a unique solution σ.

Something funny is happening, though, and we find out about it when wondering whether solutions to (5.18) and (5.20) necessarily give solutions to the maximum likelihood problem. If this is actually true, then $\nabla^2 L_n(\theta)$ should be positive-definite, assuming its existence. For $\mu \neq X_i$, $i = 1, 2, \cdots, n$, the function $L_n(\theta)$ is indeed twice continuously differentiable, and we get

$$\frac{\partial^2 L_n(\theta)}{\partial \mu^2} = -\frac{2}{n} \sum_{i=1}^{n} \frac{\left(\text{sign}(\mu - X_i)\right)^2}{\left(\sigma + |X_i - \mu|\right)^2} \ ,$$

and there is no denying that this is always negative !? The conclusion must be that although (5.18) and (5.20) may have solutions (μ, σ) with $\mu \neq X_i$ for all i, these solutions will not give (local) minima of $L_n(\theta)$. In other words, the optimal μ is equal to one of the X_i. Common sense, backed up by computational experience, suggests that μ should be the median (or thereabouts, because (5.20) has some weighting in it). Now, it is obvious how one computes an mle for the parametric family (5.16): For each i, set $\mu = X_i$, and solve (5.20) for σ. This defines the estimator θ^i, and the one for which $L_n(\theta^i)$ is smallest is the mle, assuming there are no ties. An interesting consequence of μ being one of the X_i and the weighting involved is the *counterintuitive* behavior of μ: It is actually possible for μ to decrease as the largest order statistic $X_{n,n}$ increases, while the other order statistics $X_{1,n}, X_{2,n} \cdots, X_{n-1,n}$ remain fixed. This happens, for instance, when one starts out with a nearly symmetric sample $X_1, X_2, \cdots, X_n$, but just slightly skewed to the right, and such that without the largest order statistic, it is skewed to the left. For the complete sample, this would give $\mu_n = X_{m,n}$, and for the "incomplete" sample $\mu_n = X_{m-1,n}$, where $m = \lceil n/2 \rceil + 1$. Now, as $X_{n,n}$ increases, the sample becomes even more skewed to the right, so that μ_n increases (or, at least, does not decrease). However, when $X_{n,n}$ becomes large enough that it really becomes a "gross" error, then, somehow, this procedure ignores it, and proceeds with $X_{1,n}, X_{2,n}, \cdots, X_{n-1,n}$, so that μ_n actually decreases, and, of course, decreases discontinuously. It follows that $\mu_n = T(F_n)$ is not a continuous function of F_n in any reasonable topology. Phrased differently, the mle for μ is not robust with respect to small errors!

Robust estimators. All of this does not answer the question of the robustness of the mle's with respect to a few gross errors. Simulations suggest that they are. Anyway, the point is not whether the mle's for estimating the location and scale of a Cauchy-like distribution are robust. Rather, we are interested in how this might be used to estimate the mean and variance of a normal distribution when the iid sample $X_1, X_2, \cdots, X_n$ is possibly contaminated with a few gross errors. One solution is to fix (yes, they are broken) the tails of the normal, as follows. Let $\phi(x)$ be the standard normal distribution and for $k > 0$ ($k \approx 2$?), let

$$(5.21) \qquad \phi_k(x) = \begin{cases} \phi(x) & , \quad |x| \leqslant k , \\ \dfrac{\phi(k)(1+k)^2}{(1+|x|)^2} & , \quad |x| > k . \end{cases}$$

Then, ϕ_k is continuous. Now, ϕ_k is not a pdf, but it is integrable, so we may choose c_k such that $c_k \phi_k$ is indeed a pdf. Now, define the parametric family $\mathcal{F}_k = \{ \psi_k(x ; \theta) : \theta \in \mathbb{R} \times \mathbb{R}_+ \}$ by

$$(5.22) \qquad \psi_k(x ; \theta) = c_k \, \sigma^{-1} \, \phi_k\big(\sigma^{-1}(x - \mu)\big) , \qquad x \in \mathbb{R} ,$$

and consider the maximum likelihood estimation problem for this parametric family

$$(5.23) \qquad \text{minimize} \quad \Lambda_n(\theta) \overset{\text{def}}{=} \log \sigma - \frac{1}{n} \sum_{i=1}^{n} \log \psi_k\big(\sigma^{-1}(X_i - \mu)\big) .$$

The minimizer is denoted by $\theta_{nk} = (\mu_{nk}, \sigma_{nk})$. Estimators like this are usually called M-estimators because they are similar to maximum likelihood estimators. Here, though, the likelihood has been modified in a special way, viz. by modifying the parametric family of pdfs.

What are the properties of θ_{nk}? Does θ_{nk} exist, and is it unique? And what about robustness? Before answering these questions, it is worth noting that θ_{nk} has the correct invariance properties with respect to the location and scale of $X_1, X_2, \cdots, X_n$. Specifically, if we make explicit the dependence of θ_{nk} on $X_1, X_2, \cdots, X_n$ by $\theta_{nk} \equiv \theta_{nk}(X_1, \cdots, X_n)$, then for all $m \in \mathbb{R}$, and $s > 0$, and all $X_1, X_2, \cdots, X_n$,

(5.24) $\qquad \dfrac{\theta_{nk}(X_1, \cdots, X_n) - m}{s} = \theta_{nk}\left(\dfrac{X_1 - m}{s}, \cdots, \dfrac{X_n - m}{s}\right).$

To discuss the existence, uniqueness, and robustness of θ_{nk}, we rewrite $\Lambda_n(\theta)$. To that end, let

(5.25) $\qquad I = \{\, i \,:\, |X_i - \mu| \leqslant k\sigma \,\}, \qquad J = \{\, i \,:\, |X_i - \mu| > k\sigma \,\}.$

Note, though, that I and J depend on the unknown μ, σ. We obtain

(5.26) $\Lambda_n(\theta) = \log \sigma + \dfrac{1}{n} \sum_{i \in I} \dfrac{(X_i - \mu)^2}{2\sigma^2} + \dfrac{2}{n} \sum_{i \in J} \log\left(1 + \sigma^{-1}|X_i - \mu|\right).$

The observations regarding the existence of absolute minima of the negative log-likelihood for the Cauchy-like distribution, see (5.16), apply here as well, so we consider the question of the existence of absolute minima of $\Lambda_n(\theta)$ settled. Note that the above is essentially an iterative procedure, because I and J depend on μ and σ. However, we ignore this aspect of the problem. If $J = \varnothing$, then μ_{nk} and σ_{nk}^2 are just the mle's for the mean and the variance of a normal distribution and are obviously unique. This gives an indication that the solution ought to be unique for $|J|$, the number of X_i belonging to J, not too large. The following derivation of robustness (of sorts) can be modified to also show the uniqueness of the mle's.

It is clear that $\Lambda_n(\theta)$ is differentiable. Computing the derivatives and setting them equal to 0 gives

(5.27) $\qquad \dfrac{1}{n} \sum_{i \in I} \dfrac{\mu - X_i}{\sigma^2} + \dfrac{2}{n} \sum_{i \in J} \dfrac{\operatorname{sign}(\mu - X_i)}{\sigma + |X_i - \mu|} = 0 \,,$

(5.28) $\qquad \sigma^{-1} - \dfrac{1}{n} \sum_{i \in I} \dfrac{(X_i - \mu)^2}{\sigma^3} - \dfrac{2}{n} \sum_{i \in J} \dfrac{|X_i - \mu|}{\sigma\,(\sigma + |X_i - \mu|)} = 0 \,.$

We may rewrite these equations as

(5.29) $\qquad \mu - m_I = -\dfrac{2\sigma^2}{|I|} \sum_{i \in J} \dfrac{\operatorname{sign}(\mu - X_i)}{\sigma + |X_i - \mu|} \,,$

(5.30) $\qquad \sigma^2 - \dfrac{|I|}{n}\left\{ s_I^2 + (\mu - m_I)^2 \right\} = \dfrac{2\sigma^2}{n} \sum_{i \in J} \dfrac{|X_i - \mu|}{\sigma + |X_i - \mu|} \,,$

where

(5.31) $\qquad m_I = \tfrac{1}{|I|} \sum_{i \in I} X_i \,, \qquad s_I^2 = \tfrac{1}{|I|} \sum_{i \in I} (X_i - m_I)^2$

are the sample mean and variance of $\{\, X_i \,:\, i \in I \,\}$.

Assume now that μ, σ, I, and J satisfy the above equations, with I and J as in (5.25). We assume that $|J| = \varepsilon n$ with $\varepsilon \leqslant 0.10$, say, and that

$k \geqslant 2$. Then, $|J|/n = \varepsilon$. Straightforward bounds give

$$(5.32) \qquad |\mu - m_I| \leqslant \frac{2\,|J|\,\sigma}{(k+1)\,|I|} \leqslant \varepsilon\sigma \,,$$

$$(5.33) \qquad |\sigma^2 - (1-\varepsilon)\,s_I^2| \leqslant (1-\varepsilon)\,(\mu - m_I)^2 + \frac{2\,k\,|J|\,\sigma^2}{(k+1)\,n} \,.$$

The last right-hand side may be further bounded by $(\mu - m_I)^2 + 2\,\varepsilon\,\sigma^2$. Substituting the first bound into the second inequality gives

$$|\sigma^2 - (1-\varepsilon)\,s_I^2| \leqslant (\varepsilon^2 + 2\varepsilon)\,\sigma^2 \leqslant 3\,\varepsilon\,\sigma^2 \,.$$

It follows that

$$(5.34) \qquad \sigma^2 \leqslant \frac{(1-\varepsilon)\,s_I^2}{1-3\varepsilon} \,, \qquad |\sigma^2 - s_I^2| \leqslant \frac{4\,\varepsilon\,s_I^2}{1-3\varepsilon} \,.$$

Substituting the first one of these estimates back into (5.32) gives

$$(5.35) \qquad |\mu - m_I| \leqslant \varepsilon\,s_I\,\sqrt{\frac{1-\varepsilon}{1-3\,\varepsilon}} \,.$$

So for reasonable constants c_1, c_2, e.g., for $\varepsilon \leqslant 0.10$, we may take $c_1 = 1.14$, $c_2 = 6$,

$$(5.36) \qquad |\mu - m_I| \leqslant c_1\,\varepsilon\,s_I \,, \qquad |\sigma^2 - s_I^2| \leqslant c_2\,\varepsilon\,s_I^2 \,.$$

It would be hard not to conclude that the method at hand is robust with respect to a few gross errors.

(5.37) EXERCISE. (a) Consider the functional

$$T(F) = \int_{\mathbb{R}} b(x)\,dF(x) \,,$$

for absolutely continuous F, i.e., F' is an integrable function. Assuming that $b(x)$ is a bounded continuous function, show that $T(F)$ is a robust functional in the sense of (5.10).
(b) Show that for $k > 0$ (large), the functionals

$$\mu_k(F) = \int_{\mathbb{R}} \frac{x\,dF(x)}{1+(x/k)^2} \,, \qquad \sigma^2 = \int_{\mathbb{R}} \frac{(x-\mu_k)^2\,dF(x)}{1+((x-\mu_k)/k)^2} \,,$$

are robust estimators of $\mu(F)$ and $\sigma(F)$.
(c) Consider the implicitly defined estimators

$$0 = \int_{\mathbb{R}} \frac{(x-\mu_k)\,dF_n(x)}{1+(x-\mu_k)^2/(k^2\,\sigma_k^2)} \,, \qquad \sigma_k^2 = \int_{\mathbb{R}} \frac{(x-\mu_k)^2\,dF_n(x)}{1+(x-\mu_k)^2/(k^2\,\sigma_k^2)} \,,$$

for the mean and variance of a normal distribution. Does this indeed define the estimators (uniquely), and are they robust estimators with respect to gross errors?

(5.38) EXERCISE. Would (5.36) change drastically if in (5.21) we used Cauchy tails, that is, replaced (5.21) by

$$\phi_k(x) = \begin{cases} \phi(x) & , & |x| \leqslant k , \\ \dfrac{\phi(k)\,(1+k^2)}{(1+|x|^2)} & , & |x| > k . \end{cases}$$

(5.39) EXERCISE. (a) Study the multivariate analogue of (5.16)–(5.22), that is, maximum likelihood estimation for the parametric family

$$f(x;\theta) = \sigma^{-d}\,f\!\left(\sigma^{-1}(x-\mu)\right) ,$$

with

$$f(x) = \frac{c_d\,|x|^{d-1}}{(1+|x|)^{d+1}} , \qquad x \in \mathbb{R}^d .$$

Here, $\theta = (\mu,\sigma)' \in \mathbb{R}^d \times \mathbb{R}_+$.
(b) Study the analogue of (5.22)–(5.36) for the multivariate normal (with covariance matrix $\sigma^2\,I$).

Properties of robust estimators. How is one to judge the quality of robust estimators? Obviously, the criteria are consistency, efficiency, and robustness. Consistency and efficiency refer to the case in which $X_1, X_2, \cdots, X_n$ are indeed iid random variables from $f(x;\theta_o)$: How good (or how bad) is the performance of the robustified estimator when there are no "errors" in the data. Robustness applies to data with "errors". For an estimator $\theta_n = T(F_n)$, reasonable measures would be expressions like

(5.40) $\text{inconsistency}(\theta_n) = \| E_{\theta_o}[\theta_n - \theta_o] \|$,

(5.41) $\text{inefficiency}(\theta_n) = \| \text{Var}_o(\theta_n) - [n\,\Phi(\theta_o)]^{-1} \|$,

(5.42) $\text{sensitivity}(\theta_n) = \sup\{\, \| T(F_n) - T(F_{n-1}) \| \; : \; X_n \in \mathbb{R}^d \}$.

Note that in (5.40), the expectation is over $X_1, X_2, \cdots, X_n$, which are iid with pdf $f(x;\theta_o)$, and that in (5.41), the covariance $\Phi(\theta_o)$ is the Fisher information matrix

(5.43) $\Phi(\theta) = \mathbb{E}_\theta[-\nabla_\theta^2\{\log f(X;\theta)\}]$.

An alternative for measuring the inefficiency is

(5.42′) $\text{inefficiency}(\theta_n) = \| n\,[\Phi(\theta_o)]^{1/2}\,\text{Var}_o(\theta_n)\,[\Phi(\theta_o)]^{1/2} \|$.

The emphasis in robustness studies is usually on the tradeoff between efficiency and robustness, consistency (of the location estimator, say) being guaranteed under minimal conditions. By way of example, for "our" location estimator μ_{nk}, consistency is assured if the true pdf f_r belongs to any symmetric location/scale parametric family $f_r(x) = \sigma^{-1}f_1\!\left(\sigma^{-1}(x-\mu)\right)$, with f_1 an even function. Then, it is clear that for robustness of μ_{nk}, we need $k < \infty$, whereas for efficiency, $k = \infty$ is best. So, some happy middle ground has to be found. This is typical for robust estimators in general, although there exist estimators that are robust under gross errors and that are efficient under the assumed model, see BERAN (1977). The numerical measures for efficiency and robustness allow the comparison of estimators under limited circumstances. If θ_n and $\widetilde{\theta}_n$ are two consistent estimators with

$$\text{sensitivity}(\theta_n) \approx \text{sensitivity}(\widetilde{\theta}_n) , \quad \text{inefficiency}(\theta_n) \ll \text{inefficiency}(\widetilde{\theta}_n) ,$$

then θ_n is obviously better than $\widetilde{\theta}_n$. We will not determine expressions for the robustness of estimators, but content ourselves with discussing their efficiency. As usual, very little can be said about the finite sample situation. The asymptotical efficiency is within easy reach, it being a nice application of the material in § 2.

Asymptotic efficiency of M-estimators. We now switch gears somewhat and consider general M-estimators. Let $X_1, X_2, \cdots, X_n$ be iid with pdf f_r, which is close to the paramatric family $\mathcal{F} = \{ f(x;\theta) : \theta \in \Theta \}$. This covers the case that the X_i are iid from a pdf $f(x;\theta_o)$, with a few gross errors (or many small errors). We estimate the parameter θ_o by solving the *maximum-likelihood-like* problem

(5.44)
$$\text{minimize} \quad L_{n\gamma}(\theta) \stackrel{\text{def}}{=} -\frac{1}{n} \sum_{i=1}^{n} \log \gamma(X_i;\theta)$$
$$\text{subject to} \quad \theta \in \Theta .$$

Here, $\gamma(x;\theta)$ belongs to a parametric family $\Gamma = \{ \gamma(x;\theta) : \theta \in \Theta \}$, which is a modification of $\mathcal{F}$, but the $\gamma(x;\theta)$ need not be pdfs. Any solution $\theta_{n\gamma}$ of (5.44) is called an M-estimator. The asymptotic properties of $\theta_{n\gamma}$ may be studied by repeating the material of § 2, appropriately modified and interpreted. We give a brief outline and state the results.

It seems obvious that if $\theta_{n\gamma}$ is going to converge to anything, it will be to the solution θ_γ of the problem

(5.45)
$$\text{minimize} \quad -\int_{\mathbb{R}} \log\{ \gamma(x;\theta) \} f_r(x) \, dx$$
$$\text{subject to} \quad \theta \in \Theta ,$$

and we must assume that this problem has a unique solution. This is the analogue of assumption (2.A5). Equivalently, if the $\gamma(x;\theta)$ are pdfs, θ_γ is such that $\gamma(x;\theta_\gamma)$ is the unique element from the parametric family Γ that is closest to f_r in the sense that

(5.46)
$$\text{KL}\big(f_r, \gamma(\cdot;\theta_\gamma)\big) \leqslant \min \{ \text{KL}\big(f_r, \gamma(\cdot;\theta)\big) : \theta \in \Theta \} ,$$

with $\text{KL}(\cdot, \cdot)$ the Kullback-Leibler divergence.

It is now straightforward to modify the assumptions and arguments of § 2 to show that asymptotically, local minimizers θ_n of $L_{n\gamma}(\theta)$ in a neighborhood of θ_γ exist and satisfy $\theta_{n\gamma} \longrightarrow_{as} \theta_\gamma$ for $n \to \infty$. Moreover, if the integrals

$$(5.47) \quad \Psi_\gamma(\theta) \stackrel{\text{def}}{=} \int_{\mathbb{R}} \nabla_\theta \{\log \gamma(x\,;\theta)\} \big[\nabla_\theta \{\log \gamma(x\,;\theta)\} \big]^T f_r(x)\, dx \ ,$$

$$(5.48) \quad \Phi_\gamma(\theta) \stackrel{\text{def}}{=} \int_{\mathbb{R}} \nabla_\theta^2 \{ -\log \gamma(x\,;\theta) \}\, f_r(x)\, dx \ ,$$

$$(5.49) \qquad \int_{\mathbb{R}} \| \nabla_\theta^3 \{ -\log \gamma(x\,;\theta) \} \|\, f_r(x)\, dx$$

converge uniformly for all θ in a neighborhood of θ_γ, and if $\Psi_\gamma(\theta_\gamma)$ and $\Phi_\gamma(\theta_\gamma)$ both are positive-definite, then

$$(5.50) \qquad\qquad \sqrt{n}\,(\theta_n - \theta_\gamma) \longrightarrow_d Y \sim N\big(0, V(\theta_\gamma, \gamma)\big) \ ,$$

with

$$(5.51) \qquad\qquad V(\theta, \gamma) = \big[\Phi_\gamma(\gamma) \big]^{-1} \Psi_\gamma(\theta) \big[\Phi_\gamma(\theta) \big]^{-1} \ .$$

If $\gamma(x\,;\theta) = f(x\,;\theta)$, then $\Phi_\gamma(\theta) = \Psi_\gamma(\theta)$ is just the original Fisher information matrix, and the asymptotic variance in (5.51) simplifies accordingly.

Assuming consistency, the asymptotic variance $V(\theta, \gamma)$ tells us all we are likely to find out about $\theta_{n\gamma}$. Of course, the consistency and efficiency of θ_{nk} do not really matter when the data indeed have a few gross errors. So then the question is how the robustness parameter k should be selected. Just about all methods of doing so must consider the tradeoff between minimizing the numerical measures for inefficiency and sensitivity at the expense of the other. We do not dwell on this, but merely observe that selecting k is similar to selecting window/smoothing/regularization parameters in nonparametric maximum penalized likelihood estimation, and that we discuss at length in Part II.

Misspecification of the model. Up to this point, we have been concerned with location/scale estimation. We now discuss "general" parameter estimation. In §2 and on, we have assumed that the data $X_1, X_2, \cdots, X_n$ are iid with pdf from some known parametric family, modulo a few gross errors. One could rephrase this by saying that the parametric family "adequately" models the phenomenon under investigation. Surely, this is not always the case. In many practical situations, so little is known about the phenomenon being modeled that only very few *qualitative* statements can be made regarding the distribution of the X_i. An often used approach is then to select a "reasonable" parametric family to which these qualitative properties apply. Parameters may then be selected using the maximum likelihood approach, or some other method. Ideally, but an often forgotten step, the investigator(s) now evaluate the appropriateness of the parametric family chosen, modify the parametric model if it is not, and repeat the experiment. Finally, one arrives at a parametric model that is more or less correct. If this evaluation of the parametric model is omitted, what can one say about the results of the estimation process?

To phrase the question a little more precisely, assume that we are considering the parametric family $\mathcal{F} = \{ f(\,\cdot\,;\theta) \,:\, \theta \in \Theta \}$, but the $X_1, X_2, \cdots, X_n$ are iid from the pdf f_r, which does not belong to $\mathcal{F}$. This differs from the previous setting of a few gross errors only in degree: f_r need not be close to $\mathcal{F}$ at all. If we now estimate θ, say, by means of maximum likelihood estimation, what are we then in fact estimating? Of course, it depends on what the purpose of it all is. Let $F(\,\cdot\,;\theta)$ and F_r denote the respective cdfs, and suppose we want to estimate some nice functional $T(F_r)$ of F_r. Recall that the mle of $T(F)$ is $T\big(F(\,\cdot\,;\theta_n)\big)$, where θ_n is the mle of θ [Exercise (2.41)]. Now, as before, if θ_n converges at all, it converges to θ_r, the solution of the *large sample asymptotic problem*

$$
\text{(5.52)} \qquad \text{minimize} \quad L(\theta) \stackrel{\text{def}}{=} - \int_{\mathbb{R}^d} \log\{ f(x\,;\theta) \}\, f_r(x)\, dx
$$
$$
\text{subject to} \quad \theta \in \Theta .
$$

Assuming that $h(\theta) = T\big(F(\,\cdot\,;\theta)\big)$ is a nice enough function of θ, then the above says that $\theta_n \longrightarrow_{\text{as}} \theta_r$, and

$$
\text{(5.53)} \qquad \sqrt{n}\,\big(h(\theta_n) - h(\theta_r)\big) \longrightarrow_{\text{d}} Y \sim N(0, V_r) ,
$$

with the appropriate V_r. Now, we are faced with serious problems all around. First of all, it is not clear what we would mean by the consistency of θ_n. However, if $T(F)$ is the median, or some such *nonparametric* quantity, then we know perfectly well what is meant by consistency, but it is unrealistic to assume that it actually holds, even approximately. By way of example, in Chapter 3, §§ 4 and 5, we demonstrate on some data sets (real and simulated) that choosing the Weibull parametric model rather than the log-normal distribution has a tremendous effect on the mle of the median, say. (These parametric families are interesting for other reasons as well, see §7.) Secondly, even if consistency pertains, the efficiency will surely be pretty bad when the wrong model has been chosen. However, there are ways to detect gross misspecification by testing for goodness-of-fit. Some of these questions have been answered by HUBER (1967).

EXERCISE : (5.8), (5.13), (5.37), (5.38), (5.39)

6. Ridge regression

Another standard example in which "errors" cause serious difficulties concerns the estimation problem in the multivariate linear model (1.16)

$$
\text{(6.1)} \qquad Y \sim N_n(X\beta_o, \sigma_o^2\, I) ,
$$

where $X \in \mathbb{R}^{n \times p}$ has full column rank and β and σ need to be estimated based on a sample Y. For simplicity, we assume that σ_o is known, so only β needs estimating. When the matrix X is well conditioned in the sense that the *condition number* $\mathrm{cond}(X)$,

$$(6.2) \qquad \mathrm{cond}(X) \stackrel{\mathrm{def}}{=} \{ \| (X^{\scriptscriptstyle T} X)^{-1} \| \, \| X^{\scriptscriptstyle T} X \| \}^{1/2} ,$$

is not too large, say $\leqslant 10$, then the mle of β_o, the solution of (1.18),

$$(6.3) \qquad \beta_n = (X^{\scriptscriptstyle T} X)^{-1} X^{\scriptscriptstyle T} Y$$

has all the desirable properties an estimator could have and is in fact a globally minimum variance unbiased estimator of β_o. Moreover, its mean square error

$$(6.4) \qquad \mathbb{E} \left[\| \beta_n - \beta_o \|^2 \right] = \sigma_o^2 \, \mathrm{trace}\bigl((X^{\scriptscriptstyle T} X)^{-1} \bigr)$$

is then within reason.

The fact that β_n is a globally minimum variance unbiased estimator depends very much on the model (6.1) being correct, but it turns out that β_n may still be the "best" estimator even when (6.1) is violated. In particular, one only need to assume the Gauss-Markov model,

$$(6.5) \qquad \mathbb{E} \left[Y \right] = X \beta_o \quad , \qquad \mathbb{E} \left[(Y - X \beta_o)(Y - X \beta_o)^{\scriptscriptstyle T} \right] = \sigma_o^2 \, I .$$

Under this assumption, β_n is optimal for the class of linear unbiased estimators, i.e., all estimators of the form $\widehat{\beta}_n = AY$, where $A \in \mathbb{R}^{n \times p}$, with $\mathbb{E}[\widehat{\beta}_n] = \beta_o$, (so $AX = I$). The sense in which β_n is optimal is that for all linear unbiased estimators $\widehat{\beta}_n$,

$$(6.6) \qquad \mathbb{V}\mathrm{ar}\bigl[\beta_n\bigr] \preccurlyeq \mathbb{V}\mathrm{ar}\bigl[\widehat{\beta}_n\bigr] ,$$

see RAO and MITRA (1971) or SEBER (1977). This should be compared with Theorems (2.6) and (2.7). For this reason, β_n is called a best linear unbiased estimator (BLUE).

The situation arises, and not infrequently, that X is ill-conditioned in the sense that $\mathrm{cond}(X)$ is large, say $\sim 10^{10}$. In order to elaborate on this, it is helpful to introduce the singular value decomposition of X, see RAO and MITRA (1971) or LAWSON and HANSON (1995),

$$(6.7) \qquad X = \sum_{q=1}^{p} \lambda_q \, u_q \, v_q^{\scriptscriptstyle T} ,$$

where $\{ u_1, u_2, \cdots, u_p \} \subset \mathbb{R}^n$ and $\{ v_1, v_2, \cdots, v_p \} \subset \mathbb{R}^p$ are orthonormal sets, and with the ordering

$$(6.8) \qquad \lambda_1 \geqslant \lambda_2 \geqslant \cdots \geqslant \lambda_p > 0 .$$

Note that then $\mathrm{cond}(X) = \lambda_1 / \lambda_p$, so that X is ill-conditioned if the ratio λ_1 / λ_p is large. The mle β_n of (6.3) may now be written as

$$(6.9) \qquad \beta_n = \sum_{q=1}^{p} \lambda_q^{-1} \, (u_q^{\scriptscriptstyle T} Y) \, v_q ,$$

and as we show later, its mean square error is given by

$$(6.10) \qquad \mathbb{E}\big[\,\|\,\beta_n - \beta_o\,\|^2\,\big] = \sum_{q=1}^{p} \sigma_o^2\,\lambda_q^{-1}\,.$$

So, if λ_1/λ_p is large, then the overall mean square error is large compared with the individual terms of the mle. Is this the best we can do, or can better estimators be found? One case is when the ratios λ_1/λ_q, $q = 2, 3, \cdots, p - 1$ are all reasonable, say $\leqslant 10$, but $\lambda_1/\lambda_p \sim 10^{10}$. Then, the obvious thing to do is to drop λ_p and replace β_n by

$$(6.11) \qquad \tilde{\beta}_n = \sum_{q=1}^{p-1} \lambda_q^{-1}\,(u_q^{\scriptscriptstyle T}Y)\,v_q\,.$$

Of course, one should take a peek at the value of $u_p^{\scriptscriptstyle T}Y$, and panic when this is large compared with the other values $u_q^{\scriptscriptstyle T}Y$. This situation may be summarized by saying that the column rank of X is really $p - 1$ and not p. More details may be found in SEBER (1977).

A much more difficult situation comes about when the ratios λ_1/λ_q, $q = 1, 2, \cdots, p$, increase at a steady pace, to $\sim 10^{10}$, say, without a clear break in the values λ_q/λ_{q+1}. An equivalent situation arises when X is well-conditioned, but the variance of Y is "large". A common practice, and entirely in the spirit of robustification as in Exercise (5.37), is to replace (6.9) respectively (6.11) by

$$(6.12) \qquad \beta_{n,h} = \sum_{q=1}^{p} \frac{\lambda_q}{\lambda_q^2 + h}\,(u_q^{\scriptscriptstyle T}Y)\,v_q\,,$$

with h a small positive number. If $\lambda_q \gg \sqrt{h}$, then $\lambda_q/(\lambda_q^2 + h) \approx 1/\lambda_q$, but as $\lambda_q \to 0$, the added h in the denominator prevents blowup.

It is easy to see that $\beta_{n,h}$ solves the penalized likelihood problem

$$(6.13) \qquad \text{minimize} \quad \|\,X\beta - Y\,\|^2 + h\,\|\,\beta\,\|^2\,.$$

This is an example of ridge regression, see HOERL and KENNARD (1970) and references therein, RAO and MITRA (1971), and SEBER (1977). See also WAHBA (1990). Usually, the term $\|\,\beta\,\|^2$ is replaced by a weighted version $(\beta - \beta_{\mathrm{ap}})^{\scriptscriptstyle T} D(\beta - \beta_{\mathrm{ap}})$, where β_{ap} is an a priori guess for β, e.g., $\beta_{\mathrm{ap}} = 0$, and D is a symmetric positive-definite matrix that must be chosen in accordance with the particular problem at hand. However, here, we concentrate on (6.13).

The above seems reasonable enough, based as it is on deep statistical insight. There are in fact solid statistical arguments that justify it. At first, it does not seem that way, though. In particular, for fixed positive h,

$$(6.14) \qquad \beta_h \overset{\text{def}}{=} \mathbb{E}\big[\,\beta_{n,h}\,\big] = \beta_o - h\,(X^{\scriptscriptstyle T}X + h\,I)^{-1}\beta_o\,,$$

so that $\beta_{n,h}$ is a *biased* estimator of β_o. Then, there does not appear to be much point in computing its mean square error, but let us do it anyway,

$$(6.15) \qquad \mathbb{E}\left[\|\beta_{n,h} - \beta_o\|^2\right] = \|\beta_h - \beta_o\|^2 + \mathbb{E}\left[\|\beta_{n,h} - \beta_h\|^2\right].$$

The first term is the bias and may by rewritten by means of (6.14). The second term equals $\sigma^2 \operatorname{trace}(AA^T)$, where $A = (X^TX + hI)^{-1}X^T$. The trace of a symmetric matrix is the sum of its eigenvalues, and the singular value decomposition (6.7) gives a way to compute them. The result is that

$$(6.16) \qquad \mathbb{E}\left[\|\beta_{n,h} - \beta_o\|^2\right] = \sum_{q=1}^{p} \frac{h\,(v_q^T\beta_o)^2 + \sigma^2\,\lambda_q^2}{(\lambda_q^2 + h)^2}.$$

For $h = 0$, this gives

$$(6.17) \qquad \mathbb{E}\left[\|\beta_n - \beta_o\|^2\right] = \sum_{q=1}^{p} \frac{\sigma^2}{\lambda_q^2}.$$

The claim is that for an appropriate $h > 0$, the mean square error in (6.16) is less than in (6.17). So the bias term in (6.15) is more than compensated for by the reduction in the variance of the estimator. This is easily proven in the present case. Differentiation of $\mathbb{E}\left[\|\beta_{n,h} - \beta_o\|^2\right]$ with respect to h gives

$$(6.18) \qquad \frac{d}{dh}\,\mathbb{E}\left[\|\beta_{n,h} - \beta_o\|^2\right] = \sum_{q=1}^{p} \frac{2\left(h\,(v_q^T\beta_o)^2 - \sigma^2\right)\lambda_q^2}{(\lambda_q^2 + h)^3},$$

so that

$$(6.19) \qquad \frac{d}{dh}\,\mathbb{E}\left[\|\beta_{n,h} - \beta_o\|^2\right]\Big|_{h=0} = \sum_{q=1}^{p} \frac{-2\,\sigma^2}{\lambda_q^4} < 0.$$

Thus, the mean square error is a decreasing function of h close to 0 and, in effect, smaller when h is positive and small.

So why are such estimators not used all of the time? The problem is in selecting the proper value for h. This is similar to selecting the robustness parameter k in §5. Presumably, it would make sense to minimize the mean square error, which means setting the right-hand side of (6.18) equal to zero and solving the resulting equation for h. However, the solution would depend on knowing β_o. If σ must be estimated as well, the problem becomes even more difficult. Nevertheless, this problem can be solved, see WAHBA (1990) and EUBANK (1999), and references therein. We come back to this in Volume II. As a final comment, note that $\beta_{n,h}$ is the solution to a modified likelihood problem and so is an M-estimator. The modification to the negative log-likelihood function was the addition of a penalty term. It is easy to show that $\beta_{n,h}$ is robust with respect to small changes in Y. "Gross" errors in the regression setting is a different problem, which we do not explore.

(6.20) EXERCISE. In the linear problem, the predicted, or filtered, data would be

$$Y_{n,h} = X\left(X^T X + h I\right)^{-1} X^T Y ,$$

with expected value $Y_h = X\left(X^T X + h I\right)^{-1} X^T X \beta_o$. Is there any advantage in selecting a (small) positive h? In other words, is $\mathbb{E}\left[\| Y_{n,h} - Y_h \|^2\right]$ an increasing or a decreasing function of h near $h = 0$?

(6.21) EXAMPLE. Consider the following spline interpolation problem. Given $x_1 \leqslant x_2 \leqslant \cdots \leqslant x_n \in \mathbb{R}$ (design points) and data $Y_1, Y_2, \cdots, Y_n \in \mathbb{R}$, determine the function $f(x)$ such that

(a) $f(x)$ is a cubic polynomial on the intervals (x_i, x_{i+1}), for $1 \leqslant i \leqslant n-1$, and is twice continuously differentiable on (x_1, x_n);

(b) $f(x_i) = Y_i$, $i = 1, 2, \cdots, n$;

(c) $f''(x_1) = f''(x_n) = 0$.

The first two conditions define the notion of a cubic spline. The third one makes the solution a "natural" spline. It can be shown that the solution of the above problem is also the solution of

(6.22)
$$\text{minimize} \quad \int_{x_1}^{x_n} |f''(x)|^2\, dx$$

$$\text{subject to} \quad f(x_i) = Y_i, \ i = 1, 2, \cdots, n .$$

Now, typically, the Y_i are subject to error, and a common model would be to assume normal errors, i.e., the Y_i are independent and

$$Y_i \sim N(f_o(x_i), \sigma^2) . \quad i = 1, 2, \cdots, n ,$$

where $f_o(x)$ is the function we are trying to estimate. We assume that σ is known in advance. (This is certainly not true in all cases, nor is the covariance matrix always diagonal.) Under these circumstances, it does not make a whole lot of sense to insist on condition (b) in the spline interpolation problem. Instead, one considers

(6.23)
$$\text{minimize} \quad \sum_{i=1}^{n} |f(x_i) - Y_i|^2 + h \int_{x_1}^{x_n} |f''(x)|^2\, dx ,$$

with the implicit constraint that f must be twice continuously differentiable. The solution is known as a "smoothing" spline, because it can be shown that conditions (a) and (c) still hold. This problem may be formulated in the form of (6.13), so that smoothing splines are an example of ridge regression. The problem is how to choose the parameter h, and to estimate σ if this is unknown as well. See WAHBA (1990) and EUBANK (1999), and also Volume II. A general reference on spline interpolation is AHLBERG, NILSON and WALSH (1967).

EXERCISE: (6.21).

7. Right-skewed distributions with heavy tails

The title of this section conveys only half of the story. The part it covers is that many data sets consist of observations (iid, we hope) that are positive, but that contain a nonnegligible fraction of relatively really large observations. In particular, the situation is such that it does not pay to consider these large observations as outliers. It is not uncommon that it is precisely these outliers that are of interest. Some of the parametric families traditionally used to model such data sets (or random variables) are the three-parameter log-normal, Weibull, and Gamma distributions. Of course, we would now like to sing the praise of maximum likelihood estimation to estimate the parameters of these distributions, but all is not well. This is the other part of the story. Maximum likelihood estimation as is does not (always) work. Not only may there fail to be a global maximum of the likelihood, but there may not be any local maxima either. Here, we summarize what is known about maximum likelihood estimation in this context, and to what extent the beautiful theory of § 2 is relevant. In Chapter 3, §§ 4 and 5, we discuss small sample behavior of the mle's as well as some of the practical aspects of the computation of these estimators.

For convenience, we state the parametric families under discussion (again). The log-normal pdf is given by

$$(7.1) \quad f(x\,;\theta) = \frac{1}{(x-\gamma)\,\sigma\,\sqrt{2\,\pi}}\,\exp\left[-\frac{\big(\log(x-\gamma)-\mu\big)^2}{2\,\sigma^2}\right]\,, \quad x > \gamma\,,$$

and $f(x\,;\theta) = 0$ for $x \leqslant \gamma$, with $\theta = (\gamma, \mu, \sigma) \in \mathbb{R} \times \mathbb{R} \times \mathbb{R}_{++}$. For the Weibull and Gamma distributions, the parameter θ is $\theta = (\gamma, \sigma, \alpha) \in \mathbb{R} \times \mathbb{R}_{++} \times \mathbb{R}_{++}$. For the Weibull model, the pdf is given as

$$(7.2) \qquad f(x\,;\theta) = \sigma^{-1}\,\varphi_\alpha\big(\sigma^{-1}(x-\gamma)\big)\,, \quad -\infty < x < \infty\,,$$

where

$$(7.3) \qquad \varphi_\alpha(x) = \alpha\,x^{\alpha-1}\,\exp(-x^\alpha)\,, \quad x > 0\,,$$

and $\varphi_\alpha(x) = 0$ for $x \leqslant 0$. Finally, the Gamma distribution is given by

$$(7.4) \qquad f(x\,;\theta) = \sigma^{-1}\,\psi_\alpha\big(\sigma^{-1}(x-\gamma)\big)\,, \quad -\infty < x < \infty\,,$$

where

$$(7.5) \qquad \psi_\alpha(x) = \big[\Gamma(\alpha)\big]^{-1}\,x^{\alpha-1}\,e^{-x}\,, \quad x > 0\,,$$

and, again, $\psi_\alpha(x) = 0$ for $x \leqslant 0$. For all three families, the negative log-likelihood given a sample $X_1, X_2, \cdots, X_n$ is denoted as $L_n(\theta)$. Also, for all three families, γ is a location parameter. For the Weibull and Gamma distribution, σ is a scale parameter and α determines the shape. For the log-normal pdf, μ is the scale parameter and σ influences the shape. The various shapes can be quite different, but the qualitative tail behavior is relatively insensitive with respect to the parameter choices for each family.

(7.6) EXERCISE. Use your favorite computing environment to graph the pdfs for various values of the parameters. Take $\gamma = 0$ always.
(a) For the Gamma and Weibull distributions, take $\sigma = 1$ and α from 0.5 to 2.5 in steps of 0.5, as well as $\alpha = 2$ and $\sigma = 0.5, 1.0, 1.5$.
(b) For the log-normal family, consider $\mu = 0$ and $\sigma = 0.5, 1.0, 1.5$, as well as $\sigma = 0.5$ and $\mu = \pm 1$ and 0.

In most applications, the value of the location parameter γ is supposed to be known. The parametric families are then referred to as the two-parameter families of the distributions in question. In this case, the mle's for the remaining parameters are well behaved: For all three two-parameter families, they exist and are unique, and the asymptotic considerations of § 2 apply. Moreover, they are easily computed. The following exercise deals with the log-normal distribution. The Weibull family we save for later.

(7.7) EXERCISE. (a) Suppose that the location parameter γ is known for the log-normal distribution. Show that the mle's of μ and σ are given by

$$\mu_n(\gamma) = \tfrac{1}{n} \sum_{i=1}^{n} \log(X_i - \gamma) ,$$

$$[\sigma_n(\gamma)]^2 = \tfrac{1}{n} \sum_{i=1}^{n} \big(\log(X_i - \gamma) - \mu_n \big)^2 .$$

(b) Show that these estimators are distributed as

$$\sqrt{n}\,(\mu_n - \mu_o) \sim N(0, \sigma^2) , \quad (n/\sigma_o^2)\,\sigma_n^2 \sim \chi_{n-1}^2 ,$$

a chi-squared distribution with $n - 1$ degrees of freedom.
(c) Determine minimum variance unbiased estimators for μ_o and σ_o.

For the full three-parameter families, that is, when the location parameter must be estimated as well, things become much more interesting and worrisome for dogmatic practitioners of maximum likelihood estimation, like the authors of this text. First of all, for the log-normal family, taking $\gamma_n = X_{1,n}$ makes the negative log-likelihood $L_n(\theta) = -\infty$, so that $L_n(\theta)$ admits no global minimum, cf. Exercise (2.40). Moreover, this makes the other parameters "invisible" in the sense that their values do not matter. As we shall see later, the same is also true for the other two distributions. However, there are statistical objections to choosing $\gamma_n = X_{1,n}$. In particular, for all data sets, $X_{1,n} > \gamma_o$. LOCKHART and STEPHENS (1994) suggest bias reduction methods for the Weibull distribution, but this is not the approach taken here. To be able to proceed, we must consider the three parametric families separately.

For the log-normal distribution, it is easy but tedious to show that the conditions of Theorems (2.23) and (2.24) are satisfied. In particular, the information matrix $\Phi(\theta_o)$ is positive-definite. Consequently, as $n \to \infty$, the negative log-likelihood has a local minimizer θ_n in a neighborhood of the true θ_o, and

(7.8) $$\sqrt{n}\,(\theta_n - \theta_o) \longrightarrow_d Y \sim N\big(0, [\,\Phi(\theta_o)\,]^{-1}\big)\ .$$

The difficulty is of course that for fixed sample size, the existence of local minimizers cannot be proven. It turns out, however, that with the proper interpretation, everything works out. Under the condition

$$m_3 = \sum_{i=1}^{n} (X_i - \overline{X})^3 \geqslant 0,$$

MUNROE and WIXLEY (1970) have shown that a local minimum exists, which then has the requisite asymptotic optimality properties. In addition, if $m_3 < 0$, there is a local minimum at $\gamma = -\infty$, which corresponds to a normal pdf with mean and standard deviation given by the sample mean and sample standard deviation, see EASTHAM, LARICCIA, and SCHUENE-MEYER (1987). It is clear that in this case the asymptotic theory of §2 does not apply. On the other hand, $\mathbb{P}\big[\,m_3 < 0\,\big] \longrightarrow 0$ as $n \to \infty$, so this occurrence does not affect the asymptotic theory. Of course, it is easily detected in the data analysis that $m_3 < 0$, and under these circumstances, a normal distribution with the sample mean and variance provides a reasonable fit for the underlying pdf. Even when a good local minimum exists, it may still be hard to compute, despite Exercise (7.7), which reduces the problem to that of estimating a single parameter. This is not only due to the fact that there is still a global minimum at the boundary of the parameter space, but also to the fact that $L_n(\gamma)$ changes extremely rapidly as a function of γ around its local minimum. This is illustrated in §3.4. Nevertheless, for the log-normal distribution, maximum likelihood estimation is under control.

For the Weibull and Gamma distributions, things are much less savory. As in the case for the log-normal distribution, there are no closed formulas for the mle's. Indeed, we must bend the rules somewhat to be able to talk about mle's. Some nice references for the material under discussion are PIKE (1966), ROCKETTE, ANTLE, and KLIMKO (1974), CHENG and AMIN (1983), LOCKHART and STEPHENS (1994), and HIROSE (1995), where of course more references may be found. To get our ducks in a row, we observe that for a given sample $X_1, X_2, \cdots, X_n$, the negative log-likelihood assuming a Weibull model is

(7.9) $L_n(\theta) = -\log\alpha + \alpha\log\sigma +$
$$\frac{1}{n} \sum_{i=1}^{n} \big[\,\sigma^{-1}(X_i - \gamma)\,\big]^\alpha - \frac{\alpha - 1}{n} \sum_{i=1}^{n} \log(X_i - \gamma)\ .$$

As preparation for the full three-parameter Weibull distribution, we look first at the setting where some of the parameters are assumed known.

Let us first suppose that the shape parameter α and location parameter γ are known. This case is easy, and we formulate its treatment as an exercise.

(7.10) EXERCISE. In the Weibull distribution, assume that $\alpha > 0$ and γ are known. Show that the mle σ_n of σ satisfies

$$[\sigma_n]^\alpha = \frac{1}{n} \sum_{i=1}^{n} (X_i - \gamma)^\alpha ,$$

and that

$$\sqrt{n}\,(\sigma_n - \sigma) \longrightarrow_d Y \sim N(0, V) ,$$

for a certain V, and determine V.

Next consider the case in which only one of the parameters is known. The case where the scale parameter σ is known is not quite relevant to the preparation for the full three-parameter model, so it is relegated to an exercise, see Exercise (2.40)(a). For the remaining cases, in which either the location parameter or the shape parameter is known, it is convenient to consider

(7.11) $$L_n^*(\gamma, \alpha) = \min_\sigma L_n(\gamma, \sigma, \alpha) .$$

By Exercise (7.10), we have

(7.12) $L_n^*(\gamma, \alpha) = 1 - \log \alpha +$

$$\alpha \log \left[\frac{1}{n} \sum_{i=1}^{n} (X_i - \gamma)^\alpha \right] - (\alpha - 1) \frac{1}{n} \sum_{i=1}^{n} \log(X_i - \gamma) .$$

The case of a known location parameter is nicely treated by PIKE (1966). We formulate it as an exercise.

(7.13) EXERCISE. [PIKE (1966)] Assume that γ is known in the Weibull distribution.
(a) Show that $L_n^*(\gamma, \alpha) \longrightarrow +\infty$ for $\alpha \to 0$ and for $\alpha \to \infty$.
(b) Show that $L_n^*(\gamma, \alpha)$ is a convex function of α.
(c) Conclude that the mle for α exists and is unique.
[Hint for (a): Observe that

$$\log \left[\frac{1}{n} \sum_i (X_i - \gamma)^\alpha \right] = \alpha \log(X_{n,n} - \gamma) + \log \left[1 + \left\{ \frac{X_{i,n} - \gamma}{X_{n,n} - \gamma} \right\}^\alpha \right] ,$$

and that the last term is $\mathcal{O}(1)$ as $\alpha \to \infty$. Also, observe that

$$\frac{1}{n} \sum_{i=1}^{n} \log(X_i - \gamma) < \log(X_{n,n} - \gamma) \quad \text{almost surely .}$$

Hint for (b): Show that the second derivative with respect to α is positive.]

Now, consider the situation in which the shape parameter α is known. It turns out that the cases $0 < \alpha < 1$, $\alpha = 1$, and $\alpha > 1$ behave quite differently from each other. Actually, when $\alpha = 1$, we have the shifted exponential model, which was discussed in Exercise (2.40)(b). So maximum likelihood estimation works well here, even though the theory of §2 does not cover it! The case $0 < \alpha < 1$ is simple enough to be treated as yet another exercise. For the interesting limiting case $\alpha \longrightarrow 0$, see LOCKHART and STEPHENS (1994).

(7.14) EXERCISE. Let $0 < \alpha < 1$ be fixed in the Weibull distribution. Show that

$$\lim_{\gamma \to X_{1,n}} L_n^*(\gamma, \alpha) = -\infty \quad \text{and} \quad \frac{d}{d\gamma} L_n^*(\gamma, \alpha) < 0 \quad \text{for all } \gamma < X_{1,n} .$$

In view of Exercise (7.14), we are thus forced to admit either that the mle's for γ and σ do not exist or that the mle's are γ_n and $\sigma_n = \sigma_n(\gamma_n, \alpha)$, with

$$(7.15) \qquad \gamma_n = X_{1,n} , \quad \sigma_n(\gamma_n, \alpha) = \left[\frac{1}{n} \sum_{i=1}^{n} (X_i - \gamma_n)^\alpha \right]^{1/\alpha} ,$$

the last formula by virtue of Exercise (7.10). Although it is perhaps not unreasonable to take $\gamma_n = X_{1,n}$, in that case, $L_n(\gamma, \sigma.\alpha) = -\infty$ regardless of the choice of σ. But it turns out that γ_n and σ_n are good estimators. See Exercises (7.16), (7.17) and (7.18).

(7.16) EXERCISE. It is helpful to introduce the notation $X \sim W(\gamma, \sigma, \alpha)$ to indicate that X is a random variable with the three-parameter Weibull distribution (7.2)–(7.3). Now, let $X_1, X_2, \cdots, X_n$ be iid random variables with $X_1 \sim W(\gamma, \sigma, \alpha)$. Show that

$$X_{1,n} \sim W(\gamma, \sigma\, n^{-1/\alpha}, \alpha) , \quad \text{and} \quad n^{1/\alpha}(\gamma_n - \gamma) \sim W(0, \sigma, \alpha) .$$

(7.17) EXERCISE. Under the conditions of Exercise (7.16) with $0 < \alpha < 1$, show that $\sigma_n = \sigma_n(\gamma_n, \alpha)$ satisfies

$$\sqrt{n}\,(\sigma_n - \sigma) \longrightarrow_d Z \sim N(0, s) ,$$

for a certain s, and determine s. [Hint: σ_n^α is a sum of random variables, but they are not independent. However, note that for $X_i \neq X_{1,n}$,

$$(X_i - \gamma_n)^\alpha = (X_i - \gamma)^\alpha \left\{ 1 + \frac{\gamma - \gamma_n}{X_i - \gamma} \right\}^\alpha .$$

Now, for $0 < \alpha < 1$, use the elementary inequality $|(1 + r)^\alpha - 1| \leqslant |r|^\alpha$ to conclude that one may just as well assume $\gamma_n = \gamma$.]

(7.18) EXERCISE. Repeat Exercise (7.17) for the case where $\alpha > 1$. [Hint: Now, use the mean value theorem to obtain that there exists a random λ_n

with $\gamma < \lambda_n < \gamma_n$ such that

$$\sum_{i=1}^n (X_i - \gamma_n)^\alpha = \sum_{i=1}^n \left\{ (X_i - \gamma)^\alpha - \alpha\,(X_i - \lambda_n)^{\alpha-1}(\gamma - \gamma_n) \right\}.$$

Because $X_i - \lambda_n < X_i - \gamma$, it now suffices to show that there exists a universal constant such that

$$\limsup_{n\to\infty}\ \frac{1}{n}\sum_{i=1}^n (X_i - \gamma)^{\alpha-1} \leqslant_{\text{as}}\ \text{constant .}]$$

Finally, consider the case $\alpha > 1$. It is clear from (7.12) that $L_n^*(\gamma, \alpha) \longrightarrow +\infty$ for $\gamma \to X_{1,n}$ and also for $\gamma \to -\infty$. So $L_n^*(\gamma, \alpha)$ admits an absolute minimum for which $\frac{d}{d\gamma}L_n^*(\gamma, \alpha) = 0$. Showing the uniqueness of the absolute minimum could get ugly, but ROCKETTE, ANTLE, and KLIMKO (1974) keep it edifying.

(7.19) EXERCISE. [ROCKETTE, ANTLE, and KLIMKO (1974)] (a) Show that

$$\frac{d}{d\gamma}L_n^*(\gamma, \alpha) = 0 \quad \text{implies} \quad \frac{d^2}{d\gamma^2}L_n^*(\gamma, \alpha) > 0 .$$

This says that any stationary point of $L_n^*(\gamma, \alpha)$ is a local minimum.
(b) Show that if $L_n^*(\gamma, \alpha)$ has two distinct local minima, then it must have a local maximum.
(c) Conclude that $L_n^*(\gamma, \alpha)$ has only one local minimum, which is the absolute minimum.
[Hint: For part (a), a peek at convexity in Chapter 9 might be in order.]

So the previous exercise says that mle's γ_n and σ_n exist and are unique. Note that σ_n is (still) as in (7.15). What about (asymptotic) optimality of γ_n and σ_n? CHENG and AMIN (1983) have investigated this. It is a tedious exercise to show for $\alpha \geqslant 2$ that the conditions of Theorem (2.24) are satisfied, so that γ_n and σ_n are best asymptotically normal. For $1 < \alpha < 2$, this is not the case. Just as for the shifted exponential distribution we get here better rates of convergence than those predicted by Theorem (2.24). CHENG and AMIN (1983) show that $n^{1/\alpha}\,(\gamma_n - \gamma) = \mathcal{O}_P(1)$. Then, Exercises (7.17) and (7.18) give the asymptotic normality of σ_n. Note that even $\gamma_n = X_{1,n}$ satisfies $n^{1/\alpha}\,(\gamma_n - \gamma) = \mathcal{O}_P(1)$, but the corresponding estimator σ_n has larger variance.

Finally, we are ready for the full three-parameter Weibull distribution. To some extent, it consists of interpreting the above observations in the correct manner. For one thing, we can always make $L_n(\theta) \longrightarrow -\infty$ by selecting $\alpha = \frac{1}{2}$, say ($\alpha < 1$ is what matters), and letting $\gamma \to X_{1,n}$. So here maximum likelihood fails. The only hope is that there should/could exist local minima of the negative log-likelihood in a neighborhood of θ_o. What actually happens is governed by the value of α_o. CHENG and AMIN

(1983) show that if $\alpha_o > 1$, then asymptotically (or, in probability) local maxima indeed exist, and if $\alpha_o \geqslant 2$, they are best asymptotically normal. If $1 < \alpha_o < 2$, then γ_n converges to γ_o at the rate of $n^{-1/\alpha}$, and σ_n and α_n are asymptotically normal. However, if $0 < \alpha_o < 1$, then $\gamma_n = X_{1,n}$ in probability, and we have trouble estimating σ and α, because $L_n(X_{1,n}, \sigma, \alpha) = -\infty$ for any choice of σ and $\alpha < 1$. To complicate matters even more, if we consider minimizing $L_n^*(\gamma, \alpha)$ over α first, then the minimizing α is the solution to the equation

$$(7.20) \qquad \frac{1}{\alpha} = \frac{\sum_{i=1}^n (X_i - \gamma)^\alpha \log(X_i - \gamma)}{\sum_{i=1}^n (X_i - \gamma)^\alpha} - \frac{1}{n} \sum_{i=1}^n \log(X_i - \gamma) .$$

However, the minimizing α tends to 0 and $\sigma(\gamma, \alpha)$ tends to a positive limit as $\gamma \to X_{1,n}$. (Actually, this is not all that obvious, but we will let it pass.) In effect, then, the Weibull distribution tends to a point mass distribution at $x = X_{1,n}$, which is clearly not acceptable. This is actually a combined effect of the heavy tails and the unboundedness at $x = \gamma$ of the pdf. Even though the pdf $f(x; \theta)$ tends to a delta function, its tails are heavy enough to not give very large positive contributions to $L_n(\theta)$. See also Exercise (7.26). So, when $0 < \alpha_o < 1$, maximum likelihood estimation fails in probability. These results are summarized in the following theorems. It turns out the same conclusions hold for the three-parameter Gamma distribution, see CHENG and AMIN (1983).

(7.21) EXISTENCE THEOREM. [CHENG and AMIN (1983)]
Let $X_1, X_2, \cdots, X_n$ be iid with pdf $f(x; \theta_o)$ a Weibull or a Gamma distribution, with $\theta = (\gamma, \sigma, \alpha)$, and let $L_n(\theta)$ denote the negative log-likelihood.

(a) If $\alpha_o > 1$, then in probability $L_n(\theta)$ has a local minimum in a neighborhood of θ_o.

(b) If $0 < \alpha_o \leqslant 1$, then in probability $L_n(\theta) \longrightarrow -\infty$ for $\gamma \to X_{1,n}$, and $f(x; \theta)$ tends to a point mass.

(7.22) ASYMPTOTICS THEOREM. [CHENG and AMIN (1983)]
With $X_1, X_2, \cdots, X_n$ as in Theorem 7.21, assume that $\alpha_o > 1$, and that a local minimizer θ_n of $L_n(\theta)$ exists.

(a) If $\alpha_o > 2$, then $\sqrt{n}\,(\theta_n - \theta_o) \longrightarrow_d Y \sim N(0, [\Phi(\theta_o)]^{-1})$.

(b) If $1 < \alpha_o \leqslant 2$, then $n^{1/\alpha_o}(\gamma_n - \gamma_o) = \mathcal{O}_P(1)$ and

$$\sqrt{n}\,(\vartheta_n - \vartheta_o) \longrightarrow_d Y \sim N(0, [\Phi_\vartheta(\theta_o)]^{-1}) ,$$

where $\vartheta = (\sigma, \alpha)$, and $\Phi_\vartheta(\theta)$ is the submatrix of $\Phi(\theta)$ corresponding to (σ, α).

So what is one to do? The standard method is first to estimate γ by some other method, and then proceed by minimizing $L_n(\gamma, \sigma, \alpha)$ over σ and

α. One such method is based on the fact that $n^{1/\alpha_o}\left(X_{1,n}-\gamma_o\right)=\mathcal{O}_P(1)$. Thus, we take

(7.23) $\gamma_n = X_{1,n} - c(1/n)$,

for a suitable constant c (to get the scaling right), and let σ_n, α_n be the mle's for σ_o and α_o, assuming that $\gamma_o = \gamma_n$. Is this really permissible? In particular, in view of the comments surrounding (7.20), would not α_n be really close to 0? The answer is: Not really. The reason is that in this range of values of γ_n (even if in (7.23) we replace $(1/n)$ by n^β for some fixed β, $0 < \beta < 1$), the resulting α_n is relatively insensitive to the particular value of γ_n. So this appears to be a reasonable approach. Note that for $\alpha_o \leqslant 1$, it follows that

(7.24) $n^{1/\alpha_o}\left(\gamma_n - \gamma_o\right) = \mathcal{O}_P(1)$,

(7.25) $\sqrt{n}\left(\vartheta_n - \vartheta_o\right) \longrightarrow_d Y \sim N\left(0, [\Phi_\vartheta(\theta_o)]^{-1}\right)$,

with ϑ_o and $\Phi_\vartheta(\vartheta_o)$ as in Theorem (7.22). However, this estimator is not optimal. CHENG and AMIN (1983) have exhibited estimators based on spacings for which (7.24) may be improved to $n\left(\gamma_n - \gamma_o\right) = \mathcal{O}_P(1)$, and for which (7.25) holds. See § 8.

(7.26) EXERCISE. Are there nonstandard methods for handling the case $0 < \alpha_o < 1$? The authors are glad you asked! We know how to avoid point masses when estimating pdfs nonparametrically, viz. by including a roughness penalization functional. This is, of course, very unusual in the context of parametric estimation, and the difficulty is how to choose the smoothing parameter. Investigate it!

To finish this section, a word and exercise about heavy tails are provided. The tails of the log-normal, Weibull, and Gamma distributions are not very heavy, e.g., they all have moments of arbitrary high order. If we are really interested in heavy tails, then distributions with polynomial decay come to mind. Also, if the tails are really heavy, then the structure "in the finite region" becomes hard to estimate as well (because everything is in the tail). The following exercise is instructive.

(7.27) EXERCISE. Consider the three-parameter family

$$f(x\,;\theta) = \sigma^{-1}\,\varphi_\alpha\left(\sigma^{-1}(x-\gamma)\right), \quad -\infty < x < \infty,$$

with $\theta = (\gamma, \sigma, \alpha)$, and

$$\varphi_\alpha(x) = \alpha\,(1+x)^{-\alpha-1}, \quad x > 0,$$

and $= 0$ otherwise. Investigate the maximum likelihood estimation of θ.
(a) Show that $\gamma_n = X_{1,n}$ and that $n\left(\gamma_n - \gamma\right) = \mathcal{O}_P(1)$.
(b) Show that for known γ, the conditions of Theorems (2.23) and (2.24)

are satisfied.

(c) Show that for known σ and $\gamma > X_{1,n}$, the negative log-likelihood achieves its minimum for $\alpha = \alpha_n$, where α_n is given by

$$\alpha_n^{-1} = \frac{1}{n} \sum_{i=1}^{n} \log\left[1 + \sigma^{-1}\left(X_i - \gamma\right)\right] .$$

(d) Show that for each $\gamma > X_{1,n}$, the negative log-likelihood $L_n(\gamma, \sigma, \alpha)$ has a unique minimizer (σ, α).

(e) Verify that as $\gamma \longrightarrow X_{1,n}$, the estimators for the rate of decay α and the scale parameter σ behave properly.

[Hint for (b): Integrability and differentiability conditions are easily verified. The real kicker is to show that the information matrix is positive-definite in a neighborhood of the true θ. Because of continuity, it suffices to show the positive-definiteness at the true θ. Parts (d) and (e) seem true, but the authors have not actually verified them.]

EXERCISES : (7.6), (7.7), (7.10), (7.13), (7.14), (7.16), (7.17), (7.18), (7.19), (7.26), (7.27).

8. Additional comments

Ad §2: In Theorems (2.6) and (2.7), the extra generality provided by the function $h(\theta)$ is helpful, but is in fact not more general than the standard case $h(\theta) = \theta$. This may be seen as follows. The assumption that $\nabla h(\theta)$ is continuous and has full row rank at $\theta = \theta_o$ implies that $\nabla h(\theta)$ has full row rank for all θ close enough to θ_o. Then, surely $q \leqslant p$. It is then possible to find a continuously differentiable function $k : \theta \longrightarrow \Xi' \subset \mathbb{R}^{p-q}$ such that the mapping $\theta \longmapsto \alpha \overset{\text{def}}{=} \left(h(\theta), k(\theta)\right)$ has a continuously differentiable inverse in a neighborhood of θ_o. For instance, choose $A \in \mathbb{R}^{(p-q) \times p}$ such that $\left(\left[\nabla h(\theta_o)\right]^T \mid A^T\right)$ is a nonsingular matrix, and define $k(\theta) = A\theta$. We may then reparametrize the family $f(x;\theta)$ as $\phi(x;\alpha)$, and the original theorems apply to $\phi(x;\alpha)$ and to the estimation of α, which subsumes estimating $h(\theta)$.

The critical condition in WALD's Theorem (2.10)

$$\limsup_{n \to \infty} L_n(\theta_n) - L_n(\theta_o) \leqslant_{\text{as}} 0$$

was originally phrased as

$$\liminf_{n \to \infty} \frac{\prod_{i=1}^{n} f(X_i; \theta_n)}{\prod_{i=1}^{n} f(X_i; \theta_0)} \geqslant_{\text{as}} c ,$$

for some positive constant c. In fact, it may be replaced by $\exp(-c_n)$, where $c_n \geqslant 0$ and $c_n = o(n)$ (Little Oh). In the language of the averaged negative log-likelihood the constant c disappears.

Maximum likelihood estimation (and M-estimation) with constraints is discussed by GEYER (1994). The treatment relies heavily on concepts from nonsmooth optimization and is very much worthwhile, but lies outside the scope of this text.

Ad §4: The discussion of the EM algorithm for mixtures largely follows REDNER and WALKER (1984).

Ad §5: The estimators (5.12) and the bounds (5.36) suggest the use of the "trimmed sums" m_t and s_t^2 as alternative robust estimators for the mean and variance of the normal. A slightly different implementation would use only the order statistics $X_{i,n}$, $i = [pn], [pn]+1, \cdots, n-[pn]$, where $p = 0.05$ or 0.10, say. The resulting estimators are then linear functions of the order statistics, which are known as L-estimators. See HUBER (1977) and HUBER (1981).

The usual definition of robustness is not what we make the reader believe it is. It is still based on the "closeness" of cumulative distribution functions, but now "closeness" is defined in terms of the so-called Skorokhod metric or dog-leash metric being small. This metric is defined as follows. For two distributions F and G on the line

$$d(F, G) = \inf \| F \circ h - G \|_\infty + \| h - \mathrm{id} \|_\infty \, ,$$

where $\mathrm{id}(x) = x$ for all x and the infimum is over all homeomorphisms h of $\mathbb{R}$ onto itself. (A homeomorphism of $\mathbb{R}$ is a function such that both it and its inverse are continuous in the usual sense.) One of the things this does is that two empirical distribution functions are close if the samples are close. By way of example, if F and G are the empirical distributions for the sample X_1 and Y_1, then $d(F, G) = \min(1, | X_1 - Y_1 |)$ (Exercise!). For the full treatment of robust statistics, see HUBER (1981), FERNHOLZ (1983), and RIEDER (1994) .

Ad §7: There are other ways to estimate parameters, such as the method of moments or percentiles. That not all of these methods extend to multivariate parametric families of pdfs is not of great concern because there are not many true and tested multivariate parametric distributions in actual use. However, most of the alternative methods only yield Opie-One or asymptotically normal estimators rather than best asymptotically normal estimators. Following ideas of PYKE (1965), CHENG and AMIN (1983) and RANNEBY (1984) came up with the following suggestion of more than passing interest. Suppose $X_1, X_2, \cdots, X_n$ are iid random variables with distribution function $F(x; \theta_o)$ belonging to some parametric family of distributions. For θ in the parameter space, let $Y_i = F(X_i; \theta)$, for all i. If θ was the correct value, then $Y_1, Y_2, \cdots, Y_n$ would be iid Uniform (0,1) random variables. How can we characterize uniform random variables? One way is to say that the order statistics (with $Y_{0,n} = 0$ and $Y_{n+1,n} = 1$

added) $0 = Y_{0,n} < Y_{1,n} \leqslant Y_{2,n} \leqslant \cdots \leqslant Y_{n,n} < Y_{n,n} = 1$ should be uniformly spaced in the sense that the product of the spacings

$$\prod_{i=0}^{n} \left(Y_{i+1,n} - Y_{i,n} \right)$$

should be large (close to its maximum). This leads to the idea of estimating θ by maximizing the product of spacings, or minimizing its negative logarithm, with the additional factor $(1/n)$,

$$(8.1) \quad \text{minimize} \quad -\frac{1}{n} \sum_{i=0}^{n} \log\big(F(X_{i+1,n}\,;\theta) - F(X_{i,n}\,;\theta) \big) \quad \text{over all } \theta \;.$$

Here, $X_{0,n} = -\infty$, $X_{n+1,n} = +\infty$. So this is the standard maximum likelihood method, but with $f(X_i\,;\theta)$ approximated by

$$f(X_{i,n}\,;\theta) \approx \frac{F(X_{i+1,n}\,;\theta) - F(X_{i,n}\,;\theta)}{X_{i+1,n} - X_{i,n}} \;.$$

The denominator just adds a (random) constant to the objective function in (8.1). CHENG and AMIN (1983) and RANNEBY (1984) show that this method works under the conditions of §2, say, but sometimes also when maximum likelihood estimation fails. The reason seems to be that probability density functions are fickle objects, but probabilities, such as those appearing in (8.1), are much more "real" or "stable". For the latest on this, see GOSH and RAO JAMMALAMADAKA (2001).

We also mention here minimum distance methods, in which the pdf is estimated nonparametrically, say, with a kernel density estimator denoted as $f^{nh} = A_h * dF_n$, see §1.1 or Chapter 4, as well as parametrically, by solving

$$(8.2) \qquad \text{minimize} \quad \| f^{nh} - f(\,\cdot\,;\theta) \|_1 \quad \text{over all } \theta,\, h > 0 \;,$$

see, e.g., CAO and DEVROYE (1995). BERAN (1977) replaces the L^1 distance with the Hellinger distance $\| \sqrt{f^{nh}} - \sqrt{f(\,\cdot\,;\theta)} \|_2^2$, but his point of view is robustness of the estimator of θ, cf. §5. From BERKSON (1980), one may distill the suggestion of minimum χ^2 distance, which in this case takes the form of Pearson's φ^2 distance

$$(8.3) \qquad \text{PHI}\big(f^{nh},\, f(\,\cdot\,;\theta) \big) \overset{\text{def}}{=} \int_{\mathbb{R}} \frac{|\, f^{nh}(x) - f(x\,;\theta)\,|^2}{f(x\,;\theta)} \, dx \;.$$

For "general" minimum distance estimation of this type, see, e.g., CAO, CUEVAS and FRAIMAN (1995) and references therein.

3

Parametric Maximum Likelihood Estimation in Action

1. Introduction

The title of the chapter is a bit bombastic, of course, but so be it. In this chapter, we consider the parametric estimation problem from a practical point of view. Thus, we are given a data set, and we assume/hope/pray that it is a realization of iid random variables $X_1, X_2, \cdots, X_n$, with a common pdf belonging to a specified parametric family. The goal is to estimate this pdf or, perhaps, more practically oriented, certain "properties" of the population. In the context of parametric estimation, this boils down to estimating the parameters, as well as verifying the correctness of the parametric model. The first part we have discussed in the previous chapter. It constitutes what may be called the successful side of parametric density estimation using maximum likelihood. That is, under the true parametric model, asymptotically local maximum likelihood estimators exist in a neighborhood of the true parameter, they are best asymptotically normal, and asymptotically are easy to compute. A textual analysis of the previous statements suggests what can go wrong. First of all, it makes asymptotic statements only, so for any finite sample, you are on your own. The knowledge of the existence of local maximizers of the likelihood in a neighborhood of the true parameter is of little help if there are many local maxima. Moreover, local maxima may be hard to find if the likelihood can tend to ∞ on the boundary of the parameter domain. Lastly, it is crucial that the correct parametric family be specified. Note that the advantage of parametric modeling is that relatively few observations are needed for obtaining good estimators for the parameters (by maximum likelihood estimation, say), but that the very paucity of observations is at odds with the requirements of model validation.

In this chapter, we explore some of these questions by analyzing some "real" data sets, in which the true model is unknown, and by analyzing simulated data sets (Monte Carlo simulations, or, as the authors like to call them, Atlantic City simulations). The advantage of simulated data sets is that the true parametric family as well as the true parameter are known. By way of mixtures-of-normals models, we investigate the finite sample behavior of some best asymptotically normal estimators discussed in § 2.3. There, we exhibited a multitude of best asymptotically normal estimators by means of the Newton-Raphson method or the method of scores. How well do these estimators perform in finite sample problems? (And where do the estimators $\widehat{\theta}_n$ for which $\sqrt{n}\,(\widehat{\theta}_n - \theta_o) = \mathcal{O}_P(1)$ come from? In the previous chapter, we referred to such estimators as Opie-One estimators.) The first case under investigation is that of an unknown mixture of known normals. This is relatively straightforward, because there is only one local maximum, which is then also the global maximum. However, even here practical difficulties remain. See § 2. The unknown-mixture-of-unknown-normals model is quite a bit more complicated. The likelihood tends to ∞ at some of the boundary points of the parameter domain, there exist many local maxima of the likelihood, and, finally, choosing the number of components in the mixture is worrisome. This is the subject matter of § 3. In § 6, we explore this question further, as a lead-in to nonparametric estimation. This last question also touches on the problem of how one chooses the correct parametric family of pdfs. In § 5, we explore this in some detail by way of estimating heavy-tailed distributions in a case in which the interest is precisely in the tail behavior. It will transpire that the estimated tail behavior is heavily influenced by the choice of the parametric family. This does not exactly solve the misspecification problem, it merely illustrates the severity of it.

Perhaps we should point out explicitly that the purpose of the chapter is to show the ins and outs of *some* practical estimation problems for real data sets. In particular, the emphasis is put on general considerations rather than on those specific to the parametric family at hand, although these are not completely avoidable.

2. Best asymptotically normal estimators and small sample behavior

Yes, the price we pay for banning all but a few acronyms is the rather lengthy section header. In § 2.3, a bunch of best asymptotically normal estimators were exhibited. Asymptotically, these estimators are indistinguishable from the maximum likelihood estimator. What about their behavior for small sample sizes? The ideal answer to this question would be to figure out the distribution of each estimator or, if we are comparing two or more estimators, the *joint* distribution of the estimators and to draw the appropriate conclusions. In all but a few very special cases, this can only be done approximately by means of Atlantic City simulations.

As our example, we consider mixtures of (known) normal distributions. The parametric family is

$$(2.1) \qquad f(x; w) = \sum_{j=1}^{m} w_j \phi_j(x) , \qquad x \in \mathbb{R} ,$$

where $\phi_j(r) = \sigma_j^{-1} \phi(\sigma_j^{-1}(x - \mu_j))$, with the μ_j and σ_j known. Here, $\phi(x)$ (without subscript) is the standard normal pdf. For $f(x; w)$ to be a pdf, the unknown w_j must be nonnegative and add up to 1. We assume $X_1, X_2, \cdots, X_n$ are iid with common pdf $f(\cdot; w^o)$, and our job is to estimate w^o. The maximum likelihood estimator solves

$$(2.2) \qquad \begin{aligned} &\text{minimize} \quad L_n(w) = -\frac{1}{n} \sum_{i=1}^{n} \log\Big\{ \sum_{j=1}^{m} w_j \phi_j(X_i) \Big\} + \sum_{j=1}^{m} w_j \\ &\text{subject to} \quad w \geqslant 0 \text{ component wise} , \end{aligned}$$

see Exercise (2.1.25). This version of the problem has computational advantages over the original formulation, because there is no need to enforce the pdf constraint. We show later in Chapter 9 that the solution of (2.2) exists and is unique. So theoretically all is well.

We (attempt to) solve (2.2) using the sequential quadratic programming approach, that is, Newton-Raphson, taking the nonnegativity constraints into account, see (2.3.14). It should be noted that the gradient $\nabla L_n(w)$ and Hessian $H_n(w)$ are easily computed. We also consider the EM algorithm (2.4.26). For the EM algorithm, the "uniform" starting point, $w_j = 1/m$ for $j = 1, \cdots, m$, is OK, but for the sequentially quadratic programming method, a really good starting point is needed. In §2.3, the theoretical point was made for taking any Opie-One estimator.

How can we get an Opie-One estimator? One of the simplest ways is by setting the empirical cdf and the parametric cdf equal to one another at various order statistics and solving the resulting system of equations for the estimator $\widehat{w}^n$. For this case, the system of equations is

$$(2.3) \qquad \sum_{j=1}^{m} w_j \Phi\big(\sigma_j^{-1}(X_{i,n} - \mu_j)\big) = \frac{i}{n} ,$$

for various values of i. Here, $\Phi(x)$ is the cdf for the standard normal distribution. Taking the values of i such that

$$(2.4) \qquad i/n = (p + k)/(2p + 1) , \qquad \text{for } k = 0, \pm 1, \cdots, \pm p ,$$

with $p = m$ would seem to make sense. It is clearly a good idea to take more than m order statistics and to solve the corresponding system of equations in the least-squares sense. The resulting estimator is denoted by $\widehat{w}^n$. This procedure is easily implemented on a computer.

(2.5) EXERCISE. Show that $\sqrt{n}\,(\widehat{w}_j^n - w_j^o) = \mathcal{O}_P(1)$.

Table 2.1. Means (standard deviations in parentheses) of $\sqrt{n}\,(w_1^n - w_1^o)$ for the various estimators for the weights in the mixture (2.6) based on 1000 replications, for sample sizes $n = 15, 25, 50$. The Opie-One estimator used three order statistics.

$n =$	15	25	50
Opie-One	$-.0712$ (1.01)	$-.0196$ (1.22)	$.1592$ (1.48)
One iter	$-.3670$ (0.55)	$-.3824$ (0.50)	$-.3578$ (0.48)
Two iters	$-.0341$ (0.58)	$-.0278$ (0.59)	$.1659$ (0.58)
Ten iters	$.0418$ (0.55)	$.7546$ (0.60)	$.0444$ (0.61)
ML	$.0078$ (0.51)	$.0123$ (0.51)	$-.0089$ (0.49)

Table 2.2. Means (standard deviations in parentheses) of $\sqrt{n}\,(w_1^n - w_1^o)$ for the various estimators for the weights in the mixture (2.6) based on 1000 replications, for sample sizes $n = 15, 25, 50$. The Opie-One estimator used five order statistics.

$n =$	15	25	50
Opie-One	$.0064$ (.929)	$.0688$ (.972)	$.0820$ (.947)
One iter	$-.2973$ (.520)	$-.2032$ (.480)	$-.1316$ (.471)
Two iters	$-.0104$ (.565)	$.0175$ (.544)	$.0016$ (.520)
Ten iters	$.0417$ (.556)	$.0414$ (.557)	$-.0048$ (.529)
ML	$.0078$ (.513)	$.0123$ (.511)	$-.0089$ (.492)

For the Atlantic City experiment, we took sample sizes $n = 15$, 25, and 50 and a mixture of $m = 2$ normals, with

$$
(2.6) \qquad
\begin{aligned}
(w_1^o,\, \mu_1^o,\, \sigma_1^o) &= (0.65,\, 5.0,\, 0.5) \quad \text{and} \\
(w_2^o,\, \mu_2^o,\, \sigma_2^o) &= (0.35,\, 7.0,\, 0.5) .
\end{aligned}
$$

The experiment consists of drawing n iid samples from the pdf (2.1) with the parameters as specified in (2.6), computing the Opie-One estimator, and then applying one, two and ten steps of the sequential quadratic programming procedure for the maximum likelihood problem. The maximum likelihood solution was also computed. The Opie-One estimator was "improved" somewhat by scaling the w_j to sum to 1 and by bounding them away from 0 and 1 (by a distance $1/n$). This was done because when the

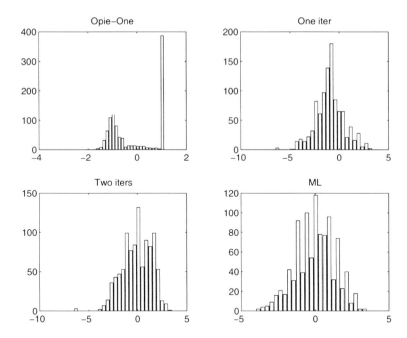

Figure 2.1. Histograms of $\sqrt{n}\left(w_1^n - w_1^o\right)$ for the various estimators, based on sample size $n = 15$ and 1000 replications, for the unknown mixture of two known normals given by (2.6). The Opie-One estimator used three order statistics.

Opie-One estimators are $(1, 0)$ or $(0, 1)$, severe roundoff problems occur in the quadratic programming algorithm to the effect that these initial guesses are never changed, even though they are not maximum likelihood solutions. The degenerate Opie-One estimators (1,0) correspond to the spike on the right in the Opie-One histograms in Figures 2.1 and 2.2. The maximum likelihood estimator *can* be computed with the EM algorithm (2.4.26), but it converges quite slowly. This is a common (and legitimate) complaint about EM algorithms. Instead, to speed things up, for each data set, the maximum likelihood estimator was computed by ten steps of the EM algorithm (starting from the uniform distribution $w_j = 1/m$ for all j), followed by ten steps of the sequential quadratic programming method. The procedure usually "converged" in fewer steps. In Table 2.1 and Table 2.2, this estimator is denoted by the ML entry.

The above was repeated 1000 times. In Tables 2.1 and 2.2, we report the sample mean and sample standard deviation of $\sqrt{n}(w_1^n - w_1^o)$ for the various estimators of w_1^o. In Figures 2.1 and 2.2, histograms of the 1000 replications are given. Some conclusions may be drawn. Obviously, the Opie-One estimator, whether based on three or five order statistics, is not very good when compared with the other estimators. Two steps of the

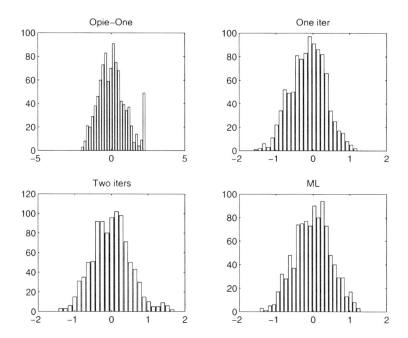

Figure 2.2. Histograms of $\sqrt{n}\left(w_1^n - w_1^o\right)$ for the various estimators, based on sample size $n = 50$ and 1000 replications, for the unknown mixture of two known normals given by (2.6). The Opie-One estimator used five order statistics.

sequential quadratic programming method is better than one step. Ten steps is generally a little bit worse since, in some cases, the iterates seem to converge to the boundary point $(0,1)$, a defect that the ML estimator does not seem to suffer from. The ML solution (rather, the estimator obtained by 10 steps of the EM algorithm, followed by at most 10 steps of the sequential quadratic programming method) is the best by far.

Exercise: (2.5).

3. Mixtures of normals

A much more challenging problem is to estimate the parameters in an unknown mixture of unknown normals, that is, when $X_1, X_2, \cdots, X_n$ are iid, with distribution,

$$(3.1) \qquad f(x\,;\,\theta) = \sum_{j=1}^{m} w_j\,\phi(x\,;\,\theta_j)\,, \quad x \in \mathbb{R}\,,$$

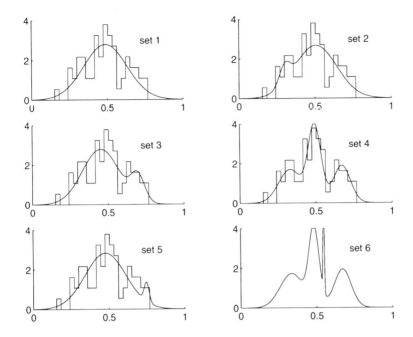

Figure 3.1. Estimated pdfs for the Buffalo snow fall data, as mixtures of unknown normals, as computed by the EM algorithm. The initial and final values of p, μ, and σ are given in Table 3.1. For Set 6, the estimated pdf is truncated to values $\leqslant 4$.

with $\theta = (w^{T}, \theta_1^{T}, \cdots, \theta_m^{T})^{T}$ and $\theta_j = (\mu_j, \sigma_j)^{T}$. Typically, m is assumed known. The estimation problem is then

$$\text{minimize} \quad L_n(\theta) = -\tfrac{1}{n} \sum_{i=1}^{n} \log f(X_i\,;\,\theta)$$

(3.2)

$$\text{subject to} \quad w \geqslant 0,\ \sum_j w_j = 1,\ \sigma_j > 0 \text{ for all } j\ .$$

In §2.4, we derived an EM algorithm for the mle of θ. This algorithm will be our main computational tool for analyzing some simulated and "real" data sets.

The first observation that needs to be made is that maximum likelihood estimators do not exist: e.g., for $m = 2$, taking μ_1 to be arbitrary, $\sigma_1 > 0$ fixed, and $\mu_2 = X_1$, $\sigma_2 \to 0^+$ gives $L_n(\theta) \to -\infty$. So global minima of $L_n(\theta)$ do not exist. However, local minima exist, and plenty of them, usually. To avoid having some σ_j tend to 0, the EM algorithm was modified by setting $w_j = 0$ when $\sigma_j < 10^{-6}$ (relative to the observations being scaled to the interval $[0, 1]$).

Table 3.1. Initial values and corresponding limit points of the EM algorithm for the Buffalo snow fall data and the negative log-likelihood. In set 5, the final mixture has only two components.

set	initial			final			neg-log-likelih'd
	w	μ	σ	w	μ	σ	
1	1.00	0.50	1.0	1.00	0.48	0.14	-0.52457
2	0.60	0.30	0.05	0.087	0.30	0.037	-0.53669
	0.40	0.50	0.05	0.913	0.51	0.137	
3	0.40	0.50	0.05	0.85	0.45	0.122	-0.55552
	0.60	0.60	0.05	0.15	0.70	0.046	
4	0.33	0.25	0.01	0.324	0.331	0.0763	-0.59599
	0.34	0.50	0.01	0.406	0.493	0.0421	
	0.33	0.75	0.01	0.271	0.671	0.0571	
5	0.33	0.10	0.01	0			-0.54015
	0.34	0.50	0.50	0.961	0.478	0.1357	
	0.33	0.75	0.01	0.039	0.749	0.0161	
6	0.30	0.30	0.01	0.324	0.484	0.0333	-0.63717
	0.20	0.45	0.01	0.340	0.336	0.0784	
	0.20	0.55	0.01	0.061	0.547	0.0038	
	0.30	0.75	0.01	0.275	0.670	0.0564	

(3.3) EXERCISE. Perform the actual experiments for the above, simulating data from a mixture of two normals, with $w = (0.9, 0.1)$, $\theta_1 = (5, 0.5)$, and $\theta_2 = (7, 0.5)$. Do this also for a mixture of three normals. Pay special attention to the (estimated ?) number of components in the mixture.

(3.4) EXERCISE. Recall Huber's model for errors, see Exercise (2.4.40),

$$f(x\,;\theta) = (1 - \varepsilon)\,\phi(x;\mu,\sigma) + \varepsilon\,\phi(x;\mu,3\,\sigma)\,,$$

with $\varepsilon \approx 0.05$ and $\phi(x;\mu,\sigma)$ the normal (μ,σ^2) distribution, and the corresponding EM algorithm, see (2.4.40). Simulate data, and run the EM

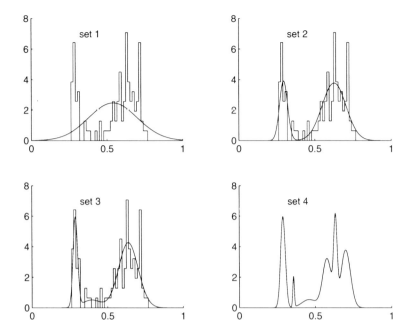

Figure 3.2. Estimated pdfs for the Old Faithful geyser data, as mixtures of unknown normals, as computed by the EM algorithm. The initial and final values of p, μ, and σ are given in Table 3.2.

algorithm to determine the mle's. Is there a unique mle? How sensitive is the behavior of the algorithm with respect to the initial point?

Compared with the above two exercises, the situation is quite a bit more complicated for the Buffalo snow fall data (see Appendix 1). In Figure 3.1, a frequency plot of the data is shown, scaled to the interval $[0, 1]$, by dividing all observations by 1.3 times the largest observation. It is not a priori clear that this data set could be modeled as an iid sample from a mixture of normals, but if we do so, then a mixture of three normals seems the most plausible, with a single normal as the second choice. A mixture of two normals does not seem realistic, but as a matter of fact, each of the above possibilities corresponds to at least one local minimum of the negative log-likelihood. In Table 3.1, we report on initial points, limit points (of the EM algorithm) and the value of $L_n(\theta)$. In Figure 3.1, the corresponding pdfs are shown, together with a frequency plot of the data. To the authors, the mixture of three normals in Set 4 of Table 3.1 seems the most reasonable fit, not because it has the largest likelihood (it hasn't), but because of the "naturalness" of the fit, see Figure 3.1.

What is the moral of all this? The lesson seems to be that one gets precisely what one wants; if a single normal is suggested, that is what you

Table 3.2. Initial values and corresponding limit points of the EM algorithm for the Old Faithful geyser data and the negative log-likelihood. In case 4 the initial mixture contained nine normals, the final solution contained only six normals.

set	initial w μ σ			final w μ σ			neg-log-likelih'd
1	1.00	0.50	1.0	1.00	0.54	0.166	-0.40396
2	0.60	0.30	0.05	0.262	0.30	0.027	-0.84702
	0.40	0.80	0.05	0.738	0.63	0.078	
3	0.33	0.25	0.01	0.224	0.286	0.0153	-0.92596
	0.34	0.50	0.01	0.077	0.393	0.0603	
	0.33	0.75	0.01	0.699	0.637	0.0654	
4	$w_j = \frac{1}{9}$ $\mu_j = (j+1)/11,$ $j = 1:9$ $\sigma_j = 0.005$			0.234	0.286	0.0158	-0.99487
				0.027	0.356	0.0053	
				0.064	0.459	0.0477	
				0.235	0.576	0.0293	
				0.146	0.632	0.0110	
				0.295	0.699	0.0312	

compute; if you want a mixture of two normals, that too can be done, and there is a choice of two reasonable ones, and one not so reasonable, or so it seems, and so on. More frustrating is the fact that the value of the negative log-likelihood is not really a useful indicator, because the more components in the mixture, the smaller $L_n(\theta)$ will be (modulo the fact that we can make it $-\infty$ whenever $m \geqslant 2$). So, how would one choose the number of components in a mixture? We do not address this problem, but merely note that it is somewhat similar to the selection of the smoothing parameter in nonparametric density estimation, see Chapter 7.

We next consider the Old Faithful geyser data, see Appendix 1. This data set is a record of the time between eruptions of Old Faithful, and it may be analyzed similarly to the Buffalo snowfall data. Again, the question arises whether this data set is accurately modeled by a mixture of normals, and if so, what the appropriate number of components is. In Table 3.2, we give initial values together with the limit points of the EM algorithm, and the negative log-likelihood. In Figure 3.2, we show the corresponding estimated pdfs and the frequency plot of the data. Again, the data were scaled to the interval $[0, 1]$ by dividing every observation by 1.3 times the largest observation.

(3.5) EXERCISE. Analyze the rubber abbrasion data set using mixtures of normals. (See Appendix 1.)

EXERCISES : (3.3), (3.4), (3.5).

4. Computing with the log-normal distribution

We now consider the computational aspects of maximum likelihood estimation for the three-parameter log-normal distribution. The practical aspects of estimation for the three-parameter Weibull and Gamma distributions are similar, but they are not discussed.

The difficulty with maximum likelihood estimation for the three parameter log-normal distribution is the determination of γ, the location parameter. As elucidated in § 2.7, the problem reduces to the minimization of

$$(4.1) \qquad L_n^*(\gamma) = \min_{\sigma, \mu} L_n(\gamma, \sigma, \mu) .$$

The minimizing σ and μ on the right of (4.1) are given in closed form in terms of γ. Because $L_n^*(\gamma)$ is a function of just one scalar parameter, finding the local minimum ought to be easy, but it turns out to be harder than expected. The reason is that for most data sets, the minimum occurs near $\gamma = X_{1,n}$, the smallest order statistic, and that $L^*(\gamma)$ changes quite rapidly in this region. Then, with Newton's method, say, or with tabulation of $L_n^*(\gamma)$ on a grid of points γ, one can easily miss this small region of interest. This tends to be worse for small sample sizes. A nice illustration is given in Figure 4.1. A random sample of size $n = 26$ was generated from a log-normal distribution with parameters

$$(4.2) \qquad \gamma = 0 , \quad \sigma = 2.397 , \quad \mu = 1.006 .$$

In this particular sample, $X_{1,n} = 0.058835$, whereas the local minimum of $L_n^*(\gamma)$ occurs at $\gamma_n = 0.058415$. Figure 4.1 shows two graphs of $L_n^*(\gamma)$ vs γ. Well, this does not quite mean what it says. In the graph on the left, the values of $L_n^*(\gamma)$ were computed for

$$(4.3) \qquad \gamma = \gamma_n \pm j \times 10^{-5} , \quad j = 0, 1, \cdots, 100 ,$$

as well as some smaller values, after which, the graph was plotted by linear interpolation. In the right half of Figure 4.1, the values of $L_n^*(\gamma)$ were computed for

$$(4.4) \qquad \gamma = X_{1,n} - j * 10^{-3} , \quad j = 0, 1, \cdots, 2000 ,$$

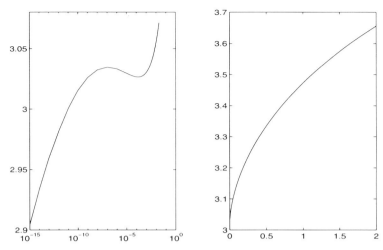

Figure 4.1. Two graphs of $L_n^*(\gamma)$ vs $X_{1,n} - \gamma$ for a random sample from the log-normal distribution with parameters (4.2). In the graph on the left, the grid of γ-values (4.3) was used. The graph on the right corresponds to the grid given by (4.4).

and the graph of $L_n^*(\gamma)$ was plotted using linear interpolation.

Of course, when analyzing one data set, this is not a problem. Presumably, one can afford to be careful. However, if *many* data sets like these must be analyzed, then it becomes a tedious task of finding out what is what. This becomes also apparent when conducting simulation studies, and it explains the extensive literature on the subject. The standard way to proceed is to determine an Opie-One estimator, and use one or two iterations of Newton's method to get best asymptotically normal estimators. A nice Opie-One estimator is determined in the following exercise.

(4.5) EXERCISE. (a) For the log-normal distribution, an Opie-One estimator for γ can be obtained as follows. If we knew the correct value of γ, then $\log(X - \gamma)$ would be normal. So its mean and median would be equal, and the sample mean and sample median of $\log(X_i - \gamma)$, $i = 1, 2, \cdots, n$, should be close as well. Thus, it seems reasonable to estimate γ by that value of γ for which the sample mean and sample median of $\log(X - \gamma)$ are equal. Thus, $\gamma = \gamma_n$ solves the equation

$$\frac{1}{n} \sum_{i=1}^{n} \log(X_i - \gamma) = \log(X_{\text{med}} - \gamma) ,$$

where X_{med} is the sample median. This equation can be solved using Newton-Raphson, with a starting value just a teensy bit less than $X_{1,n}$, the first order statistic.

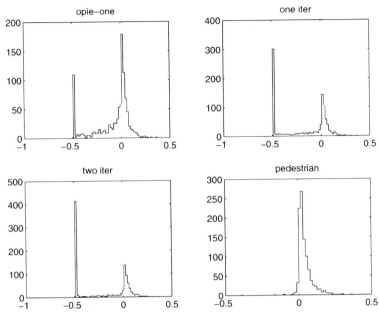

Figure 4.2. Histograms of the various estimators of γ, from 1000 replications of a sample of size $n = 26$ from a log-normal distribution with parameters given in (4.2).

(b) Show that this Opie-One estimator exists if and only if $X_{\mathrm{med}} < \overline{X}$.

(c) Show that $\sqrt{n}\left(\gamma_n - \gamma_o\right) = \mathcal{O}_P(1)$.

In Figure 4.2, some results of a simulation study are shown. One thousand random samples of size $n = 26$ were drawn from the log-normal distribution with parameters given by (4.2). For each sample, the Opie-One estimator discussed in Exercise (4.5) was determined, and Newton's method was applied with the Opie-One estimator as the starting point. In Figure 4.2, we show histograms of the estimators of γ, with the proviso that any estimator less than -0.5 was set equal to -0.5. Typically, when this happened, the actual value of the estimator was less than -1.0×10^6, say. In other words, $-\infty$ for all intents and purposes. The figure shows that the Opie-One estimator does a reasonable job, perhaps, but "improving" this estimator using Newton's method (or also the method of scores) is not a good idea. In fact, two steps of Newton's method gives worse results than does only one step. Running Newton's method to convergence (to $-\infty$, as a matter of fact) is even worse, but is not shown. As a solution, a pedestrian but extremely effective alternative is considered, see KÜBLER (1979). This solution still employs Newton's method, but with a careful choice of the starting point to the iteration as well as a careful monitoring of the iteration process. In particular, when there is no convergence, then the iterates either converge to $-\infty$ or to $X_{1,n}$. The first initial guess is $\gamma = X_{1,n} - 1$, say. If the Newton iteration has not converged in a certain number of steps, say, 20, then the iteration is restarted, with the initial guess whose distance to $X_{1,n}$ is 1/10-th of the distance of the previous starting guess. This procedure almost always works. In fact, in the simulation study, it never failed to converge for any of the 1000 replications.

(4.6) EXERCISE. Repeat the above for the Weibull and/or Gamma distribution.

EXERCISES: (4.5), (4.6)

5. On choosing parametric families of distributions

The log-normal, Gamma and Weibull distributions have been used extensively to model right-skewed, heavy-tailed distributions, e.g., in probabilistic models in fracture mechanics, see, e.g., SOBCZYK and SPENCER (1992), in oil resource estimation, see NAIR and WANG (1989) and SCHUENEMEYER and DREW (1991), in weather modification experiments, see SIMPSON, OLSEN and EDEN (1975), and for flood levels, see DUMONCEAUX and ANTLE (1973). In this section, we touch on how one might choose between these three families, and the effect the wrong choice might have. Typically, the literature in question is rather taciturn about the reasons for preferring one of these parametric families over the others, or else it just repeats the "model" of the founding paper on the subject. Of course, if just one single data set is being analyzed, there is not very much that can be done. If over a period of time a sequence of data sets have been gathered and analyzed, then it should be viable to give statistical reasons for choosing one particular family.

The questions of concern are not the same in all areas of applications. In oil-field exploration and cloud seeding (to produce rain), one has to be satisfied with a few huge successes: Many trials will be only moderately successful, or even be outright failures. In these areas, one is interested in estimating the tails of the relevant distributions. This could take the form of estimating the 90th or 95th percentiles. Because we are talking about distributions with heavy tails, the observations will be spaced far apart in this region. In time-to-failure problems, on the contrary, one would perhaps like to estimate the 5th or 10th percentile, but one would expect lots of data in this region.

We explore these issues by analyzing two data sets, one dealing with oil-field exploration, the other with weather modification experiments (cloud seeding).

The Texo oil-field data. This data set consists of recorded sizes of oil fields found after 318 successful drillings in the exploration "play" of

Table 5.1. Maximum likelihood estimators of the three-parameter log-normal, Gamma, and Weibull distributions for the Texo oil-field data.

	γ	μ or α	σ
log-normal	0.0190	2.400	1.01
Gamma	0.0169	0.297	95.64
Weibull	0.0169	0.447	8.90

Table 5.2. Estimated and empirical quantiles for the Texo oil-field data, based on the various parametric models. The last row reports the negative log-likelihood corresponding to the maximum likelihood estimators of the parameters.

u	log-normal	Gamma	Weibull	$X_{[un],n}$
0.05	0.07	0.02	0.03	0.05
0.10	0.15	0.05	0.07	0.11
0.20	0.38	0.31	0.33	0.33
0.30	0.80	1.18	0.90	0.90
0.40	1.51	3.13	2.00	1.86
0.50	2.75	6.81	3.93	3.27
0.60	5.04	13.20	7.34	5.25
0.70	9.62	24.10	13.50	10.00
0.80	20.60	43.40	25.80	19.00
0.90	58.90	83.70	57.50	58.40
0.95	144.00	1310.00	105.00	146.00
$L_n(\theta)$	3.30	3.47	3.30	

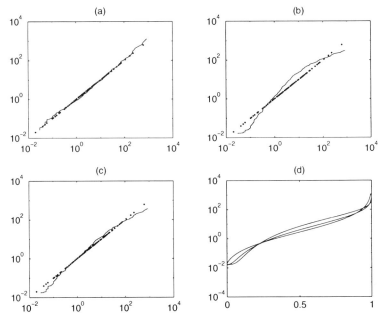

Figure 5.1. Plots of the estimated quantile functions $F^{\mathrm{inv}}(\cdot\,;\theta_n)$ (vertical axis) for the Texo data vs the empirical quantile function (horizontal axis). (a) log-normal, (b) Gamma, and (c) Weibull. In these graphs, every other fifth order statistic is plotted vs itself as well. In (d), the quantile functions $F^{\mathrm{inv}}(u\,;\theta_n)$ vs u are plotted. At $u = 0.5$, the ordering of the quantile function from smallest to largest is log-normal, Weibull, and Gamma.

the Frio strand in the central coastal plane of Texas. See Appendix 1. Note that in fact there were 695 drillings. This data set is an example of sampling proportional to size (if in some way we equate the size of oil fields in barrels of oil with "geological" size) from a finite population without replacement. As such, it provides interesting modeling problems. In particular, the data set is definitely not an iid sample from some (unknown) distribution. Nevertheless, we shall consider it as an iid sample from a heavy-tailed distribution and proceed to model the data set with the log-normal, Weibull, or Gamma distributions. An alternative model is briefly discussed below. It should be noted that in EUBANK, LARIC-CIA, and SCHUENEMEYER (1995), the hypothesis that the data set is from a log-normal could not be rejected by the goodness-of-fit techniques proposed there, and indeed, the Q-Q plot shown in Figure 5.3 bears this out. The log-normal distribution has traditionally been used in modeling oil discovery and resources modeling, but the theoretical foundation is tenuous. From a practical point of view, strong cases can be made for the Weibull and Gamma distributions as well.

Maximum likelihood estimation for the parametric families at hand was discussed in § 2.7 following Theorems (7.21) and (7.22). For the Weibull distribution, it seemed clear from the calculation of several Opie-One estimators for α that the true $\alpha < 1$, and so we are in the "impossible" regime of maximum likelihood estimation. Thus, the location parameter γ was estimated using (7.23) and the function $L_n^*(\theta)$ was minimized using the Newton-Raphson procedure applied to

$$\frac{\partial}{\partial \alpha} L_n^*(\gamma, \alpha) = 0.$$

The considerations for the Gamma distribution are similar.

In Table 5.1, the maximum likelihood estimators (well, suitably interpreted) of the three-parameter log-normal, Weibull, and Gamma distributions for the Texo data are given, for reference purposes. In Table 5.2, we report on the estimated quantiles corresponding to these three parametric models with the parameters from Table 5.1. In Figure 5.1, the corresponding graphs of the estimated quantile functions vs the order statistics are given. That is, for each pdf, we plot the points $\left(X_{i,n}, F^{\mathrm{inv}}(i/(n+1); \theta_n) \right)$ and complete the graph by linear interpolation. The data are represented as well, by plotting the points (X_i, X_i), $i = 1, 2, \cdots, n$. Let us briefly consider Table 5.1 and look at $F^{\mathrm{inv}}(0.5; \theta_n)$. The estimated value ranges from 2.75 to 6.81. Also note that for values close to 0 and close to 1, the quantile estimator based on the log-normal distribution is larger than for the other two families, whereas in the middle region, the opposite holds. Apparently, we must have good reason to believe the correctness of a particular model before we can believe the estimated quantiles. So which parametric model best fits the data? Based on the negative log-likelihood, the Weibull distribution would be chosen. Based on the Q-Q plots, there is no doubt that the log-normal would be it.

The view of the Texo oil-field data as an iid sample of a random variable (let alone of a log-normal or Weibull random variable) is of course not even remotely correct. A much more reasonable model would be as follows. The oil exploration "play" consists of N oil fields, of size $Y_1, Y_2, \cdots, Y_N$, with N unknown. One would like to estimate N, or perhaps more relevant, the number of "big" oil fields. What does the data consist of? The data set is actually growing with time. Here, time, denoted by t, is the number of completed oil-drilling operations. With each t is associated the random variable $X(t)$, the size of the oil field. The number of oil fields discovered so far is the number of positive $X(t)$. What is the distribution of the underlying finite population $Y_1, Y_2, \cdots, Y_N$? NAIR and WANG (1989) derive an EM algorithm for this problem. What one would really like to estimate are the probable number of oil fields, or perhaps more importantly, the number of "large" oil fields remaining.

The cloud seeding data. [SIMPSON, OLSEN and EDEN (1975)] The cloud seeding data set, see Appendix 1, consists of two random samples

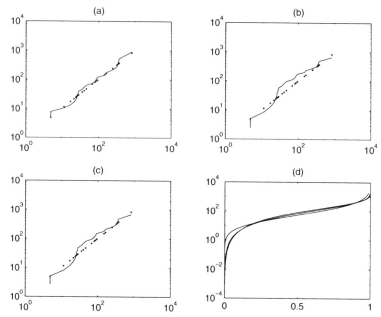

Figure 5.2. Plots of the estimated quantile functions $F^{\text{inv}}(\,\cdot\,;\theta_n)$ (vertical axis) for the control clouds set of the rain data vs the empirical quantile function (horizontal axis). (a) log-normal (b) Gamma (c) Weibull. In these graphs every other fifth order statistic is plotted vs itself as well. In (d) the quantile functions $F^{\text{inv}}(\,u\,;\theta_n)$ vs u are plotted. At $u = 0.5$, the ordering of the quantile function from smallest to largest is log-normal, Weibull and Gamma.

from a well-designed experiment to investigate the effects on rain fall of seeding (with silver-oxide) of a specific type of cloud. The variable of interest is the total rain volume from clouds as measured by calibrated radar. One data set consists of the rain volume from seeded clouds, whereas the other one corresponds to a control group. Each data set has a sample size of 26. The $ 64,000 question is whether there is a statistically significant difference between the populations represented by the two data sets. It should be observed that classical procedures, such as analysis of variance (ANOVA), even after an appropriate transformation of variables, are not applicable to this type of data. SIMPSON, OLSEN and EDEN (1975) propose to model each one of the two data sets by the two-parameter Gamma distribution. Why this distribution should be the Gamma is not really addressed. The design of the experiment assures that both data sets are random samples from some distribution.

We are again interested in estimating the tail of the distribution of the measured rain volume, and implement this goal by estimating the quantiles of the distribution. As in the previous example, we take as the parametric

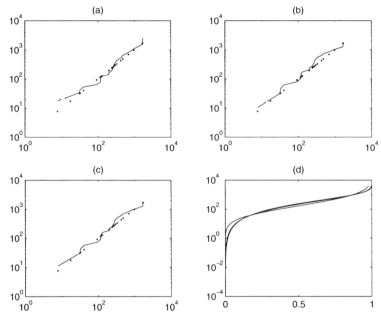

Figure 5.3. Plots of the estimated quantile functions $F^{\text{inv}}(\,\cdot\,;\theta_n)$ (vertical axis) for the seeded clouds set of the rain data vs the empirical quantile function (horizontal axis). (a) log-normal, (b) Gamma, and (c) Weibull. In these graphs every other fifth order statistic is plotted vs itself as well. In (d), the quantile functions $F^{\text{inv}}(\,u\,;\theta_n)$ vs u are plotted. At $u = 0.5$, the ordering of the quantile function from smallest to largest is log-normal, Weibull, and Gamma.

model the Gamma, the log-normal, or the Weibull family. We also assume that the location parameter $\gamma = 0$ for all three cases, so that the maximum likelihood has a unique minimizer (and a unique stationary point), see § 2.7.

The maximum likelihood estimators of the parameters for the two sets of rain data are given in Table 5.3. In Table 5.4, the estimated quantiles for the two data sets are given, assuming the parametric model in question. To see which parametric family best fit the data, in Figures 5.2 and 5.3, we give the Q-Q plot, the graphs of $F^{\text{inv}}(\,\cdot\,;\theta_n)$ vs F_n^{inv} for the three models. These figures indicate that the log-normal appears to give the best fit for the control data set, whereas either the Gamma or Weibull seems the most appropriate for the seeded-cloud data set. However, for such small data sets, it is hard to draw conclusions regarding the relevance of any of the parametric models considered. Again, note that the relative ordering of the quantile estimators observed for the Texo oil field data occurs here too.

In the above analysis of the oil data and the rain data, we observed a definite ordering of the quantile estimators based on the three parametric

Table 5.3. Maximum likelihood estimators of the two-parameter log-normal, Gamma, and Weibull distributions for the rain data.

		control	seeded
log-normal	μ	3.99	5.134
	σ	1.61	1.568
Gamma	α	0.561	0.640
	σ	294.0	691.0
Weibull	α	0.669	0.734
	σ	119.0	395.0

Table 5.4. Estimated quantiles $Q(u)$ for the rain data. Columns 2-5 correspond to the control data, columns 5-7 correspond to the seeded cloud data. ("Wei" is short for Weibull, and Γ stands for Gamma.)

u	log	Γ	Wei	log	Γ	Wei
0.05	3.87	1.15	1.42	12.7	5.31	6.31
0.10	6.88	3.98	4.14	22.8	16.3	16.8
0.20	14.0	14.0	12.7	45.4	49.4	46.6
0.30	23.2	29.7	25.6	74.6	97.0	88.2
0.40	36.0	52.0	43.7	114.	160.	144.
0.50	54.1	82.4	69.0	170.	243.	218.
0.60	91.3	124.	104.	252.	353.	318.
0.70	126.	182.	158.	386.	503.	462.
0.80	209.	271.	243.	635.	728.	686.
0.90	425.	434.	415.	1260	1130	1120
0.95	762.	606.	614.	2230	1550	1600
$L_n(\theta)$	5.87	5.96	5.66	7.00	7.01	6.99

Table 5.5. Estimated means (and standard deviations in parentheses) of the quantile estimators, when the data are drawn from the log-normal, with parameters as in Table 5.3.

u	true	log	Weibull	Gamma
0.05	12.9	13.5 (3.3)	4.23 (1.8)	3.51 (2.0)
0.10	22.7	23.6 (5.0)	12.3 (4.0)	11.9 (4.6)
0.20	45.3	46.5 (8.5)	38.0 (8.9)	42.5 (9.7)
0.30	74.5	76.0 (12.)	77.4 (15.)	92.9 (15.)
0.40	114.	116. (19.)	134. (22.)	168. (27.)
0.50	170.	172. (27.)	213. (34.)	271. (51.)
0.60	252.	255. (41.)	328. (55.)	416. (93.)
0.70	386.	386. (67.)	500. (92.)	625. (163)
0.80	635.	640. (122)	784. (164)	949. (285)
0.90	1266	1279 (286)	1368 (342)	1540 (533)
0.95	2237	2264 (583)	2065 (587)	2200 (813)

Table 5.6. Estimated means (and standard deviations in parentheses) of the quantile estimators, when the data are drawn from the Gamma distribution, with parameters as in Table 5.3.

u	true	Gamma	log	Weibull
0.05	5.45	6.41 (3.1)	10.2 (4.6)	8.06 (3.2)
0.10	16.2	18.0 (17.)	19.0 (7.4)	20.3 (6.7)
0.20	49.4	52.2 (14.)	40.3 (13.)	53.6 (14.)
0.30	97.0	100. (20.)	69.8 (18.)	98.6 (21.)
0.40	160.	164. (28.)	112. (24.)	157. (29.)
0.50	243.	247. (36.)	174. (31.)	235. (37.)
0.60	353.	357. (47.)	273. (42.)	338. (48.)
0.70	504.	507. (64.)	441. (63.)	484. (63.)
0.80	729.	731. (90.)	777. (124)	710. (88.)
0.90	1133	1134 (143)	1717 (390)	1140 (145)
0.95	1555	1553 (202)	3330 (1017)	1615 (221)

Table 5.7. Estimated means (and standard deviations in parentheses) of the quantile estimators, when the data are drawn from the Weibull, with parameters as in Table 5.3.

u	true	Weibull	log	Gamma
0.05	6.28	6.72 (2.8)	10.2 (4.3)	5.49 (2.7)
0.10	16.7	17.4 (5.6)	18.7 (6.8)	15.9 (5.8)
0.20	46.5	47.4 (12.)	39.1 (11.)	47.7 (12.)
0.30	88.1	88.9 (18.)	66.9 (16.)	93.4 (18.)
0.40	144.	144. (26.)	107. (22.)	155. (26.)
0.50	218.	217. (34.)	164. (28.)	235. (35.)
0.60	318.	317. (46.)	253. (39.)	343. (48.)
0.70	462.	459. (63.)	404. (58.)	491. (167)
0.80	687.	681. (92.)	702. (110)	712. (99)
0.90	1118	1109 (160)	1517 (320)	1113 (163)
0.95	1601	1589 (249)	2882 (782)	1531 (235)

Table 5.8. Estimated means (and standard deviations in parentheses) of the quantile estimators, when the data are drawn from the exceptional Weibull distribution, with $\gamma = 0$, $\sigma = 3.0$, and $\alpha = 2.1$.

u	true	Weibull	log	Gamma
0.05	.729	.740 (.10)	.849 (.12)	.818 (.12)
0.10	1.03	1.04 (.12)	1.06 (.13)	1.07 (.12)
0.20	1.50	1.48 (.13)	1.37 (.14)	1.44 (.13)
0.30	1.84	1.84 (.13)	1.66 (.14)	1.76 (.13)
0.40	2.18	2.18 (.14)	1.95 (.14)	2.08 (.13)
0.50	2.52	2.51 (.14)	2.28 (.14)	2.40 (.14)
0.60	2.89	2.87 (.15)	2.65 (.14)	2.75 (.14)
0.70	3.28	3.26 (.16)	3.12 (.16)	3.18 (.15)
0.80	3.76	3.74 (.17)	3.78 (.20)	3.71 (.18)
0.90	4.43	4.94 (.21)	4.55 (.34)	1540 (.23)
0.95	5.06	5.02 (.26)	6.16 (.52)	5.34 (.29)

families. Here, we try to investigate whether this is a property of the data sets or of the parametric families. We ran a small Atlantic City experiment with 1000 replications, in which $n = 100$ iid observations were drawn from one of the three distributions in question, with the location parameter $\gamma = 0$, and the other parameters from those given in Table 5.3, that is, with the mle's of the parameters for the seeded-cloud rain data. An "exceptional" Weibull distribution with $\gamma = 0$, $\sigma = 3.0$, and $\alpha = 2.1$ was considered also. For each data set, all three two-parameter families ($\gamma = 0$) were fitted using maximum likelihood estimation and a few quantiles were estimated accordingly.

The observed expected values and standard deviations of the estimators are summarized in Tables 5.5-5.8. As expected, the quantiles are estimated reasonably well when the correct model is chosen and miserably when the wrong model is postulated. Rather surprisingly, the relative ordering observed for the Texo data and cloud seeding data is observed for the simulated data as well, even for the "exceptional" Weibull model. So the quantiles based on the log-normal fit are relatively quite a bit larger for the lower and upper tails, but quite a bit smaller in the middle region.

6. Toward nonparametrics : mixtures revisited

How can one avoid model misspecification? One method is the nonparametric approach to estimation, which as the name implies, attempts estimation under minimal assumptions on the population model. The point is that one avoids making the *wrong* assumptions about the model, which would lead to the wrong conclusions in the estimation process. The alternative to nonparametrics is to consider parametric families with more model parameters, with the hope that now most distributions found in practice can be adequately approximated. Even so, the number of scalar parameters is seldom more than five. It is pretty much received knowledge that increasing the number of parameters in this fashion is not the way to proceed, and we agree: Otherwise, the text would have ended here! The objections are mainly concerned with the *ad hoc* fashion in which many-parameter families are constructed. In particular, a scientific foundation for the model is often lacking. For an interesting historical account of some of these attempts, see THOMPSON and TAPIA (1990) and SILVERMAN (1986).

Interesting parametric families with many parameters can be constructed as mixture. For brevity, we restrict attention to the univariate case. We consider two such models. The mixtures of normals model was already introduced before. It is given by

$$(6.1) \qquad f(x; \theta) = \sum_{j=1}^{m} w_j \, \phi(x; \theta_j) \, , \quad x \in \mathbb{R} \, ,$$

where $\theta = (w^\top, \theta_1, \cdots, \theta_m)^\top \in \Theta_m$, in which $\theta_j = (\mu_j, \sigma_j)$, and $\phi(x; \theta_j)$ is the normal (μ_j, σ_j^2) density. Moreover, the w_j are nonnegative and add up to one. The parameter space Θ_m is implicitly described by this. Now, for fixed $m \geqslant 5$, say, it is a rare experimental situation in which there is any reason to believe that this would be an accurate model for the pdf, although such settings do exist, of course, see, e.g., MCLACHLAN and BASFORD (1988). However, it is easy to show that for any pdf f_o,

$$(6.2) \quad \inf \{ \| f_o - \sum_{j=1}^{m} w_j \, \phi(\,\cdot\,; \theta_j) \|_1 \, : \, \theta \in \Theta_m \} \longrightarrow 0 \quad \text{for } m \to \infty \, .$$

Thus, a proper selection of m and θ is all that matters. Indeed, taking m very large would seem the best policy: One can always set some of the w_j equal to zero. The problem is how to estimate θ. We have seen in §3 that maximum likelihood estimation is full of hazards, because we can make $L_n(\theta) \longrightarrow -\infty$. Moreover, lots of local minima exist. One can even argue that for $m \geqslant n$, the mle would be a sum of delta peaks, one at each observation. Then, we are definitely "overfitting" the data. So the large mixture model is not appropriate, at least if the σ_j can be arbitrarily small. On the other hand, for the Buffalo snow fall data and for the Old Faithful geyser data, a mixture of a few normals seems to provide a good fit. Indeed, if the true pdf was of this form, one could reasonably expect to observe data similar to the one obtained. See ROEDER (1992) for an elaboration of these points and much more.

REMARK. We will see in the next chapter that if the number of components is large, then selecting a proper lower bound for the σ_j goes a long way toward providing good estimators for the density. In particular, with ϕ as above, estimators of the form

$$(6.3) \qquad f_n(x) = \frac{1}{n} \sum_{i=1}^{n} \phi(x; X_i, \sigma) \, , \quad x \in \mathbb{R} \, ,$$

are constructed. So the number of components in the mixture equals the sample size, and each observation X_i acts as a location parameter. Finally, the standard deviation σ is the same for each component. For the appropriate choice of σ, this provides for good estimators, but it is precisely choosing σ appropriately that provides the difficult problem.

The second mixture model we consider is that of mixtures of disjoint, uniform pdfs. What we mean by disjointness is that the supports of the uniform distributions do not overlap. This parametric family may be defined as follows. Let $-\infty < \theta_1 < \theta_2 < \cdots < \theta_{m+1} < \infty$, and let

$$(6.4) \qquad f(x; \theta) = \sum_{j=1}^{m} w_j \, \mathrm{U}(x; \theta_j, \theta_{j+1}) \, , \quad x \in \mathbb{R} \, ,$$

where U is a uniform density

(6.5) $$U(x; a, b) = \begin{cases} (b-a)^{-1}, & a < x < b , \\ 0 & , & \text{otherwise} . \end{cases}$$

For this parametric family too, one can make $L_n(\theta) \longrightarrow -\infty$, and the analogue of (6.2) holds. For fixed θ, things are a lot better. Now, only the w_j need to be estimated. The mle w^n for w, based on the iid sample $X_1, X_2, \cdots, X_n$, is given by

(6.6) $$w_j^n = \int_{\theta_j}^{\theta_{j+1}} dF_n(x) , \quad j = 1, 2, \cdots, m .$$

(6.7) EXERCISE. Verify (6.6).

In this case, the density estimator is the histogram estimator (the frequency diagram of the data), the oldest and most used method for discovering the nature of the pdf underlying the data. As different as it seems from the mixture of normals model, putting a lower bound on the widths (narrows?) $\theta_{j+1} - \theta_j$ is similar to putting a lower bound on the σ_j, and when the number of components is large, say, $m \approx n$, the difference between the two models is not that impressive.

These two examples, together with the observation (6.2), suggest that a mixtures-of-normals or a mixture-of-uniforms model would be a good choice for a parametric family when few assumptions about the shape of the pdf can be made. In the next chapters, this claim is substantiated, in the sense that mixtures of pdfs from any reasonable location/scale parametric family will work.

This ends (our treatment of) parametrics.

EXERCISE : (6.7).

Part II:

Nonparametric Density Estimation

4

Kernel Density Estimation

1. Introduction

We continue the study of density estimation, but now from the nonparametric point of view. The problem is the same as in Chapters 2 and 3 :
Given an iid sample $X_1, X_2, \cdots, X_n \in \mathbb{R}^d$ with common pdf f_o, estimate
f_o. The difference with the previous chapters is that here, we make as
few assumptions as possible. In particular, a parametric model is not postulated, whence the designation *nonparametric* estimation. At times, we
refer to the present problem as the direct estimation problem, as opposed
to the indirect estimation of, say, nonparametric deconvolution, see § 1.2.
Also, we restrict attention to the univariate case. Regarding the results
we have in mind, the multivariate case is harder mostly from the practical
point of view.

We begin with a general introduction to univariate nonparametric density estimation and are again interested in the maximum likelihood approach. The nonparametric maximum likelihood density estimation problem is

$$\text{minimize} \quad L_n(f) \stackrel{\text{def}}{=} -\tfrac{1}{n} \sum_{i=1}^{n} \log f(X_i)$$

(1.1)

$$\text{subject to} \quad f \text{ is a pdf} ,$$

with the implicit assumption that f be continuous, or at least continuous
at each X_i. The continuity requirement is necessary to make sense out of
the point evaluations $f(X_i)$. But from the start, we are in the trouble,
because we can make $L_n(f) \longrightarrow -\infty$ by letting some (or all) $f(X_i) \longrightarrow \infty$,
e.g., as follows. Let $A(x) = (1 - |x|)_+$ for $x \in \mathbb{R}$ be the tent pdf (we set
$x_+ = x \vee 0$). For positive weights w_i, that add up to 1, and $h > 0$, let

$$f(x) = \sum_{i=1}^{n} w_i A_h(x - X_i) , \quad -\infty < x < \infty ,$$

(1.2)

in which $A_h(x) = h^{-1}A(h^{-1}x)$. If the X_i's are distinct, then we have for all h small enough

$$(1.3) \qquad L_n(f) = -\frac{1}{n} \sum_{i=1}^{n} \log\{w_i h^{-1}\} ,$$

so that $L_n(f) \longrightarrow -\infty$ for $h \to 0$. Note that the estimator given by (1.2) becomes extremely spiky as $h \to 0$. Thus, nonparametric maximum likelihood estimation seems pretty hopeless.

The above may be looked at in a slightly different (equally hopeless) way. First, the problem (1.1) may be equivalently phrased as

$$(1.4) \qquad \text{minimize} \quad \widetilde{L}_n(f) \overset{\text{def}}{=} -\frac{1}{n} \sum_{i=1}^{n} \log f(X_i) + \int_{\mathbb{R}} f(y)\, dy$$

$$\text{subject to} \quad f \text{ is continuous}, \ f \geqslant 0 .$$

The equivalence is of course somewhat nonsensical, because neither problem has a solution. The following exercise makes some sense out of it, though.

(1.5) EXERCISE. Let $f \in L^1(\mathbb{R})$ be nonnegative and continuous, and suppose that $f(X_i) > 0$, $i = 1, 2, \cdots, n$. Show that
$$\widetilde{L}_n(f/t) \leqslant \widetilde{L}_n(f) ,$$
where $t = \int_{\mathbb{R}} f(y)\, dy$, with equality if and only if f is a pdf.

Now, consider the following. Let f be a pdf, with $L_n(f)$ finite. We are going to make $L_n(f)$ smaller by performing surgery on f. For arbitrary $\delta > 0$, define

$$S_\delta = \bigcup_{i=1}^{n} \{ x \in \mathbb{R} : |x - X_i| \leqslant \delta \} .$$

Thus, S_δ is the set of points within distance δ from an observation X_i. Now, set f equal to 0 outside S_δ, i.e., consider

$$(1.6) \qquad \varphi_\delta(x) = f(x)\, \mathbb{1}(x \in S_\delta) , \quad x \in \mathbb{R} .$$

Then, φ_δ agrees with f at each X_i, and

$$(1.7) \qquad \int_{\mathbb{R}} \varphi_\eta(y)\, dy \leqslant \int_{\mathbb{R}} \varphi_\delta(y)\, dy \leqslant \int_{\mathbb{R}} f(y)\, dy , \quad \text{for all } 0 < \eta < \delta .$$

It follows that $\widetilde{L}(\varphi_\delta)$ is an increasing function of δ. The same is true for the scaled case, i.e., $L_n(\psi_\delta)$ is increasing in δ, where $\psi_\delta = \varphi_\delta / \int_{\mathbb{R}} \varphi_\delta$. (Actually, the φ_δ need not be continuous, but never mind.)

All of this purports to show that "solutions" of (1.1) or (1.4) must be zero everywhere. The conclusion is that the corresponding distribution functions Ψ_δ must consist of point masses located at the X_i. That being settled, the problem now is to estimate these point masses. So let

$$(1.8) \qquad \Psi(x) = \sum_{i=1}^{n} w_i\, \mathbb{1}(x \geqslant X_i) , \quad x \in \mathbb{R} ,$$

where the w_i are nonnegative and add up to 1. The likelihood of Ψ given $X_1, X_2, \cdots, X_n$ is the probability of observing $X_1, X_2, \cdots, X_n$ under the model Ψ, which equals $\prod_{i=1}^{n} w_i$. Hence, the maximum likelihood problem for estimating the distribution function is

$$
\begin{aligned}
\text{minimize} \quad & - \tfrac{1}{n} \sum_{i=1}^{n} \log w_i \\
\text{subject to} \quad & w \geqslant 0 \text{ component wise, } \sum_{i=1}^{n} w_i = 1 .
\end{aligned}
$$
(1.9)

One verifies that the unique solution is given by

$$
(1.10) \qquad\qquad w_i = \tfrac{1}{n} , \quad i = 1, 2, \cdots, n .
$$

Thus, the maximum likelihood estimator of the distribution function is precisely the empirical distribution function of the observations

$$
(1.11) \qquad\qquad F_n(x) = \tfrac{1}{n} \sum_{i=1}^{n} \mathbb{1}(x \geqslant X_i) ,
$$

with $\mathbb{1}(x \geqslant y) = 1$ for $x \geqslant y$, and $= 0$ otherwise. This is nice, of course, but does not help in estimating the pdf.

(1.12) EXERCISE. Verify that (1.10) indeed is the unique solution of (1.9).

So what is to be done ? Somehow, we must remove from contention the spiky estimators like those given by (1.2).

The simplest way of not admitting spiky solutions to the problem (1.1) is by restricting the minimization in (1.1) to sets of smooth pdfs. This is the basic idea of sieves, briefly discussed in § 1.6. We leave it at that.

Another way to eliminate the spiky estimators is by modifying the likelihood. This is similar to the treatment of robust estimators in § 2.5. Thus, we add a "roughness penalization" term to the likelihood function in (1.1) and replace it by

$$
\begin{aligned}
\text{minimize} \quad & - \tfrac{1}{n} \sum_{i=1}^{n} \log f(X_i) + h^2 \, R(f) \\
\text{subject to} \quad & f \text{ is a pdf} ,
\end{aligned}
$$
(1.13)

where $h > 0$ must be appropriately chosen and, e.g.,

$$
(1.14) \qquad\qquad R(f) = \int_{\mathbb{R}} \frac{|f'(x)|^2}{f(x)} \, dx .
$$

The logic behind this is that rough densities will have a large value of $R(f)$, which presumably eliminates them from being solutions to (1.13). The functional $R(f)$ of (1.14) is the roughness penalty functional proposed by GOOD (1971), see GOOD and GASKINS (1971). The solution of the problem (1.13)–(1.14) exists and is unique, and it is given implicitly by

$f = u^2$, where u satisfies

(1.15)
$$u(x) = \frac{1}{n} \sum_{i=1}^{n} \frac{\mathcal{B}_h(x - X_i)}{u(X_i)} , \qquad x \in \mathbb{R} ,$$

in which

(1.16)
$$\mathcal{B}_h(x) = (2h)^{-1} \exp(-h^{-1} |x|) , \qquad x \in \mathbb{R} ,$$

is the scaled two-sided exponential kernel. The representation (1.15) serves as the basis for an efficient way to compute this estimator. A detailed study of the GOOD estimator is given in the next chapter. It turns out to be a remarkably GOOD estimator.

Another way to modify the likelihood is by smoothing it out using kernel smoothers. Thus, $\log f$ is replaced by $A_h * \log f$, defined as

(1.17)
$$[\, A_h * \log f \,](x) = \int_{\mathbb{R}} A_h(x - y) \log f(y) \, dy , \qquad x \in \mathbb{R} .$$

It is perhaps not obvious how this should be interpreted, but proceeding without fear, we obtain the maximum *smoothed* likelihood problem

(1.18)
$$\text{minimize} \quad L_{nh}(f) \stackrel{\text{def}}{=} -\frac{1}{n} \sum_{i=1}^{n} [\, A_h * \log f \,](X_i)$$

$$\text{subject to} \quad f \text{ is a pdf} .$$

To find out what the solution of (1.18) is, if it exists at all, note that for symmetric A,

(1.19)
$$\frac{1}{n} \sum_{i=1}^{n} [\, A_h * \log f \,](X_i) = \int_{\mathbb{R}} [\, A_h * dF_n \,](x) \log f(x) \, dx .$$

Here, similar to (1.17), for any distribution function F,

(1.20)
$$A_h * dF(x) = \int_{\mathbb{R}} A_h(x - y) \, dF(y) ,$$

in which the integral is in the sense of Lebesgue-Stieltjes. (If F is absolutely continuous, so that $dF(y) = f(y) \, dy$ with f integrable, then the standard notation is to write $A_h * dF = A_h * f$.) To return to (1.18), it follows that for all pdfs f,

(1.21)
$$L_{nh}(f) = \mathrm{KL}(A_h * dF_n , f) + L_{nh}(A_h * dF_n) ,$$

and so the solution of (1.18) is the minimizer of $\mathrm{KL}(A_h * dF_n , f)$. But it is clear that the minimizer of *this* is $f^{nh} = A_h * dF_n$, written alternatively as

(1.22)
$$f^{nh}(x) = \frac{1}{n} \sum_{i=1}^{n} A_h(x - X_i) , \qquad x \in \mathbb{R} .$$

The estimator (1.22) is called a kernel density estimator, with A being the kernel. It is in fact the density estimator in general use. This goes back to

AKAIKE (1954), ROSENBLATT (1956), WHITTLE (1958), PARZEN (1962), and WATSON and LEADBETTER (1963). The standard kernels encountered in theory and practice are the normal and uniform densities, as well as the Epanechnikov kernel

$$(1.23) \qquad A(x) = \tfrac{3}{4}\left(1 - x^2\right)_+ , \quad x \in \mathbb{R} ,$$

which is "optimal" in a well-defined sense, see § 7. We must also mention the two-sided exponential kernel, which is anything but optimal (it is really bad), but still features prominently, see, e.g., (1.15).

Although the kernel estimator is standard, the motivation given above is certainly not the standard one. In § 1.1, we gave a least-squares interpretation for the kernel estimator with the two-sided exponential kernel, which is nonstandard also. Here, we give a least-squares interpretation for more general kernels. If one is willing to live with the squared L^2 norm as a good way to measure the distance between two pdfs, then the best one can do is determine an estimator φ so as to

$$(1.24) \qquad \begin{aligned} &\text{minimize} \quad \|\varphi - f\|_2^2 \\ &\text{subject to} \quad \varphi \text{ is a pdf} . \end{aligned}$$

Using an idea of STONE (1984), we expand the objective function as

$$(1.25) \qquad \|\varphi - f\|_2^2 = \|\varphi\|_2^2 + \|f\|_2^2 - 2 \int_{\mathbb{R}} \varphi(y)\, dF(y)$$

and observe that the last integral may be estimated by $\int_{\mathbb{R}} \varphi(y)\, dF_n(y)$. Also, observe that the term $\|f\|_2^2$ on the right of (1.25) is unknown, but that it does not influence the minimization problem. It can be shown similar to our previous exploits that the "solution" of the resulting least-squares problem is the empirical distribution function F_n. True to the motto of this text (when in trouble penalize!), we add a roughness penalization functional to the objective function and obtain the problem

$$(1.26) \qquad \begin{aligned} &\text{minimize} \quad \|\varphi\|_2^2 - 2 \int_{\mathbb{R}} \varphi(y)\, dF_n(y) + h^2\, R(\varphi) \\ &\text{subject to} \quad \varphi \text{ is a pdf} . \end{aligned}$$

In view of the L^2 binge we are on, let us consider

$$(1.27) \qquad R(\varphi) = \|\varphi'\|_2^2 .$$

Solving problems like (1.26)–(1.27) is a standard exercise in the calculus of variations. The solution of (1.26) exists and is unique, and it is given by

$$(1.28) \qquad f^{nh}(x) = \tfrac{1}{n} \sum_{i=1}^{n} \mathfrak{B}_h(x - X_i) , \quad x \in \mathbb{R} .$$

Thus, we get a kernel estimator as the solution to this least penalized squares method.

(1.29) EXERCISE. It is tempting to consider the penalization
$$R(\varphi) = \| \varphi'' \|_2^2 \ ,$$
but explicitly solving (1.26) with this penalization is complicated. Determine the solution of (1.26) with the pdf constraint omitted. Is the solution always positive?

In a formal way, one can get more general kernel estimators by least penalized squares problems, as follows. Let A be a pdf such that for all $\varphi \in L^2(\mathbb{R})$

(1.30) $A_h * \varphi(x) = 0$ for all x $\implies$ φ vanishes everywhere .

Consider the operator $T_h : L^2(\mathbb{R}) \longrightarrow L^2(\mathbb{R})$ defined by

(1.31) $$T_h \varphi = A_h * \varphi \ .$$

By the assumption on A, the operator T_h is one-to-one. Now, define the roughness functional R_h by

(1.32) $$R_h(\varphi) = \begin{cases} h^{-2} \{ \| T_h^{-1} \varphi \|_2^2 - \| \varphi \|_2^2 \} \ , & \text{if } \varphi \in \text{range}(T_h) \ , \\ + \infty & , \quad \text{otherwise} \ . \end{cases}$$

With $\varphi = A_h * \psi$, the problem (1.26)–(1.32) may then be written as

(1.33) $$\begin{array}{c} \text{minimize} \quad - 2 \int_{\mathbb{R}} A_h * \psi(y) \, dF_n(y) + \| \psi \|_2^2 \\[2mm] \text{subject to} \quad \psi \text{ is a pdf} \ . \end{array}$$

Now, the solution ψ is given by $\psi = B_h * dF_n$, where $B_h(x) = A_h(-x)$. (If A is symmetric, then $B_h = A_h$.) Thus, the solution φ of (1.26)–(1.32) is

(1.34) $$\varphi = A_h * B_h * dF_n \ .$$

The conclusion is that we can get kernels of a particular form. This is made more precise in the following exercise.

(1.35) EXERCISE. Let A be symmetric. We say that A is a positive-definite kernel if
$$\int_{\mathbb{R}} \varphi(x) \, [\, A_h * \varphi \,](x) \, dx > 0 \quad \text{for all } \varphi \in L^2(\mathbb{R}) \ , \ \varphi \text{ not identically } 0 \ .$$

(a) Let A satisfy (1.30), and define B_h by $B_h(x) = A_h(-x)$ for all x. Show that $A_h * B_h$ and $B_h * A_h$ are positive-definite kernels.
(b) Let A be positive-definite. Show that $\varphi = A_h * dF_n$ is the unique solution to the least penalized squares problem (1.26) with $R = R_h$
$$R_h(\varphi) = h^{-2} \int_{\mathbb{R}} \varphi(x) \{ [\, T_h^{-1} \varphi \,](x) - \varphi(x) \} \, dx \ .$$

The above are some ways in which kernel density estimators arise as solutions to variational problems. This interpretation of kernel estimators comes in handy when qualitative information regarding the unknown density is available. It may be desirable and even advantageous to require that the estimator conform to this information. In the maximum likelihood context, this may be achieved by modifying (1.18) to

(1.36)
$$\text{minimize} \quad -\frac{1}{n} \sum_{i=1}^{n} \left[A_h * \log f \right] (X_i)$$

$$\text{subject to} \quad f \in \mathcal{F} ,$$

where $\mathcal{F}$ is the appropriate set of densities. We refer to (1.36) as constrained density estimation. Some interesting shape constraints are monotonicity and convexity of pdfs on $(0, \infty)$, or unimodality and log-concavity of pdfs on $(-\infty, \infty)$. An important alternative is to first determine an estimator for the "true" distribution function, e.g., by maximum likelihood estimation, and then use kernel smoothing to estimate the density. See GROENEBOOM and WELLNER (1992) for some impressive examples of this. Our version of constrained density estimation is discussed in the next two chapters. In Chapter 6, monotone and unimodal density estimation is discussed in detail. Log-concave estimation, together with maximum penalized likelihood estimation, is the topic of Chapter 5.

This ends our general introduction to nonparametric estimation and, in particular, maximum penalized likelihood estimation.

We continue with a general introduction to this chapter on kernel density estimation. To make kernel estimation work in practice, two decisions must be made: What kernel should one use, and how to choose the smoothing parameter h. When all is said and done, it turns out that the choice of the kernel does not matter much, see §7, and, for practical comparisons, §8.5. The choice of the smoothing parameter is much more critical, as already illustrated in Figure 1.1.1. It is clear that the smoothing parameter should be chosen in a rational manner, i.e., the smoothing parameter h should be a function of the data

(1.37)
$$h = H_n(X_1, X_2, \cdots, X_n) .$$

Of course, in constrained density estimation (1.36), it makes sense to let the choice of H_n also be influenced by the class of densities $\mathcal{F}$ under consideration. Note that the smoothing parameter (1.37) is now *random*, and that tremendously complicates the analysis of the kernel estimators. In this chapter, we avoid these complications and study kernel estimators for *deterministic* smoothing parameters only. Thus, the smoothing parameters are allowed to deterministically vary with the sample size n, that is, independently of $X_1, X_2, \cdots, X_n$. This serves as preparation for the random case studied in Chapter 7, but we are not able to get a completely satisfactory treatment there.

How does one go about analyzing kernel density estimators for deterministic h? Recall that f^{nh} may be written as a functional of the empirical distribution function F_n. The expected value of f^{nh} for fixed x is

$$(1.38) \qquad f_h(x) = \mathbb{E}\left[\, f^{nh}(x)\,\right] = A_h * dF_o(x) \, ,$$

if indeed the common pdf of $X_1, X_2, \cdots, X_n$ is f_o with cdf F_o. Note that for most x, one will have $f_h(x) \neq f_o(x)$, so that $f^{nh}(x)$ is a biased estimator of $f_o(x)$. We will come back to this in § 2. The "error" in the estimator may then be decomposed as

$$(1.39) \qquad f^{nh} - f_o = \{\, f^{nh} - f_h \,\} + \{\, f_h - f_o \,\} \, ,$$

where $f_h(x) = \mathbb{E}[\, f^{nh}(x)\,] = A_h * f_o(x)$. The second term on the right is the bias term, and the first term gives rise to the variance term. The task at hand is to show that the bias and variance are small. For reasons explained in § 1.3, the authors agree with DEVROYE and GYÖRFI (1985) that the L^1 distance,

$$(1.40) \qquad \|\, f^{nh} - f_o \,\|_1 = \int_{\mathbb{R}} \left|\, f^{nh}(x) - f_o(x)\,\right| dx \, ,$$

is a more than reasonable choice. In particular, its scale invariance is a very desirable feature. This may even be strengthened to invariance under monotone transformations of x. See Exercise (1.44). At times, we also need the Hellinger, Kullback-Leibler, and Pearson's φ^2 distances, see (1.3.18)–(1.3.20). The L^1 error does not admit an exact decomposition into bias and variance, but the triangle inequality gives us something close

$$(1.41) \qquad \|\, f^{nh} - f_o \,\|_1 \leqslant \|\, f^{nh} - f_h \,\|_1 + \|\, f_h - f_o \,\|_1 \, .$$

The first term on the right is the variance part, and the second one is the bias. In the following sections, various upper bounds are given for the bias and variance terms of kernel density estimators for some of these distances. The study of the bias is essentially a part of approximation theory in that no stochastic elements are present. The study of the variance part is (applied) probability. The main tool used is the notion of discrete time martingales, which results in the exponential inequalities of DEVROYE (1991), viz.

$$(1.42) \qquad \mathbb{P}[\,|\, e_{n,h} - \mathbb{E}[\, e_{n,h}\,]\,| > t\,] \leqslant 2\,e^{-\frac{1}{2}n t^2} \, , \qquad \text{for all } t > 0 \, ,$$

where either $e_{n,h} = \|\, f^{nh} - f_h \,\|_1$ or $e_{n,h} = \|\, f^{nh} - f_o \,\|_1$. Discrete time submartingales are extremely useful also, and lead to DOOB's submartingale inequality. This is all spelled out in detail in § 4.

It will transpire that in general it is impossible to achieve the parametric rate of convergence $n^{-1/2}$, and that $n^{-2/5}$ is typical for the L^1 error (but we shall not actually be concerned with lower bounds on the L^1 error). In contrast, it is sometimes possible to estimate functionals of f at the parametric rate, the mean and variance being the standard examples, but others are possible. Both for theoretical and practical reasons we are interested in the estimation of the (negative) entropy

$$(1.43) \qquad \mathcal{E}(f_o) = \int_{\mathbb{R}} f_o(x) \log f_o(x) \, dx$$

by the natural estimator $\mathcal{E}(f^{nh})$. This estimator not only achieves the parametric rate under reasonable conditions, but also is even best asymptotically normal. See § 6.

(1.44) EXERCISE. By a monotone transformation of $\mathbb{R}$, we mean any continuously differentiable function $t\,(x)$, $x \in \mathbb{R}$, with $t'(x) > 0$ everywhere. This induces a transformation T on the set of all pdfs by

$$Tf(x) = f\big(t\,(x)\big)\, t'(x)\,, \quad x \in \mathbb{R}\,.$$

(a) Show that Tf is a pdf if f is a pdf. In fact, if f is the pdf of the random variable X, then Tf is the pdf of the random variable $t\,(X)$.
(b) Show that $\|\,Tf - Tg\,\|_1 = \|\,f - g\,\|_1$ for any two pdfs f and g, and likewise for the Hellinger, Kullback-Leibler, and Pearson's φ^2 distances.
(c) Show that $\mathcal{E}(f)$ is not scale invariant.

To prove that the bias and variance terms of the kernel estimator tend to 0 at appropriate rates, some assumptions about f_o and the kernel A have to be made. These are of two kinds, to wit smoothness conditions and tail conditions. We are not too worried about requiring as much smoothness as seems desirable; we are more meticulous about minimal assumptions regarding the tail behavior. For the concise description of smoothness conditions, it is helpful to introduce the Sobolev spaces $W^{m,p}(\mathbb{R})$, see ADAMS (1975). For $1 \leqslant p \leqslant \infty$ and $m = 0, 1, 2, \cdots$, let

$$(1.45) \qquad W^{m,p}(\mathbb{R}) = \big\{\, f \in L^p(\mathbb{R}) \,:\, f^{(m)} \in L^p(\mathbb{R}) \,\big\}\,,$$

where $f^{(m)}$ denotes the m-th derivative of f. The norm on $W^{m,p}(\mathbb{R})$ is taken to be

$$(1.46) \qquad \|\, f \,\|_{W^{m,p}(\mathbb{R})} = \|\, f \,\|_p + \|\, f^{(m)} \,\|_p\,,$$

with $\|\cdot\|_p$ the L^p norm, see (1.3.8). The cases $p = 1$, $p = 2$, and $p = \infty$ are the most interesting.

Regarding the smoothness of f_o, we make the usual nonparametric assumption, viz. that f_o is smooth in the sense that

$$(1.47) \qquad f_o \in W^{2,1}(\mathbb{R})\,.$$

The assumption (1.47) is used in studying the bias term. For the variance term, we need conditions on the tails of the pdf. One type of constraint concerns the existence of moments of the pdf. The minimal requirement is that some moment of order > 1 exists. For the Kullback-Leibler distance $\mathrm{KL}(f^{nh}, f_h)$, a moment of order > 2 is required. So the typical assumption is that for a suitable $\lambda > 1$ (or > 2 as the need arises),

$$(1.48) \qquad \mathbb{E}\big[\,|\,X\,|^{\lambda}\,\big] = \int_{\mathbb{R}} |\,x\,|^{\lambda}\, f_o(x)\, dx < \infty \ .$$

In the limiting case in which f_o has moments of all orders, we sometimes also assume that it has a finite exponential moment, i.e., for some $t > 0$,

$$(1.49) \qquad \mathbb{E}\big[\,e^{\,t\,|\,X\,|}\,\big] < \infty \ .$$

Regarding the kernel function A, we only require that it is symmetric and has a moment of order 2, but most kernels in practical use have finite exponential moments.

The conditions (1.47) and (1.48) are somewhat weaker than the assumptions of the following type. The condition is that there exists a positive constant c and nonnegative constants α and β such that

$$(1.50) \quad f\big(\,F^{\text{inv}}(t)\,\big) \geqslant c\,\big\{\,t\,(1-t)\,\big\}^{\alpha}\,\big|\log\{\,t\,(1-t)\,\}\,\big|^{\beta} \ , \quad 0 < t < 1 \ .$$

where F^{inv} is the inverse of F if F is strictly increasing. Since one should not exclude that f is zero on an interval, in general, F^{inv} is defined as

$$(1.51) \qquad F^{\text{inv}}(t) = \inf\{\,u\,:\,F(u) \geqslant t\,\} \ , \quad 0 < t < 1 \ .$$

Strictly speaking, the behavior near $t = 0$ could be different from that at $t = 1$, but this hardly widens the scope of the results. It should be realized that condition (1.50) involves the regularity (slow variation) as well as the rate of the decay of f. In contrast, assumptions (1.47) and (1.48) involve decay only. Also, the crucial information in (1.50) is contained in the values of α and β and not so much in the value of c. If $\alpha \leqslant 1$, we speak of light tails, otherwise, we have heavy tails. However, for $\alpha = 1$, the further refinements embodied in the parameter β become important. For more on this, see PARZEN (1979). The case $\alpha < 1$ is much less interesting.

(1.52) EXERCISE. Suppose the pdf f satisfies (1.50) for some $\alpha < 1$. Let $\mathbb{1}_f$ denote the indicator function of the support of f, i.e., of the set $\text{support}(\,f\,) = \text{closure}\{\,x : f(x) > 0\,\}$, so

$$\mathbb{1}_f(x) = \begin{cases} 1 \ , & x \in \text{support}(\,f\,) \ , \\ 0 \ , & \text{otherwise} \ . \end{cases}$$

Show that $\text{support}(\,f\,)$ has finite measure, i.e.,

$$\int_{\mathbb{R}} \mathbb{1}_f(x)\, dx < \infty \ .$$

This almost says that f has compact support, but not quite, as the example

$$f(x) = 1 \quad \text{for } x \in [\,n\,,\,n+2^{-n}\,] \ , \quad \text{for } n = 1, 2, \cdots ,$$

(and $= 0$ otherwise) illustrates. Verify this.

We finish with some additional remarks on the selection of the smoothing parameter h. In Chapter 1, Figure 1.1, we illustrated the importance of this. How would one go about selecting the smoothing parameter? The first and foremost requirement is that the choice of h should be based on the data $X_1, X_2, \cdots, X_n$ only, and not on the experimenter's intuition. In other words, selecting h should be done by a rational procedure. Note, though, that then h is a random variable. However, whatever the method for choosing h, one must have that $h \longrightarrow_{as} 0$, $nh \longrightarrow_{as} \infty$, as the following well-known theorem states. The theorem is due to DEVROYE (1983), who credits ABOU-JAOUDÉ (1977) with some of the key ideas. We prove it for densities f_o with a finite moment of order > 1, using the normal and double exponential kernel, but not until Chapter 7.

(1.53) THEOREM. [DEVROYE (1983), ABOU-JAOUDÉ (1977)] *Let A be a kernel with finite moment of order > 1, and let $H = H_n$ be the smoothing parameter, deterministically or randomly varying with n. Then,*

$$\| A_H * dF_n - f_o \|_1 \longrightarrow_{as} 0$$

if and only if

$$H \longrightarrow_{as} 0 \ , \quad nH \longrightarrow_{as} \infty \ .$$

Nice as the theorem is, it does not really have practical implications. So, how should one go about selecting h? It seems clear that this should be done such that the resulting density estimator has "nice" properties. Finite sample optimality would be ideal, but it appears to be unachievable. Even asymptotic optimality is hard to achieve, although some recent efforts by DEVROYE and LUGOSI (1996), (1997) are coming very, very close. In Chapters 7 and 8, we give a brief introduction to some current practical procedures for selecting the smoothing parameter and show some simulation results illustrating their performance for moderate sample sizes.

In the remainder of this chapter, we discuss in detail the convergence of kernel density estimators in various distances, but only for deterministic choices of the smoothing parameter. We also discuss "optimal" kernels that are not necessarily densities. Maximum penalized likelihood estimation, in particular, the GOOD estimator, is studied in Chapter 5. Constrained density estimation is discussed there as well. The special cases of monotone and unimodal density estimation get their own chapter, Chapter 6. Window parameter selection (random smoothing parameters) is considered in Chapter 7, and practical examples and simulation experiments are reported in Chapter 8.

EXERCISES: (1.5), (1.12), (1.29), (1.35), (1.44), (1.52).

2. The expected L^1 error in kernel density estimation

In this section, we begin the study of the convergence of kernel density estimators. To keep things simple, the smoothing parameter is only allowed to vary deterministically with the sample size n. We start with the expected L^1 error. Let $X_1, X_2, \cdots, X_n$ be iid random variables on the real line, with pdf f. (We use f_o to denote the density of the X_i only in the maximum likelihood context). The kernel estimator of f with kernel A is then given by $f^{nh} = A_h * dF_n$ or, explicitly,

$$(2.1) \qquad f^{nh}(x) = \tfrac{1}{n} \sum_{i=1}^{n} A_h(x - X_i) , \qquad x \in \mathbb{R} .$$

We assume that

(2.2) A is a bounded pdf, symmetric about 0, and

$$(2.3) \qquad \int_{\mathbb{R}} x^2 A(x) \, dx < \infty .$$

Some results require smoothness of the kernel, in the form

$$(2.4) \qquad\qquad A \in W^{1,1}(\mathbb{R}) .$$

Recall the definition (1.45) of the Sobolev spaces $W^{m,p}(\mathbb{R})$. Of the usual kernels (normal, two-sided exponential, Epanechnikov, uniform), only the uniform kernel does not satisfy this condition.

(2.5) EXERCISE. Suppose A satisfies (2.2) and (2.3).
(a) Show that $\| A \|_2^2 \leqslant \| A \|_1 \| A \|_\infty$,
(b) and

$$\int_{\mathbb{R}} |x| \, A(x) \, dx < \infty , \quad \int_{\mathbb{R}} x \, A(x) \, dx = 0 , \quad \int_{\mathbb{R}} x^2 \, \{ A(x) \}^2 \, dx < \infty .$$

(c) If A satisfies only (2.4), show that $\| A \|_\infty \leqslant \| A' \|_1$.

Regarding f, we make the usual nonparametric assumptions

$$(2.6) \qquad\qquad f \in W^{2,1}(\mathbb{R}) ,$$

$$(2.7) \qquad \int_{\mathbb{R}} |x|^\lambda f(x) \, dx < \infty , \quad \text{for some } \lambda > 1 .$$

Initially, the assumption (2.7) will be used with $\lambda > 2$.

The starting point in the study of f^{nh} is the decomposition (1.39) into bias and variance components

$$(2.8) \qquad f^{nh} - f = A_h * (dF_n - dF) + \{ f_h - f \} ,$$

where $f_h = A_h * dF = \mathbb{E}[f^{nh}]$. Note that estimating the bias term $f_h - f$ is a deterministic problem, even if h is random, and of a type studied in approximation theory, see SHAPIRO (1969) for a beautiful little introduction. See also (the inexpensive reprint of) ACHIESER (1956). We state the result here, but reserve the proof for the end of this section.

(2.9) THEOREM. *Suppose f satisfies the assumptions (2.6) and (2.7). If A satisfies (2.2) and (2.3), then*

(a) $$\| A_h * f - f \|_1 \leqslant c_2(A) h^2 \| f'' \|_1 ,$$

(b) $$\| A_h * f - f \|_\infty \leqslant c_1(A) h \| f' \|_\infty \leqslant c_1(A) h \| f'' \|_1 .$$

If, in addition, A satisfies (2.4), then

(c) $$\| (A_h * f)' - f' \|_1 \leqslant c_1(A) h \| f'' \|_1 ,$$

(d) $$\| (A_h * f)' - f' \|_1 \leqslant c_2(A) h^2 \| f^{(3)} \|_1 , \quad \text{if } f \in W^{3,1}(\mathbb{R}) .$$

Here, $c_j(A) = (1/j!) \int_\mathbb{R} |x|^j A(x)\, dx, \ j = 1, 2$.

Observe that part (a) of the theorem implies that the bias term goes to 0 for $h \to 0$. The bound (a) is sharp in the sense that the power h^2 cannot be improved on, in general. The same applies to the bounds (c) and (d). See, e.g., SHAPIRO (1969). The bound (b) is sharp also, as is made clear in Exercise (2.34) below.

We next consider the variance term at a fixed point x, written as

(2.10) $$f^{nh}(x) - f_h(x) = \frac{1}{n} \sum_{i=1}^{n} Z_i ,$$

with

(2.11) $$Z_i = A_h(x - X_i) - f_h(x) , \quad i = 1, 2, \cdots, n .$$

Thus, because h is deterministic, the Z_i are iid random variables with zero mean and variance

$$\sigma_Z^2 = \mathbb{E}\left[\{A_h(x - X_1)\}^2 \right] - \{f_h(x)\}^2 ,$$

and so, with $B(x) = \{A(x)\}^2$, and $B_h(x) = h^{-1} B(h^{-1}x)$ as usual,

(2.12) $$\sigma_Z^2 = h^{-1} B_h * f(x) - \{f_h(x)\}^2 \leqslant h^{-1} B_h * f(x) .$$

Now, the kernel B_h is not a pdf, but it is nonnegative and integrable, see Exercise (2.5)(c), so that $\widetilde{B}_h = B_h/(\int_\mathbb{R} A^2)$ is a pdf. Since $h \to 0$, Theorem (2.9) implies that $B_h * f \longrightarrow c_0(B) f$, with $c_j(\cdot)$ as in Theorem (2.9), and so in the last inequality, not much is lost. Since $\mathbb{E}[\,|Y|\,] \leqslant \{\mathbb{E}[Y^2]\}^{1/2}$, it follows that

(2.13) $$\mathbb{E}\left[|f^{nh}(x) - f_h(x)| \right] \leqslant (nh)^{-1/2} \left\{ B_h * f(x) \right\}^{1/2} .$$

For all intents and purposes, this is a sharp upper bound for $nh \to \infty$. For $nh \to 0$, it is not sharp, see the following exercise.

(2.14) EXERCISE. (a) Show that $\mathbb{E}\big[\,|\,f^{nh}(x) - f_h(x)\,|\,\big] \leqslant 2\,f_h(x)$.

(b) Show that this is bounded as $h \to 0$, under the usual assumptions on f and A.

(c) Note that the right-hand side of (2.13) tends to ∞ for $nh \to 0$. Since then $h \to 0$, the left-hand side is bounded. Thus, equality is far from being attained in (2.13).

As an aside, we note that the fundamental features of kernel density estimation are beginning to take shape. From Theorem (2.9), we gather that $h \to 0$ is required for the bias term to go to 0, whereas (2.13) suggests that $nh \to \infty$ is required for the variance term to tend to 0. So we may conclude that h should tend to 0 as $n \to \infty$, but not too fast.

(2.15) REMARK. For fixed x, the Central Limit Theorem may be applied to $f^{nh}(x) - f_h(x)$, and yields for $n \longrightarrow \infty$,

$$\sqrt{n}\,\{f^{nh}(x) - f_h(x)\} \longrightarrow_{\mathrm{d}} Y \sim N(0, \sigma_Z^2)\,.$$

The law of the iterated logarithm, see, e.g., SHORACK and WELLNER (1986), gives

$$(2.16) \qquad \limsup_{n \to \infty} \{\, n^{-1} \log\log n\,\}^{1/2}\,|\,f^{nh}(x) - f_h(x)\,| =_{\mathrm{as}} \sqrt{2}\,\sigma_Z\,.$$

However, these hold for fixed h only, so their use in the present context is limited.

The inequality (2.13) provides useful information for fixed x, but the interest is in the *random function* $f^{nh} - f_h$, i.e., in $f^{nh}(x) - f_h(x)$ for all values of x. However, functions are infinite-dimensional objects, and they are much too hard to handle. So, instead, the essence (for us) is captured by a single number, such as $\|\,f^{nh} - f_h\,\|_1$ or $\mathbb{E}\big[\,\|\,f^{nh} - f_h\,\|_1\,\big]$. Fortunately, the bound (2.13) applies directly, after an appeal to Fubini's theorem to justify the change in the order of integration,

$$(2.17) \qquad \mathbb{E}\big[\,\|\,f^{nh} - f_h\,\|_1\,\big] \leqslant (nh)^{-1/2}\,\|\,\{\,B_h * f\,\}^{1/2}\,\|_1\,.$$

We are now forced to deal with the L^1 norm on the right. The following lemma can be found in DEVROYE and GYÖRFI (1985), with a nice proof in DEVROYE (1987), Lemma 7.1.

(2.18) LEMMA. [DEVROYE and GYÖRFI (1985)] *If the random variable Y has pdf φ, then*

$$\|\,\sqrt{\varphi}\,\|_1 \leqslant \sqrt{2\pi}\,\{\,\mathbb{V}\mathrm{ar}(Y)\,\}^{1/4}\,.$$

The elementary proof of DEVROYE (1987) is outlined in Exercise (2.24) below. How do we apply this to $B_h * f$? As before, let $\widetilde{B}_h = B_h/(\int_{\mathbb{R}} A^2)$, and let X, Y be independent random variables with pdfs f and $\widetilde{B}$. Then, the random variable $X + hY$ has pdf $\widetilde{B}_h * f$, and so Lemma (2.18) yields

$$(2.19) \qquad \left\| \{ \widetilde{B}_h * f \}^{1/2} \right\|_1 \leqslant \sqrt{2\pi} \left\{ \operatorname{Var}(X) + h^2 \operatorname{Var}(Y) \right\}^{1/4} .$$

Now,

$$\operatorname{Var}(Y) = \int_{\mathbb{R}} x^2 \{A(x)\}^2 \, dx \bigg/ \int_{\mathbb{R}} \{A(x)\}^2 \, dx ,$$

and this is finite, see Exercise (2.5). Since $\operatorname{Var}(X)$ is finite as well, by assumption (2.7), with $\lambda > 2$ for now, then

$$(2.20) \qquad \left\| \{ B_h * f \}^{1/2} \right\|_1 \leqslant \text{Constant} , \quad \text{uniformly in } h \to 0 .$$

Combining this with Theorem (2.9) proves the following theorem, with the strengthened version of assumption (2.7), i.e., $\lambda > 2$. The general case $\lambda > 1$ is worked out in the exercises below.

(2.21) THEOREM. *Under the conditions (2.2)–(2.3) on A and (2.7) on f, there exist constants c_i such that for all n,h,*

(a) $$\mathbb{E}\left[\| f^{nh} - f_h \|_1 \right] \leqslant c_1 \, (nh)^{-1/2} ,$$

(b) $$\mathbb{E}\left[\| f^{nh} - f \|_1 \right] \leqslant c_2 \, h^2 + c_3 \, (nh)^{-1/2} .$$

If, in addition, f satisfies (2.6), then

(c) $$\inf_{h>0} \mathbb{E}\left[\| f^{nh} - f \|_1 \right] \leqslant c_4 \, n^{-2/5} .$$

(2.22) EXERCISE. Fill in the details of the proof of Theorem (2.21), under the assumption (2.7) with $\lambda > 2$.

(2.23) EXERCISE. Formulate and prove the analogue of Theorem (2.21) for the derivatives, that is, for $\mathbb{E}\left[\| (f^{nh})' - (f_h)' \|_1 \right]$, with (2.2) replaced by $f \in W^{3,1}(\mathbb{R})$, and whatever assumptions on A you desire (as long as such an A exist). Do it also under the same assumptions on f as before.

(2.24) EXERCISE. Prove Lemma (2.18). [Hint: First assume that the mean equals 0. Use Cauchy-Schwarz to show that for all $a > 0$,

$$\left\{ \int_{\mathbb{R}} \sqrt{\varphi} \right\}^2 \leqslant \left\{ \int_{\mathbb{R}} \frac{dy}{a^2 + y^2} \right\} \left\{ \int_{\mathbb{R}} (a^2 + y^2) \, \varphi(y) \, dy \right\} ,$$

compute the first factor on the right, and minimize over $a > 0$ using calculus.]

(2.25) EXERCISE. (a) Prove that for every $\lambda > 1$ there exists a constant $c(\lambda)$ such that for the random variable X with density f,

$$\| \sqrt{f} \|_1 \leqslant c(\lambda) \left\{ \mathbb{E}\left[|X|^\lambda \right] \right\}^{1/(2\lambda)} .$$

This is the second part of Lemma 1 in Chapter 5 of DEVROYE and GYÖRFI (1985). [Hint: In (2.24), replace $a^2 + y^2$ by $a^\lambda + |y|^\lambda$.]

(b) Show that for all $p > 1$ and for all $\lambda > p - 1$ there exists a constant $c(\lambda, p)$ such that for the random variables X with density f,

$$\| f^{1/p} \|_1 \leq c(\lambda, p) \{ E[|X|^\lambda] \}^{1/(\lambda q)},$$

with $(1/p) + (1/q) = 1$.

(2.26) EXERCISE. Let the random variables Y and X have densities A and f. Show that for all $\lambda > 1$ there exists a constant $c(\lambda)$ such that

$$\| \sqrt{A_h * dF_n} \|_1 \leq c(\lambda) \{ h\, E[|Y|^\lambda] + \tfrac{1}{n} \sum_{i=1}^n |X_i|^\lambda \}^{1/(2\lambda)}.$$

(2.27) EXERCISE. Prove Theorem (2.21) as stated (i.e., with the assumption (2.7) for some $\lambda > 1$.)

The above Theorem (2.21) is nice! But what about the almost sure behavior of $\| f^{nh} - f_h \|_1$, or the pointwise almost sure behavior (2.16) if we want h (deterministically) changing with n? The case of the expected values is somewhat of a misdirection, because of Fubini's theorem leading to the inequality (2.17). Even for fixed $h > 0$, we are not allowed to integrate (2.16) to get a statement about the almost sure behavior of $\| f^{nh} - f_h \|_1$! Example (1.3.23) and Exercise (1.3.26) show what could go wrong. In the next sections, we provide two approaches to the problem.

We finish this section with a proof still owed to the reader.

PROOF OF THEOREM (2.9). Considering A_h is a pdf, we may rewrite $f_h(x) - f(x) = [A_h * f](x) - f(x)$ as

$$f_h(x) - f(x) = \int_\mathbb{R} A_h(y) \{ f(x - y) - f(x) \}\, dy.$$

By the symmetry of A about 0, and thus that of A_h, then $\int_\mathbb{R} y\, A_h(y)\, dy = 0$, so that we may subtract $y f'(x)$ from the expression in curly brackets in the integrand above. Then,

$$(2.28) \qquad f_h(x) - f(x) = \int_\mathbb{R} y^2\, A_h(y)\, f(x, y)\, dy,$$

where

$$(2.29) \qquad f(x, y) = \frac{f(x - y) - f(x) - y f'(x)}{y^2}, \qquad y \neq 0.$$

Roughly, $f(x, y) \approx \frac{1}{2} f''(y)$, as we make precise. Taylor's theorem with exact remainder gives

$$(2.30) \qquad f(x, y) = y^{-2} \int_{x}^{x-y} (x - y - t) f''(t) \, dt$$

$$= \int_{0}^{1} (1 - t) f''(x - y t) \, dt$$

$$= \int_{\mathbb{R}} w(t) f''(x - y t) \, dt \; ,$$

with $w(t) = 1 - t$ for $0 < t < 1$, and $= 0$ otherwise. From (2.28), we get after changing the order of integration,

$$\| f_h - f \|_1 \leqslant \int_{\mathbb{R}} y^2 A_h(y) \left\{ \int_{\mathbb{R}} | f(x, y) | \, dx \right\} dy$$

$$(2.31) \qquad \leqslant \sup_{y} \left\{ \int_{\mathbb{R}} | f(x, y) | \, dx \right\} \int_{\mathbb{R}} y^2 A_h(y) \, dy \; .$$

The last integral equals $2 \, c_2(A) \, h^2$. For the first integral, we have by (2.30),

$$\int_{\mathbb{R}} | f(x, y) | \, dx \leqslant \int_{\mathbb{R} \times \mathbb{R}} w(t) | f''(x - y t) | \, dt \, dx =$$

$$\leqslant \int_{\mathbb{R}} w(t) \left\{ \int_{\mathbb{R}} | f''(x - y t) | \, dx \right\} dt =$$

$$\leqslant \| w \|_1 \| f'' \|_1 = \tfrac{1}{2} \| f'' \|_1 \; .$$

It follows that

$$\| f_h - f \|_1 \leqslant c_2(A) \, h^2 \, \| f'' \|_1 \; .$$

This proves Part (a) of Theorem (2.9). Part (d) is a one-liner now. The proof of part (c) is analogous to that of part (a), as is the first inequality of (b). The second inequality of (b) follows from Exercise (2.5)(c).   Q.e.d.

(2.32) EXERCISE. Prove parts (b) and (c) of Theorem (2.9).

(2.33) EXERCISE. Any smoothness condition of Theorem (2.9) of the form $f \in W^{m,1}(\mathbb{R})$ may be replaced by the condition that $f^{(m-1)}$ has bounded variation. In particular, prove the following: If f' has bounded variation, then under the stated conditions on the kernel A,

$$\| A_h * f - f \|_1 \leqslant c_2(A) \, h^2 \, \| f' \|_{\mathrm{BV}} \; .$$

Here, for any function g on the line,

$$\| g \|_{\mathrm{BV}} = \int_{\mathbb{R}} | dg(x) | \; .$$

[Hint : First, prove it under the usual smoothness condition, that is, assume that $f \in W^{2,1}(\mathbb{R})$, and show that

$$\| f'' \|_1 = \| f' \|_{\mathrm{BV}} \, .$$

To get rid of the smoothness, first *apply* extra smoothness and consider

$$\| A_h * \phi_\lambda * f - \phi_\lambda * f \|_1 \leqslant c_2(A) \, h^2 \, \| (\phi_\lambda)' * (f') \|_1 \, ,$$

where ϕ is the standard normal density. Now, show that

$$\| \phi_\lambda * (f') \|_{\mathrm{BV}} \leqslant \| f' \|_{\mathrm{BV}} \, ,$$

and let $\lambda \longrightarrow 0$.]

(2.34) EXERCISE. Show that the factor h in the bound (b) of Theorem (2.9) cannot be improved in general, as follows. Let $f(x) = (1 - |x|)_+$.
(a) Show that $[A_h * f](0) - f(0) = -h \, c_1(A)$, where $c_1(A)$ is defined in Theorem (2.9).
(b) Show that $f \notin W^{2,1}(\mathbb{R})$, but that $\| f' \|_{\mathrm{BV}} = 2$.
(c) Conclude that $\| A_h * f - f \|_\infty \geqslant \frac{1}{2} \, c_1(A) \, h \, \| f' \|_{\mathrm{BV}}$.
(d) Now, smooth out f a little so that $f \in W^{2,1}(\mathbb{R})$, as per Exercise (2.33).

EXERCISES : (2.5), (2.14), (2.22), (2.23), (2.24), (2.25), (2.26), (2.27), (2.32), (2.33), (2.34).

3. Integration by parts tricks

One of the main difficulties in the study of kernel estimators, alluded to in the previous section, amounts to the decoupling of the smoothing factor h from the sample $X_1, X_2, \cdots, X_n$. Here, we study integration by parts tricks. Although they do not give optimal results, they do provide for an interesting view of the problem, which turns out helpful in later chapters.

So how can the decoupling be done ? A reasonable first attempt is the straightforward integration by parts trick, see PRAKASA RAO (1983) and references therein,

$$f^{nh}(x) - f_h(x) = \int_{\mathbb{R}} A_h(x - y)\{ \, dF_n(y) - dF(y) \, \}$$

(3.1)
$$= h^{-1} \int_{\mathbb{R}} C_h(x - y) \, \{ \, F_n(y) - F(y) \, \} \, dy \, ,$$

where $C(x) = A'(x)$ and $C_h(x) = h^{-1} C(h^{-1}x)$ as usual. So, concisely, $f^{nh} - f_h = h^{-1} \, C_h * \{ \, F_n - F \, \}$. We thus obtain

(3.2)
$$\| f^{nh} - f_h \|_1 = h^{-1} \| C_h * \{ \, F_n - F \, \} \|_1$$
$$\leqslant h^{-1} \| C_h \|_1 \| F_n - F \|_\infty \, .$$

Note that $\| C_h \|_1 = \| A' \|_1 < \infty$, if $A \in W^{1,1}(\mathbb{R})$.

(3.3) EXERCISE. The CHUNG (1949) law of the iterated logarithm, see, e.g., SHORACK and WELLNER (1986), says that

$$\limsup_{n\to\infty} \left\{ \tfrac{1}{2}\, n^{-1} \log\log n \right\}^{-1/2} \| F_n - F \|_\infty =_{as} 1 \ .$$

(a) Use this to show that

$$\| f^{nh} - f_h \|_p =_{as} \mathcal{O}\left(h^{-1} \left(n^{-1} \log\log n \right)^{1/2} \right) \ ,$$

uniformly in $h > 0$, deterministic or random.

(b) Under the assumption (2.6) on f, show that for $h = (n^{-1} \log\log n)^{1/6}$,

$$\| f^{nh} - f \|_1 =_{as} \mathcal{O}\left((n^{-1} \log\log n)^{1/3} \right) \ .$$

All of this is fast and easy, or, rather, quick and dirty. The nice part is that Exercise (3.3) would hold for random h, but in comparison with the expected value, see Theorem (2.21), we have lost a factor $h^{1/2}$ in the process, and because $h \to 0$, that is significant. It is not so easy to pinpoint the exact difficulty in the above. It would be hard to improve on (3.2), so our attention turns to (3.1). One approach, which proved successful in DEHEUVELS and MASON (1992), is to replace $F_n(y) - F(y)$ by $F_n(y) - F(y) - \{F_n(x) - F(x)\}$ and study moduli of continuity of empirical processes. However, we study instead fractional integration by parts. This is a well-known topic in analysis, with a hallowed history, such as WEYL (1917) and HARDY and LITTLEWOOD (1928). See GORENFLO and VESSELLA (1991) for the complete scoop.

When all is said and done, the fractional integration by parts trick leads to the following result.

(3.4) THEOREM. *Let $\tfrac{1}{2} < \kappa < 1$, and assume that $A \in W^{1,1}(\mathbb{R})$. Then, for every distribution function Φ,*

$$A_h * d\Phi = A_{h,\kappa} * g_\kappa * d\Phi \ ,$$

where $A_{h,\kappa} \in L^1(\mathbb{R})$ satisfies

$$\| A_{h,\kappa} \|_1 \leqslant c\, h^{-\kappa}$$

and $g_\kappa \in L^1(\mathbb{R}) \cap L^2(\mathbb{R})$ is a pdf, symmetric about 0, with

$$g_\kappa(x) \sim c_\kappa\, |x|^{\kappa-1} \ , \quad |x| \to 0 \ ,$$
$$g_\kappa(x) \sim \gamma_\kappa\, |x|^{-\kappa-1} \ , \quad |x| \to \infty \ ,$$

for certain constants c_κ, γ_κ. Moreover, $|x|^{1-\kappa} g_\kappa(x)$ is bounded on $\mathbb{R}$.

We leave the proof for later. With $f^{nh} - f_h = A_h * (dF_n - dF)$, the theorem implies that

$$(3.5) \qquad \| f^{nh} - f_h \|_1 \leqslant c\, h^{-\kappa} \| g_\kappa * (dF_n - dF) \|_1 \ .$$

How does this behave as $n \to \infty$? To get good behavior of $g_\kappa * (dF_n - dF)$, we need $g_\kappa \in L^2(\mathbb{R})$, so that in view of Theorem (3.4), we need $\kappa > \frac{1}{2}$. In view of the factor $h^{-\kappa}$, it is clear that we want κ close to $\frac{1}{2}$. For deterministic h, the inequality (3.5) does not improve on (2.13)–(2.17), as a matter of fact, it gives slightly worse results ($h^{-\kappa}$ vs $h^{-1/2}$). It should be possible to coax a.s. convergence rates out of (3.5), but unfortunately, a law of the iterated logarithm or some such thing for $\| k * (dF_n - dF) \|_1$ with fixed $k \in L^1 \cap L^2$ does not appear to be known. Note, though, that it should be easier than with k replaced by A_h with deterministically changing or even random h.

How can we get around this problem? The probabilists' solution would be to derive so-called exponential inequalities, and like good school boys, we do so in the next section. This gives quite sharp bounds. We shall also take the simple(r) route to asymptotically optimal convergence rates by means of submartingale inequalities. The submartingale inequalities result in a.s. rates that are the same as for the expected values, give or take a few factors $\log n$, which the authors do not mind. Surprisingly, the submartingale approach not only works for $\| A_h * (dF_n - dF) \|_1$ with deterministically varying h, but also for the Hellinger and Kullback-Leibler distances.

THE SKETCH OF THE PROOF OF THEOREM (3.4). The easiest way to explain the fractional integration by parts trick is by means of Fourier transforms, at least informally. For $f \in L^1(\mathbb{R})$, the Fourier transform $f^\wedge$ (the notation $\widehat{f}$ is also common) of f is defined as

$$(3.6) \qquad f^\wedge(\omega) = \int_{\mathbb{R}} f(x) \, e^{-2\pi i \omega} \, dx \, , \quad \omega \in \mathbb{R} \, .$$

We mention here that $\left(A_h \right)^\wedge(\omega) = A^\wedge(h\omega)$, $\omega \in \mathbb{R}$.

Let Φ be a distribution function. Then, $A_h * d\Phi$ is a pdf, and its Fourier transform equals

$$(A_h * d\Phi)^\wedge(\omega) = A^\wedge(h\omega) \, (d\Phi)^\wedge(\omega) \, , \quad \omega \in \mathbb{R} \, .$$

Informally, integration by parts of the convolution $A_h * d\Phi$ corresponds to

$$(A_h * d\Phi)^\wedge(\omega) = 2\pi i \omega \, A^\wedge(h\omega) \, \frac{(d\Phi)^\wedge(\omega)}{2\pi i \omega} \, , \quad \omega \in \mathbb{R} \, ,$$

since $(\psi')^\wedge(\omega) = 2\pi i \omega \, \psi^\wedge(\omega)$ for nice functions ψ. This gives us a good idea of what fractional integration by parts should look like. It ought to be

$$(A_h * d\Phi)^\wedge(\omega) = (2\pi i \omega)^\kappa \, A^\wedge(h\omega) \, \frac{(d\Phi)^\wedge(\omega)}{(2\pi i \omega)^\kappa} \, , \quad \omega \in \mathbb{R} \, .$$

with $0 < \kappa < 1$, and as a not-so-wild guess, $\kappa = \frac{1}{2}$. The following modification looks a bit more reasonable:

$$(A_h * d\Phi)^\wedge(\omega) = \{1 + (2\pi i\omega)^\kappa\} A^\wedge(h\omega) \frac{(d\Phi)^\wedge(\omega)}{1 + (2\pi i\omega)^\kappa}, \quad \omega \in \mathbb{R},$$

because $\{1 + (2\pi i\omega)^\kappa\}^{-1}$ looks like it might be the Fourier transform of a function. To avoid the complex fractional power functions, it is in fact more convenient to implement the formula

$$(3.7) \quad (A_h * d\Phi)^\wedge(\omega) = \{1 + |2\pi\omega|^\kappa\} A^\wedge(h\omega) \frac{(d\Phi)^\wedge(\omega)}{1 + |2\pi\omega|^\kappa}, \quad \omega \in \mathbb{R}.$$

This leads to the definition of the functions g_κ and $A_{h,\kappa}$ by means of their Fourier transforms

$$(3.8) \qquad\qquad (g_\kappa)^\wedge(\omega) = \{1 + |2\pi\omega|^\kappa\}^{-1},$$

$$(3.9) \quad (A_{h,\kappa})^\wedge(\omega) = A^\wedge(h\omega) + h^{-\kappa}\{|2\pi h\omega|^\kappa A^\wedge(h\omega)\}, \quad \omega \in \mathbb{R},$$

and their properties advertised in the theorem must be shown. The function g_κ is studied in Appendix 2, § 4. For the function $A_{h,\kappa}$, it is useful to observe that

$$A_{h,\kappa} = A_h + h^{-\kappa}\mathfrak{A}_h,$$

where $\mathfrak{A}_h = h^{-1}\mathfrak{A}(h^{-1}x)$, and $\mathfrak{A}$ has Fourier transform

$$\mathfrak{A}^\wedge(\omega) = |2\pi\omega|^\kappa A^\wedge(\omega), \quad \omega \in \mathbb{R}.$$

In Appendix 2, § 4, we show that $\mathfrak{A} \in L^1(\mathbb{R})$ if $A \in W^{1,1}(\mathbb{R})$.

EXERCISE: (3.3).

4. Submartingales, exponential inequalities, and almost sure bounds for the L^1 error

In this section, we study martingales and submartingales from a very simple point of view, with an eye toward obtaining inequalities for the L^1 error $\| A_h * (dF_n - dF) \|_1$. In the next section, the application of submartingales to other distances is considered. This section follows some of the material of WILLIAMS (1991) and DEVROYE (1991). Parts of it rely heavily on convexity considerations, see Part III.

(4.1) EXAMPLE. Let $S_n = n \| g_\kappa * (dF_n - dF) \|_1$, with g_κ from Theorem (3.4), with $\frac{1}{2} < \kappa < 1$. It is easily shown that

$$\mathbb{E}[S_{n+1} \mid X_1, \cdots, X_n] \geq \| \mathbb{E}[(n+1) g_\kappa * (dF_{n+1} - dF) \mid X_1, \cdots, X_n] \|_1.$$

Now,

$$\mathbb{E}[\,(n+1)\,g_\kappa * (dF_{n+1} - dF)\,|\,X_1, X_2, \cdots, X_n\,] =$$
$$n\,g_\kappa * (dF_n - dF) + \mathbb{E}[\,g_\kappa(\,\cdot\, - X_{n+1}) - g_\kappa * dF\,] =$$
$$n\,g_\kappa * (dF_n - dF) \, ,$$

and so

$$\mathbb{E}[\,S_{n+1}\,|\,X_1, X_2, \cdots, X_n\,] \geqslant S_n \, .$$

This is the property that makes $\{\,S_n\,\}_{n\geqslant 1}$ a *submartingale*.

(4.2) DEFINITION. *Let $\{\psi_n\}_{n\geqslant 1}$ be a sequence of (deterministic) functions, where $\psi_n : \mathbb{R}^n \longrightarrow \mathbb{R}$ are continuous. Let $X_1, X_2, \cdots, X_n$ be a sequence of independent random variables, and define $\{S_n\}_{n\geqslant 1}$ by*

$$S_n = \psi_n(X_1, X_2, \cdots, X_n) \, , \quad n = 1, 2, \cdots \, .$$

Then, $\{S_n\}_{n\geqslant 1}$ is a submartingale if

$$\mathbb{E}[\,S_{n+1}\,|\,X_1, X_2, \cdots, X_n\,] \geqslant S_n \, , \quad n = 1, 2, \cdots \, .$$

If there is equality for every n, then $\{S_n\}_{n\geqslant 1}$ is called a martingale.

Often, we say that $\{S_n\}_{n\geqslant 1}$ is a submartingale, without explicitly mentioning the underlying random variables $X_1, X_2, \cdots, X_n$. The context should make it clear what is happening. We should also mention that in the precise, measure-theoretic treatment of (sub)martingales, the sequence of σ-fields generated by the observations $X_1, X_2, \cdots, X_n$ for $n \geqslant 1$ plays a key role, but we shall not go to such lengths. See, e.g., WILLIAMS (1991).

The above is a quite pedestrian description of submartingales, but it is good enough for our purposes. Note that positivity of the S_n is not a requirement, the example in (4.1) just happens to be that way.

(4.3) EXERCISE. Show : If $\{S_n\}_{n\geqslant 1}$ is a submartingale, then

$$\mathbb{E}[\,S_n\,|\,X_1, \cdots, X_k\,] \geqslant S_k \, , \quad \text{for all } n > k \, .$$

[Hint : Use induction on n. The result holds for $n = k + 1$.]

(4.4) LEMMA. *Let φ be a convex function.*
(a) If $\{S_n\}_{n\geqslant 1}$ is a martingale, then $\{\varphi(S_n)\}_{n\geqslant 1}$ is a submartingale.
(b) If φ is also increasing over $\mathbb{R}$ and $\{S_n\}_{n\geqslant 1}$ is a submartingale, then $\{\varphi(S_n)\}_{n\geqslant 1}$ is a submartingale.

PROOF. By the convexity of φ, Jensen's inequality gives

$$\mathbb{E}[\,\varphi(S_{n+1})\,|\,X_1, X_2, \cdots, X_n\,] \geqslant \varphi\big(\mathbb{E}[\,S_{n+1}\,|\,X_1, X_2, \cdots, X_n\,]\big) \, .$$

This proves part (a) by the martingale property. If φ is also increasing, then the submartingale property does the trick.   Q.e.d.

(4.5) EXERCISE. (a) Let $\{S_n\}_n$ be a submartingale. Show that $\{\exp(S_n)\}_n$ and $\{\max(S_n, 0)\}_n$ are submartingales also.
(b) Let $\{S_n\}_n$ be a positive submartingale. Show that $\{-\log(S_n)\}_n$ and $\{(S_n)^p\}_n$ (for all $p > 1$) are submartingales.

The following is a very devious way to associate a martingale with a random function, i.e., a function that depends on independent random variables $X_1, X_2, \cdots, X_n$.

(4.6) EXERCISE. Let $X_1, X_2, \cdots, X_n$ be independent random variables, and let $S = \psi(X_1, X_2, \cdots, X_n)$, with ψ continuous, say. Assume that $\mathbb{E}[|S|] < \infty$, and define

$$S_k = \mathbb{E}[S \mid X_1, \cdots, X_k], \quad k = 1, 2, \cdots, n.$$

Show that $S_n = S$ and that $S_1, S_2, \cdots, S_n$ is a martingale.

(4.7) EXERCISE. If $\{S_n\}_{n \geqslant 1}$ is a submartingale, then $\mathbb{E}[S_{n+1}] \geqslant \mathbb{E}[S_n]$.

On to submartingale inequalities. First note that if $\{S_n\}_{n \geqslant 1}$ is a submartingale, then $\{S_n\}_{n \geqslant 1}$ should increase on average, and one would expect that the largest one of $S_1, S_2, \cdots, S_n$ is the last one, S_n. Recalling the Tchebyshev inequality, Exercise (4.18) below,

$$\mathbb{P}[S_n > t] \leqslant \frac{1}{t} \mathbb{E}[|S_n|], \quad t > 0,$$

should both lower and raise the level of wonderment at the following inequality.

(4.8) THEOREM (DOOB'S SUBMARTINGALE INEQUALITY). If $\{S_n\}_{n \geqslant 1}$ is a positive submartingale, then for all n, $t > 0$,

$$\mathbb{P}[\max_{1 \leqslant k \leqslant n} S_k > t] \leqslant \frac{1}{t} \mathbb{E}[S_n].$$

Moreover, for all $p > 1$,

$$\mathbb{E}[\max_{1 \leqslant k \leqslant n} (S_k)^p] \leqslant \left(\frac{p}{p-1}\right)^p \mathbb{E}[(S_n)^p].$$

PROOF. Let $M_n = \max_{1 \leqslant k \leqslant n} S_k$. The event $\{M_n \geqslant t\}$ is thought of as a subset of $\mathbb{R}^n$. The first step is to partition this event into a disjoint union

$$\{M_n \geqslant t\} = \mathcal{G}_1 \cup \mathcal{G}_2 \cup \cdots \cup \mathcal{G}_n,$$

where

$$\mathcal{G}_1 = \{S_1 > t\}$$

and

$$\mathcal{G}_k = \{S_1 \leqslant t\} \cap \cdots \cap \{S_{k-1} \leqslant t\} \cap \{S_k > t\},$$

for $k = 2, 3, \cdots, n$. Here, each event $\{ S_k > t \}$ is shorthand for

$$\{ S_k > t \} = \{ (X_1, X_2, \cdots, X_n) \in \mathbb{R}^n : S_k = \psi_k(X_1, \cdots, X_k) > t \} ,$$

and likewise for $\{ S_k \leqslant t \}$. Actually, our pedestrian approach forces us to consider $\mathcal{G}_k$ also as a subset of $\mathbb{R}^k$. To do this properly, we define $\mathcal{F}_k \subset \mathbb{R}^k$ such that

$$\mathcal{G}_k = \mathcal{F}_k \times \mathbb{R}^{n-k+1}$$

or, more precisely,

$$\mathcal{F}_k = \{ (X_1, \cdots, X_k) : (X_1, X_2, \cdots, X_n) \in \mathcal{G}_k \}.$$

Now, apparently,

$$\mathbb{P}[M_n > t] = \sum_{k=1}^{n} \mathbb{P}[\mathcal{G}_k] = \sum_{k=1}^{n} \mathbb{P}[\mathcal{F}_k \times \mathbb{R}^{n-k+1}] .$$

Now, let P_k be the joint cdf for the random variables $X_1, X_2, \cdots, X_k$, and Q_k the joint cdf for $X_{k+1}, \cdots, X_n$. Then,

$$\mathbb{P}[\mathcal{F}_k \times \mathbb{R}^{n-k+1}] = \int_{\mathcal{F}_k} dP_k \leqslant t^{-1} \int_{\mathcal{F}_k} S_k \, dP_k ,$$

since $S_k > t$ on $\mathcal{F}_k$. (This looks suspiciously like the trick used to prove Tchebyshev's inequality!) By the submartingale property,

$$S_k \leqslant \mathbb{E}[S_n \,|\, X_1, \cdots, X_k] = \int_{\mathbb{R}^{n-k}} S_n \, dQ_k ,$$

so that

$$\mathbb{P}[\mathcal{F}_k \times \mathbb{R}^{n-k+1}] \leqslant t^{-1} \int_{\mathcal{F}_k \times \mathbb{R}^{n-k+1}} S_n \, dP_k \, dQ_k = t^{-1} \int_{\mathcal{G}_k} S_n \, dP_n .$$

Finally, by the disjointness of the $\mathcal{G}_k$,

$$\sum_{k=1}^{n} \mathbb{P}[\mathcal{F}_k \times \mathbb{R}^{n-k+1}] \leqslant t^{-1} \int_{\bigcup \mathcal{G}_k} S_n \, dP_n \leqslant t^{-1} \mathbb{E}[S_n] ,$$

and that does it. The proof of the inequality for the expected values is left as a daunting exercise.   Q.e.d.

(4.9) EXERCISE. What if the submartingale is not positive? Show that if $\{S_n\}_{n \geqslant 1}$ is any submartingale, then

$$\mathbb{P}[\max_{1 \leqslant k \leqslant n} S_k > t] \leqslant t^{-1} \mathbb{E}[|S_n|] .$$

(4.10) EXERCISE. If $\{S_n\}_{n \geqslant 1}$ is a submartingale, show that for all $t > 0$ and $p \geqslant 1$,

$$\mathbb{P}[\max_{1 \leqslant k \leqslant n} S_k > t] \leqslant t^{-p} \mathbb{E}[|S_n|^p] .$$

[Hint: In (4.9) and (4.10), use Theorem (4.8), rather than prove it again.]

(4.11) EXERCISE. Let $X_1, X_2, \cdots, X_n$ be independent random variables, with $\mathbb{E}[X_i] = 0$, and $\mathbb{E}[X_i^2] = \sigma_i^2 < \infty$. Let $S_n = X_1 + X_2 + \cdots + X_n$ and $V_n = \sum_i \sigma_i^2$. Show that

$$\mathbb{P}[\max_{1 \leqslant k \leqslant n} |S_k| > t] \leqslant t^{-2} V_n .$$

This is due to KOLMOGOROV, but he did not use (sub)martingales.

We have the following application of the submartingale inequality to the business of getting almost sure bounds.

(4.12) THEOREM. Let $\{S_n\}_{n \geqslant 1}$ be a submartingale. Then, for all $p \geqslant 1$ and all $\varepsilon > 0$,

$$S_n =_{\text{as}} \mathcal{O}\left(\left\{ (\log n)^{1+\varepsilon} \mathbb{E}[|S_{2n}|^p] \right\}^{1/p} \right) .$$

Note that the bound for S_n is in terms of S_{2n}.

PROOF. The proof uses the Borel-Cantelli lemma and a well-known blocking technique. Let

$$M_n = \max_{2^{n-1} < k \leqslant 2^n} S_k .$$

From the submartingale inequality (4.8), we get that for all $t > 0$,

$$\mathbb{P}[M_n > t] \leqslant t^{-p} \mathbb{E}[|S_{2^n}|^p] .$$

Now, let $\varepsilon > 0$ and take

$$t = t_n = \left\{ n^{1+\varepsilon} \mathbb{E}[|S_{2^n}|^p] \right\}^{1/p} ,$$

whence

$$\sum_{n=1}^{\infty} \mathbb{P}[M_n > t_n] \leqslant \sum_{n=1}^{\infty} n^{-1-\varepsilon} < \infty ,$$

and so, by the Borel-Cantelli lemma,

$$\mathbb{P}[M_n > t_n \text{ infinitely often}] = 0 .$$

Then, M_n / t_n is almost surely bounded. Since for $2^{n-1} < k \leqslant 2^n$, the ratios $n^{1+\varepsilon} / (\log_2 k)^{1+\varepsilon}$ are bounded, uniformly in n and k, the theorem follows.   Q.e.d.

(4.13) THEOREM. Let $\frac{1}{2} < \kappa < 1$, and let $\int_{\mathbb{R}} |x|^\lambda f(x) \, dx < \infty$ for some $\lambda > 1$. Then, for all $\varepsilon > 0$,

$$\| g_\kappa * (dF_n - dF) \|_1 =_{\text{as}} \mathcal{O}\left(n^{-1/2} (\log n)^{1+\varepsilon} \right) . \quad /$$

PROOF. Let $S_n = n \| g_\kappa * (dF_n - dF) \|_1$. Then, by Exercise (4.1), $\{S_n\}_{n \geqslant 1}$ is a submartingale, and as in (2.17),

$$\mathbb{E}[S_n] \leqslant n^{1/2} \| (g_\kappa^2 * f)^{1/2} \|_1 .$$

Now, from Exercise (2.25), and in view of the asymptotic behavior of g_κ for $\kappa > \frac{1}{2}$, see Theorem (3.4),

$$\| (g_\kappa^2 * f)^{1/2} \|_1 \leqslant \text{const} \int_R |x|^\lambda g_\kappa^2 * f(x)\, dx$$

$$\leqslant \text{other const} \int_{\mathbb{R}} |x|^\lambda \{ (g_\kappa(x))^2 + f(x) \}\, dx$$

$$\leqslant \text{yet another const} ,$$

by assumption. Theorem (4.12) then implies $S_n =_{\text{as}} \mathcal{O}\big(n^{1/2} (\log n)^{1+\varepsilon} \big)$, and the theorem follows.   Q.e.d.

The above theorem is not too bad if one is willing to live with extra factors of $\log n$. The application of Theorem (4.12) to $\| A_h * (dF_n - dF) \|_1$ and $\mathrm{KL}(A_h * dF_n , A_h * dF)$ is a bit more involved and is postponed to the next section. In the remainder of this section, we get a much more precise result.

Exponential inequalities. Martingales have a very nice connection with exponential inequalities, and this may be fruitfully applied in the study of $\| g_\kappa * (dF_n - dF) \|_1$ and $\| A_h * (dF_n - dF) \|_1$. This was all spelled out beautifully in DEVROYE (1991). But wait a minute! Did we not show that the $n \| A_h * (dF_n - dF) \|_1$ formed a submartingale? Exercise (4.6) provides the clue: Under appropriate conditions, every random function may be associated with a martingale.

Exponential inequalities for a scalar random variable Y come about when $\mathbb{E}[\exp(\lambda Y)]$ is suitably bounded for all λ. The following lemma due to HOEFFDING (1963) seems to be the prototype.

(4.14) LEMMA. [HOEFFDING (1963)] *Let Y be a bounded random variable, with $a \leqslant Y \leqslant b$ and $\mathbb{E}[Y] = 0$. Then, for all $\lambda > 0$,*

$$\mathbb{E}[\exp(\lambda Y)] \leqslant \exp\big(\tfrac{1}{8} \lambda^2 (b - a)^2 \big) .$$

PROOF. Despite the generality of the lemma, it suffices to prove it for $\lambda = 1$. (Exercise!) We obviously have, for all t,

$$\mathbb{E}[\exp(Y)] = \mathbb{E}[-t Y + \exp(Y)] \leqslant \inf_t \ \sup_{a < y < b} \ \{ -t y + e^y \} .$$

Now, the function $y \longmapsto -t y + e^y$ is convex, so the supremum occurs at either $y = a$ or $y = b$, or both. Consequently,

$$\mathbb{E}[\exp(Y)] \leqslant \inf_t \ \max\{ -t a + e^a , -t b + e^b \} .$$

Graphical considerations show that the minimum over t occurs when the values at the two endpoints are equal, so $t = (e^b - e^a)/(b - a)$. Whether this value of t actually provides the minimum does not matter, the resulting bound is not going to be smaller than the infimum. Thus,

$$(4.15) \qquad \mathbb{E}[\exp(Y)] \leqslant \frac{b\,\mathrm{e}^a}{b-a} - \frac{a\,\mathrm{e}^b}{b-a} .$$

So far, so good. Now, we need to properly bound the right-hand side of (4.15). Let us rewrite this right-hand side as

$$\text{rhs} = p\,\mathrm{e}^{-qc} + q\,\mathrm{e}^{pc} ,$$

where

$$c = b - a , \quad p = \frac{b}{b-a} , \quad q = \frac{-a}{b-a} .$$

Note that c, p, and q are all positive, and $p+q = 1$. Since we want a bound in terms of c, one could try to maximize over p, and pretend c does not change when p changes. This, however, we are not able to do. Instead, consider the logarithm of "rhs" as a function of c, so

$$\varphi(c) = \log(p\,\mathrm{e}^{-qc} + p\,\mathrm{e}^{pc}) .$$

Graphing this for various values of p and q suggests that it is bounded above by a parabola. This is easily verified using differentiation. One verifies that

$$\varphi'(c) = \frac{pq\,(-\mathrm{e}^{-qc} + \mathrm{e}^{pc})}{p\,\mathrm{e}^{-qc} + q\,\mathrm{e}^{pc}} = \frac{pq\,(\mathrm{e}^c - 1)}{q\,\mathrm{e}^c + p}$$

and

$$\varphi''(c) = \frac{pq\,\mathrm{e}^c}{(p + q\,\mathrm{e}^c)^2} \leqslant \tfrac{1}{4} ,$$

the last inequality by the elementary inequality $4\,x\,y \leqslant (x + y)^2$.

Now, one verifies that $\varphi(0) = \varphi'(0) = 0$, so that from Taylor's theorem with exact remainder, for $c > 0$,

$$\varphi(c) = \varphi(0) + c\,\varphi'(0) + \int_0^c (c - \gamma)\,\varphi''(\gamma)\,d\gamma \leqslant \tfrac{1}{8}c^2 ,$$

and this is it. Q.e.d.

(4.16) EXERCISE. Show that the inequality (4.15) is in fact sharp, i.e., construct a random variable Y for which there is equality.

(4.17) EXERCISE. Come up with more intuitive proofs for bounds for (4.15) when $a = -b$. The inequality then is $\cosh(c) \leqslant \exp(\tfrac{1}{2}c^2)$.

On to exponential inequalities. We need a simple device, which is just Tchebyshev's inequality in disguise.

(4.18) EXERCISE. Let Y be a random variable. (a) Prove the Tchebyshev inequality, that is, $\mathbb{P}[\, Y > t \,] \leqslant t^{-1} \mathbb{E}[\, Y \vee 0 \,] \leqslant t^{-1} \mathbb{E}[\, |\, Y\,|\,]$, for all $t > 0$. (b) Also show the following inequality

$$\mathbb{P}[\, Y > t \,] \leqslant \mathrm{e}^{-\lambda t} \, \mathbb{E}[\, \mathrm{e}^{\lambda Y} \,] \,, \quad \text{for all } t > 0, \ \lambda > 0 \,.$$

[Hint: The event $\{\, Y > t \,\}$ is the same as $\{\, \exp(\lambda Y) > \exp(\lambda t) \,\}$.]

(4.19) THEOREM. [HOEFFDING (1963)] Let $\{S_n\}_{n \geqslant 1}$ be a martingale based on the independent random variables $X_1, X_2, \cdots$. Suppose that there exist random variables $T_n = \varphi_n(X_1, \cdots, X_n)$ and constants T_0 and c_n, such that

$$T_{n-1} \leqslant S_n \leqslant T_{n-1} + c_n \,, \quad n = 1, 2, \cdots \,.$$

Let $s_n^2 = \sum_{i=1}^{n} c_i^2$. Then, for all $n \geqslant 1$,

$$\mathbb{P}[\, |\, S_n - \mathbb{E}[S_1]\,| > s_n t \,] \leqslant 2\, \mathrm{e}^{-2t^2} \,.$$

PROOF. Let $S_0 = \mathbb{E}[\, S_1 \,]$. We may assume that $S_0 = 0$, since otherwise we proceed with the martingale $\{S_n - S_0\}_{n \geqslant 1}$. Let $Y_n = S_n - S_{n-1}$, $n \geqslant 1$. The martingale property then gives that $\mathbb{E}[\, Y_n \mid X_1, \cdots, X_{n-1} \,] = 0$, and $T_{n-1} - S_{n-1} \leqslant Y_n \leqslant T_{n-1} - S_{n-1} + c_n$, for all n. The starting point is Exercise (4.18), which gives

$$\mathbb{P}[\, S_n > t \,] \leqslant \mathrm{e}^{-\lambda t} \, \mathbb{E}[\, \mathrm{e}^{\lambda S_n} \,] \,.$$

Now, write $S_n = S_{n-1} + Y_n$, so that

$$\mathbb{E}[\, \mathrm{e}^{\lambda S_n} \,] = \mathbb{E}\Big[\, \mathrm{e}^{\lambda S_{n-1}} \, \mathbb{E}[\, \mathrm{e}^{\lambda Y_n} \mid X_1, \cdots, X_{n-1} \,] \,\Big] \,.$$

Since $\mathbb{E}[\, Y_n \mid X_1, \cdots, X_{n-1} \,] = 0$ and $0 \leqslant Y_n - T_{n-1} + S_{n-1} \leqslant c_n$, we may use HOEFFDING's Lemma (4.14) to bound the conditional expectation of Y_n, so that

$$\mathbb{E}[\, \mathrm{e}^{\lambda S_n} \,] \leqslant \mathrm{e}^{\frac{1}{8}\lambda^2 c_n^2} \, \mathbb{E}[\, \mathrm{e}^{\lambda S_{n-1}} \,] \,.$$

By induction, it follows that

$$\mathbb{E}[\, \mathrm{e}^{\lambda S_n} \,] \leqslant \mathrm{e}^{\frac{1}{8}\lambda^2 s_n^2} \,,$$

whence

$$\mathbb{P}[\, S_n > t \,] \leqslant \mathrm{e}^{-\lambda t + \frac{1}{8}\lambda^2 s_n^2} \,.$$

Minimization of the right-hand side of this inequality over λ gives

$$\mathbb{P}[\, S_n > t \,] \leqslant \mathrm{e}^{-2t^2/s_n^2} \,.$$

By symmetry, we get the same bound for $\mathbb{P}[\, -S_n > t \,]$. The theorem follows.   Q.e.d.

This is nice stuff, but how does it apply to $\| A_h * (dF_n - dF) \|_1$ or $\| g_\kappa * (dF_n - dF) \|_1$? The good news is of course that they are bounded random variables. The following theorem does the trick.

(4.20) THEOREM. [MCDIARMID (1989)] For $n \geqslant 1$, let $\psi : \mathbb{R}^n \longrightarrow \mathbb{R}$ be such that

$$\sup_{(i)} \big| \psi(x_1, x_2, \cdots, x_n) - \psi(y_1, y_2, \cdots, y_n) \big| \leqslant c_i ,$$

where the supremum is over all $x_1, \cdots, x_n$ and $y_1, \cdots, y_n$ with $x_j = y_j$ for all $j \neq i$. Let $X_1, X_2, \cdots, X_n$ be independent random variables, and set $S = \psi(X_1, X_2, \cdots, X_n)$. Set $s^2 = \sum_{i=1}^n c_i^2$. Then, for all $n \geqslant 1$,

$$\mathbb{P}[\,|S - \mathbb{E}[S]| > s\,t\,] < 2\,e^{-2t^2} .$$

PROOF. Define $S_k = \mathbb{E}[S \mid X_1, \cdots, X_k]$, for $k = 0, 1, \cdots, n - 1$, and set $S_n = S$. One verifies that $S_1, S_2, \cdots, S_n$ is a martingale, see Exercise (4.6). Finally, we need a bound on S_k. It suffices to get a bound on $\sup S_k - \inf S_k$, where the supremum and infimum are over X_k, since

$$\inf_{X_k} S_k - S_{k-1} \leqslant S_k - S_{k-1} \leqslant \Big\{ \inf_{X_k} S_k - S_{k-1} \Big\} + \gamma_k ,$$

in which

$$\gamma_k = \sup_{X_k} S_k - \inf_{X_k} S_k.$$

So here it goes. It is convenient to denote X_k in the supremum as Y and in the infimum as Z. We have

$$\gamma_k = \sup_{X_k} S_k - \inf_{X_k} S_k$$

$$= \sup_{Y,Z} \mathbb{E}[S \mid X_1, \cdots, X_{k-1}, Y] - \mathbb{E}[S \mid X_1, \cdots, X_{k-1}, Z]$$

$$= \int \big\{ \psi(X_1, \cdots, X_{k-1}, Y, y_{k+1}, \cdots, y_n) -$$

$$\psi(X_1, \cdots, X_{k-1}, Z, y_{k+1}, \cdots, y_n) \big\} \, dP(y_{k+1}, \cdots, y_n) ,$$

where P is the joint cdf of $X_{k+1}, \cdots, X_n$. It follows that $\gamma_k \leqslant c_k$. Now, we may apply Theorem (4.19) to prove the conclusion of the present theorem. Q.e.d.

How do we get almost sure bounds from this? The following theorem and proof answer the question. The first part is just an ever-so-slight reformulation of Theorem (4.20) that simplifies later applications.

(4.21) THEOREM. [MCDIARMID (1989)] *For $n \geqslant 1$, let $\psi_n : \mathbb{R}^n \longrightarrow \mathbb{R}$ be such that*

$$\sup_{(i)} \left| \psi_n(x_1, x_2, \cdots, x_n) - \psi_n(y_1, y_2, \cdots, y_n) \right| \leqslant c_{i,n} \, ,$$

where the supremum is over all $x_1, \cdots, x_n$ and $y_1, \cdots, y_n$ with $x_j = y_j$ for all $j \neq i$. Let $X_1, X_2, \cdots$ be independent random variables, and set $S_n = \psi_n(X_1, X_2, \cdots, X_n)$. Set $s_n^2 = \sum_{i=1}^{n}(c_{i,n})^2$. Then,

$$\mathbb{P}[\, |S_n - \mathbb{E}[S_n]| > s_n t \,] < 2\,e^{-2t^2}$$

and

$$\limsup_{n \to \infty} (2s_n \log n)^{-1/2} \left| S_n - \mathbb{E}[S_n] \right| \leqslant_{as} 1 \, .$$

PROOF. Evidently, only the last statement needs proving. Let $\varepsilon > 0$ be arbitrary. Taking

$$t = t_n = \left\{ \tfrac{1}{2}(1 + \varepsilon) \log n \right\}^{1/2}$$

in Theorem (4.20) gives

$$\mathbb{P}[\, |S_n - \mathbb{E}[S_n]| > s_n t_n \,] < 2\,n^{-1-\varepsilon} \, .$$

By the Borel-Cantelli lemma, then

$$\mathbb{P}[\, |S_n - \mathbb{E}[S_n]| > s_n t_n \text{ infinitely often}\,] = 0 \, ,$$

and so

$$\limsup_{n \to \infty} (2s_n \log n)^{-1/2} \left| S_n - \mathbb{E}[S_n] \right| \leqslant_{as} (1 + \varepsilon)^{1/2} \, .$$

Since $\varepsilon > 0$ was arbitrary, this proves the last statement of the theorem. Q.e.d.

   We are almost there. If we want to apply this theorem to obtain almost sure bounds for $\| g_\kappa * (dF_n - dF) \|_1$ and $\| A_h * (dF_n - dF) \|_1$, the conditions must be verified. Let k be an arbitrary pdf and define $\psi_n : \mathbb{R}^n \to \mathbb{R}$ by

$$\psi_n(X_1, X_2, \cdots, X_n) = \| k * (dF_n - dF) \|_1 \, .$$

Then, ψ_n is a symmetric function of $X_1, X_2, \cdots, X_n$ (we may scramble the order of the X_i without changing the value ψ_n), so to verify MCDIARMID's condition, it suffices to just vary X_1, say. Again, we replace X_1 by Y and Z in MCDIARMID's condition. Let G_n be the empirical distribution of

$Y, X_2, \cdots, X_n$, and $\widetilde{G}_n$ the empirical distribution of $Z, X_2, \cdots, X_n$. Then,

$$\left| \psi_n(Y, X_2, \cdots, X_n) - \psi_n(Z, X_2, \cdots, X_n) \right|$$
$$= \left| \, \| k * (dG_n - dF) \|_1 - \| k * (d\widetilde{G}_n - dF) \|_1 \, \right|$$
$$\leqslant \| k * (dG_n - d\widetilde{G}_n) \|_1 = \tfrac{1}{n} \| k(\cdot - Y) - k(\cdot - Z) \|_1$$
$$\leqslant \tfrac{1}{n} \left(\| k(\cdot - Y) \|_1 + \| k(\cdot - Z) \|_1 \right)$$
$$\leqslant \tfrac{2}{n} \, .$$

Upon application of Theorem (4.21), we get the following.

(4.22) THEOREM. [DEVROYE (1991)] *If k is a pdf, then*

$$\mathbb{P}\left[\left| \, \| k * (dF_n - dF) \|_1 - \mathbb{E}[\, \| k * (dF_n - dF) \|_1\,] \, \right| > t \right] \leqslant 2\mathrm{e}^{-\frac{1}{2}n t^2} \, ,$$

$$\limsup_{n \to \infty} \, (n/\log n)^{1/2} \left| \, \| k * (dF_n - dF) \|_1 - \mathbb{E}[\, \| k * (dF_n - dF) \|_1\,] \, \right| \leqslant_{\mathrm{as}} 1 \, .$$

(4.23) EXERCISE. Prove the above theorem.

(4.24) EXERCISE. What happens in the above if $k \in L^1(\mathbb{R})$ is not necessarily a pdf?

This is the last piece of the puzzle. We now use Theorem (4.21) to obtain bounds on $\| g_\kappa * (dF_n - dF) \|_1$ and $\| A_h * (dF_n - dF) \|_1$, but only for deterministic h. Actually, it suffices to do so for the latter. So consider deterministic h varying with the sample size n, say, $h \asymp n^{-1/3}$ or $\asymp n^{-1/5}$. Recall the notations $f^{nh} = A_h * dF_n$ and $f_h = A_h * dF$. From the triangle inequality,

$$\| f^{nh} - f \|_1 \leqslant \| f_h - f \|_1 + \mathbb{E}[\, \| f^{nh} - f_h \|_1\,] +$$
$$\left| \, \| f^{nh} - f_h \|_1 - \mathbb{E}[\, \| f^{nh} - f_h \|_1\,] \, \right| \, ,$$

and with Theorem (2.9) to bound the bias term, Theorem (2.21) to bound the expected value, and Theorem (4.22) to bound the deviation from the expected value, we get under suitable conditions that

$$\| f^{nh} - f \|_1 =_{\mathrm{as}} \mathcal{O}\left(h^2 + (nh)^{-1/2} + (n^{-1} \log n)^{1/2} \right) \, .$$

Finally, the (deterministic) choice $h \asymp n^{-1/5}$ gives $\| f^{nh} - f \|_1 =_{\mathrm{as}} \mathcal{O}(n^{-2/5})$. We have proven the following result.

(4.25) THEOREM. *Suppose A satisfies (2.2) and (2.4).*
(a) *If f has a finite moment of order > 1 [condition (2.7)], then*

$$\| f^{nh} - f_h \|_1 =_{\mathrm{as}} \mathcal{O}\left((nh)^{-1/2} + (n^{-1} \log n)^{1/2} \right) \, .$$

(b) *If, in addition, $f \in W^{2,1}(\mathbb{R})$ [condition (2.6)], then for $h \asymp n^{-1/5}$ (deterministic),*

$$\| f^{nh} - f \|_1 =_{as} \mathcal{O}(n^{-2/5}) .$$

(4.26) EXERCISE. Repeat the above for $\| g_\kappa * (dF_n - dF) \|_1$, with $0 < \kappa < 1$.

(4.27) EXERCISE. Use Exercise (4.24) to prove the almost sure L^1 rates of convergence for kernel density estimation with random smoothing parameter h_n, which satisfies $c_1 n^{-1/5-\varepsilon} \leqslant_{as} h_n \leqslant_{as} c_2 n^{-1/5+\varepsilon}$ for all $\varepsilon > 0$ for suitable positive constants c_1, c_2.

Theorem (4.25) is not subject to improvement, in the sense that the rate $n^{-2/5}$ cannot be improved (after all, it is the exact rate for the expected value, see DEVROYE and GYÖRFI (1985)). Can MCDIARMID's Theorem (4.20) be made to work for the other distances? For Kullback-Leibler and Pearson's φ^2 distances, the answer appears to be no. The reason is that these distances are not *a priori* bounded (let alone finite), but MC-DIARMID's condition on the ψ function implies boundedness. See Exercise (4.28). For the Hellinger distance, the corresponding ψ function satisfies a MCDIARMID condition, since Hellinger distance is bounded by L^1 distance, but that is not good enough, because then the quadratic character is lost. See Exercise (4.29).

(4.28) EXERCISE. Let $(y_1, y_2, \cdots, y_n) \in \mathbb{R}$ be fixed. If $\psi(y_1, y_2, \cdots, y_n)$ is finite, and ψ satisfies the MCDIARMID condition, then for all realizations of $X_1, X_2, \cdots, X_n$,

$$| \psi(X_1, X_2, \cdots, X_n) | \leqslant | \psi(y_1, y_2, \cdots, y_n) | + \sum_{i=1}^n c_{i,n} .$$

(4.29) EXERCISE. (a) Apply Theorem (4.20) to the Hellinger distance

$$\psi(X_1, X_2, \cdots, X_n) = \| (f^{nh})^{1/2} - (f_h)^{1/2} \|_2^2 .$$

(b) Assuming that $\mathbb{E}[\| (f^{nh})^{1/2} - (f_h)^{1/2} \|_2^2] = \mathcal{O}((nh)^{-1} \log(nh)^{-1})$, use the results of part (a) to give a.s. bounds on the Hellinger distance. (The bound on the expected values holds under suitable conditions on f, as shown in §6 on entropy estimation.)
(c) Does it help to consider the square root of the Hellinger distance?

As a final comment, we note that Theorem (4.22) implies that the kernel estimator satisfies

$$\| A_h * dF_n - f \|_1 =_{as} \mathbb{E}[\| A_h * dF_n - f \|_1] + \mathcal{O}((n^{-1} \log n)^{-1/2}) .$$

and so, if the expected value is asymptotically larger than $(n^{-1} \log n)^{-1/2}$, this proves the asymptotic stability of the kernel estimator, i.e.,

$$(4.30) \qquad \limsup_{n \to \infty} \frac{\| A_h * dF_n - f \|_1}{\mathbb{E}[\| A_h * dF_n - f \|_1]} = 1 .$$

For a precise treatment under minimal conditions, see DEVROYE (1988).

This concludes our treatment of martingales and submartingales, and the L^1 distance.

EXERCISES : (4.3), (4.5), (4.6), (4.7), (4.9), (4.10), (4.11), (4.16), (4.17), (4.18), (4.23), (4.24), (4.26), (4.27), (4.28), (4.29)

5. Almost sure bounds for everything else

Well, everything else means just about everything else having to do with kernel estimators that has not been done in the previous section. The main effort there was in getting bounds for $\| f^{nh} - f_h \|_1$. Here, we want to get bounds for other distances between f^{nh} and f_h, the main reason being that this will be useful later on. The other distances of interest are the Hellinger, Kullback-Leibler, and Pearson's φ^2 distances, see (1.3.18)–(1.3.20). The Hellinger distance is not covered well by the exponential inequalities framework, and Kullback-Leibler and Pearson's φ^2 distances are not covered at all. Recourse must be had to submartingales and submartingale inequalities. The funny thing is that the "everything else" now comes down to the study of the Kullback-Leibler distance, because we have trouble with the expected values of the Pearson's φ^2 and Hellinger distances. We are not really able to handle the Hellinger distance, other than by bounding it with Kullback-Leibler, but since Kullback-Leibler is not even *a priori* bounded, surely something is lost in the process. The Pearson's φ^2 distance is a different story. Of course, with $(A^2)_h(x) = h^{-1}\{A(h^{-1}x)\}^2$,

$$(5.1) \qquad \mathbb{E}[\, \text{PHI}(f^{nh}, f_h)\,] = (nh)^{-1} \int_{\mathbb{R}} \frac{(A^2)_h * f - h(A_h * f)^2}{A_h * f} ,$$

but that is where it ends. We really do not know how to bound this under reasonable conditions on the the kernel and the pdf, because the tails are quite influential due to the division by $A_h * f$. See Exercise (5.25).

So in the remainder of this section, we study the Kullback-Leibler distance for deterministic smoothing parameter, by means of submartingales and submartingale inequalities. But immediately, there is a snag because of the occurrence of the smoothing parameter h. If there are submartingales to be found, then surely only *with h fixed*! But we definitely want h to vary with n. Fortunately, the true and tested blocking technique already used in the proof of Theorem (4.12) provides the solution.

We formulate the general approach. For every (fixed) $h > 0$, suppose that $S_n(h)$, $n = 1, 2, \cdots$, is a positive submartingale, and let

$$(5.2) \qquad M_n(h) = \max_{2^{n-1} < k \leqslant 2^n} S_k(h) .$$

Now, DOOB's submartingale inequality (4.8) says that

$$\mathbb{P}[\, M_n(h) > t \,] \leqslant t^{-1} \, \mathbb{E}[\, S_{2^n}(h) \,] ,$$

for all n and h. Note that we may take h dependent on n, denoted as $h = H_n$, but the effect is that $h = h_k$ is constant on blocks $2^{n-1} < k \leqslant 2^n$. In other words, for these k, we have $h_k = H_n$. Be that as it may, similar to the proof of Theorem (4.12), it follows with $t = t_n = n^{1+\varepsilon} \, \mathbb{E}[\, S_{2^n}(H_n) \,]$ that

$$\mathbb{P}[\, M_n(H_n) > t_n \text{ infinitely often} \,] = 0 ,$$

and so $M_n(H_n) =_{\text{as}} \mathcal{O}(t_n)$. In terms of $S_k(h)$, this reads as

$$(5.3) \qquad S_k(h_k) =_{\text{as}} \mathcal{O}\big(n^{1+\varepsilon} \, \mathbb{E}[\, S_{2^n}(h_{2^n}) \,] \big) .$$

This is as far as it goes in general. But we now observe that it is no great loss to assume that $\{H_n\}_n$ is a decreasing (nonincreasing) sequence, at least when considering deterministic sequences. To proceed, we make what seems an outrageous assumption:

$$(5.4) \qquad \begin{array}{l} \text{For every realization of } X_1, X_2, \cdots, X_n \text{ , the function} \\ S_n(h) = \psi_n(h\,;\, X_1, X_2, \cdots, X_n) \\ \text{is decreasing in } h > 0 . \end{array}$$

Now, for decreasing $\{H_n\}$, redefine $\{h_k\}_k$ to be any decreasing sequence such that $h_{2^n} = H_n$. Then, $S_k(h_k) \leqslant S_k(H_{n-1})$, for $2^{n-1} < k \leqslant 2^n$, by the decreasing assumption (5.4), and we get from (5.3) that

$$S_k(h_k) =_{\text{as}} \mathcal{O}\big(n^{1+\varepsilon} \, \mathbb{E}[\, S_{2^n}(h_{2^{n-1}}) \,] \big) ,$$

for every deterministic, decreasing sequence $\{h_k\}_k$. With the observation that the expected values are increasing with h, we have proved the following theorem.

(5.5) THEOREM. *For each fixed $h > 0$, let $\{S_n(h)\}_n$ be a submartingale, and suppose $S_n(h)$ is a decreasing function of h, in the sense of (5.4). Then, for all $\varepsilon > 0$ and for every deterministic, decreasing sequence $\{h_n\}_n$,*

$$S_n(h_n) =_{\text{as}} \mathcal{O}\big((\log n)^{1+\varepsilon} \, \mathbb{E}[\, S_{2n}(h_n) \,] \big) .$$

Note that the bound on $S_n(h_n)$ in terms of $S_{2n}(h)$, with $h = h_n$, and not with $h = h_{2n}$. However, since in the applications we typically have

$$\mathbb{E}[\, S_n(h) \,] = \mathcal{O}\big(n^{-\alpha} \big) \quad \text{for} \quad h \asymp n^{-\beta} ,$$

for suitable α, β between 0 and 1, this is hardly a drawback. Also note that the decreasing assumption could be replaced by the assumption that there exists a constant c such that for every realization of $X_1, X_2, \cdots, X_n$, and every $\lambda < h$,

$$(5.6) \qquad\qquad S_n(h) \leqslant c\, S_n(\lambda) \; .$$

This actually seems reasonable. Moreover, the constant should depend on the kernel A only.

(5.7) EXERCISE. Prove Theorem (5.5) if the assumption (5.4) is replaced by the condition (5.6).

Kullback-Leibler and submartingales. We now apply Theorem (5.5) to the Kullback-Leibler distance. For reasons to do with assumption (5.4), we choose the scaled two-sided exponential kernel $A_h = \mathcal{B}_h$, where

$$(5.8) \qquad\quad \mathcal{B}_h(x) = (2h)^{-1} \exp(-h^{-1}|x|) \; , \quad x \in \mathbb{R} \; .$$

Apparently, we must show that the Kullback-Leibler distance "is" a submartingale and verify the other conditions of the theorem. The Kullback-Leibler distance $\mathrm{KL}(f^{nh}, f_h)$ is a convex function of f^{nh}, but unfortunately not an increasing one, cf. Lemma (4.4). However, $\mathrm{KL}(f^{nh}, f_h)$ may be "decomposed" as

$$(5.9) \qquad \mathrm{KL}(f^{nh}, f_h) = \mathrm{KL}(f^{nh} \wedge f_h \,,\, f_h) + \mathrm{KL}(f^{nh} \vee f_h \,,\, f_h) \;,$$

which saves the day. Here, for any two functions φ and ψ on the line, $\varphi \vee \psi(x) = \max(\varphi(x), \psi(x))$ and $\varphi \wedge \psi(x) = \min(\varphi(x), \psi(x))$. We will prove the following result by elementary means.

(5.10) LEMMA. *For fixed $h > 0$, the sequence $\{\, n^2\, \mathrm{KL}(f^{nh} \vee \bar{f}_h \,,\, f_h)\,\}_n$ is a submartingale.*

The other part $\mathrm{KL}(f^{nh} \wedge f_h \,,\, f_h)$ is not a submartingale even after suitable scaling, it appears. But we have the inequality $\mathrm{KL}(\varphi, \psi) \leqslant \mathrm{PHI}(\varphi, \psi)$, with PHI the Pearson's φ^2 distance, and, indeed, again by elementary means, we have the following lemma.

(5.11) LEMMA. *For fixed $h > 0$, the sequence $\{\, n^2\, \mathrm{PHI}(f^{nh} \wedge f_h \,,\, f_h)\,\}_n$ is a submartingale.*

We postpone the proofs until the end of the section. Next, given the lemmas, we must find the expected values of these submartingales. Obviously, $\mathrm{KL}(f^{nh} \vee f_h \,,\, f_h) \leqslant \mathrm{KL}(f^{nh}, f_h)$, in view of (5.9), so it suffices to compute $\mathbb{E}[\,\mathrm{KL}(f^{nh}, f_h)\,]$. The other submartingale seems to be much more troublesome, because of its relation to the plain Pearson's φ^2 distance, which as discussed earlier is a nasty customer. However, for any two nonnegative functions φ, ψ, we have

$$\text{PHI}(\varphi \wedge \psi, \psi) = \int_{\mathbb{R}} \frac{|\varphi \wedge \psi - \psi|^2}{\psi} = \int_{\mathbb{R}} |(\varphi \wedge \psi)^{1/2} - \psi^{1/2}|^2 W^2 ,$$

where

$$W = \frac{(\varphi \wedge \psi)^{1/2} + \psi^{1/2}}{\psi^{1/2}} .$$

Obviously, $1 \leqslant W \leqslant 2$, so that

(5.12) $\text{H}(\varphi \wedge \psi, \psi) \leqslant \text{PHI}(\varphi \wedge \psi, \psi) \leqslant 4 \text{H}(\varphi \wedge \psi, \psi) \leqslant 4 \text{KL}(\varphi \wedge \psi, \psi) .$

The very last inequality was already mentioned in (1.3.21). Since

$$\text{PHI}(f^{nh} \wedge f_h, f_h) \leqslant 4 \text{KL}(f^{nh}, f_h) ,$$

bounding $\mathbb{E}[\text{PHI}(f^{nh} \wedge f_h, f_h)]$ can be done by computing $\mathbb{E}[\text{KL}(f^{nh}, f_h)]$. Moreover, not much has been lost during all of these bounding steps. The original sequence $\{n^2 \text{KL}(f^{nh}, f_h)\}_n$ (not apparently a submartingale) is dominated by two submartingales, each of which is dominated by a fixed multiple of the original sequence. So the expected values are off by at most a constant factor (a factor of 4, actually), and nothing is lost in the way of *rates* of convergence.

The expected value of the Kullback-Leibler distance. Write the Kullback-Leibler distance $\text{KL}(f^{nh}, f_h)$ as

$$\text{KL}(f^{nh}, f_h) = \tfrac{1}{n} \sum_{i=1}^{n} \int_{\mathbb{R}} A_h(x - X_i) L \, dx ,$$

where L is shorthand for $\log\{f^{nh}(x)/f_h(x)\}$. Then, by conditioning,

$$\mathbb{E}[\text{KL}(f^{nh}, f_h)] = \tfrac{1}{n} \sum_{i=1}^{n} \int_{\mathbb{R}} \mathbb{E}[A_h(x - X_i) \mathbb{E}[L \mid X_i]] \, dx$$

$$= \int_{\mathbb{R}} \mathbb{E}[A_h(x - X_1) \mathbb{E}[L \mid X_1]] \, dx ,$$

the last equality by the fact that the X_i are iid. Now, since the logarithm is a concave function, Jensen's inequality gives

$$\mathbb{E}[L \mid X_1] \leqslant \log \mathbb{E}[\{f^{nh}(x)/f_h(x)\} \mid X_1] =$$

$$\leqslant \log \frac{(n-1) f_h(x) + A_h(x - X_1)}{n f_h(x)}$$

$$\leqslant \log\Big(1 + \frac{\varepsilon}{f_h(x)}\Big) ,$$

where $\varepsilon = (nh)^{-1} \|A\|_\infty$. This yields the nice bound

(5.13) $\mathbb{E}[\text{KL}(f^{nh}, f_h)] \leqslant \int_{\mathbb{R}} f_h \log\Big(1 + \frac{\varepsilon}{f_h}\Big) ,$

with $\varepsilon = (nh)^{-1} \| A \|_\infty$. We proceed to bound it further, using the elementary inequality

(5.14)
$$\log(1 + t) \leqslant 2\sqrt{t}, \quad t > 0.$$

Applying (5.14) to the right-hand side of (5.13) gives

$$\mathbb{E}[\,\mathrm{KL}(f^{nh}, f_h)\,] \leqslant 2\,\varepsilon^{1/2} \, \| f_h^{1/2} \|_1 .$$

Now, recall our dealings with $\| f_h^{1/2} \|_1$ from §2, Exercise (2.25), and the following result emerges.

(5.15) LEMMA. *If f and A have moments of order $\lambda > 1$, then there exists a constant $c(\lambda)$ such that*

$$(nh)^{1/2}\,\mathbb{E}[\,\mathrm{KL}(f^{nh}, f_h)\,] \leqslant c(\lambda)\,\{\,\mathbb{E}[\,|X|^\lambda\,] + h^2\,\mathbb{E}[\,|Y|^\lambda\,]\,\}^{1/(2\lambda)} ,$$

where X has pdf f and Y has pdf A.

Is this bound for $\mathbb{E}[\,\mathrm{KL}(f^{nh}, f_h)\,]$ any good? In view of the quadratic nature of the Kullback-Leibler distance, perhaps not, but it has its uses. Either way, the lemma can be improved on under a general moment condition, or even an exponential moment condition.

(5.16) LEMMA. *Let $\alpha > 1$. If f and A have finite moments of some order $> 1/(\alpha - 1)$, then for some constant $c(\alpha)$ (also dependent on f and A),*

$$\mathbb{E}[\,\mathrm{KL}(f^{nh}, f_h)\,] \leqslant c(\alpha)\,(nh)^{-1/\alpha} .$$

If there exist a $t > 0$ such that $\mathbb{E}[e^{t\,|X|}] < \infty$, and the kernel A has a finite exponential moment, then for some constant c (dependent on t),

$$\mathbb{E}[\,\mathrm{KL}(f^{nh}, f_h)\,] \leqslant c\,(nh)^{-1}\,\log(nh) .$$

(5.17) EXERCISE. Let $1 < p < \infty$, and define q by $(1/p) + (1/q) = 1$.
(a) Show that $\log(1 + t) \leqslant p\,t^{1/p}$ for all $t > 0$.
(b) Show that $\mathbb{E}[\,\mathrm{KL}(f^{nh}, f_h)\,] \leqslant p\,\varepsilon^{1/p}\,\| (f_h)^{1/q} \|_1$, with ε as before.
(c) Show that if the random variable X has pdf f, then for all $\lambda > q/p$, there is a (finite) constant $c(\lambda)$ such that

$$\| f^{1/q} \|_1 \leqslant c(\lambda)\,\{\,\mathbb{E}[\,|X|^\lambda\,]\,\}^{1/(\lambda p)} .$$

(d) Show that $\| f^{1/q} \|_1^q \leqslant (p - 1)^{q/p}\,\mathbb{E}[e^{|X|}]$ for all $q > 1$.
[Hint: Use Hölder's inequality.]
(e) Prove Lemma (5.16). [Hint: Compare with Lemma (2.18) and Exercises (2.24), (2.25). For the exponential moment condition, use the bound from (d), and define p such that $(p - 1)^{-1} = \log(nh)$.]

Would it be possible to get a bound $\mathbb{E}[\,\mathrm{KL}(f^{nh}, f_h)\,] \leqslant c\,(nh)^{-1}$ under suitable conditions? The answer is yes, but under extreme conditions on the tail behavior of f. See § 6.

The monotonicity as a function of the smoothing parameter. Before we can actually apply Theorem (4.12), we must verify the monotonicity condition (5.4). It is here that the choice of the two-sided exponential smoother comes in. Actually, assumption (5.4) turns out to be true also for the normal and the Cauchy smoother. See Exercise (5.22). The main ingredient of the proof is the well-known inequality

$$(5.18) \qquad \mathrm{KL}(A * \varphi, A * \psi) \leqslant \mathrm{KL}(\varphi, \psi), \quad \text{for every pdf } \varphi, \psi, A.$$

This depends (solely) on the convexity of $\mathrm{KL}(\varphi, \psi)$ in φ and ψ jointly, see Exercise (9.1.25). The other ingredient is expressed by (5.20) below, which says that for $\lambda > h$, the pdf $\mathfrak{B}_\lambda$ is divisible by $\mathfrak{B}_h$.

(5.19) LEMMA. *Let F and G be two cdfs. If $\mathrm{KL}(\mathfrak{B}_h * dF, \mathfrak{B}_h * dG)$ is finite for $h = h_o$, then it is finite for all $h > h_o$ and is a decreasing function of h.*

PROOF. It is easily verified using Fourier transforms that

$$(5.20) \qquad \mathfrak{B}_\lambda = (h/\lambda)^2\, \mathfrak{B}_h + [\,1 - (h/\lambda)^2\,]\, \mathfrak{B}_\lambda * \mathfrak{B}_h.$$

For $\lambda > h$, the right-hand side is a convex combination of $\mathfrak{B}_h$ and $\mathfrak{B}_\lambda * \mathfrak{B}_h$, and it follows by Jensen's inequality and (5.18) that

$$
\begin{aligned}
\mathrm{KL}(\mathfrak{B}_\lambda * dF, \mathfrak{B}_\lambda * dG) \leqslant & \\
\leqslant\ & (h/\lambda)^2\, \mathrm{KL}(\mathfrak{B}_h * dF, \mathfrak{B}_h * dG) \\
& + [\,1 - (h/\lambda)^2\,]\, \mathrm{KL}(\mathfrak{B}_\lambda * \mathfrak{B}_h * dF, \mathfrak{B}_\lambda * \mathfrak{B}_h * dG) \\
\leqslant\ & (h/\lambda)^2\, \mathrm{KL}(\mathfrak{B}_h * dF, \mathfrak{B}_h * dG) \\
& + [\,1 - (h/\lambda)^2\,]\, \mathrm{KL}(\mathfrak{B}_h * dF, \mathfrak{B}_h * dG) = \\
\leqslant\ & \mathrm{KL}(\mathfrak{B}_h * dF, \mathfrak{B}_h * dG),
\end{aligned}
$$

thus, proving the lemma.   Q.e.d.

With Theorem (4.12) and Exercise (5.17), we now get a.s. bounds for the Kullback-Leibler distance.

(5.21) THEOREM. *Let $\alpha > 1$. If f has a finite moment of order $> \dfrac{1}{\alpha - 1}$, then for all $\varepsilon > 0$,*

$$\mathrm{KL}(f^{nh}, f_h) =_{\mathrm{as}} \mathcal{O}\big((nh)^{-1/\alpha} (\log n)^{1+\varepsilon}\big),$$

$$\mathrm{H}(f^{nh}, f_h) =_{\mathrm{as}} \mathcal{O}\big((nh)^{-1/\alpha} (\log n)^{1+\varepsilon}\big).$$

Moreover, if f and the kernel A have a finite exponential moment, then for all $\varepsilon > 0$,

$$\mathrm{KL}(f^{nh}, f_h) =_{\mathrm{as}} \mathcal{O}\big((nh)^{-1} \log((nh)^{-1}) (\log n)^{1+\varepsilon} \big) .$$

The monotonicity property (5.4) does not appear to be true for arbitrary pdfs A, such as the Epanechnikov kernel, but it seems reasonable that a relaxed version of (5.4) should hold (but the authors do not know how to show it).

(5.22) EXERCISE. (a) Let A be the normal kernel or the Cauchy kernel and as usual, set $A_h(x) = h^{-1}A(h^{-1}x)$, for all x. Show for all $\lambda > h$ that there exists a $\mu > 0$ such that $A_\lambda = A_\mu * A_h$.
(b) Prove Lemma (5.19) for the normal kernel and for the Cauchy kernel.

(5.23) EXERCISE. For the analysis of the estimator of GOOD (1971) in Chapter 5, we also need $\mathrm{KL}(T_h * dF_n , T_h * dF)$, where T_h is the one-sided exponential distribution, that is, $T_h(x) = h^{-1}\exp(-h^{-1}x)$ for $x > 0$ and $= 0$ for $x < 0$. Reformulate Theorem (5.21) to cover this case. In particular, prove the analogue of (5.20): For $\lambda, h > 0$,

$$T_\lambda = (h/\lambda)\, T_h + [\,1 - (h/\lambda)\,]\, T_\lambda * T_h .$$

The above covers the Kullback-Leibler distance $\mathrm{KL}(f^{nh} , f_h)$ as well as one could hope. What about $\mathrm{KL}(f^{nh} , f)$ or $\mathrm{KL}(f , f^{nh})$? In the next section, we give bounds for $\mathrm{KL}(f , f_h)$, but since Kullback-Leibler does not satisfy anything resembling a triangle inequality, this is not sufficient to get bounds on either one of the two Kullback-Leibler distances between f^{nh} and f. We shall not explore this further. As said before, we cannot really deal with Hellinger distance, other than the bound $\mathrm{H}(f^{nh}, f_h) \leqslant \mathrm{KL}(f^{nh}, f_h)$. Exercise (5.24) deals with a.s. bounds on $\mathrm{H}(f^{nh} , f)$, but they are "almost surely" not the best. For Pearson's φ^2 distance, we are definitely in trouble. Perhaps one should not expect to get reasonable bounds for this. At any rate, the authors cannot get them. See Exercise (5.25).

(5.24) EXERCISE. (a) State conditions on f such that $\mathrm{H}(f_h, f) = \mathcal{O}(h^4)$.
(b) If, in addition to the conditions of part (a), the density f has a finite moment of order $> 1/(\alpha - 1)$, show that

$$\mathrm{H}(f^{nh}, f) = \mathcal{O}\big((nh)^{-1/\alpha}(\log n)^{1+\varepsilon} + h^4 \big) .$$

(c) Determine the asymptotic order of $\min_h \mathrm{H}(f^{nh}, f)$.

(5.25) EXERCISE. (a) Show that (for arbitrary kernel A)

$$\mathbb{E}[\,\mathrm{PHI}(f^{nh}, f_h)\,] = (nh)^{-1} \int_{\mathbb{R}} \frac{(A^2)_h * f - h(A_h * f)^2}{A_h * f} .$$

(x) And now what? The authors do not know!
(b) To show how bad (a) is, let A be the uniform pdf, and suppose $f > 0$ everywhere. Show that the expected value is ∞.

We finish with the promised proofs of the two lemmas.

PROOF OF LEMMA (5.10). Let $S_n = \text{KL}(\varphi_n, f_h)$, where $\varphi_n = f^{nh} \vee f_h$. We need to show that $\{n^2 S_n\}_n$ is a submartingale, that is,

$$(n+1)^2 \, \mathbb{E}[\, S_{n+1} \mid X_1, X_2, \cdots, X_n \,] \geqslant n^2 \, S_n \ .$$

To simplify notation, let $\mathbb{E}_n \equiv \mathbb{E}[\, \cdot \mid X_1, X_2, \cdots, X_n \,]$. By the convexity of $\text{KL}(\varphi, f_h)$ as function of φ, Jensen's inequality yields

$$\mathbb{E}_n[\, S_{n+1} \,] \geqslant \text{KL}(\mathbb{E}_n[\, \varphi_{n+1} \,], f_h) \ .$$

Now,

$$\mathbb{E}_n[\, \varphi_{n+1} \,] = \mathbb{E}_n[\, f^{n+1,h} \vee f_h \,] \geqslant \left(\mathbb{E}_n[\, f^{n+1,h} \,] \right) \vee f_h = g_n \vee f_h \ ,$$

where

$$g_n \stackrel{\text{def}}{=} \theta \, f^{nh} + (1-\theta) \, f_h \ ,$$

with $\theta = n/(n+1)$. Since $\text{KL}(p, q)$ is an increasing function of p for $p > q$, it follows that

$$\mathbb{E}_n[\, S_{n+1} \,] \geqslant \text{KL}\big(\mathbb{E}_n[\, \varphi_{n+1} \,], f_h\big) \geqslant \text{KL}(g_n \vee f_h, f_h) \ .$$

Next, we observe that $\big(\theta \, f^{nh} + (1-\theta) \, f_h \big) > f_h$ precisely on the set where $f^{nh} > f_h$, so that

$$\mathbb{E}_n[\, S_{n+1} \,] \geqslant \int_{\{\, f^{nh} > f_h \,\}} g_n \log \frac{g_n}{f_h} + f_h - g_n \ .$$

Now, use the elementary inequality, for $r > 1$,

(5.26)
$$\frac{(\theta r + 1 - \theta) \, \log(\theta r + 1 - \theta) + 1 - (\theta r + 1 - \theta)}{r \, \log r + 1 - r} \geqslant \theta^2 \ ,$$

with $r = f^{nh}/f_h$, to obtain

$$\mathbb{E}_n[\, S_{n+1} \,] \geqslant \theta^2 \int_{\{\, f^{nh} > f_h \,\}} f^{nh} \log \frac{f^{nh}}{f_h} + f_h - f^{nh} = \theta^2 \, \text{KL}(f^{nh} \vee f_h, f_h) \ ,$$

which concludes the proof.   Q.e.d.

NOTE. Inequality (5.26) fails for $0 < r < 1$, or alternatively, the lower bound occurs for $r = 0^+$, which gives a bound of $\theta + (1-\theta) \log(1-\theta)$, but this is not usable, it appears. This necessitated the decomposition (5.9).

PROOF OF LEMMA (5.11). With $\mathbb{E}_n$ as in the previous proof, the convexity of $\mathrm{PHI}(\varphi, f_h)$ as function of φ gives that

$$\mathbb{E}_n[\mathrm{PHI}(f^{n+1,h} \wedge f_h, f_h)] \geq \mathrm{PHI}(\mathbb{E}_n[f^{n+1,h} \wedge f_h], f_h) .$$

Now, we have

$$\mathbb{E}_n[f^{n+1,h} \wedge f_h] \leq (\mathbb{E}_n[f^{n+1,h}]) \wedge f_h = g_n \wedge f_h ,$$

with $g_n \stackrel{\text{def}}{=} \theta f^{nh} + (1 - \theta) f_h$ and $\theta = n/(n+1)$. Next, observe that for $x < y < z$, we have $|x - z| > |y - z|$, to conclude that

$$\mathrm{PHI}(\mathbb{E}_n[f^{n+1,h} \wedge f_h], f_h) \geq \mathrm{PHI}(g_n, f_h) .$$

Again, the set where $g_n < f_h$ is precisely the set where $f^{nh} < f_h$, so that

$$\mathrm{PHI}(g_n, f_h) = \int_{\{f^{nh} < f_h\}} \frac{|\theta f^{nh} + (1-\theta) f_h - f_h|^2}{f_h}$$

$$= \theta^2 \int_{\{f^{nh} < f_h\}} \frac{|f^{nh} - f_h|^2}{f_h} = \theta^2 \, \mathrm{PHI}(f^{nh} \wedge f_h, f_h) .$$

Putting all this together shows that

$$\mathbb{E}_n[\mathrm{PHI}(f^{n+1,h} \wedge f_h, f_h)] \geq \theta^2 \, \mathrm{PHI}(f^{nh} \wedge f_h, f_h) ,$$

and we are done.   Q.e.d.

(5.27) EXERCISE. Prove inequality (5.26). [Hint: Convexity!]

EXERCISES: (5.7), (5.17), (5.22), (5.23), (5.24), (5.25), (5.27).

6. Nonparametric estimation of entropy

We study the estimation of the entropy, as an example of the estimation of a nonlinear functional of the density. This is of interest in, e.g., discrimination problems and information theory, see TSYBAKOV and VAN DER MEULEN (1996) and references therein. For us, an added incentive is that the results are needed to justify a particular procedure for choosing the smoothing parameter for the deconvolution problem in Volume II.

The negative entropy of a pdf f is defined as

$$(6.1) \qquad \mathcal{E}(f) = \int_{\mathbb{R}} f(x) \log f(x) \, dx .$$

The objective is to estimate $\mathcal{E}(f)$ based on an iid sample $X_1, X_2, \cdots, X_n$ from the pdf f. There are many nonparametric estimators around, but here we consider the natural estimator $\mathcal{E}(f^{nh})$, where f^{nh} is a kernel density estimator. In the entropy context, it is unusual to impose smoothness conditions, but if statements about rates of convergence are to be made, then some smoothness seems indispensable. We assume that

$$(6.2) \qquad \int_{\mathbb{R}} \frac{|f'|^4}{f^3} < \infty \quad , \qquad \int_{\mathbb{R}} \frac{|f''|^2}{f} < \infty \; ,$$

which is actually equivalent to the condition

$$(6.3) \qquad\qquad f^{1/2} \in W^{2,2}(\mathbb{R}) \; .$$

(6.4) EXERCISE. (a) Prove the inequality

$$\int_{\mathbb{R}} \frac{|f'|^4}{f^3} < \frac{9}{4} \int_{\mathbb{R}} \frac{|f''|^2}{f} \; .$$

[Hint: Integration by parts.]
(b) Show that

$$8 \, \| (f^{1/2})'' \|_2 \leqslant 7 \left\{ \int_{\mathbb{R}} \frac{|f''|^2}{f} \right\}^{1/2} \leqslant 56 \, \| (f^{1/2})'' \|_2 \; .$$

(The numerical constants are not guaranteed !)
(c) Find the best constants in (a) and (b). [The authors do not know how to do this.]

It is again convenient to choose the two-sided exponential kernel. This time, the usefulness of the two-sided exponential comes about by the fact that it is the Green's function for a boundary value problem for a second-order differential equation. That is, if φ is a continuous pdf, then $\psi = \mathfrak{B}_h * \varphi$ is also a pdf that satisfies

$$(6.5) \qquad \begin{aligned} -h^2 \, \psi'' + \psi &= \varphi & \text{on } \mathbb{R} \; , \\ \psi &\longrightarrow 0 & \text{at } \pm\infty \; . \end{aligned}$$

(6.6) EXERCISE. Show that $\psi = \mathfrak{B}_h * \varphi$ solves (6.5) if φ is a continuous pdf.

We are somewhat prejudicing the proceedings by setting

$$(6.7) \qquad \mathcal{E}(f^{nh}) = \frac{1}{n} \sum_{i=1}^{n} \log f(X_i) + \delta_{nh} \; ,$$

but this is precisely what is required in view of the following results.

(6.8) THEOREM. *Suppose f satisfies (6.2). Let $\alpha > 1$. If f has a moment of order $> 1/(\alpha - 1)$, then for $h \asymp n^{-1/(2\alpha+1)}$ (deterministic and constant on blocks $2^{k-1} < n \leqslant 2^k$),*

$$\delta_{nh} =_{as} \mathcal{O}\big(n^{-2/(2\alpha+1)} \big) \; .$$

In general, if f has a moment of order > 0, and if h (still constant on blocks) satisfies $nh \longrightarrow \infty$, $h \longrightarrow 0$, then

$$\delta_{nh} \longrightarrow_{\text{as}} 0 .$$

(6.9) COROLLARY. If f satisfies (6.2), and has a finite moment of some order > 2 and $\sigma^2 = \int_{\mathbb{R}} f \, |\log f|^2 < \infty$, then for $h \asymp n^{-1/3}$ (deterministic and constant on blocks),

$$\sqrt{n} \left\{ \mathcal{E}(f^{nh}) - \mathcal{E}(f) \right\} \longrightarrow_{d} Y \sim N(0, \sigma^2) .$$

It is beyond the scope of this text, but it can be shown that $\mathcal{E}(f^{nh})$ is in fact a best asymptotically normal estimator of $\mathcal{E}(f)$. See LEVIT (1978) and TSYBAKOV and VAN DER MEULEN (1996).

(6.10) EXERCISE. Prove Corollary (6.9) assuming Theorem (6.8).

The limiting case $\alpha \to 1$ in Theorem (6.8) is interesting, but handling it requires a few extra conditions. We postpone it until after the proof of Theorem (6.8). Also, the constancy of h on blocks is somewhat of a nuisance. We come back to this in (6.20) and (6.21).

Before going on to prove Theorem (6.8), a convexity warning must be issued. In the remainder of this section, convexity arguments can and will be used at the drop of a hat, at which time, a peak at Chapters 9 and 10 might be in order.

To begin the proof of Theorem (6.8), we observe that

(6.11) $$\delta_{nh} = \text{KL}(f^{nh}, f_h) + \left\{ \mathcal{E}(f_h) - \mathcal{E}(f) \right\} + S_{n,h} ,$$

where

(6.12) $$S_{n,h} = \int_{\mathbb{R}} (\mathcal{B}_h * \log f_h - \log f)(dF_n - dF) .$$

What about the various terms in (6.11)? The term $\text{KL}(f^{nh}, f_h)$ has been discussed ad nauseam in the previous section, see Theorem (5.21). The last term $S_{n,h}$ is a sum of iid random variables with zero mean. So, it is a martingale [Exercise (6.31)]. The only complicating circumstance is that it depends on the parameter h, so that there are some problems concerning the blocking phenomenon. The middle term $\mathcal{E}(f_h) - \mathcal{E}(f)$ is entirely deterministic and can be elegantly handled by virtue of the double exponential kernel being the Green's function. It is convenient and instructive to begin with this middle term.

(6.13) LEMMA. Under the smoothness conditions (6.2) on f, for $h \to 0$,

$$0 \leqslant \mathcal{E}(f) - \mathcal{E}(f_h) = \mathcal{O}(h^2) .$$

PROOF. We rewrite $\mathcal{E}(f) - \mathcal{E}(f_h)$ as the sum of two integrals, each of which turns out to be positive, in particular,

$$(6.14) \qquad \mathcal{E}(f) - \mathcal{E}(f_h) = \int_{\mathbb{R}} f \, \log \frac{f}{f_h} + \int_{\mathbb{R}} (f - f_h) \, \log f_h \ .$$

The first integral on the right equals $\mathrm{KL}(f, f_h)$, which is nonnegative. Moreover, by the convexity of $\mathrm{KL}(f, f_h)$ with respect to the second argument (Chapter 10!),

$$\mathrm{KL}(f, f_h) \leqslant \int_{\mathbb{R}} \frac{|f - f_h|^2}{f_h} = h^4 \int_{\mathbb{R}} \frac{|(f_h)''|^2}{f_h} \ ,$$

where the last equality comes from the Green's function business (6.5). Finally, since $(u, v) \longmapsto u^2/v$ is convex in $(u, v) \in \mathbb{R} \times (0, \infty)$, then

$$\frac{|(f_h)''|^2}{f_h} = \frac{|\mathcal{B}_h * (f'')|^2}{\mathcal{B}_h * f} \leqslant \mathcal{B}_h * \frac{|f''|^2}{f} \ ,$$

so that with assumption (6.2),

$$\mathrm{KL}(f, f_h) \leqslant h^4 \int_{\mathbb{R}} \frac{|f''|^2}{f} = \mathcal{O}(h^4) \ .$$

The second integral in (6.14) equals, again by (6.5) and integration by parts, and the convexity of u^2/v,

$$-h^2 \int_{\mathbb{R}} (f_h)'' \, \log f_h = h^2 \int_{\mathbb{R}} (f_h)' \, (\log f_h)' = h^2 \int_{\mathbb{R}} \frac{|(f_h)'|^2}{f_h} \leqslant h^2 \int_{\mathbb{R}} \frac{|f'|^2}{f} \ .$$

So this term is $\mathcal{O}(h^2)$, by assumption (6.2).   Q.e.d.

(6.15) LEMMA. *Under the conditions (6.2) on f, for all $\varepsilon > 0$, and for $h = h_n$ constant on blocks $2^{k-1} < n \leqslant 2^k$,*

$$S_{n,h} =_{\mathrm{as}} \mathcal{O}\left(h^2 \, n^{-1/2} \, (\log n)^{(1+\varepsilon)/2} \right) \ .$$

Since $\{ S_{n,h} \}_{n \geqslant 1}$ is a martingale, we first need the expected value.

(6.16) LEMMA. *Under the assumptions (6.2), we have, uniformly in $h > 0$,*

$$\mathbb{E}\left[|S_{n,h}|^2 \right] = \mathcal{O}(h^4 \, n^{-1}) \ .$$

PROOF. Since $S_{n,h}$ is the sum of iid random variables with zero mean, the above expected value equals

$$(6.17) \qquad n^{-1} \int_{\mathbb{R}} f \, |\mathcal{B}_h * \log f_h - \log f|^2 \ .$$

Writing

$$\big|\, \mathcal{B}_h * \log f_h - \log f \,\big|^2 = \big|\, \mathcal{B}_h * \log f_h - \log f_h + \log(f_h/f) \,\big|^2$$

$$\leqslant 2\,\big|\, \mathcal{B}_h * \log f_h - \log f_h \,\big|^2 + 2\,\big|\, \log(f_h/f) \,\big|^2 \,,$$

we have

(6.18) $$\int_{\mathbb{R}} f \,\big|\, \mathcal{B}_h * \log f_h - \log f \,\big|^2$$

$$\leqslant 2 \int_{\mathbb{R}} f \,\big|\, \mathcal{B}_h * \log f_h - \log f_h \,\big|^2 + 2 \int_{\mathbb{R}} f \,\big|\, \log(f_h/f) \,\big|^2 \,.$$

Again, using the Green's function property (6.5), we see that

$$\mathcal{B}_h * \log f_h - \log f_h = h^2 \big(\mathcal{B}_h * \log f_h \big)'' = h^2\, \mathcal{B}_h * \big(\log f_h \big)'' ,$$

so that the first integral in the right-hand side of (6.18) is dominated by

(6.19) $$h^4 \int_{\mathbb{R}} f \,\big|\, \mathcal{B}_h * \{ (\log f_h)'' \} \,\big|^2 \leqslant$$

$$h^4 \int_{\mathbb{R}} f \, \{ \mathcal{B}_h * \big|\, (\log f_h)'' \,\big|^2 \} = h^4 \int_{\mathbb{R}} f_h \,\big|\, (\log f_h)'' \,\big|^2 \,,$$

the inequality by the convexity of $t \longmapsto t^2$, and the last equality by changing the order of integration. But since

$$(\log f_h)'' = \frac{(f_h)''}{f_h} - \left\{ \frac{(f_h)'}{f_h} \right\}^2 ,$$

the expression in (6.19) is dominated by

$$2\, h^4 \, \{ \int_{\mathbb{R}} \frac{\big|\, (f_h)'' \,\big|^2}{f_h} + \int_{\mathbb{R}} \frac{\big|\, (f_h)' \,\big|^4}{\big|\, f_h \,\big|^3} \} \,.$$

By the convexity of $(u, v) \longmapsto u^2/v$ and $(u, v) \longmapsto u^4/v^3$ on $\mathbb{R} \times (0, \infty)$, this in turn is dominated by

$$2\, h^4 \, \{ \int_{\mathbb{R}} \frac{\big|\, f'' \,\big|^2}{f} + \int_{\mathbb{R}} \frac{\big|\, f' \,\big|^4}{f^3} \} \,,$$

which is $\mathcal{O}(h^4)$, by assumption (6.2).

For the second integral in (6.18), we use the elementary inequality

$$\big|\, \log t \,\big| = 2\,\big|\, \log \sqrt{t} \,\big| \leqslant 2\, \frac{\big|\, t - 1 \,\big|}{\sqrt{t}} \,, \qquad t > 0 \,,$$

to get, with $t = f/f_h$,

$$\int_{\mathbb{R}} f \,\big|\, \log(f_h/f) \,\big|^2 \leqslant 4 \int_{\mathbb{R}} \frac{\big|\, f - f_h \,\big|^2}{f_h} \,,$$

and in the proof of Lemma (6.14), we already showed this is $\mathcal{O}(h^4)$. The lemma follows.   Q.e.d.

The proof of Lemma (6.15) is now a simple application of Theorem (4.12), and it is omitted.

PROOF OF THEOREM (6.8). The starting point is (6.11). Let $\varepsilon > 0$. From Theorem (5.21), and Lemmas (6.13) and (6.15), we have with h_n constant on blocks,

$$\delta_{nh} =_{as} \mathcal{O}\left((nh_n)^{-1/\alpha} + h_n^2 + h_n^2 n^{-1/2} (\log n)^{(1+\varepsilon)/2} \right).$$

Now, take h_n constant on blocks such that $nh_n \longrightarrow \infty$, $h_n \longrightarrow 0$, and it follows that $\delta_{nh} \longrightarrow_{as} 0$. Taking $h_n \asymp n^{-1/(2\alpha+1)}$ (but still constant on blocks) gives $\delta_{nh} =_{as} \mathcal{O}\left(n^{-2/(2\alpha+1)} \right)$, completing the proof.   Q.e.d.

(6.20) REMARK. The constancy on blocks of $h = h_n$ seems to be a nuisance, because it only matters for the term $S_{n,h}$, which is then smaller by an order of magnitude compared with the other two terms. So one should be able to sacrifice "half an order" of magnitude to get rid of the blocking.

(6.21) EXERCISE. In Lemma (6.15) and Theorem (6.8), get rid of the constancy on blocks condition.

(6.22) EXERCISE. What is the correct formulation of Theorem (6.8) if f is any pdf with a finite exponential moment $\mathbb{E}[e^{t|X|}] < \infty$ for some $t > 0$? [Hint : Lemma (5.16) !]

The above is a pretty concise treatment of entropy estimation. The conditions on the pdf f having moments of a certain order are not too bad, but they are not sharp in the sense that to get the rate $n^{-2/(2\alpha+1)}$, the pdf f needs to have a moment of order $> 1/(\alpha - 1)$. There is a way around it, but at the cost of imposing regular tail behavior. Yes, the cure may be worse than the disease, but it fits together so nicely that there must be more to it than meets the eye. Specifically, we replace the moment condition with the following PARZEN condition on f, viz.

(6.23) $f\left(F^{inv}(t)\right) \geqslant c\,[t\,(1-t)]^\alpha$, $0 < t < 1$,

for some positive constant c and some $\alpha \geqslant 1$. See (1.50). Moreover, in addition, the pdf f is required to be approximately bell shaped, in the following precise sense. We assume that there exist numbers $A < B$ and $b > 0$ such that

$(6.24a)$ $f(x) \geqslant b \cdot f(y)$,   for all $x \in [A, B]$, $y \notin [A, B]$,

$(6.24b)$ $f(x) \geqslant b \cdot f(y)$,   for all $B < x < y$, and all $y < x < A$.

For $b = 1$, this says that everywhere on $[A, B]$, the density f is larger than everywhere outside $[A, B]$, and that f is monotone in the tails $(-\infty, A)$ and (B, ∞). For $b < 1$, these statements are fudged a bit. The conditions

(6.24) are surely not independent of (6.2) and (6.23) combined, but it hardly seems worth while to sort this out.

In the study of entropy estimation, the new assumptions only affect the treatment of the Kullback-Leibler distance.

(6.25) LEMMA. *Under the conditions* (6.2), (6.23), *and* (6.24) *on* f,

$$\mathbb{E}\big[\,\mathrm{KL}(f^{nh},\,f_h)\,\big] = \begin{cases} \mathcal{O}\big(\,(nh)^{-1/\alpha} + h^2\,\big) & , \quad \alpha > 1\,, \\ \mathcal{O}\big(\,(nh)^{-1}\log(nh) + h^2\,\big) , & \alpha = 1\,, \end{cases}$$

provided $nh \to \infty$, $h \to 0$.

PROOF. The starting point is the bound (5.13) with the right-hand side written in the form

$$\int_{\mathbb{R}\times\mathbb{R}} f(y)\,\mathfrak{B}_h(x-y)\,\log\Big(1 + \frac{\mathfrak{B}_h(x-y)}{n\,f_h(x)}\Big)\,dx\,dy =$$

$$\int_{\mathbb{R}\times\mathbb{R}} f(y)\,\mathfrak{B}_h(x-y)\,\log\big(\,f_h(x) + (1/n)\,\mathfrak{B}_h(x-y)\,\big)\,dx\,dy - \int_{\mathbb{R}} f_h\,\log f_h\ .$$

Since $t \longmapsto \log t$ is concave, the first integral is dominated by

$$\int_{\mathbb{R}} f(y)\,\log\Big(\int_{\mathbb{R}} \mathfrak{B}_h(x-y)\,\{\,f_h(x) + (1/n)\,\mathfrak{B}_h(x-y)\,\}\,dx\,\Big)\,dy =$$

$$\int_{\mathbb{R}} f\,\log(\,p_h * f + \varepsilon\,)\,dy\,,$$

where $p_h = \mathfrak{B}_h * \mathfrak{B}_h$ and $\varepsilon = (nh)^{-1}\int_{\mathbb{R}} |\,\mathfrak{B}(x)\,|^2\,dx$. So we have

$$\mathbb{E}\big[\,\mathrm{KL}(f^{nh},\,f_h)\,\big] \leqslant \int_{\mathbb{R}} f\,\log(p_h * f + \varepsilon) - \mathcal{E}(f_h) =$$

(6.26)
$$\leqslant \int_{\mathbb{R}} f\,\log\frac{p_h * f}{f} + \mathcal{E}(f) - \mathcal{E}(f_h) +$$

$$\int_{\mathbb{R}} f\,\log\Big(1 + \frac{\varepsilon}{p_h * f}\Big)\,.$$

The first integral on the right of (6.26) equals $-\mathrm{KL}(f,\,p_h * f)$ and is thus negative. The next term $\mathcal{E}(f_h) - \mathcal{E}(f) = \mathcal{O}(h^2)$ by Lemma (6.13). To handle the last integral, we need to show that

(6.27)
$$p_h * f(x) \geqslant c\,f(x)\,, \quad x \in \mathbb{R}\,,$$

for some positive constant c depending on f. This is where the bell shape of f embodied by assumptions (6.24ab) comes into play. (It does not really depend on the double exponential kernel.) First, consider the case $x > B$. Then,

$$p_h * f(x) \geqslant \int_A^x p_h(x-y)\,f(y)\,dy \geqslant b\,f(x)\int_A^x p_h(x-y)\,dy\,,$$

by assumptions (2.3a,b). Now, the last integral equals

$$\int_0^{x-A} p_h(y)\, dy \;,$$

and this tends to $\frac{1}{2}$ as $h \to 0$. This shows that (6.27) holds for $x > B$. The proof for $x < A$ is similar. Finally, for $A \leqslant x \leqslant B$, both $f(x)$ and $p_h * f$ are bounded and bounded away from 0, so here too (6.27) holds.

So the last integral in (6.26) is dominated by

$$I(\delta) \stackrel{\text{def}}{=} \int_{\mathbb{R}} f \log\left(1 + \frac{\delta}{f} \right) \;,$$

with $\delta = \varepsilon/c$, with c the constant of (6.27). With the PARZEN assumption (6.23), we get

$$I(\delta) = \int_0^1 \log\left(1 + \frac{\delta}{f F^{\text{inv}}(t)} \right) dt \;\leqslant\; \int_0^1 \log\left(1 + c\,\delta\,[\,t\,(1-t)\,]^{-\alpha} \right) dt \;,$$

with c from (6.23). It is now an advanced calculus exercise to show that for $\delta \to 0$,

(6.28) $$I(\delta) = \begin{cases} \mathcal{O}\!\left(\delta^{1/\alpha} \right) & , \quad \alpha > 1 \;, \\ \mathcal{O}\!\left(\delta \, \log(1/\delta) \right) , & \alpha = 1 \;, \end{cases}$$

and the lemma follows. Q.e.d.

(6.29) REMARK. Consider the case $\alpha = \frac{3}{2}$ in condition (6.23). This corresponds to slowly varying decay of f like $|x|^{-3}$ as $|x| \to \infty$, so that f has moments of order < 2 and not of order 2. Then, Lemma (6.25) gives a rate of $(nh)^{-2/3}$ for $\mathbb{E}[\,\mathrm{KL}(f^{nh}, f_h)\,]$. However, in Theorem (5.20), to get a rate of $(nh)^{-2/3}$ requires a moment of order > 2. So all this was not completely in vain.

(6.30) EXERCISE. Prove (6.28). [Hint : Do integration by parts, and split the resulting integral in half. On $(0, \frac{1}{2})$, you may replace $t\,(1-t)$ by just t (at the cost of a constant factor), and compute the integral for $\alpha = 1$. For $\alpha > 1$, estimate the integral.]

(6.31) EXERCISE. (a) Verify (6.11).
(b) Show that $\{\, S_{n,h} \,\}_{n \geqslant 1}$, as defined by (6.12), is a martingale.

(6.32) EXERCISE. Extend Theorem (6.8) to the multivariate case $X \in \mathbb{R}^d$, with the kernel

$$\prod_{\ell=1}^d \mathcal{B}_h(x_\ell) \;, \quad x = (x_1, x_2, \cdots, x_d) \in \mathbb{R}^d \;,$$

with $\mathfrak{B}_h$ as before, under the smoothness conditions

$$\int_{\mathbb{R}^d} \frac{\|\nabla f(x)\|^4}{|f(x)|^3} \, dx < \infty \,, \qquad \int_{\mathbb{R}^d} \frac{|\Delta f(x)|^2}{f(x)} \, dx < \infty \,,$$

where ∇f is the gradient of f and Δ is the Laplacian, and the moment condition

$$\mathbb{E}\big[\, \|X\|^{\lambda d}\,\big] < \infty \,,$$

for some $\lambda > 2$.

(6.33) EXERCISE. State the appropriate conditions, and prove the best asymptotic normality of the nonparametric entropy estimator of AHMAD and LIN (1976) (see also HALL and MORTON (1993))

$$\frac{1}{n} \sum_{i=1}^{n} \log f^{nh}(X_i) \,.$$

EXERCISES : (6.4), (6.6), (6.10), (6.21), (6.22), (6.30), (6.31), (6.32), (6.33).

7. Optimal kernels

Statisticians generally love to do things optimally, so the question arises as to the optimal implementation of kernel density estimators. There are two "parameters" that one can choose optimally: the kernel and the smoothing parameter. It is customary to separate the two, because considering them together makes the problem intractable. Exceptions are WATSON and LEADBETTER (1963) and, more recently, DEVROYE and LUGOSI (2000). Furthermore, it is an act of self-preservation to let the choice of the kernel be governed by general considerations and to let only the choice of the smoothing parameter depend on the data $X_1, X_2, \cdots, X_n$. In practice, the kernel is not very important. That is why its choice gets one section (two actually, see § 11.2), whereas choosing h gets a whole chapter, Chapter 7.

To determine "optimal" kernels, some criterion is needed to judge them by. It seems beyond question that such a criterion should involve the error $f^{nh} - f$ and, as argued before, to measure that error by $\|f^{nh} - f\|_1$. However, in doing so, we let the smoothing parameter and the data sneak back into the picture. A good way around this is first to take the expected value with respect to the data and then choose the "optimal" h. (This could perhaps be done better the other way around, but that is quite a bit harder.) There is still the dependence on the underlying f, but we let that pass. (If this really bothers you, see DEVROYE (1987), Chapter 7). To summarize, we would like to choose the kernel A so as to solve

(7.1)
$$
\begin{aligned}
&\text{minimize} && \min_{h} \mathbb{E}\big[\, \|f^{nh} - f\|_1\,\big] \\
&\text{subject to} && A \text{ is a pdf} \,.
\end{aligned}
$$

This problem is still untractable not only because of the dependence on f, but also because of the dependence on the sample size n. However, asymptotically, (7.1), can almost be solved. The starting point is the following result.

(7.2) THEOREM. Let $f \in W^{2,1}(\mathbb{R})$, and assume that the pdf A has zero mean, finite standard deviation. Then,

$$\mathbb{E}\left[\| f^{nh} - f \|_1\right] \leqslant |c_2(A)| \| f'' \|_1 h^2 + (nh)^{-1/2} d_o(A) \| \sqrt{f} \|_1 ,$$

where $d_o(A) = \| A \|_2$ and $c_j(A)$ is as in Theorem (2.9).

(7.3) EXERCISE. Prove Theorem (7.2).

Actually, this is only a slight improvement over Theorems (2.9) and (2.21). Now, minimizing (over h) the upper bound for $\mathbb{E}\left[\| f^{nh} - f \|_1\right]$ of Theorem (7.2) yields

(7.4) $\limsup\limits_{n \to \infty} \min\limits_{h} n^{2/5} \mathbb{E}\left[\| f^{nh} - f \|_1\right] \leqslant$

$$\text{const} \{ |c_2(A)| (d_o(A))^4 \}^{1/5} \{ \| f'' \|_1 \| \sqrt{f} \|_1^4 \}^{1/5}$$

for some universal constant. It is not at all easy to show (or in fact clear) that there is equality in (7.4), and we refrain from doing so, but we shall proceed on the assumption that minimizing the right-hand side of (7.4) is the same asymptotically as minimizing the left-hand side.

So to find "optimal" kernels, we must minimize

(7.5) $J(A) \overset{\text{def}}{=} c_2(A) (d_o(A))^4 = \left| \int_{\mathbb{R}} x^2 A(x) \, dx \right| \left| \int_{\mathbb{R}} \{A(x)\}^2 \, dx \right|^2 ,$

over all pdfs A with zero mean. The first thing to note is that there is a trivial kind of nonuniqueness about minimizing $J(A)$ (but maybe a nontrivial nonuniqueness as well), as the following exercise makes clear.

(7.6) EXERCISE. (a) Show that for any pdf A, one has $J(A_h) = J(A)$, that is, $J(A)$ is scaling invariant.
(b) Let A be a square integrable pdf with finite variance, i.e., $c_2(A) < \infty$. Show that for every $t > 0$, there exists an $h > 0$ such that $c_2(A_h) = t$.

It now turns out that the nonuniqueness illustrated by Exercise (7.6) is the only kind of nonuniqueness found in minimizing $J(A)$. The following lemma gives an optimal kernel, but any scaled version is optimal also.

(7.7) LEMMA. The minimum of $J(A)$ over all pdfs A is achieved by

$$A_o(x) = \tfrac{3}{4} (1 - x^2)_+ , \quad x \in \mathbb{R} .$$

PROOF. Let B be any pdf. Without loss of generality, one may assume that B is square integrable and has finite variance. Then, for a scaled version B_h, we have $c_2(B_h) = c_2(A_o)$. Again, without loss of generality, assume that $h = 1$. Now, observe that $B^2 = (B - A_o)^2 + A_o^2 + 2(B - A_o)A_o$ and that

$$\frac{4}{3}\int_{\mathbb{R}}(B - A_o)A_o = \int_{-1}^{1}\big(B(x) - A_o(x)\big)(1 - x^2)\,dx$$
$$= -\int_{|x|>1}B(x)\,dx + c_2^2(A_o) - \int_{-1}^{1}x^2\,B(x)\,dx$$
$$\geqslant c_2^2(A_o) - \int_{\mathbb{R}}x^2\,B(x)\,dx$$
$$= c_2^2(A_o) - c_2^2(B) = 0 \ .$$

In the second line, we used that A_o and B are both pdfs, and in the last line that B was scaled to have the same variance as A_o. It now follows that

$$\int_{\mathbb{R}}B^2 = \int_{\mathbb{R}}(B - A_o)^2 + \int_{\mathbb{R}}A_o^2 + 2\int_{\mathbb{R}}(B - A_o)A_o \geqslant \int_{\mathbb{R}}A_o^2 \ ,$$

with equality for $B = A_o$. Thus, A_o minimizes $J(A)$.   Q.e.d.

The above considerations may now be summarized as follows.

(7.8) THEOREM. *The kernel that achieves asymptotically the fastest rate of convergence for twice continuously differentiable pdfs f with a finite moment of order > 1 is the Epanechnikov kernel*

$$A_o(x) = \tfrac{3}{4}(1 - x^2)_+ \ , \quad x \in \mathbb{R} \ .$$

It is one of life's coincidences that if one considers $\min_h \mathbb{E}\big[\,\|\,f^{nh} - f\,\|_2^2\,\big]$, instead of (7.1), then the problem again reduces to minimizing $J(A)$ over all pdfs A with mean zero, see BARTLETT (1963) and EPANECHNIKOV (1969). Even earlier than that, HODGES and LEHMANN (1956) solved the same minimization problem, but in a different context.

So the above theorem shows that the Epanechnikov kernel is optimal. However, both in theory and practice, it is hardly noticeable. The asymptotic order of convergence is $\mathcal{O}(n^{-2/5})$, just as for any other symmetric kernel A with finite variance. The only gain is that the constant in the order term is smaller. Exercise (7.9) shows that this improvement over some reasonable (and popular) alternatives is not very impressive. Many simulation studies indicate that this is also true for small-to-moderate sample sizes. We illustrate this in § 8.5.

(7.9) EXERCISE. Calculate $J(A)$ for each of the kernels: the standard normal, the double exponential, the uniform, and the biweight $(15(1-x^2)_+^2/16)$ densities. Compare this with the value for the Epanechnikov kernel.

The previous observations naturally lead to the question of whether it is possible to improve on the order of convergence $n^{-2/5}$. To settle this, let us go back to the starting point of Theorem (7.2). The first term in the upper bound of Theorem (7.2) is the variance term and cannot be avoided: $\int_{\mathbb{R}} \{A(x)\}^2 \, dx > 0$, whether we jump high or low. The second term is the bias term, and $\int_{\mathbb{R}} x^2 A(x) \, dx > 0$ for any density A. This is true, but herein lies the rub. The condition that A is a pdf can be relaxed. In particular, suppose that A is symmetric and that

$$\int_{\mathbb{R}} A(x)\,dx = 1 \quad , \qquad \int_{\mathbb{R}} x^2 A(x)\,dx = 0 \ ,$$

then the bias term is $o(h^2)$ at most. In general, assume that A is a kernel of order $2k$ (a $2k$-th order kernel), that is, A is symmetric, $(1 + x^{2k})A(x)$ is integrable, and

(7.10)
$$\int_{\mathbb{R}} A(x)\,dx = 1 \quad , \qquad \int_{\mathbb{R}} x^{2k} A(x)\,dx \neq 0 \ ,$$
$$\int_{\mathbb{R}} x^j A(x)\,dx = 0 \ , \quad j = 1, \cdots, 2k-1 \ .$$

Kernels of odd order may be defined similarly, but we restrict attention to kernels of even order.

Under these assumptions, the behavior of the bias term is described by the following refinement of Theorem (2.9).

(7.11) LEMMA. Let $k \geqslant 1$ (integer). If $f \in W^{2k,1}(\mathbb{R})$ and A is a kernel of order $2k$, then

$$\| A_h * f - f \|_1 \leqslant c_{2k}(A)\, h^{2k} \, \| f^{(2k)} \|_1 \ ,$$

where $c_k(A) = 1/k! \int_{\mathbb{R}} |x^k A(x)| \, dx$.

(7.12) EXERCISE. Prove the lemma.

Before proceeding, it might be prudent to make sure that $2k$-th order kernels actually exist. The following exercise shows how they may be constructed.

(7.13) EXERCISE. [STUETZLE and MITTAL (1979), DEVROYE (1989)]
(a) Suppose A is a symmetric density having a finite fourth moment. Then, $B = 2A - A * A$ is a 4-th order kernel.
(b) If A is a $2k$-th order kernel with a finite moment of order $2k + 2$, then $B = 2A - A * A$ is a kernel of order $2k + 2$.

(c) If A is a $2\,k$-th order kernel with a finite moment of order $2k + 2\ell$, and B is an $2\,\ell$-th order kernel with a finite moment of order $2k + 2\ell$, then $C = A + B - A * B$ is a kernel of order $2\,k + 2\,\ell$.

With Lemma (7.11) in hand, bounds for $\| f^{nh} - f \|_1$ may be derived similarly to the way it was done for second-order kernels. The result is as follows.

(7.14) THEOREM. *If $f \in W^{2k,1}(\mathbb{R})$, and if A is a $2\,k$-th order kernel, then*

$$\limsup_{n \to \infty} \inf_{h > 0} n^{2k/(4k+1)} \mathbb{E}\big[\, \| f^{nh} - f \|_1 \,\big] \leqslant$$

$$\gamma_k \left\{ c_{2k}(A) \, \| A \|_2^{4k} \, \| f^{(2k)} \|_1 \, \| \sqrt{f} \|_1^{4k} \right\}^{1/(4k+1)},$$

where γ_k depends on k only.

(7.15) EXERCISE. (a) Prove Theorem (7.14).
(b) Derive almost sure bounds for $\| f^{nh} - f \|_1$.

(7.16) EXERCISE. Keeping in mind the opening line of this section, determine the "optimal" $2\,k$-th order kernel. [This amounts to minimizing

$$J_k(A) = \big| c_{2k}(A) \big| \, \{ d_o(A) \}^{4k}$$

over all functions A satisfying (7.10).]

Theorem (7.14) tells us that for $k \to \infty$, kernel estimators exist that approach the parametric rate of $n^{-1/2}$. Of course, one should keep in mind that this requires that $f \in W^{2k,1}(\mathbb{R})$. It depends on one's point of view if this is a serious drawback or not. Another drawback is that with higher order kernels A, the density estimator is negative in some places. If this bothers you, as it does the authors, then the *ad hoc* modification

$$g^{n,h}(x) = \frac{\max\{\, 0 \,, \, A_h * dF_n(x) \,\}}{\int_{\mathbb{R}} \max\{\, 0, \, A_h * dF_n(y) \,\} \, dy}$$

is a quick fix. In fact, $g^{n,h}$ is a better approximation to f than is f^{nh}, as is made clear in Exercise (7.17). In Chapter 8, we give some illustrations of the performance of these higher order kernels for small-to-moderate sample sizes. Again, the improvement is not very impressive.

(7.17) EXERCISE. Let $\varphi \in L^1(\Omega)$, with $\int_{\mathbb{R}} \varphi(x) \, dx = 1$. Define ψ and θ by $\psi(x) = \max\{0, \varphi(x)\}$ and $\theta(x) = \psi(x)/\int_{\mathbb{R}} \psi(y) \, dy$.
(a) If φ is negative on a set with positive measure, show that for any pdf f,

$$\| \theta - f \|_1 < \| \psi - f \|_1 < \| \varphi - f \|_1 \ .$$

[DEVROYE and GYÖRFI (1985)].
(b) Show the improved version $\| \theta - f \|_1 \leqslant \big(\| \psi \|_1 \big)^{-1} \| \varphi - f \|_1$. [GAJEK (1987). For more on this, see KALUSZKA (1998).]

So, by increasing the order of the kernel for very smooth pdfs, we can get a rate of convergence approaching $n^{-1/2}$, but never quite getting there. Which brings us to the final question in this section: Is it possible to achieve the rate $n^{-1/2}$? IBRAGIMOV and HAS'MINSKII (1982) have shown the way, see also DEVROYE (1987). A quick perusal of Theorem (7.2) shows that it would suffice if the bias term vanished. So the question becomes whether it is possible to find a kernel A and a reasonable class of pdfs f such that

(7.18)
$$A_h * f = f .$$

This is a problem that screams out for Fourier transforms, see Appendix 2. Assuming A has a Fourier transform, then $(A_h)^\wedge(\omega) = A^\wedge(h\omega)$ for $\omega \in \mathbb{R}$, and so (7.18) is equivalent to

(7.19)
$$\left(\widehat{A}(h\omega) - 1 \right) \widehat{f}(\omega) = 0 , \quad \text{for all } \omega .$$

Now, the following will make it work. Let A satisfy

(7.20)
$$\widehat{A}(\omega) = 1 , \quad \text{for all } |\omega| \leqslant 1 ,$$

(and "arbitrary" for $|\omega| > 1$), and then (7.18) holds for all f with $\widehat{f}(\omega) = 0$ for $|\omega| > h^{-1}$. It is useful to introduce the classes of band-limited functions $\mathcal{B}(t)$, defined for $t > 0$ by

(7.21)
$$\mathcal{B}(t) \overset{\text{def}}{=} \{ f \text{ is a pdf} : \widehat{f}(\omega) = 0 \text{ for } |\omega| > t \} .$$

Then, (7.18) holds for all $f \in \mathcal{B}(h^{-1})$, provided A satisfies (7.20). We thus obtain the following optimal result.

(7.22) THEOREM. *Suppose f has a finite moment of order > 1, and that $f \in \mathcal{B}(t)$ for some $t > 0$. Then, with the pdf A having a finite moment of order > 1, and satisfying (7.20),*

$$\min_h \mathbb{E}\left[\| f^{nh} - f \|_1 \right] = \mathcal{O}\left(n^{-1/2} \right) .$$

It should be observed that the optimal h in this case does not tend to 0 as $n \to \infty$: If it did, we would *not* achieve the $n^{-1/2}$ rate, in view of the $(nh)^{-1/2}$ rate for $\mathbb{E}\left[\| f^{nh} - f_h \|_1 \right]$.

(7.23) EXERCISE. Prove Theorem (7.22). [Observe that the error satisfies $\| f^{nh} - f \|_1 = \| f^{nh} - f_h \|_1$, and use Theorem (2.21)(a).]

(7.24) EXERCISE. Under the conditions of Theorem (7.22), apply the exponential inequalities from §4 to obtain a.s. rates of convergence for the L^1 error.

Before we can consider the job done, we had better make sure that kernels A satisfying (7.20) actually exist, and that the classes $\mathcal{B}(t)$ do contain pdfs that have moments of order > 1. The following kernel is at the heart of it all. Consider the (L^2) function

(7.25)
$$A(x) = \frac{\sin(\pi x)}{\pi x} , \quad x \in \mathbb{R} .$$

This function is the (inverse) Fourier transform of the function

(7.26)
$$\widehat{A}(\omega) = \begin{cases} 1 , & |\omega| < 1/2 , \\ 0 , & \text{otherwise} . \end{cases}$$

Now, this kernel A satisfies (7.20), but, unfortunately, it does not belong to $L^1(\mathbb{R})$. We may fix this by extending $\widehat{A}$ smoothly to the whole line by

(7.27a)
$$\widehat{A}(\omega) = \begin{cases} 1 , & |\omega| \leqslant 1 , \\ 0 , & |\omega| \geqslant 2 , \end{cases}$$

and for $1 < |\omega| < 2$ by

(7.27b)
$$\widehat{A}(\omega) = \left\{ 1 - \exp\left(-\frac{1}{\omega^2 - 1}\right) \right\} \exp\left(\frac{1}{3} - \frac{1}{4 - \omega^2}\right) .$$

Then, $\widehat{A} \in C_{\downarrow}^{\infty}(\mathbb{R})$, the space of infinitely often continuously differentiable functions, all of whose derivatives decay faster than any power function at ∞. Consequently, $\widehat{A}$ is indeed the Fourier transform of a function A, and in fact, $A \in C_{\downarrow}^{\infty}(\mathbb{R})$. In particular, then A has finite moments of all orders. See Appendix 2 for the details.

It is now also obvious that the classes $\mathcal{B}(t)$ contain lots of pdfs with finite moments of all orders, as follows.

(7.28) EXERCISE. Let A be defined by (7.27). Show that if the pdf φ has a finite moment of all orders, then for $h > 0$, so does $A_h * \varphi$, and $A_h * \varphi \in \mathcal{B}((2h)^{-1})$.

EXERCISES: (7.3), (7.6), (7.9), (7.12), (7.13), (7.15), (7.16), (7.17), (7.23), (7.24), (7.28).

8. Asymptotic normality of the L^1 error

In §4, we discussed (sharp) bounds on the L^1 error for kernel density estimators, based on the exponential inequalities of DEVROYE (1991). The bounds took the form

(8.1)
$$Y_{n,h} =_{as} \mathcal{O}\left((\log n)^{1/2}\right) ,$$

where

(8.2)
$$Y_{n,h} \overset{\text{def}}{=} \sqrt{n} \left\{ \| f^{nh} - \mathbb{E}[f^{nh}] \|_1 - \mathbb{E}[\| f^{nh} - \mathbb{E}[f^{nh}] \|_1] \right\} ,$$

and $h = h_n$ (deterministic). The next step in this program is to determine the asymptotic distribution of $Y_{n.h}$, the final step being the consideration of *random* h. Here, we present a proof of the asymptotic normality of the L^1 error for kernel estimators, following material kindly provided by David Mason [MASON (2000)] (There is a quantum jump in prerequisites here.) This also provides a glimpse at some other useful tools, such as Poissonization, not used elsewhere in the text.

BEIRLANT and MASON (1995) introduced a general method for deriving the asymptotic normality of norms of empirical functionals. Here, their method is applied to the special case of the L^1 error of kernel density estimators. Throughout this section, we let Υ be a standard normal random variable,

(8.3) $$\Upsilon \sim N(0,1) \; ,$$

but sometimes also a multivariate standard normal. We also need Poisson random variables. If p is a Poisson random variable with mean λ, this is denoted as

(8.4) $$p \sim \text{Poisson}(\lambda) \; .$$

However, most of the time, we remind the reader of this.

First, we need some basic assumptions. Let $h = h_n$ be such that

(8.5) $$h \longrightarrow 0 \; , \quad \sqrt{n}\, h \longrightarrow \infty \; .$$

We assume that A is bounded,

(8.6a) $$\kappa = \| A \|_\infty < \infty \; ,$$

and that A has compact support

(8.6b) $$A(x) = 0 \; , \quad \text{for } |x| > \tfrac{1}{2} \; .$$

So then $A \in L^1(\mathbb{R})$. Of course, we also assume that

(8.6c) $$\int_{\mathbb{R}} A(x)\, dx = 1 \; ,$$

so that kernels of arbitrary order are allowed. We set $A_h(x) = h^{-1} A(h^{-1}x)$. The data $X_1, X_2, \cdots$ are independent and distributed as X with pdf f. The kernel estimator is denoted as $f^{nh} = A_h * dF_n$, and its expected value as $f_h = \mathbb{E}[A_h * dF_n] = A_h * f$. Define the function ϱ by

(8.7) $$\varrho(t) = (\| A \|_2)^{-2} \int_{\mathbb{R}} A(t+y) A(y)\, dy \; , \quad t \in \mathbb{R} \; .$$

Let $\Upsilon = (\Upsilon_1, \Upsilon_2) \sim N_2(0, I)$, and set

(8.8) $$\sigma^2(A) = \| A \|_2^2 \int_{-1}^{1} \text{Cov}\left(\left| \sqrt{1 - \varrho^2(t)}\, \Upsilon_1 + \varrho(t)\, \Upsilon_2 \right|, \left| \Upsilon_2 \right| \right) dt \; .$$

The following theorem is remarkable in that there are no assumptions on the density f, and that the asymptotic variance is independent of f.

(8.9) ASYMPTOTIC NORMALITY THEOREM. [MASON (2000)] *For any pdf* f, *under the assumptions* (8.5)–(8.6abc),

$$\frac{Y_{n,h}}{\sqrt{\mathrm{Var}[Y_{n,h}]}} \longrightarrow_d \Upsilon \sim N(0,1)$$

and $\lim_{n\to\infty} \mathrm{Var}[Y_{n,h}] = \sigma^2(A)$.

GINÉ, MASON and ZAITSEV (2001) show, among other things, that the theorem holds with the boundedness assumption on the kernel (8.6a) replaced by the less restrictive assumption of square integrability $\|A\|_2 < \infty$.

The proof consists of three steps: truncation, Poissonization, and dePoissonization. We develop each step separately and in the end put everything together in an actual proof. An underlying theme in the proof is essentially Lusin's theorem, which implies that for every pdf f,

> there exists a sequence $\{C_k\}_k$ of compact subsets of $\mathbb{R}$ such that f is strictly positive and continuous relative to each C_k, and

(8.10) $$\lim_{k\to\infty} \int_{C_k} f(x)\,dx = 1 \ .$$

However, the sets C_k must be chosen more carefully.

(8.11) APPROXIMATION LEMMA. *Suppose* $\mathcal{A}$ *is a finite set of bounded kernels with compact support. For* $A \in \mathcal{A}$, *set* $I(A) = \int_{\mathbb{R}} A(x)\,dx$. *Then,*

(a) $$\lim_{h\to 0} A_h * f - I(A)\,f = 0 \quad \text{almost everywhere} \ .$$

Moreover, for all $0 < \varepsilon < 1$ *there exist positive constants* M, ν, *and* α, *with* $0 < \alpha < \frac{1}{2}$, *and a Borel set* C *with finite Lebesgue measure such that*

(b) $$C \subset [-M + \nu\,, M - \nu] \ ,$$

(c) $$\tfrac{1}{2} > \int_{|x|>M} f(x)\,dx = \alpha > 0 \ ,$$

(d) $$\int_C f(x)\,dx > 1 - \varepsilon \ ,$$

(e) f *is continuous, bounded, and bounded away from* 0 *on* C ,

and for each $A \in \mathcal{A}$,

(f) $$\lim_{h\to 0} \|(A_h * f - I(A)\,f)\,\mathbb{1}_C\|_\infty = 0 \ ,$$

where $\mathbb{1}_C$ *is the indicator function of the set* C.

PROOF. For (a), see DEVROYE and GYÖRFI (1985), Chapter 2, Theorem 3.

Let $0 < \varepsilon < 1$. Using the continuity of the measure associated with f, one can find an interval $[-M, M]$ and a positive number ν such that

$$\alpha = \int_{|x|>M} f(x)\,dx = \tfrac{1}{8}\varepsilon$$

and

$$\int_{|x|>M-\nu} f(x)\,dx = \tfrac{1}{4}\varepsilon \ .$$

The rest of the proof is inferred from Lusin's theorem followed by Egorov's theorem. For both of these theorems, see, e.g., WHEEDEN and ZYGMUND (1977). By Lusin's theorem, we can find a Borel set F such that f is continuous relative to F and

$$\int_F f(x)\,dx > 1 - \tfrac{1}{4}\varepsilon \ .$$

Now, we may extract a compact subset D of $\mathbb{R}$ such that f restricted to $D \cap F$ is continuous, bounded, and bounded away from zero, and

$$\int_{D \cap F} f(x)\,dx > 1 - \tfrac{1}{2}\varepsilon \ .$$

Finally, using Egorov's theorem, which is justified by (a), we can find a Borel set $C \subset [-M+\nu, M-\nu] \cap D \cap F$ such that (d) and (e) are satisfied, and (f) holds for all $A \in \mathcal{A}$.   Q.e.d.

For any Borel set B, let

(8.12) $$F(B) = \int_B f(x)\,dx \ .$$

The truncation step is based on the following result.

(8.13) LEMMA. Let $B \subset \mathbb{R}$ be a Borel set. Then, for any $\{a_n\}_n \subset L^1(\mathbb{R})$,

$$\limsup_{n \to \infty} n \operatorname{Var}\big[\,\|(f^{nh} - a_n)\,\mathbb{1}_B\,\|_1\big] \leqslant 4\kappa^2\,F(B) \ .$$

PROOF. Applying the theorem in PINELIS (1994), cf. Theorem 2.1 of DE ACOSTA (1981), and also DEVROYE (1991), we get that the limsup in the lemma is bounded from above by

$$4\,\mathbb{E}\left[\left|\int_B |A_h(x-X)|\,dx\right|^2\right] \leqslant 4\,\|A_h\|_1\,\mathbb{E}\left[\int_B |A_h(x-X)|\,dx\right] \ .$$

The first two factors on the right equal just $4\,\|A\|_1$. By Fubini's theorem, the expected value equals

$$\int_B [\,|A_h| * f\,](y)\,dy \ .$$

Since $|A| \in L^1(\mathbb{R})$, then $|A_h| * f$ converges in $L^1(\mathbb{R})$ to $\|A\|_1 f$, so that the above integral tends to $\|A\|_1 F(B)$. Thus, the bound for the limsup is $4 \|A\|_1^2 F(B)$, but by (8.6b), we have $\|A\|_1 \leqslant \|A\|_\infty$.   Q.e.d.

The next step in the proof of the Asymptotic Normality Theorem is what is known as *Poissonization*. Let p be a Poisson random variable with mean n, and independent of the $X_1, X_2, \cdots$, and let

$$(8.14) \qquad \varphi^{ph}(x) = \frac{1}{n} \sum_{i=1}^{p} A_h(x - X_i) , \quad x \in \mathbb{R} ,$$

with $\varphi^{ph}(x) = 0$ if $p = 0$. Then,

$$\mathbb{E}[\varphi^{ph}(x)] = \mathbb{E}[f^{nh}(x)] = f_h(x) ,$$

$$n \operatorname{Var}[f^{nh}(x)] = h^{-1} (A^2)_h * f(x) - (f_h(x))^2 ,$$

and $\qquad\qquad n \operatorname{Var}[\varphi^{ph}(x)] = h^{-1} (A^2)_h * f(x) .$

The variances of f^{nh} and φ^{ph} are nearly equal, in the following sense.

(8.15) LEMMA. *Let C be a Borel subset of $\mathbb{R}$ with finite Lebesgue measure satisfying* (e) *and* (f) *of the Approximation Lemma* (8.11) *for the collection* $\mathcal{A} = \{A, A^2, |A|^3\}$. *Then,*

$$\sup_{x \in C} \left| \sqrt{n \operatorname{Var}[\varphi^{ph}(x)]} - \sqrt{n \operatorname{Var}[f^{nh}(x)]} \right| = \mathcal{O}(\sqrt{h}) .$$

PROOF. By the square root trick, we have

$$\left| \sqrt{n \operatorname{Var}[\varphi^{ph}(x)]} - \sqrt{n \operatorname{Var}[f^{nh}(x)]} \right|$$

$$\leqslant \frac{|f_h(x)|^2}{\sqrt{n \operatorname{Var}[\varphi^{ph}(x)]}} = \frac{|f_h(x)|^2}{\sqrt{h^{-1} (A^2)_h * f(x)}} .$$

Now, on C, the function f is bounded and bounded away from 0, so that f_h is bounded and $(A^2)_h * f$ is bounded away from 0. Thus, the above is $\mathcal{O}(h^{1/2})$ uniformly on C.   Q.e.d.

We require the following basic inequalities, which follow from Theorem 1 in SWEETING (1977).

(8.16) LEMMA. [SWEETING (1977)] *Let $\zeta, \zeta_1, \zeta_2, \zeta_3, \cdots$ be a sequence of iid mean zero random vectors, with $\zeta_i = (\zeta_{i,1}, \zeta_{i,2}, \zeta_{i,3})^T \in \mathbb{R}^3$. Assume that*

$$\mathbb{E}[|\zeta_{i,1}|^3 + |\zeta_{i,2}|^3 + |\zeta_{i,3}|^3] < \infty ,$$

and let

$$\mathbb{E}[\,\zeta\,\zeta^{\,T}\,] = \Sigma^2 \overset{\text{def}}{=} \begin{bmatrix} 1 & \varrho_1 & \varrho_2 \\ \varrho_1 & 1 & \varrho_3 \\ \varrho_2 & \varrho_3 & 1 \end{bmatrix} .$$

Let $Z = (Z_1, Z_2, Z_3)^{\,T} \sim N_3(0, \Sigma^2)$. Finally, define the sums

$$S_{n,j} = n^{-1/2} \sum_{i=1}^{n} \zeta_{i,j} , \quad j = 1, 2, 3 .$$

Then, there exists a universal constant K, not depending on the distribution of ζ, such that

$$\left| \mathbb{E}[\,|\,S_{n,1}\,|\,] - \mathbb{E}[\,|\,Z_1\,|\,] \right| \leqslant K\,n^{-1/2}\,\mathbb{E}[\,|\,\zeta_{1,1}\,|^3\,] ,$$

$$\left| \mathbb{E}[\,|\,S_{n,1}\,S_{n,2}\,|\,] - \mathbb{E}[\,|\,Z_1\,Z_2\,|\,] \right| \leqslant K\,n^{-1/2}\,\mathbb{E}[\,|\,\zeta_{1,1}|^3 + |\,\zeta_{1,2}|^3\,] ,$$

$$\left| \mathbb{E}[\,S_{n,1}\,|\,S_{n,2}\,|\,] \right| \leqslant K\,n^{-1/2}\,\mathbb{E}[\,|\,\zeta_{1,1}|^3 + |\,\zeta_{1,2}|^3\,]$$

and

$$\left| \mathbb{E}[\,|\,S_{n,1}\,S_{n,2}\,S_{n,3}\,|\,] - \mathbb{E}[\,|\,Z_1\,Z_2\,Z_3\,|\,] \right| \leqslant K\,n^{-1/2}\,\mathbb{E}[\,|\,\zeta_{1,1}|^3 + |\,\zeta_{1,2}|^3 + |\,\zeta_{1,3}|^3\,] .$$

(8.17) LEMMA. *Let C be a Borel subset of $\mathbb{R}$ with finite Lebesgue measure satisfying (e) and (f) of the Approximation Lemma (8.11) for the collection $\mathcal{A} = \{\,A\,,\,A^2\,,\,|\,A\,|^3\,\}$. Then,*

$$\lim_{n \to \infty} \sqrt{n}\,\mathbb{E}[\,\|\,(\varphi^{ph} - f_h)\,\mathbb{1}_C\,\|_1\,] - \mathbb{E}[\,|\,\Upsilon\,|\,]\,\|\,\sqrt{n\,\mathbb{V}\text{ar}[\,\varphi^{ph}\,]}\,\mathbb{1}_C\,\|_1 = 0$$

and

$$\lim_{n \to \infty} \sqrt{n}\,\mathbb{E}[\,\|\,(f^{nh} - f_h)\,\mathbb{1}_C\,\|_1\,] - \mathbb{E}[\,|\,\Upsilon\,|\,]\,\|\,\sqrt{n\,\mathbb{V}\text{ar}[\,\varphi^{ph}\,]}\,\mathbb{1}_C\,\|_1 = 0 ,$$

where $\Upsilon \sim N(0, 1)$.

PROOF. Let $p_1 \sim \text{Poisson}(1)$, independent of the X_i, and define

$$W_n(x) = \frac{\{\,\sum_{i=1}^{p_1} A_h(x - X_i)\,\} - \mathbb{E}[\,A_h(x - X)\,]}{\sqrt{\mathbb{E}[\,|\,A_h(x - X)\,|^2\,]}} .$$

Then, $\mathbb{V}\text{ar}[\,W_n(x)\,] = 1$, and for constants K_r, uniformly in $x \in C$,

$$\mathbb{E}[\,|\,W_n(x)\,|^3\,] \leqslant K_1\,\frac{h^{-2}\,\mathbb{E}[\,|\,(A^3)_h(x - X)\,|\,]}{h^{-3/2}\,(\,\mathbb{E}[\,(A^2)_h(x - X)\,]\,)^{3/2}} \leqslant K_2\,h^{-1/2} ,$$

where the last inequality follows by an argument similar to the one at the end of the proof of Lemma (8.15).

Now, let $W_{n,1}(x), W_{n,2}(x), \cdots, W_{n,n}(x)$ be iid random variables, distributed as $W_n(x)$. Then, clearly,

$$\frac{\sqrt{n}\,[\,\varphi^{ph}(x) - f_h(x)\,]}{\sqrt{\mathbb{E}[\,|\,A_h(x - X)\,|^2\,]}} \quad \text{and} \quad \frac{1}{\sqrt{n}} \sum_{i=1}^{n} \{\,W_{n,i}(x) - \mathbb{E}[\,W_n(x)\,]\,\}$$

are equal in distribution. Therefore, by Lemma (8.16), there exists a constant K such that for all x,

$$\left|\,\frac{\mathbb{E}[\,\sqrt{n}\,|\,\varphi^{ph}(x) - f_h(x)\,|\,]}{\sqrt{\mathbb{E}[\,|\,A_h(x - X)\,|^2\,]}} - \mathbb{E}[\,|\,\Upsilon\,|\,]\,\right| \leqslant K\,n^{-1/2}\,\mathbb{E}[\,|\,W_n(x)\,|^3\,] \overset{\text{def}}{=} B_n(x)\,.$$

From the above, we see that the supremum of $B_n(x)$ over $x \in C$ is of order $(nh)^{-1/2}$, and since

$$(8.18) \quad \sup_{x \in C} \sqrt{\mathbb{E}[\,|\,A_h(x - X)\,|^2\,]} = \sup_{x \in C} \sqrt{n\,\mathbb{V}\mathrm{ar}[\,\varphi^{ph}(x)\,]} = \mathcal{O}\big(h^{-1/2}\big)\,,$$

then

$$\sup_{x \in C} \left|\,\sqrt{n}\,\mathbb{E}[\,|\,\varphi^{ph}(x) - f_h(x)\,|\,] - \mathbb{E}[\,|\,\Upsilon\,|\,]\,\sqrt{n\,\mathbb{V}\mathrm{ar}[\,\varphi^{ph}(x)\,]}\,\right| = \mathcal{O}\big((nh^2)^{-1/2}\big)\,.$$

Similarly, one obtains by Lemma (8.16),

$$\sup_{x \in C} \left|\,\sqrt{n}\,\mathbb{E}[\,|\,f^{nh}(x) - f_h(x)\,|\,] - \mathbb{E}[\,|\,\Upsilon\,|\,]\,\sqrt{n\,\mathbb{V}\mathrm{ar}[\,f^{nh}(x)\,]}\,\right| = \mathcal{O}\big((nh^2)^{-1/2}\big)\,.$$

By Lemma (8.15), this implies

$$\sup_{x \in C} \left|\,\sqrt{n}\,\mathbb{E}[\,|\,f^{nh}(x) - f_h(x)\,|\,] - \mathbb{E}[\,|\,\Upsilon\,|\,]\,\sqrt{n\,\mathbb{V}\mathrm{ar}[\,\varphi^{ph}(x)\,]}\,\right|$$

$$= \mathcal{O}\big((nh^2)^{-1/2} + h^{1/2}\big)\,,$$

thus, proving the lemma.   Q.e.d.

The next lemma and its proof provide a useful tool.

(8.18A) LEMMA. *Let the subset C of $\mathbb{R}$ have finite Lebesgue measure. Then, as $h \to 0$, the functions $\mathbb{1}_C(x + ht)$ and $f(x + ht)\,\mathbb{1}_C(x + ht)$ converge in $L^1(\mathbb{R} \times [-1, 1])$, and also in measure, to $\mathbb{1}_C(x)$ respectively $f(x)\,\mathbb{1}_C(x)$.*

PROOF. Let $g \in L^1(\mathbb{R})$, and observe that

$$\int_{-1}^{1} \int_{\mathbb{R}} |\,g(x + ht) - g(x)\,|\,dx\,dt \leqslant 2 \sup_{|\tau| \leqslant h} \int_{\mathbb{R}} |\,g(x + \tau) - g(x)\,|\,dx\,,$$

which tends to 0 as $h \to 0$ by the continuity of translation in $L^1(\mathbb{R})$. Thus, the function $g(x + ht)$ converges to $g(x)$ in $L^1(\mathbb{R} \times [-1, 1])$ and, therefore, also in measure. Now, take $g = \mathbb{1}_C$ and $g = f\,\mathbb{1}_C$.   Q.e.d.

Continuing our quest toward asymptotic normality, let

$$\sigma_n^2(C, A) = n \, \mathbb{V}\mathrm{ar}\big[\, \|\, (\varphi^{ph} - f_h)\, \mathbb{1}_C\, \|_1\, \big]\ .$$

(8.19) LEMMA. *Let C be a Borel subset of $\mathbb{R}$ with finite Lebesgue measure satisfying* (e) *and* (f) *of the Approximation Lemma* (8.11) *for the collection $\mathcal{A} = \{\, A,\, A^2,\, |A|^3\, \}$. Then,* $\lim\limits_{n\to\infty}\ \sigma_n^2(C, A) = F(C)\, \sigma^2(A)\ .$

PROOF. Observe that $|\, \varphi^{ph}(x) - f_h(x)\, |$ and $|\, \varphi^{ph}(y) - f_h(y)\, |$ are independent whenever $|x - y| > h$. Therefore,

$$n^{-1}\, \sigma_n^2(C, A) =$$

$$= \int_{C\times C} \big\{\, \mathbb{E}[\, |\, \varphi^{ph}(x) - f_h(x)\, |\, |\, \varphi^{ph}(y) - f_h(y)\, |\,] -$$

$$\mathbb{E}[\, |\, \varphi^{ph}(x) - f_h(x)\, |\,]\, \mathbb{E}[\, |\, \varphi^{ph}(y) - f_h(y)\, |\,]\, \big\}\, dy\, dx$$

$$= \int_{C\times C} \mathbb{1}(\, |x - y| \leqslant h\,)\, \mathbb{C}\mathrm{ov}\big(\, |\, \varphi^{ph}(x) - f_h(x)\, |\, ,\ |\, \varphi^{ph}(y) - f_h(y)\, |\, \big)\, dy\, dx\ .$$

Now, for each x let, $Z_n(x)$ be a standard normal, satisfying

$$\big(Z_n(x),\, Z_n(y)\big) =_d \big(\, \sqrt{1 - \varrho_n^2(x, y)}\ \Upsilon_1 + \varrho_n(x, y)\, \Upsilon_2,\, \Upsilon_2\, \big)\ ,$$

where Υ_1 and Υ_2 are independent standard normal random variables, and

$$\varrho_n(x, y) = \mathbb{E}[\, W_n(x)\, W_n(y)\,] = \frac{\mathbb{E}[\, A_h(x - X)\, A_h(y - X)\,]}{\sqrt{\mathbb{E}[\, |\, A_h(x - X)\, |^2\,]\, \mathbb{E}[\, |\, A_h(x - X)\, |^2\,]}}\ .$$

Note that $|\, \varrho_n(x, y)\, | \leqslant 1$. For later use, we introduce the abbreviation

$$\gamma_n(x, y) = \mathbb{C}\mathrm{ov}\big(\, |\, Z_n(x)\, |\, ,\ |\, Z_n(y)\, |\, \big)\ .$$

Now, by the properties (e) and (f) of Lemma (8.11), we have

$$h\, \mathbb{E}[\, |\, (A_h)^2(x - X)\, |\, \longrightarrow \|\, A\, \|_2^2\, f(x)\ \text{uniformly on}\ C\ ,$$

and likewise for

$$h\, \mathbb{E}[\, |\, (A_h)^2(y - X)\, |\, \longrightarrow \|\, A\, \|_2^2\, f(y)\ .$$

Also note that the same holds for the reciprocals.

Now, by Lemma (8.16), keeping in mind the formula for $\mathbb{V}\mathrm{ar}[\, \varphi^{ph}(x)\,]$, and the uniform boundedness of $\gamma_n(x, y)$, we see that

$$\sigma_n^2(C, A) =$$

$$h^{-1}\, \|\, A\, \|_2^2 \iint\limits_{C\times C} \mathbb{1}(\, |x - y| \leqslant h\,)\, \gamma_n(x, y)\, \sqrt{f(x)\, f(y)}\ dy\, dx + o(1)\ .$$

By a slight variation of the $L^1(\mathbb{R} \times [-1, 1])$ result of Lemma (8.18A), we see that

$$\sigma_n^2(C, A) = \|\, A\, \|_2^2\, \tau_n^2(C, A) + o(1)\ ,$$

where
$$\tau_n^2(C, A) = h^{-1} \iint\limits_{C \times C} \mathbb{1}(\,|x - y| \leqslant h\,)\,\gamma_n(x, y)\,f(x)\,dy\,dx\ .$$

It is convenient to perform the change of variable $y = x + t\,h$, which gives
$$\tau_n^2(C, A) = \int_C \int_{-1}^1 \gamma_n(x, x + t\,h)\,f(x)\,dt\,dx\ .$$

Again, by the uniform boundedness of $\gamma_n(x, y)$, the proof is completed if we can show that
$$\gamma_n(x, x + t\,h) \longrightarrow \mathbb{C}\mathrm{ov}\Big(\big|\,\sqrt{1 - \varrho^2(t)}\ \Upsilon_1 + \varrho(t)\,\Upsilon_2\big|, \big|\,\Upsilon_2\big|\Big)\ ,$$

in measure on $C \times [-1, 1]$, or, what is the same thing, that
$$\varrho_n(x, x + t\,h) \longrightarrow \varrho(t) \quad \text{in measure on } C \times [-1, 1]\ .$$

To study $\varrho_n(x, y)$, we must consider $h\,\mathbb{E}[\,A_h(x - X)\,A_h(x - X + t\,h)\,]$. This may be written as
$$f_{h, t}(x) \overset{\text{def}}{=} \int_{\mathbb{R}} h^{-1} A\big(h^{-1}(x - z)\big) A\big(h^{-1}(x - z) + t\big)\,f(z)\,dz\ .$$

By all accounts, this should converge to $J(t)\,f(x)$, where
$$J(t) = \int_{\mathbb{R}} A(u)\,A(u + t)\,du\ .$$

To make sure, note that
$$|\,f_{h, t}(x) - J(t)\,f(x)\,| \leqslant$$
$$\leqslant \int_{\mathbb{R}} h^{-1}\,\big|\,A\big(h^{-1}(x - z)\big) A\big(h^{-1}(x - z) + t\big)\,\big|\,\big|\,f(z) - f(x)\,\big|\,dz$$
$$\leqslant \|\,A\,\|_\infty^2 \int_{\mathbb{R}} h^{-1}\,\mathbb{1}(\,|x - z| \leqslant h\,)\,\big|\,f(z) - f(x)\,\big|\,dz$$
$$\leqslant \|\,A\,\|_\infty^2 \int_{-1}^1 \big|\,f(x + \tau h) - f(x)\,\big|\,d\tau\ ,$$

and so
$$\|\,f_{h, t} - J(t)\,f\,\|_{L^1(\mathbb{R} \times [-1, 1])} \leqslant$$
$$\leqslant \|\,A\,\|_\infty^2 \int_{-1}^1 \int_{\mathbb{R}} \int_{-1}^1 \big|\,f(x + \tau h) - f(x)\,\big|\,d\tau\,dx\,dt$$
$$\leqslant 4\,\|\,A\,\|_\infty^2 \sup_{|\tau| \leqslant h} \|\,f(\,\cdot\, + \tau) - f\,\|_{L^1(\mathbb{R})} \longrightarrow 0\ ,$$

the last statement by the continuity of translation in $L^1(\mathbb{R})$. Thus,
$$f_{h, t} \longrightarrow J(t)\,f \quad \text{in measure on } \mathbb{R} \times [-1, 1]\ .$$

By our earlier treatment of $h\,\mathbb{E}[|(A_h)^2(x-X)|]$ and $h\,\mathbb{E}[|(A_h)^2(x-X)|]$, we thus get

$$\varrho_n(x, x+t\,h) \longrightarrow \varrho(t) \text{ in measure on } C \times [-1, 1] .$$

The lemma follows.   Q.e.d.

Believe us, we are going in the right direction. Let

$$k_n(x) = \sqrt{\mathbb{E}[|A_h(x-X)|^2]} ,$$

and set

(8.20) $$\Delta_n = \frac{\sqrt{n}\,\{\,|\varphi^{ph}(x) - f_h(x)| - \mathbb{E}[\,|\varphi^{ph}(x) - f_h(x)|\,]\,\}}{k_n(x)} .$$

Assume that the set C satisfies the requirements of the Approximation Lemma (8.11), with M, ν, and α as stated. Assume also that n is so large that $h = h_n \leqslant \nu$ and $h \leqslant \frac{1}{2}M$. Define $m = m_n = \lfloor M/h \rfloor$, the integer part of M/h, and $h^* = M/m$. Then, $M/(2h) \leqslant m \leqslant M/h$, and so

$$h \leqslant h^* \leqslant 2h .$$

Define for all $i \in \mathbb{Z}$,

$$\delta_{i,n} = \{\sigma_n(C, A)\}^{-1} \int_{ih^*}^{(i+1)h^*} 1\!\!1_C(x)\,\Delta_n(x)\,k_n(x)\,dx .$$

(8.21) LEMMA. *Let C be a Borel subset of $\mathbb{R}$ with finite Lebesgue measure satisfying the conditions of the Approximation Lemma (8.11) for the collection $\mathcal{A} = \{A, A^2, |A|^3\}$. Then, there exists a constant K such that for all i, n,*

$$\mathbb{E}[\,|\delta_{i,n}|^3\,] \leqslant K\,h^{3/2} .$$

PROOF. Let $I_{i,n} = [ih^*, (i+1)h^*]$ and $J_{i,n} = I_{i,n} \times I_{i,n} \times I_{i,n}$. Observe that $\sigma_n^3(C, A)\,\mathbb{E}[\,|\delta_{i,n}|^3\,]$ is bounded by

$$\int_{J_{i,n}} 1\!\!1_{C^3}(x, y, z)\,\mathbb{E}[\,|\Delta_n(x)\,\Delta_n(y)\,\Delta_n(z)|\,]\,k_n(x)\,k_n(y)\,k_n(z)\,dx\,dy\,dz .$$

By Lemma (8.16), this may be bounded for a suitable constant K_1 by

$$K_1 \int_{J_{i,n}} 1\!\!1_{C^3}(x, y, z)\,M(x, y, z)\,k_n(x)\,k_n(y)\,k_n(z)\,dx\,dy\,dz +$$

$$K_1\,(nh)^{-1/2} \int_{J_{i,n}} k_n(x)\,k_n(y)\,k_n(z)\,dx\,dy\,dz ,$$

where

$$M(x, y, z) = \big(\mathbb{E}[\,|Z_n(x)|^3\,]\,\mathbb{E}[\,|Z_n(y)|^3\,]\,\mathbb{E}[\,|Z_n(z)|^3\,]\big)^{1/3} .$$

Since the volume of $J_{i,n}$ is bounded by $8\,h^3$, the statement (8.18) implies that the above is of the order $h^{3/2}$. Now, Lemma (8.19) does the trick. Q.e.d.

We need the following cute little fact.

(8.22) LEMMA. [BEIRLANT and MASON (1995)] Let $p \sim Poisson(\lambda)$. Then, for all $r \geqslant 1$, there exists a constant K_r such that

$$\mathbb{E}[\,|\,p - \lambda\,|^{2r}\,] \leqslant K_r \max(\lambda,\,\lambda^r)\,.$$

Introduce $p \sim Poisson(n)$, independent of the X_i, set $\mathcal{M} = [-M,\,M]$, and define

$$D_n = \sum_{i=-m}^{m-1} \delta_{i,n}\,,$$

$$U_n = n^{-1/2}\left\{\left(\sum_{i=1}^{p} \mathbb{1}(X_i \in \mathcal{M})\right) - n\,\mathbb{P}[\,X \in \mathcal{M}\,]\right\}\,,$$

$$V_n = n^{-1/2}\left\{\left(\sum_{i=1}^{p} \mathbb{1}(X_i \notin \mathcal{M})\right) - n\,\mathbb{P}[\,X \notin \mathcal{M}\,]\right\}\,.$$

Then, (D_n, U_n) and V_n are independent. Now, observe that

$$\mathbb{V}\mathrm{ar}[\,D_n\,] = 1\,,\quad \mathbb{V}\mathrm{ar}[\,U_n\,] = 1 - \alpha\,,$$

where $\alpha = \mathbb{P}(\,X \notin \mathcal{M}\,)$. By properties (e) and (f) of the Approximation Lemma (8.11), and Lemmas (8.16) and (8.22) applied inside the integral, then

$$|\,\mathbb{C}\mathrm{ov}(D_n, U_n)\,| \leqslant K\,(nh)^{-1/2} \int_C k_n(x)\,dx = \mathcal{O}\big((nh^2)^{-1/2}\big) \longrightarrow 0\,.$$

Thus, for any two scalars λ_1 and λ_2,

(8.23) $$\mathbb{V}\mathrm{ar}[\,\lambda_1\,D_n + \lambda_2\,U_n\,] \longrightarrow \lambda_1^2 + \lambda_2^2\,(1 - \alpha)\,.$$

All this helps in the proof of the following lemma.

(8.24) LEMMA. Let C be a Borel subset of $\mathbb{R}$ with finite Lebesgue measure satisfying the conditions of the Approximation Lemma (8.11) for the collection $\mathcal{A} = \{\,A,\,A^2,\,|\,A\,|^3\,\}$. Then,

$$(D_n, (1 - \alpha)^{-1/2}\,U_n) \longrightarrow_{\mathrm{d}} \Upsilon \sim N_2(0, I)\,,$$

where $\alpha = \mathbb{P}(\,X \notin \mathcal{M}\,)$. (Recall $0 < \alpha < \frac{1}{2}$.)

PROOF. By the Cramér-Wold device, see, e.g., BILLINGSLEY (1968), it suffices to show that for any λ_1, λ_2,

(8.25) $$\lambda_1\,D_n + \lambda_2\,U_n \longrightarrow_{\mathrm{d}} Z \sim N\big(0, \lambda_1^2 + \lambda_2^2\,(1 - \alpha)\big)\,.$$

To do this, we split U_n into little pieces

$$u_{i,n} = n^{-1/2} \left\{ \left(\sum_{j=1}^{p} \mathbb{1}(X_j \in I_{i,n}) \right) - n \, \mathbb{P}[X \in I_{i,n}] \right\} ,$$

with $I_{i,n} = [ih^*, (i+1)h^*]$ and

$$w_{i,n} = \lambda_1 \delta_{i,n} + \lambda_2 u_{i,n} .$$

By Jensen's inequality,

$$\sum_{i=-m}^{m-1} \mathbb{E}[|w_{i,n}|^3] \leqslant 4 \sum_{i=-m}^{m-1} \left\{ |\lambda_1|^3 \, \mathbb{E}[|\delta_{i,n}|^3] + |\lambda_2|^3 \, \mathbb{E}[|u_{i,n}|^3] \right\} .$$

Applying Lemma (8.21) and Lemma (8.22), this is readily shown to converge to zero as $n \to \infty$. Moreover, the sequence $w_{i,n}$, $-m \leqslant i \leqslant m-1$, is 1-dependent, and by (8.23),

$$\mathbb{V}\mathrm{ar}\Big[\sum_{i=-m}^{m-1} w_{i,n} \Big] \longrightarrow \lambda_1^2 + \lambda_2^2(1-\alpha) .$$

Thus, we can conclude (8.21) from Corollary 2 in SHERGIN (1979).   Q.e.d.

The next step is *dePoissonization*. We need another cute little fact.

(8.26) LEMMA. [BEIRLANT and MASON (1995)] *Let $q_{1,n}$ and $q_{2,n}$ be independent Poisson random variables with means $n(1-\alpha)$ and $n\alpha$, with $\alpha \in (0, \frac{1}{2})$. Set*

$$Q_n = n^{-1/2} \{ q_{1,n} - \mathbb{E}[q_{1,n}] \} , \quad R_n = n^{-1/2} \{ q_{2,n} - \mathbb{E}[q_{2,n}] \} .$$

Let $\{ S_n \}_n$ be a sequence of random variables such that
(a) for each $n \geqslant 1$, the random vector (S_n, Q_n) is independent of R_n, and
(b) for some positive constant σ, there holds

$$(S_n, (1-\alpha)^{-1/2} \sigma Q_n) \longrightarrow_{\mathrm{d}} Z \sim N_2(0, \sigma^2 I) .$$

Then, for all x,

$$\mathbb{P}[S_n \leqslant \sigma x \,|\, q_{1,n} + q_{2,n} = n] \longrightarrow \mathbb{P}[\Upsilon \leqslant x] ,$$

where $\Upsilon \sim N(0, 1)$.

Let

$$L_n(C) = \frac{n^{1/2}}{\sigma_n(C, A)} \left\{ \| (f^{nh} - f_h) \mathbb{1}_C \|_1 - \mathbb{E}[\| (f^{nh} - f_h) \mathbb{1}_C \|_1] \right\} .$$

(8.27) LEMMA. *Let C be a Borel subset of $\mathbb{R}$ with finite Lebesgue measure satisfying the conditions of the Approximation Lemma (8.11) for the collection $\mathcal{A} = \{ A, A^2, |A|^3 \}$. Then,*

$$L_n(C) \longrightarrow_{\mathrm{d}} \Upsilon \sim N(0, 1) .$$

PROOF. Note that D_n is almost the same as $L_n(C)$, but not quite:

$$D_n = \frac{n^{1/2}}{\sigma_n(C,A)} \left\{ \| (\varphi^{ph} - f_h) \mathbb{1}_C \|_1 - \mathbb{E}[\| (\varphi^{ph} - f_h) \mathbb{1}_C \|_1] \right\} ,$$

and conditioned on $p = n$, we have $D_n =_d \mathfrak{D}_n$, where

$$\mathfrak{D}_n = \frac{n^{1/2}}{\sigma_n(C,A)} \left\{ \| (f^{nh} - f_h) \mathbb{1}_C \|_1 - \mathbb{E}[\| (\varphi^{ph} - f_h) \mathbb{1}_C \|_1] \right\} .$$

By Lemmas (8.26) and (8.24), we have

$$\mathfrak{D}_n \longrightarrow_d \Upsilon \sim N(0,1) .$$

The conclusion follows upon using Lemmas (8.17) and (8.19).   Q.e.d.

Finally, we are ready for the main result.

PROOF OF THEOREM (8.9). Consider the sequence of Borel sets C_k, each with finite Lebesgue measure, and for each k, the properties of Lemma (8.11) hold, with

(8.28) $$\lim_{k \to \infty} \int_{C_k} f(x)\, dx = 1 .$$

Let Γ_k be the complement of C_k, and let $\mathbb{1}_k$ be the indicator function of Γ_k. By Lemma (8.13), we have

$$\limsup_{n \to \infty} \sqrt{n}\, \mathbb{E}\left[\left| \| (f^{nh} - f_h) \mathbb{1}_k \|_1 - \mathbb{E}[\| (f^{nh} - f_h) \mathbb{1}_k \|_1] \right|^2 \right]$$

$$\leqslant 4 \| A \|_1^2 \int_{\Gamma_k} f(x)\, dx .$$

By (8.28), this may be made arbitrarily small by choosing k large enough. Fix k. Then, by the previous lemma,

$$L_n(C_k) \longrightarrow_d \Upsilon \sim N(0,1) ,$$

and by Lemma (8.19),

$$\sigma_n^2(C_k, A) \longrightarrow F(C_k)\sigma^2(A) .$$

By a standard argument, see, e.g., Theorem 4.2 in BILLINGSLEY (1968),

(8.29) $$\sqrt{n} \left\{ \| f^{nh} - f_h \|_1 - \mathbb{E}[\| f^{nh} - f_h \|_1] \right\} \longrightarrow_d Z \sim N\left(0, \sigma^2(A) \right) .$$

Finally, by Theorem 2.3 of PINELIS (1990), we have for all $r \geqslant 2$,

$$\mathbb{E}\left[\left| n^{1/2} \left\{ \| f^{nh} - f_h \|_1 - \mathbb{E}[\| f^{nh} - f_h \|_1] \right\} \right|^r \right] < \infty ,$$

and by (8.29), this implies the variance part of the theorem, using Theorem 5.4 of BILLINGSLEY (1968).   Q.e.d.

9. Additional comments

Ad § 1 : The treatment of the maximum likelihood interpretation of F_n was inspired by WALTER and BLUM (1984).

The sentiments expressed in the introduction regarding the separation of concerns for the kernel and the smoothing parameter are not universally accepted, see, e.g., WATSON and LEADBETTER (1963).

The two-sided exponential kernel plays an important role in this text. It also goes under the name of the Picard kernel, see, e.g., SHAPIRO (1969), and the Laplace density, see, e.g., DEVROYE and GYÖRFI (1985).

Ad § 4 : We owe David Mason for pointing out to us the submartingale structure of the L^1 error and its application to the Kullback-Leibler distance. The exponential inequalities of DEVROYE (1991) provide the almost perfect treatment of the L^1 error. For more variations on martingale inequalities, see, e.g., NEVEU (1975), MILMAN and SCHECHTMAN (1986), and VAN DE GEER (2000).

The blocking technique used is well known and appears in the standard probability text books.

Ad § 5 : Regarding Pearson's φ^2 distance, everything is under control if one restricts attention to a finite interval on which the density is bounded away from 0. See ROSENBLATT (1956), BICKEL and ROSENBLATT (1973), and DEHEUVELS and MASON (1992).

Ad § 6 : This section follows EGGERMONT and LARICCIA (1999a). Some relevant references for nonparametric entropy estimation are AHMAD and LIN (1976) and HALL and MORTON (1993), who study the estimator discussed in Exercise (6.33). We also mention GYÖRFI and VAN DER MEULEN (1987), (1990), JOE (1989), MOKKADEM (1989), TSYBAKOV and VAN DER MEULEN (1996), and VAN ES (1992), as well as the review article BEIRLANT, DUDEWICZ, GYÖRFI and VAN DER MEULEN (1997).

Ad § 7 : ZHANG and FAN (2000) give a minimax justification for minimizing $J_k(A)$ and determine the optimal kernels. Somewhat different notions of optimality are explored by BERLINET (1993).

It should be noted that the classes $\mathcal{B}(t)$ of (7.21) may be made into a reproducing kernel Hilbert space and that (7.18) is a statement about the reproducing kernel. See Appendix 3, § 7.

Ad § 9 : Had there been a real § 9, it would have dealt with boundary kernels for the estimation of densities on bounded intervals or on the half line. Here we merely state some references, HALL and WEHRLY (1991), JONES (1993), and MÜLLER (1993). In Chapter 6, we use these for estimating monotone densities on $(0, \infty)$.

5

Nonparametric Maximum Penalized Likelihood Estimation

1. Introduction

Here, we begin a serious study of maximum penalized likelihood estimation. For a general introduction to nonparametric maximum likelihood estimation, see § 4.1.

Let $X_1, X_2, \cdots, X_n$ be iid random variables, with common density f_o. The maximum penalized likelihood problem for estimating f_o is

(1.1)
$$\text{minimize} \quad -\frac{1}{n} \sum_{i=1}^{n} \log f(X_i) + h^2 R(f)$$
$$\text{subject to} \quad f \text{ is a pdf} ,$$

where $R(f)$ is the roughness penalization functional. Two instances are considered, viz. the first roughness penalization of GOOD (1971), GOOD and GASKINS (1971),

(1.2)
$$R(f) = \left\| \left\{ \sqrt{f} \right\}' \right\|_2^2$$

(L^2 norm, and the prime denotes differentiation), and the penalization functional of SILVERMAN (1982)

(1.3)
$$R(f) = \left\| \left\{ \log f \right\}^{(3)} \right\|_2^2 ,$$

where "(3)" denotes the *third* derivative. Other derivatives (differential operators) here and in (1.2) suggest themselves, but are not considered. Despite a flurry of activity in the 1970's and early 1980's, maximum penalized likelihood estimation did not catch on, for obvious reasons. From a practical point of view, the computation of the mple's remains a nontrivial problem demanding close attention. Theoretically, establishing consistency and convergence rates proved remarkably hard. Yet both penalizations (1.2) and (1.3) yield remarkably accurate estimators and seem to be making a modest come back.

In this chapter, we study a.s. convergence rates for the above mple's. For the GOOD estimator, this proves spectacularly successful, due to what appears to be a lucky accident. In general, the penalization employed strongly suggests how the error in the estimator should be measured. For the GOOD estimator, the Hellinger distance recommends itself; for the log-penalization of SILVERMAN, the (symmetrized) Kullback-Leibler distance seems natural. So for these mple's, everything seems clear, except that we wish to consider the L^1 error. Actually, there is one other maximum penalized likelihood problem where everything is clear, i.e., in the case of kernel density estimation. As already mentioned in §4.1, if A is a symmetric pdf, then the kernel estimator $f^{nh} = A_h * dF_n$ is the solution to the maximum smoothed likelihood problem

$$(1.4) \qquad \text{minimize} \quad -\frac{1}{n} \sum_{i=1}^{n} [A_h * \{\log f\}](X_i) + \int_{\mathbb{R}} f(y) \, dy \ .$$

But the usual analyses of the kernel estimator do not make use of this property.

The interpretation of kernel estimators as maximum smoothed likelihood estimators is particularly useful when estimating densities under (shape) constraints. Some typical examples occur in the estimation of monotone, unimodal, or log-concave densities. To make things concrete, consider the estimation of log-concave densities. Denote the set of such densities by $\mathcal{C}$. The maximum smoothed likelihood problem of estimating a log-concave density could thus be phrased as (1.4) with minimization only over $f \in \mathcal{C}$. Equivalently, the constrained estimation problem is

$$(1.5) \qquad \begin{aligned} \text{minimize} \quad &\text{KL}(A_h * dF_n \, , \, f) \\ \text{subject to} \quad &f \in \mathcal{C} \, , \end{aligned}$$

with KL the Kullback-Leibler distance. Thus, the maximum smoothed likelihood problem results in a minimum distance estimation problem and suggests the use of other distances. Such methods are well known in parametric estimation, see §2.8. Besides the Kullback-Leibler distance, other choices are the Hellinger distance or the L^1 norm. For these distances, some of the technical difficulties with (1.5) proper seem to go away. Although the praise of minimum Hellinger distance estimation has been sung in the literature, we shall stick with maximum penalized likelihood estimation. Monotone density estimation has somewhat special properties, and we give a detailed study of (1.5) with the monotonicity constraint in Chapter 6. This is then put to good use in the study of the estimation of unimodal densities.

Be that as it may, the best thing is to let these approaches speak for themselves. One reason for discussing them here is that the setting is fairly simple, and that it gives some indication of what must be done for more complicated problems, such as generalized deconvolution or Poisson regression. See Volume II. It should be noted that the approach is different

from the WALD (1949) approach to (parametric) maximum likelihood estimation of § 2.2. The development of § 2.2 was geared toward showing (best)
asymptotic normality. Here, we are satisfied with less, but enquiring minds
are encouraged to see how distinct (similar!?) the approaches actually are.

2. Good's roughness penalization of root-densities

In this section, we study the maximum penalized likelihood density estimation problem

$$(2.1) \quad \text{minimize} \quad -\tfrac{1}{n} \sum_{i=1}^{n} \log f(X_i) + \int_{\mathbb{R}} f(x)\,dx + h^2\,R(f)$$

$$\text{subject to} \quad f \in L^1(\mathbb{R})\,, \quad f \geqslant 0\,,$$

where the roughness penalization functional is

$$(2.2) \quad R(f) = \|\,(f^{1/2})'\,\|_2^2\,,$$

see GOOD (1971) and GOOD and GASKINS (1971). It turns out that the
solution exists and is unique, see § 11.3. Denote the solution of (2.1)–(2.2)
by f^{nh}. We perform a thorough analysis of the error in f^{nh}, culminating
in the error bound under quite reasonable conditions on the true pdf f_o,
for $h \asymp n^{-1/5}$,

$$(2.3) \quad \| f^{nh} - f_o \|_1 =_{\text{as}} \mathcal{O}\big(n^{-2/5} \big)\,,$$

see Theorem (2.51) below. In other words, the same convergence rate as
for kernel density estimation is obtained. Consistency under minimal conditions is also considered, under the heading of universal consistency. The
crux in the treatment of the GOOD estimator is a fundamental, if unexpected, connection of the estimator with kernel estimators, after which we
essentially follow KLONIAS (1984) with some important technical improve-
ments. Computational issues are briefly addressed at the end of the section.
Existence and uniqueness of the estimator are considered in Chapter 10.

This section is divided into a number of subsections, with the headings:
Some preliminary observations, The Euler equations, Comparison with kernel density estimators, The convergence of the bias term, The convergence
of the variance term, Almost sure convergence rates, The computation of
the Good estimator, and Universal consistency.

Some preliminary observations. The reader will have noticed that
the pdf constraint is not enforced in (2.1), the reason being that it is awkward to do so, but in fact, it has advantages: It makes life simpler, it does
not change the rate of approximation of the true density by the estimator,
and if the solution without the pdf constraint is scaled to be a pdf, then
this is the solution with the pdf constraint for a slightly different value of
h. See Exercise (2.62).

(2.4) EXERCISE. Show that if f solves (2.1)–(2.2), then $\int_{\mathbb{R}} f = 1 - h^2 R(f)$. [Hint: Denote the objective function in (2.1) by $L_{nh}(f)$. If f is the minimizer, consider $L_{nh}(t\,f)$ as function of t, $t > 0$. Its minimum occurs for $t = 1$, so the derivative with respect to t vanishes at $t = 1$.]

The second observation is that the transformation $u = f^{1/2}$ seems to recommend itself. Then, problem (2.1)–(2.2) is equivalent to

$$\text{minimize} \quad -\tfrac{2}{n} \sum_{i=1}^{n} \log u(X_i) + \| u \|_2^2 + h^2 \| u' \|_2^2$$

(2.5)

$$\text{subject to} \quad u \in W^{1,2}(\mathbb{R}) \,, \ u \geqslant 0 \,,$$

with $W^{1,2}(\mathbb{R})$ as in (4.1.45). See Exercise (2.14). Thus, the problem has a Hilbert space setting, which has many advantages. The third observation is the following. It seems reasonable to think that if the solution u_n of (2.5) is going to converge to anything for fixed h, it must be toward w, the solution of

$$\text{minimize} \quad -2 \int_{\mathbb{R}} \log u(x)\,dF_o(x) + \| u \|_2^2 + h^2 \| u' \|_2^2$$

(2.6)

$$\text{subject to} \quad u \in W^{1,2}(\mathbb{R}) \,, \ u \geqslant 0 \,.$$

This is the large sample asymptotic problem. Showing the existence and uniqueness of the solution of (2.6) is similar to that for the small sample problem, see § 11.3. As in kernel density estimation, the error $u_n - f^{1/2}$ is then decomposed into an asymptotic bias and variance term

(2.7) $$u_n - f_o^{1/2} = \{\, u_n - w \,\} + \{\, w - f_o^{1/2} \,\} \,.$$

The study of the last term falls again under the heading of approximation theory. The study of the first term is (applied) probability. Because we are interested in $\| f^{nh} - f_o \|_1$, a reasonable norm to use for $u_n - f_o^{1/2}$ is the L^2 norm, in view of the fact that for (sub)pdfs φ and ψ,

(2.8) $$\| \varphi - \psi \|_1 \leqslant 2 \, \| \varphi^{1/2} - \psi^{1/2} \|_2 \,,$$

see (1.3.21). Here, a sub-pdf is a nonnegative, integrable function with integral $\leqslant 1$. The above would seem to be the general approach to obtaining convergence rates for the present maximum penalized likelihood estimator. However, we are in for a pleasant surprise.

The Euler equations. We first discuss the necessary and sufficient conditions for a solution of (2.1)–(2.2), known as the Euler equations. As shown in § 11.3, the necessary and sufficient condition for u_n to solve (2.5) is that it solves the Euler equations

$$-h^2\,u'' + u = \frac{dF_n}{u} \qquad \text{on } \mathbb{R} \,,$$

(2.9)

$$u \longrightarrow 0 \qquad \text{at } \pm\infty \,.$$

The difficulty is the interpretation of the differential equation with the point masses in the right-hand side, see §11.3. In short, the function u solves (2.9) if for all $\psi \in C_o^2(\mathbb{R})$, the set of all twice continuously differentiable ψ with compact support,

$$(2.10) \qquad -h^2 \int_{\mathbb{R}} u \, \psi'' + \int_{\mathbb{R}} u \, \psi = \int_{\mathbb{R}} \frac{\psi}{u} \, dF_n \; ,$$

see COURANT and HILBERT (1953). Here, u'' is a measure that satisfies

$$\int_{\mathbb{R}} \psi \, u'' = \int_{\mathbb{R}} u \, \psi'' \; , \qquad \text{for all } \psi \in C_o^2(\mathbb{R}) \; ,$$

cf. Appendix 3, Exercise (1.6)(g). The solution w of the large sample asymptotic problem (2.6) satisfies a similar equation. Recall from §4.6 that the scaled two-sided exponential density $\mathcal{B}_h(x) = (2h)^{-1} \exp(-h^{-1}|x|)$ is the Green's function for the boundary value problem (2.9), i.e., for a distribution G, the function $u = \mathcal{B}_h * dG$ solves the problem

$$(2.11) \qquad \begin{aligned} -h^2 \, u'' + u &= dG & &\text{on } \mathbb{R} \; , \\ u &\longrightarrow 0 & &\text{at } \pm\infty \; . \end{aligned}$$

Hence, the solution of (2.9) is given implicitly by

$$(2.12) \qquad u_n(x) = \frac{1}{n} \sum_{i=1}^{n} \frac{\mathcal{B}_h(x - X_i)}{u_n(X_i)} \; , \qquad -\infty < x < \infty \; ,$$

or, equivalently, with a slight abuse of notation,

$$(2.13) \qquad u_n = \mathcal{B}_h * \frac{dF_n}{u_n} \; .$$

Note that (2.13) implies that $u_n(x) > 0$ for all x. A similar derivation applies to w, the solution of the large sample asymptotic problem (2.6).

(2.14) EXERCISE. Show that problem (2.1)–(2.2) and problem (2.5) are equivalent, i.e., if u_n solves (2.5), then $f_n = (u_n)^2$ solves (2.1)–(2.2), and vice versa.

It is easily verified that $\left(\mathcal{B}_h \right)'(x) = -\mathrm{sign}(x) \, h^{-1} \mathcal{B}_h(x)$, whence (2.13) implies

$$(2.15) \qquad |u_n'(x)| \leqslant h^{-1} \left[\mathcal{B}_h * \frac{dF_n}{u_n} \right](x) = h^{-1} u_n(x) \; , \qquad \text{for all } x \; .$$

This very useful inequality is due to DE MONTRICHER, TAPIA and THOMPSON (1975).

Comparison with kernel density estimators. Euler's equation leads to the following unexpected comparison of the GOOD estimator with kernel density estimators. Let G be an arbitrary distribution function, and let v be the solution of

$$-h^2 v'' + v = \frac{dG}{v} \qquad \text{on } \mathbb{R} ,$$
(2.16)
$$v \longrightarrow 0 \qquad \text{at } \pm \infty .$$

As in (2.15), we have the inequality

(2.17)
$$|v'| \leqslant h^{-1} v .$$

From the observation that $(v^2)'' = 2 v v'' + 2 (v')^2$, it follows that v^2 satisfies the differential equation

(2.18)
$$-\tfrac{1}{2} h^2 (v^2)'' + v^2 = dG - V v^2 ,$$

where $V = h^2 (v'/v)^2$. From (2.17), we have that $0 \leqslant V \leqslant 1$. Now, on the one hand, we obtain with $\lambda = h/\sqrt{2}$ that

(2.19)
$$v^2 = \mathfrak{B}_\lambda * \{ dG - V v^2 \} \leqslant \mathfrak{B}_\lambda * dG .$$

On the other hand, rewriting (2.18) as

$$-\tfrac{1}{4} h^2 (v^2)'' + v^2 = \tfrac{1}{2} dG + \tfrac{1}{2} (1 - V) v^2$$

yields

(2.20)
$$v^2 = \tfrac{1}{2} \mathfrak{B}_{h/2} * \{ dG + (1 - V) v^2 \} \geqslant \tfrac{1}{2} \mathfrak{B}_{h/2} * dG .$$

We have thus proven the following quite remarkable comparisons:

(2.21)
$$\tfrac{1}{2} \mathfrak{B}_{h/2} * dG \leqslant v^2 \leqslant \mathfrak{B}_{h/\sqrt{2}} * dG .$$

It should be observed that these inequalities hold whether G is absolutely continuous or not. In particular, we may take $G = F_o$ as well as $G = F_n$.

(2.22) COMPARISON-WITH-KERNEL-ESTIMATION-LEMMA. *The* GOOD *estimator* u_n *and the large sample asymptotic estimator* w *satisfy*

$$\tfrac{1}{2} \mathfrak{B}_{h/2} * dF_n \leqslant (u_n)^2 \leqslant \mathfrak{B}_{h/\sqrt{2}} * dF_n ,$$
$$\tfrac{1}{2} \mathfrak{B}_{h/2} * dF_o \leqslant w^2 \leqslant \mathfrak{B}_{h/\sqrt{2}} * dF_o .$$

The *coup de grâce* is the following observation regarding Lemma (2.22). The stated lower bounds are not very accurate: u_n^2 and w^2 are estimators of pdfs, but the lower bounds integrate to $\tfrac{1}{2}$. The upper bounds are superb, however, because they are pdfs themselves. Thus, the upper bound implies

(2.23) $\| u_n^2 - \mathfrak{B}_{h/\sqrt{2}} * dF_n \|_1 = \displaystyle\int_{\mathbb{R}} \{ \mathfrak{B}_{h/\sqrt{2}} * dF_n - u_n^2 \} = h^2 \| (u_n)' \|_2^2 ,$

see Exercise (2.4). So all that must be shown is that (taking a not-so-wild guess as to what it ought to be)

$$(2.24) \qquad \| (u_n)' \|_2 \longrightarrow \| (f_o^{1/2})' \|_2 \,, \qquad \text{almost surely, say },$$

and we would have that u_n^2 behaves essentially like a kernel density estimator, with the two-sided exponential kernel.

It is possible that there is a quick way to establish (2.24), but it should be pointed out that the comparison with kernel density estimators appears to be somewhat of a coincidence. For other roughness penalty functionals, we may not be so lucky. So, in the following, we give the authorized version via the study of the bias and variance terms.

The convergence of the bias term. For the remainder, the "true" density f_o is denoted by plain f. We assume here that $f^{1/2} \in W^{2,2}(\mathbb{R})$. What happens when only $f^{1/2} \in W^{1,2}(\mathbb{R})$, or, heaven forbid, $f^{1/2} \in L^2(\mathbb{R})$ is left until later. Analogous to (2.9), the solution w of the large sample asymptotic problem satisfies

$$(2.25) \qquad \begin{aligned} -h^2 \, w'' + w &= \frac{f}{w} \qquad && \text{on } \mathbb{R} \,, \\ w &\longrightarrow 0 && \text{at } \pm \infty \,. \end{aligned}$$

Writing f/w as

$$\frac{f}{w} = f^{1/2} - \frac{f^{1/2}}{w} \, (w - f^{1/2}) \,,$$

substituting this into (2.25), and adding $h^2 (f^{1/2})''$ to both sides gives

$$(2.26) \qquad -h^2 \, (w - f^{1/2})'' + \Big(1 + \frac{f^{1/2}}{w}\Big) (w - f^{1/2}) = h^2 \, (f^{1/2})'' \,.$$

Now, multiply both sides by $w - f^{1/2}$ and integrate over $\mathbb{R}$. After integration by parts for the first term, we get

$$(2.27) \quad h^2 \, \| (w - f^{1/2})' \|_2^2 + \| w - f^{1/2} \|_2^2 +$$
$$\int_{\mathbb{R}} \frac{f^{1/2}}{w} \, | w - f^{1/2} |^2 = h^2 \int_{\mathbb{R}} (f^{1/2})'' \, (w - f^{1/2}) \,.$$

The right-hand side may be bounded with Cauchy-Schwarz as

$$\int_{\mathbb{R}} (f^{1/2})'' \, (w - f^{1/2}) \leqslant h^2 \, \| (f^{1/2})'' \|_2 \, \| w - f^{1/2} \|_2 \,.$$

Upon ignoring the ugly term on the left of (2.27), we get the beautiful inequality

$$(2.28) \quad h^2 \, \| (w - f^{1/2})' \|_2^2 + \| w - f^{1/2} \|_2^2 \leqslant h^2 \, \| (f^{1/2})'' \|_2 \, \| w - f^{1/2} \|_2 \,.$$

Now, ignore the first term on the left, and cancel a factor $\| w - f^{1/2} \|_2$, to see that

$$\| w - f^{1/2} \|_2 \leqslant h^2 \, \| (f^{1/2})'' \|_2 \,.$$

Consequently, the right-hand side of (2.28) is less than $h^4 \| (f^{1/2})'' \|_2^2$. Now, ignore the second term on the left of (2.28), and obtain

$$\| (w - f^{1/2})' \|_2 \leqslant h \| (f^{1/2})'' \|_2 .$$

We have proven the following theorem.

(2.29) THEOREM. If $f^{1/2} \in W^{2,2}(\mathbb{R})$, then

$$\| w - f^{1/2} \|_2 \leqslant h^2 \| (f^{1/2})'' \|_2 \quad , \quad \| (w - f^{1/2})' \|_2 \leqslant h \| (f^{1/2})'' \|_2 .$$

(2.30) EXERCISE. Show that the theorem implies

$$\| w - f^{1/2} \|_\infty \leqslant 3^{1/4} h^{3/2} \| (f^{1/2})'' \|_2 .$$

(2.31) EXERCISE. (a) Working directly with (2.6) show that

$$\mathrm{KL}(f , w^2) + h^2 \| w' \|_2^2 \leqslant h^2 \| (f^{1/2})' \|_2^2 .$$

(b) Conclude that $\| w' \|_2 \leqslant \| (f^{1/2})' \|_2$.
(c) Use this to show that $\| w - f^{1/2} \|_2 \leqslant h \| (f^{1/2})' \|_2$.
(d) Improve (2.4) to $1 - h^2 \| (f^{1/2})' \|_2^2 \leqslant \| w \|_2^2 \leqslant 1 - h^2 \| w' \|_2^2$.

The convergence of the variance term. From (2.9) and its analogue for w, we get that

$$-h^2 (u_n - w)'' + u_n - w = \frac{dF_n}{u_n} - \frac{dF}{w} .$$

Multiplying both sides by $u_n - w$ and integrating (by parts for the first term) gives

$$(2.32) \qquad h^2 \| u_n' - w' \|_2^2 + \| u_n - w \|_2^2 = \int_{\mathbb{R}} (u_n - w) \left(\frac{dF_n}{u_n} - \frac{dF}{w} \right) .$$

Note that the integral on the right is well defined if we admit the value $-\infty$, since

$$\int_{\mathbb{R}} (u_n - w) \frac{dF_n}{u_n}$$

is just a finite sum, and

$$\int_{\mathbb{R}} (u_n - w) \left\{ -\frac{dF}{w} \right\} = 1 - \int_{\mathbb{R}} u_n \frac{dF}{w} < 1 .$$

Now, observe that there are two equivalent ways of writing the measure on the right of (2.32), viz.

$$(u_n - w) \left(\frac{dF_n}{u_n} - \frac{dF}{w} \right) = -\frac{(u_n - w)^2}{u_n w} dF_n + \frac{u_n - w}{w} (dF_n - dF)$$

$$= -\frac{(u_n - w)^2}{u_n w} dF + \frac{u_n - w}{u_n} (dF_n - dF) .$$

It is obvious that the negative quadratic terms are going to be dropped, but there is some freedom in that it can be done in either formula. Doing this so as to maximize the denominator of the remaining term seems a good idea, so

$$\int_{\mathbb{R}} (u_n - w) \left(\frac{dF_n}{u_n} - \frac{dF}{w} \right) \leqslant \int_{\mathbb{R}} \frac{u_n - w}{u_n \vee w} \left(dF_n - dF \right)$$

(recall that u_n and w are both strictly positive) and (2.32) is replaced by

$$(2.33) \qquad h^2 \, \| u_n' - w' \|_2^2 + \| u_n - w \|_2^2 \leqslant \int_{\mathbb{R}} \frac{u_n - w}{u_n \vee w} \left(dF_n - dF \right) \, .$$

And now what? The right-hand side is of the form $\int_{\mathbb{R}} \varphi \, (dF_n - dF)$ for a nice function φ, we hope, but φ is random! The similarity with kernel estimation suggests integration by parts. We have in fact the new and improved integration by parts trick, as follows.

(2.34) LEMMA. *Let T_λ be the one-sided exponential distribution*
$$T_\lambda(x) = \lambda^{-1} \exp(-\lambda^{-1} x) \, \mathbb{1}(x \geqslant 0) \, , \quad x \in \mathbb{R} \, .$$
Then, for all $\varphi \in W^{1,1}(\mathbb{R})$ and all distribution functions Ψ,

$$\int_{\mathbb{R}} \varphi(x) \, d\Psi(x) = \int_{\mathbb{R}} \left\{ -\lambda \, \varphi'(x) + \varphi(x) \right\} [\, T_\lambda * d\Psi \,](x) \, dx \, .$$

(2.35) EXERCISE. Prove Lemma (2.34) using plain integration by parts on
$$\int_{\mathbb{R}} \varphi'(x) \, [\, T_\lambda * d\Psi \,](x) \, dx \, .$$

Applying Lemma (2.34) to the right-hand side of (2.33) gives

$$(2.36) \qquad \text{rhs} = \int_{\mathbb{R}} \left\{ (-\lambda \nabla + 1) \frac{u_n - w}{u_n \vee w} \right\} T_\lambda * \left(dF_n - dF \right) \, ,$$

where ∇ represents differentiation, so $\nabla\varphi(x) \equiv \varphi'(x)$. This uses implicitly that $(u_n - w)/(u_n \vee w)$ is bounded at $\pm\infty$ [Exercise (2.38)]. Now, one easily verifies that

$$(2.37) \quad \{ u_n \vee w \} \left\{ (-\lambda \nabla + 1) \frac{u_n - w}{u_n \vee w} \right\} =$$

$$- \lambda \, (u_n' - w') + (u_n - w) \left\{ \lambda \frac{\{ u_n \vee w \}'}{u_n \vee w} + 1 \right\} \, .$$

We need a bound for the last factor.

(2.38) EXERCISE. (a) Show that $| \{ u_n \vee w \}' | \leqslant \lambda^{-1} \, u_n \vee w$, except possibly at isolated points, where $u_n = w$. *Open intervals where $u_n = w$ cause no problems.*
(b) Give a nice bound for $| (u_n - w)/(u_n \vee w) |$ everywhere.

Using the bound of Exercise (2.38) in (2.37) yields

$$(2.39) \qquad \left| \left\{ (-\lambda \nabla + 1) \, \frac{u_n - w}{u_n \vee w} \right\} \right| \leqslant \frac{\lambda \, | \, u_n' - w' \, |}{u_n \vee w} + \frac{2 \, | \, u_n - w \, |}{u_n \vee w} \, ,$$

and the Cauchy-Schwarz inequality gives that the "rhs" of (2.33) satisfies

$$(2.40) \qquad \text{rhs} \leqslant B \, \{ \lambda \, \| \, u_n' - w' \, \|_2 + 2 \, \| \, u_n - w \, \|_2 \} \; E_{n,\lambda} \, ,$$

where

$$(2.41) \qquad E_{n,\lambda} = \left\{ \int_{\mathbb{R}} \frac{| \, T_\lambda * (\, dF_n - dF) \, |^2}{(T_\lambda * f) \vee (T_\lambda * dF_n)} \right\}^{1/2}$$

and

$$(2.42) \qquad B = \left\| \frac{(T_\lambda * dF_n) \vee (T_\lambda * dF)}{(u_n \vee w)^2} \right\|_\infty \, .$$

Before worrying about the choice of λ and getting a bound for B, we note that the complicated integral on the right of (2.41) is very close to a familiar customer.

(2.43) EXERCISE. Let φ, ψ be pdfs. With $H(\varphi, \psi)$ the Hellinger distance, show that

$$H(\varphi, \psi) \leqslant \int_{\mathbb{R}} \frac{| \, \varphi - \psi \, |^2}{\varphi \vee \psi} \leqslant 4 \, H(\varphi, \psi) \, .$$

Now on to choosing λ. In view of the Comparison Lemma (2.22), the choice $\lambda = h/2$ seems to make eminent sense. Since $\mathcal{B}_\lambda(x) \geqslant \frac{1}{2} T_\lambda(x)$, then

$$(u_n)^2 \geqslant \tfrac{1}{2} \, \mathcal{B}_{h/2} * dF_n \geqslant \tfrac{1}{4} \, T_{h/2} * dF_n,$$

and likewise for w, which gives the very nice bound $B \leqslant 4$.

Parenthetically, it was the concern about B being bounded that prompted the search for lower bounds on the GOOD estimator, and then the upper bounds of Lemma (2.22) are but a small additional step.

We can now take stock. From (2.33), (2.36), (2.39), and (2.40)–(2.42), we obtain

$$(2.44) \qquad e_{n,h}^2 \leqslant 8 \, e_{n,h} \, E_{n,h/2} \, ,$$

where

$$(2.45) \qquad e_{n,h}^2 = h^2 \, \| \, u_n' - w' \, \|_2^2 + \| \, u_n - w \, \|_2^2 \, .$$

It follows that

$$(2.46) \qquad e_{n,h} \leqslant 8 \, E_{n,h/2} \, .$$

As noted in Exercise (2.43), the right-hand side behaves like the Hellinger distance between $T_{h/2} * dF_n$ and $T_{h/2} * dF$. In view of (1.3.21), this is dominated by the Kullback-Leibler distance, which we studied in Exercise (4.5.23). The fact that T_h is not symmetric or does not have a vanishing first moment does not matter. What does matter is that the one-sided exponential distribution has finite moments of all orders. We may thus state the following result.

(2.47) THEOREM. (a) *There is a universal constant c such that for all $h > 0$,*

$$h^2 \| u_n' - w' \|_2^2 + \| u_n - w \|_2^2 \leqslant c \, \mathrm{H}(\, T_{h/2} * dF_n \,, T_{h/2} * dF \,) \,.$$

(b) *If $\mathbb{E}\big[\, |X|^m \,\big] < \infty$ for some $m > \kappa > 2$, then uniformly in $h > 0$ (deterministically) as $n \to \infty$,*

$$h^2 \| u_n' - w' \|_2^2 + \| u_n - w \|_2^2 =_{\mathrm{as}} \mathcal{O}\big((nh)^{-2\kappa/(2\kappa+1)}\big) \,,$$

$$\| u_n' \|_2 - \| (f^{1/2})' \|_2 =_{\mathrm{as}} \mathcal{O}\big(h^{-1}(nh)^{-\kappa/(2\kappa+1)}\big) \,.$$

(c) *Finally, if $\mathbb{E}\big[\, e^{r|X|} \,\big] < \infty$ for some $r > 0$, then for all $s > 1$,*

$$h^2 \| u_n' - w' \|_2^2 + \| u_n - w \|_2^2 =_{\mathrm{as}} \mathcal{O}\big((nh)^{-1} \log(nh)^{-1} (\log n)^s\big) \,,$$

uniformly in $h > 0$ (deterministically) as $n \to \infty$.

(2.48) EXERCISE. Prove Theorem (2.47).

(2.49) EXERCISE. Reformulate Theorem (2.47) for random h, and prove it. (The authors do not know how to do it, but monotonicity in h may prove helpful.)

(2.50) EXERCISE. Formulate and prove rates of convergence for $u_n \longrightarrow f^{1/2}$ by supplying bounds for

$$\| u_n - f^{1/2} \|_2 \quad \text{and} \quad h^2 \| u_n' - (f^{1/2})' \|_2^2 \,.$$

Almost sure convergence rates. As far as its implications to the convergence $u_n \longrightarrow f^{1/2}$ is concerned, the above theorem and exercise constitute somewhat of a mixed blessing. For the Hellinger distance, we get essentially the same convergence rates under the same conditions as for kernel estimators. But going from Hellinger to L^1 distance apparently is not very efficient and seems to require a finite exponential moment to get the "usual" rates of convergence. Recall that for kernel density estimation, we only needed a finite moment of order > 1. Fortunately (but unfortunately for other penalizations), the comparison-with-kernel-estimation saves our bacon. It shows that the GOOD estimator achieves the same *rate* of a.s. convergence as the kernel estimator, be it with the two-sided exponential kernel. The slight drawback is that apparently $f^{1/2}$ must be smooth, rather than just f itself.

(2.51) THEOREM. *Suppose the density f satisfies $f^{1/2} \in W^{2,2}(\mathbb{R})$ and $f \in W^{2,1}(\mathbb{R})$. If $\mathbb{E}\big[\,|\,X\,|^m\,\big] < \infty$ for some $m > 1$, then for $h \asymp n^{-1/5}$ (deterministically),*

$$\|\, u_n^2 - f \,\|_1 =_{\mathrm{as}} \mathcal{O}\big(\,n^{-2/5}\,\big) \;.$$

(2.52) EXERCISE. Prove Theorem (2.51).

(2.53) EXERCISE. The condition $f \in W^{2,1}(\mathbb{R})$ seems superfluous. Determine whether it is in fact implied by the assumption $f^{1/2} \in W^{2,2}(\mathbb{R})$ and the moment condition or by $f^{1/2} \in W^{2,2}(\mathbb{R}) \cap L^1(\mathbb{R})$, cf. Exercise (4.6.4).

(2.54) EXERCISE. (a) Prove that the GOOD estimator depends continuously on h: Let u be the solution of (2.5) for h, and v the solution for λ, with the same distribution function F_n. Then,

$$\|\, u - v \,\|_2^2 + h^2 \,\|\, u' - v' \,\|_2^2 \leqslant \{\, 1 - (h/\lambda)^2 \,\}\, \lambda^2 \,\|\, v' \,\|_2^2$$
$$\leqslant (h\lambda)^{-2}\,(\, h^2 - \lambda^2\,)^2 \;.$$

(b) Derive analogous bounds for $\|\, u^2 - v^2 \,\|_1$.

(2.55) EXERCISE. Analyze the second roughness penalty functional of GOOD and GASKINS (1971)

$$R\,(f) = \gamma \,\|\, (f^{1/2})' \,\|_2^2 + \delta \,\|\, (f^{1/2})'' \,\|_2^2 \;.$$

The choice $\gamma = 2$, $\delta = h^2$ should go just like the first roughness penalization functional. Any other choice apparently leads to trouble. [Hint: The Green's function, for γ and δ as stated, is $[\mathfrak{B}_h * \mathfrak{B}_h](x - y).$]

Universal consistency. A problem of some interest is the universal consistency of the GOOD estimator, similar to the case of Theorem (4.1.53) for density estimation, expressed in the following theorem. (We revert back to writing f_o for the true density.)

(2.56) THEOREM. *If the density f_o has a finite moment of order > 1, then*

$$\|\, f^{nh} - f_o \,\|_1 \longrightarrow_{\mathrm{as}} 0 \;,$$

provided $h \to 0$, $nh \to \infty$,

PROOF. We recall that u_n solves (2.5), and w is the solution to (2.6). In view of Theorem (2.47)(a), we have

$$H(u_n\,,\, w) \leqslant c\, H(T_\lambda * dF_n\,,\, T_\lambda * dF_o) \leqslant c\, \|\, T_\lambda * (dF_n - dF_o)\,\|_1 \;,$$

the last step by virtue of the elementary inequality $|\,\sqrt{p} - \sqrt{q}\,|^2 \leqslant |\,p - q\,|$, for all $p, q \geqslant 0$. Thus, Theorem (4.4.25)(a) implies that the variance part of the error tends to 0,

$$\| u_n - w \|_2 \longrightarrow 0 \quad \text{provided} \quad nh \longrightarrow \infty .$$

(Theorem (4.4.25) does not actually apply, because T is not symmetric, but it is easily fixed.) It thus suffices to consider the bias term $\| w - (f_o)^{1/2} \|_2$. The way we go about this is by smoothing f_o and applying the results for smooth f_o.

Let $\lambda > 0$, to be chosen later, and set $f_\lambda = (\mathcal{B}_\lambda * \mathcal{B}_\lambda * f_o)^{1/2}$. Then, $f_\lambda \in L^2(\mathbb{R})$ and

$$(f_\lambda)'' = \frac{(\mathcal{B}_\lambda * \mathcal{B}_\lambda * f_o)''}{2 (\mathcal{B}_\lambda * \mathcal{B}_\lambda * f_o)^{1/2}} - \frac{|(\mathcal{B}_\lambda * \mathcal{B}_\lambda * f_o)'|^2}{4 (\mathcal{B}_\lambda * \mathcal{B}_\lambda * f_o)^{3/2}} .$$

Now, the argument leading to (2.15) shows

$$| (\mathcal{B}_\lambda * \mathcal{B}_\lambda * f_o)'(x) | \leqslant \lambda^{-1} \mathcal{B}_\lambda * \mathcal{B}_\lambda * f_o(x) ,$$
$$| (\mathcal{B}_\lambda * \mathcal{B}_\lambda * f_o)''(x) | \leqslant \lambda^{-2} \mathcal{B}_\lambda * \mathcal{B}_\lambda * f_o(x) ,$$

and thus,

$$| (f_\lambda)''(x) | \leqslant \tfrac{3}{4} \lambda^{-2} f_\lambda(x) .$$

It follows that

(2.57) $$\| (f_\lambda)'' \|_2 \leqslant \tfrac{3}{4} \lambda^{-2} ,$$

and thus, $f_\lambda \in W^{2,2}(\mathbb{R})$. Let $f = \vartheta_{h,\lambda}$ be the solution of

$$\text{minimize} \quad - \int_{\mathbb{R}} \mathcal{B}_\lambda * \mathcal{B}_\lambda * f_o(x) \log f(x) \, dx + h^2 R(f)$$

$$\text{subject to} \quad f \in L^1(\mathbb{R}) , \ f \geqslant 0 ,$$

i.e., $\vartheta_{h,\lambda}$ is the solution to the large sample asymptotic problem (2.6) with the true density f_o replaced by $\mathcal{B}_\lambda * \mathcal{B}_\lambda * f_o$. Now, interpreting Theorem (2.47)(a) as showing the Lipschitz continuous dependence of the solution of (2.6) on F_o, or that of the solution of (2.1) on F_n, we obtain the bound on the difference between w and $\vartheta_{h,\lambda}$

$$\| w - \vartheta_{h,\lambda} \|_2^2 + h^2 \| (w - \vartheta_{h,\lambda})' \|_2^2 \leqslant$$

$$c \, H(T_{h/2} * f_o , T_{h/2} * \mathcal{B}_\lambda * \mathcal{B}_\lambda * f_o) \leqslant c \| f_o - \mathcal{B}_\lambda * \mathcal{B}_\lambda * f_o \|_1 ,$$

and this last expression tends to 0 as $\lambda \to 0$. Thus,

(2.58) $$\| w - \vartheta_{h,\lambda} \|_2 \longrightarrow 0 \quad \text{for} \ \lambda \to 0 , \text{ uniformly in } h > 0 .$$

Finally, we apply Theorem (2.29) and obtain

$$\| \vartheta_{h,\lambda} - f_\lambda \|_2 \leqslant h^2 \| \{ (f_\lambda)'' \|_2 \leqslant h^2 \lambda^{-2} ,$$

the last inequality by (2.57). Now, the triangle inequality gives

$$(2.59) \quad \| w - (f_o)^{1/2} \|_2 \leqslant \| w - \vartheta_{h,\lambda} \|_2 +$$

$$\| \vartheta_{h,\lambda} - f_\lambda \|_2 + \| f_\lambda - (f_o)^{1/2} \|_2 \, ,$$

in which the last term may be bounded as

$$\| f_\lambda - (f_o)^{1/2} \|_2 = \| (\mathcal{B}_\lambda * \mathcal{B}_\lambda * f_o)^{1/2} - (f_o)^{1/2} \|_1$$

$$\leqslant \{ \| \mathcal{B}_\lambda * \mathcal{B}_\lambda * f_o - f_o \|_1 \}^{1/2} \, .$$

Now, for $\lambda = h^{1/2}$ and $h \to 0$, each term on the right in (2.59) tends to 0.
Q.e.d.

(2.60) EXERCISE.. Prove Theorem (2.56) for arbitrary densities f_o, i.e.,
without any moment conditions. [Hint: Theorem (4.1.53) should be helpful.]

The computation of the Good estimator. How does one compute
the GOOD estimator? From the representation (2.13), it is clear that only
the $u_n(X_i)$ must be determined, and that one gets the system of equations

$$(2.61) \qquad u_n(X_j) = \frac{1}{n} \sum_{i=1}^{n} \frac{\mathcal{B}_h(X_j - X_i)}{u_n(X_i)} \, , \qquad j = 1, 2, \cdots, n \, ,$$

which can be solved using the Newton-Raphson method, say. In view
of the comparison lemma, one might take $u_n(X_j) = \mathcal{B}_{h/\sqrt{2}} * dF_n(X_j)$,
$j = 1, 2, \cdots, n$, as the initial guess. This is the authorized version of DE
MONTRICHER, TAPIA and THOMPSON (1975), and it works quite well.

Although the development in this section shows that the pdf constraint
in (2.5) has little theoretical value, in practice, one would like to enforce it.
The following exercise shows how this may be done implicitly.

(2.62) Exercise. Let f^{nh} be the solution of (2.1), and define

$$\varphi^{n,h} = \frac{f^{nh}}{\int_{\mathbb{R}} f^{nh}(x) \, dx} \, .$$

Let $f = \psi_{n,h}$ be the solution to

$$(2.63) \qquad \text{minimize} \quad -\frac{1}{n} \sum_{i=1}^{n} \log f(X_i) + \int_{\mathbb{R}} f(x) \, dx + h^2 \, R(f)$$

$$\text{subject to} \quad f \text{ is a pdf} \, .$$

Show that for each $h > 0$ there exists a $\lambda > 0$ such that

$$\psi_{n,h} = \varphi^{n,\lambda} \, ,$$

and vice versa. [Hint: With $f = u^2$, the Euler equations for (2.63) are

$$-h^2 \, u'' + (1 + \mu) \, u = \frac{dF_n}{u} \, ,$$

together with appropriate boundary conditions, where $\mu > 0$ is such that the resulting u satisfies $\|u\|_2 = 1$. Take it from there.]

We finish this section by seeing how KLONIAS (1982, 1984) looked at these things. In the following exercise, Part (a) introduces the reproducing kernel Hilbert space $\mathfrak{H}$ and shows that $\mathfrak{B}_h(x-y)$ is the *reproducing kernel* of the Hilbert space $\mathfrak{H}$. Part (c) is a very nice inequality derived by KLONIAS, but he has already lost the thread. In part (d) the "correct" inequality is stated.

(2.64) EXERCISE. Let $\mathfrak{H} = W^{1,2}(\mathbb{R})$. Define an inner product and a norm on $\mathfrak{H}$ by $\langle\, u\,,\, v\,\rangle_{\mathfrak{H}} = \langle\, u\,,\, v\,\rangle_2 + h^2\,\langle\, u'\,,\, v'\,\rangle_2$, and $\|u\|_{\mathfrak{H}}^2 = \langle\, u\,,\, u\,\rangle_{\mathfrak{H}}$, for all $u,\, v \in \mathfrak{H}$.
(a) Show that with $\mathfrak{B}_h$ as before
$$\langle\, \mathfrak{B}_h(\,\cdot\, - y)\,,\, u\,\rangle_{\mathfrak{H}} = u(y)\,, \quad y \in \mathbb{R}\,,$$
for all $u \in \mathfrak{H}$, and that
$$\|\,\mathfrak{B}_h * u\,\|_{\mathfrak{H}} = \|u\|_2 \quad \text{for all } u \in \mathfrak{H}\,.$$

(b) With u_n the solution of (2.5), show that for all $v \in \mathfrak{H}$,
$$\|\, u_n - v\,\|_{\mathfrak{H}} \leqslant \left\|\, \mathfrak{B}_h * \frac{dF_n}{u_n} - v\,\right\|_{\mathfrak{H}}\,.$$
Also, derive a similar inequality for w, the solution of (2.6).
[Hint : Start with
$$\frac{1}{n}\sum_{i=1}^n \left(\, \{u_n(X_i)\}^{-1} - \{v(X_i)\}^{-1}\right)\left(\, u_n(X_i) - v(X_i)\right) \leqslant 0\,,$$
and write $u_n(X_i) - v(X_i) = \langle\, \mathfrak{B}_h * (\,\cdot\, - X_i)\,,\, u_n - v\,\rangle_{\mathfrak{H}}$, and take it from there.]

(c) With u_n and w from (b), show under suitable conditions that
$$\|\, u_n - w\,\|_{\mathfrak{H}} \leqslant \left\|\, \mathfrak{B}_h * \frac{dF_n - dF}{w}\,\right\|_{\mathfrak{H}}\,.$$
(x) Time-out! For fixed $h > 0$, compute the expected value of the square of the right-hand side of the inequality in Part (c), and weep: Under what conditions on f is the expected value finite? Compare this with Pearson's φ^2 distance in Exercise (4.5.25).
(d) Show that
$$\|\, u_n - w\,\|_{\mathfrak{H}} \leqslant \left\|\, \mathfrak{B}_h * \frac{dF_n - dF}{u_n \vee w}\,\right\|_{\mathfrak{H}}\,.$$
(e) Derive bounds for $\|\, u_n - w\|_{\mathfrak{H}}$ based on the representation of part (d).

EXERCISES : (2.4), (2.14), (2.30), (2.31), (2.35), (2.38), (2.42), (2.47), (2.48), (2.49), (2.51), (2.52), (2.53), (2.54), (2.55), (2.60), (2.62), (2.64).

3. Roughness penalization of log-densities

We continue the investigation of (1.1), but now with the very interesting roughness penalization proposed by SILVERMAN (1982)

$$(3.1) \qquad R(f) = \int_{\mathbb{R}} |\{ \log f(x) \}^{(3)}|^2 \, dx \ .$$

Solutions of (1.1) with this $R(f)$ will have a smooth log-density, but it is not so clear what this means. A motivation for this penalization is that $R(f) = 0$ precisely when f is a Gaussian distribution, so that in this setup, densities are also penalized for not being normal. Thus, the penalization (3.1) recommends itself when the true density is (believed to be) close to normal. However, since $R(f)$ is finite even for the Cauchy distribution, it is not clear what aim is being achieved, and how. See also (3.6) below.

In this section, the problem (2.1)–(3.1) is studied under the minimal assumptions that the true density $f_o \in L^1(\mathbb{R})$ satisfies

$$(3.2) \qquad R(f_o) < \infty$$

and

$$(3.3) \qquad \mathbb{E}[\,|X|^5\,] < \infty \ .$$

Note that (3.2) implies that f_o is continuous and that $f_o > 0$ everywhere. In the next section, we restrict attention to smooth densities on a bounded interval. Condition (3.3) is a tail condition, which is minimal in view of the above motivation for the log-density penalization.

As a preliminary observation in the study of this estimator, one should note that if $\{\log f\}^{(3)} \in L^2(\mathbb{R})$, then Taylor expansion gives

$$(3.4) \qquad \log f(x) = p(x) + [\,\mathcal{V}q\,](x) \ , \qquad x \in \mathbb{R} \ ,$$

where $p(x) = p(\log f \, ; x)$ is the quadratic Taylor polynomial, $q = \{\log f\}^{(3)}$, and

$$(3.5) \qquad [\,\mathcal{V}q\,](x) = \tfrac{1}{2} \int_0^x (x - y)^2 \, q(y) \, dy \ , \qquad x \in \mathbb{R} \ .$$

Then, with Cauchy-Schwarz,

$$\left| [\,\mathcal{V}q\,](x) \right| \leqslant \frac{1}{2\sqrt{5}} \, |x|^{5/2} \, \| q \|_2 \ ,$$

and so, as $|x| \to \infty$ for a suitable constant c,

$$(3.6) \qquad |\log f(x)| \leqslant c\,|x|^{5/2} \, \| \{\log f\}^{(3)} \|_2 \ .$$

Hence, $f(x) \geqslant c_1 \exp(-c_2 |x|^{5/2})$. So, if (2.1)–(3.1) has solutions that satisfy $R(f) < \infty$, then there is a bound on how fast f can decay at ∞. This sheds an unexpected light on the preference this roughness penalization has toward the normal density.

From the theoretical as well as the practical point of view, the transformation $u = \log f$ makes sense. Then, the problem becomes

$$
\text{minimize} \quad L_{nh}(u) \overset{\text{def}}{=} -\frac{1}{n} \sum_{i=1}^{n} u(X_i) + \int_{\mathbb{R}} e^{u(x)} \, dx + h^2 \, \| \, u^{(3)} \, \|_2^2
$$

(3.7)

$$
\text{subject to} \quad u^{(3)} \in L^2(\mathbb{R}) \, .
$$

The solution u of (3.7) is denoted as $u = u_{nh}$.

(3.8) EXERCISE. Show that $\exp(u_{nh})$ is a pdf.

The following observations and statements are proved in § 11.4, making extensive use of convexity. The existence of solutions to (3.7) is shown by relating it to the existence of maximum likelihood estimators for a certain parametric exponential family. Uniqueness follows from the strong convexity of L_{nh}, that is, for all u,

$$
(3.9) \qquad h^2 \, \| \, (u - u_{nh})^{(3)} \, \|_2^2 + \text{KLL}(u, u_{nh}) \leqslant L_{nh}(u) - L_{nh}(u_{nh}) \, ,
$$

with KLL what we call the "logarithmic" Kullback-Leibler divergence,

$$
(3.10) \qquad \text{KLL}(u, w) = \text{KL}(e^u, e^w) = \int_{\mathbb{R}} e^u (u - w) + e^w - e^u \, .
$$

However, the real importance of (3.9) is for the derivation of convergence rates for $\| \exp(u_{nh}) - f_o \|_1$. We also have occasion to refer to the Euler equations for (3.7). It is convenient to write them in terms of p and q, see (3.4). So, $u_{nh} = p + \mathcal{V}q$ solves (3.7) if and only if p and q solve

$$
(3.11) \qquad 2h^2 q + \mathcal{V}^* \{ \exp[p + \mathcal{V}q] - dF_n \} = 0 \, , \quad x \in \mathbb{R} \, ,
$$

$$
(3.12) \qquad \int_{\mathbb{R}} x^\ell \exp\big[p(x) + \mathcal{V}q(x) \big] \, dx = \int_{\mathbb{R}} x^\ell \, dF_n(x) \, , \quad \ell = 0, 1, 2 \, .
$$

Here, $\mathcal{V}^*$ is the adjoint of $\mathcal{V}$ defined for distributions Ψ by

$$
(3.13) \qquad \mathcal{V}^* d\Psi(x) =
\begin{cases}
\frac{1}{2} \displaystyle\int_x^\infty (x - y)^2 \, d\Psi(y) & , \quad x > 0 \, , \\[2mm]
-\frac{1}{2} \displaystyle\int_{-\infty}^x (x - y)^2 \, d\Psi(y) \, , & x < 0 \, .
\end{cases}
$$

If Ψ has density ψ, then we write $\mathcal{V}^* d\Psi = \mathcal{V}^* \psi$.

(3.14) EXERCISE. Verify that $\mathcal{V}^*$ is indeed the adjoint of the operator $\mathcal{V}$, say, in the sense that for all densities φ and distributions Ψ with finite fifth moment,

$$
\int_{\mathbb{R}} [\mathcal{V}\varphi](x) \, d\Psi(x) = \int_{\mathbb{R}} \varphi(y) \, [\mathcal{V}^* d\Psi](y) \, .
$$

As always, the first step in the analysis is the decomposition of the error $\exp(u_{nh}) - f_o$ into asymptotic bias and variance terms. To that end, we introduce the large sample asymptotic problem for (3.7), given by

$$(3.15) \quad \text{minimize} \quad -\int_{\mathbb{R}} u(x)\, dF_o(x) + \int_{\mathbb{R}} e^{u(x)}\, dx + h^2 \, \| u^{(3)} \|_2^2$$

$$\text{subject to} \quad u^{(3)} \in L^2(\mathbb{R}) \ .$$

The objective function in (3.15) is denoted by $L_{\times h}(u)$. The solution of (3.15) exists and is unique, see §11.4, and is denoted by w_h. The analogue of (3.9) holds, viz. for all u,

$$(3.16) \qquad h^2 \, \| (u - w_h)^{(3)} \|_2^2 + \text{KLL}(u, w_h) \leqslant L_{\times h}(u) - L_{\times h}(w_h) \ .$$

(3.17) EXERCISE. Verify that $\exp(w_h)$ is a pdf.

We may now decompose the error $\exp(u_{nh}) - f_o$ as

$$(3.18) \quad \exp(u_{nh}) - f_o = \big\{ \exp(u_{nh}) - \exp(w_h) \big\} + \big\{ \exp(w_h) - f_o \big\} \ ,$$

and we study each term in turn. Because the ultimate goal is to obtain L^1 error bounds, the development that follows is justified by the inequality

$$\tfrac{1}{2} \, \| \exp(u) - \exp(v) \|_1^2 \leqslant \text{KLL}(u, v) \ .$$

The asymptotic bias term. From (3.9) with $u = w_o = \log f_o$, we get

$$(3.19) \quad h^2 \, \|(w_o - w_h)^{(3)}\|_2^2 + \text{KLL}(w_h, w_o) \leqslant$$

$$-\int_{\mathbb{R}} (w_o - w_h)\, f_o + h^2 \, \|w_o^{(3)}\|_2^2 - h^2 \, \|w_h^{(3)}\|_2^2 \ .$$

The first term on the right equals $-\text{KLL}(w_o, w_h)$ and may be moved to the left. For the remaining terms, we observe that

$$\|w_o^{(3)}\|_2^2 - \|w_h^{(3)}\|_2^2 = 2 \big\langle\, w_o^{(3)}, (w_o - w_h)^{(3)} \,\big\rangle - \|(w_o - w_h)^{(3)}\|_2^2 \ .$$

With $\langle\, \cdot\,,\, \cdot\, \rangle$ denoting the L^2 inner product, then (3.19) may be written as

$$(3.20) \quad 2\, h^2 \, \|(w_o - w_h)^{(3)}\|_2^2 + \mathcal{D}(w_h, w_o) \leqslant 2\, h^2 \big\langle\, w_o^{(3)}, (w_o - w_h)^{(3)} \,\big\rangle \ ,$$

with $\mathcal{D}$ the symmetrized version of KLL

$$(3.21) \qquad\qquad \mathcal{D}(u, w) = \text{KLL}(u, w) + \text{KLL}(w, u) \ .$$

It follows that

$$(3.22) \quad 2\, h^2 \, \|(w_o - w_h)^{(3)}\|_2^2 + \mathcal{D}(w_o, w_h) \leqslant 2\, h^2 \, \| w_o^{(3)} \|_2 \, \|(w_o - w_h)^{(3)} \|_2 \ ,$$

and we have just about proven the following theorem.

(3.23) THEOREM. *If* $\{\log f_o\}^{(3)} \in L^2(\mathbb{R})$, *then*

$$\mathcal{D}(w_o, w_h) = \mathcal{O}(h^2) \ , \qquad \| w_h^{(3)} \|_2 \leqslant \| w_o^{(3)} \|_2 \ .$$

(3.24) EXERCISE. Prove Theorem (3.23). Actually, (3.22) implies the inequality $\| w_h^{(3)} \|_2 \leqslant 2 \, \| w_o^{(3)} \|_2$, but show that the factor 2 is superfluous.

It is possible to improve on Theorem (3.23) under stronger conditions on the density f_o. We leave it as an exercise.

(3.25) EXERCISE. (a) Suppose that $w_o = \log f_o$ is six times differentiable, and that there exists a function $a_o \in L^\infty(\mathbb{R})$ such that
$$w_o^{(6)}(x) = a_o(x) f_o(x) , \quad x \in \mathbb{R} ,$$
and
$$x^\ell w_o^{(3+\ell)}(x) \longrightarrow 0 \quad \text{at } \pm\infty \text{ for } \ell = 0, 1, 2 .$$
Improve the conclusions of Theorem (3.23) to
$$\mathcal{D}(w_o, w_h) = \mathcal{O}(h^4) , \quad \| (w_o - w_h)^{(3)} \|_2 = \mathcal{O}(h) .$$
(b) Are the above conditions optimal? (The authors do not know!)

The asymptotic variance term. For the asymptotic variance term, we try to repeat the above. Now, the inequalities (3.9), with $u = w_h$, and (3.16), with $u = u_{nh}$, yield

$$(3.26) \quad 2 h^2 \|(u_{nh} - w_h)^{(3)}\|_2^2 + \mathcal{D}(u_{nh}, w_h) \leqslant \int_{\mathbb{R}} (u_{nh} - w_h)(dF_n - dF_o) .$$

With the representation (3.4), the right-hand side may be rewritten as

$$(3.27) \quad \int_{\mathbb{R}} p(u_{nh} - w; x)(dF_n(x) - dF_o(x)) +$$
$$\int_{\mathbb{R}} (u_{nh} - w_h)^{(3)} \mathcal{V}^* \{ dF_n - dF_o \} ,$$

where $\mathcal{V}^*$ is given in (3.13). The first observation is the crude bound

$$\int_{\mathbb{R}} (u_{nh} - w_h)^{(3)} \mathcal{V}^*(dF_n - dF_o) \leqslant \| (u_{nh} - w_h)^{(3)} \|_2 \| \mathcal{V}^*(dF_n - dF_o) \|_2 ,$$

(can it be improved?) and the following result.

(3.28) LEMMA. If $\mathbb{E}[\,|X|^5\,] < \infty$, then
$$\| \mathcal{V}^* (dF_n - dF_o) \|_2 =_{\text{as}} \mathcal{O}\big((n^{-1} \log\log n)^{1/2} \big) .$$

(3.29) EXERCISE. Prove the lemma. [Hint: Write the square of the left-hand side as sums of iid random variables.]

Regarding the remaining term in (3.27), we note that direct computation gives, since $p(u_{nh} - w_h ; \cdot)$ is a quadratic polynomial,

$$(3.30) \quad \left| \int_{\mathbb{R}} p(u_{nh} - w_h ; x)(dF_n(x) - dF_o(x)) \right| \leqslant_{\text{as}}$$
$$c (n^{-1} \log\log n)^{1/2} \| p(u_{nh} - w_h ; \cdot) \|_{\mathbb{P}_3} ,$$

where $\| p \|_{\mathbb{P}_3}$ is any convenient norm on the polynomials of degree < 3 (quadratic polynomials). We take

$$(3.31) \qquad \| p \|_{\mathbb{P}_3} = | p(0) | + | p'(0) | + | p''(0) | .$$

To make progress, $\| p(u_{nh} - w_h ; \cdot) \|_{\mathbb{P}_3}$ must now be bounded in terms of $\mathcal{D}(u_{nh}, w_h)$.

(3.32) LEMMA. *Under the assumptions* (3.2) *and* (3.3), *there exists a constant* c *dependent on* f_o *such that for all* h *small enough,*

$$\| p(u_{nh} - w_h ; \cdot) \|_{\mathbb{P}_3} \leqslant c \left\{ \mathcal{D}(w_h, u_{nh}) + \mathcal{D}(w_h, u_{nh})^{1/2} + \| (u_{nh} - w_h)^{(3)} \|_2 \right\} .$$

PROOF. For the purpose of this proof only, we introduce the abbreviations $u = u_{nh}$, $w = w_h$, and $\varepsilon = u - w$. Choose $\delta > 0$ fixed. Then, there exists a constant $c = c(\delta)$ such that for all quadratic polynomials p,

$$\| p \|_{\mathbb{P}_3} \leqslant c \int_{-\delta}^{\delta} | p(x) | \, dx .$$

In particular, then

$$\| p(\varepsilon ; \cdot) \|_{\mathbb{P}_3} \leqslant c \int_{-\delta}^{\delta} | p(\varepsilon ; x) | \, dx$$

$$\leqslant c(w; \delta) \int_{-\delta}^{\delta} e^{w} | p(\varepsilon ; x) | \, dx ,$$

where $c(w; \delta) = c / \min \left\{ \exp(w(x)) : |x| < \delta \right\} .$

Proceeding, the triangle inequality gives

$$\int_{-\delta}^{\delta} e^{w} | p(\varepsilon ; x) | \, dx \leqslant \int_{\mathbb{R}} e^{w} | p(\varepsilon ; \cdot) + \mathcal{V}\{(\varepsilon)^{(3)}\} | + \int_{\mathbb{R}} e^{w} | \mathcal{V}\{(\varepsilon)^{(3)}\} | .$$

Now, once more by the triangle inequality,

$$e^{w} | p(\varepsilon ; \cdot) + \mathcal{V}\{(\varepsilon)^{(3)}\} | = e^{w} | u - w |$$

$$\leqslant | e^{w} (u - w) + e^{w} - e^{u} | + | e^{u} - e^{w} | ,$$

and we may drop the absolute values signs for the first term on the far right. Putting everything together, it follows that

$$\| p(\varepsilon ; \cdot) \|_{\mathbb{P}_3} \leqslant$$

$$c(w; \delta) \left\{ \mathrm{KLL}(w, u) + \left\{ \mathrm{KLL}(w, u) \right\}^{1/2} + \| \mathcal{V}^* e^{w} \|_2 \| (\varepsilon)^{(3)} \|_2 \right\} .$$

Now, the Euler equations (3.11) for the large sample asymptotic problem yield

$$\mathcal{V}^* e^{w_h} = \mathcal{V}^* e^{w_o} + 2 \, h^2 \, w_h^{(3)} ,$$

and so

$$\| \mathcal{V}^* e^w \|_2 \leqslant \| \mathcal{V}^* f_o \|_2 + 2\, h^2 \, \| w^{(3)} \|_2$$

$$\leqslant \frac{1}{2\sqrt{5}} \left(\mathbb{E}[\,|X|^5\,] \right)^{1/2} + 2\, h^2 \, \| w^{(3)} \|_2$$

$$\leqslant \text{const}\,, \quad \text{for } h \to 0\,,$$

the last observation by Theorem (3.23). By the same theorem, we also have

$$\min \left\{ \exp(\, w(x)\,) \; : \; |x| < \delta \right\} \longrightarrow \min \left\{ f_o(x) \; : \; |x| < \delta \right\}\,,$$

which is a fixed positive constant under the assumption (3.2). The lemma follows.   Q.e.d.

Summarizing the above, we have the somewhat involved inequality, for a suitable constant c, and with $\lambda_n = (\, n^{-1} \log \log n\,)^{1/2}$,

$$2\, h^2 \, \| (u_{nh} - w_h)^{(3)} \|_2^2 + \mathcal{D}(w_h, u_{nh}) \leqslant$$
$$\| (u_{nh} - w_h)^{(3)} \|_2 \, \| \mathcal{V}^*(dF_n - dF_o) \|_2 +$$
$$c\left(\mathcal{D}(w_h, u_{nh}) + \mathcal{D}(w_h, u_{nh})^{1/2} + \| (u_{nh} - w_h)^{(3)} \|_2 \right)\,,$$

from which the following theorem may be derived using standard manipulations.

(3.33) THEOREM. *Under the assumptions* (3.2) *and* (3.3), *for all h small enough (independent of $X_1, X_2, \cdots, X_n$, but dependent on f_o),*

$$2\, h^2 \, \|(u_{nh} - w_h)^{(3)} \|_2^2 + \mathcal{D}(u_{nh}, w_h) =_{as} \mathcal{O}\big((n\, h^2)^{-1} \log \log n \big)\,.$$

(3.34) COROLLARY. *Under the assumptions* (3.2) *and* (3.3), *uniformly in* $h \to 0$,

$$\| \exp(u_{nh}) - f_o \|_1 =_{as} \mathcal{O}\big(\{ (nh^2)^{-1} \log \log n \}^{1/2} + h \big)\,,$$

and for $h \asymp (\, n^{-1} \log \log n\,)^{1/4}$,

$$\| \exp(u_{nh}) - f_o \|_1 =_{as} \mathcal{O}\big((\, n^{-1} \log \log n\,)^{1/4} \big)$$

and

$$\| u_{nh}^{\ (3)} \|_2 =_{as} \mathcal{O}(\, 1\,)\,.$$

EXERCISES: (3.8), (3.14), (3.17), (3.24), (3.25), (3.29).

4. Roughness penalization of bounded log-densities

We continue the study of the roughness penalization of log-densities, but now with the added assumption that they are bounded. This can only happen on a bounded interval, which is chosen to be $[0, 1]$. In this section, we assume that

(4.1) $$\log f_o \in W^{3,2}(0, 1) .$$

This implies that f_o is bounded and bounded away from 0. Later on, we also consider the stronger conditions

(4.2)
$$\log f_o \in W^{6,2}(0, 1) ,$$
$$\{\log f_o\}^{(\ell)} = 0 \text{ at } x = 0, 1 , \quad \text{for } \ell = 3, 4, 5 .$$

Strange as these conditions may seem, they are "natural". The goal of this section is to prove (presumably) optimal rates of convergence.

As before, we estimate $w_o = \log f_o$ by the solution to

(4.3)
$$\text{minimize} \quad L_{nh}(u) \overset{\text{def}}{=} -\frac{1}{n} \sum_{i=1}^{n} u(X_i) + \int_0^1 e^{u(x)} \, dx + h^2 \| u^{(3)} \|_2^2$$
$$\text{subject to} \quad u^{(3)} \in L^2(0, 1) .$$

The solution is denoted by u_{nh} and the solution of the corresponding large sample asymptotic problem by w_h. The approach to proving convergence rates is as in § 3, but cutting right to the chase, the boundedness of $\log f_o$ implies that $\| u_{nh} - w_h \|_2^2$ and $\mathcal{D}(u_{nh}, w_h)$, see (3.20), are essentially equivalent. This, however, must be proven! The analysis of § 3 applies verbatim to the present case and yields the following lemma.

(4.4) LEMMA. *Under the assumption (4.1), for $h \to 0$,*
$$\| w_h - w_o \|_\infty \longrightarrow 0 .$$
If, in addition, $h \geqslant c \, (\, n^{-1} \log \log n \,)^\beta$ for some $\beta \leqslant \frac{1}{2}$ and $c > 0$, then
$$\| u_{nh} - w_o \|_\infty \longrightarrow_{\text{as}} 0 .$$

(4.5) EXERCISE. Prove the lemma.

The lemma is useful, in that it implies that w_h and u_{nh} are bounded, uniformly in h, provided

(4.6) $$h \geqslant \lambda_n \overset{\text{def}}{=} c \, (\, n^{-1} \log \log n \,)^{1/4} ,$$

where c is an arbitrary positive constant. We can now prove the equivalence of the squared L^2 norm and the symmetric logarithmic Kullback-Leibler distance $\mathcal{D}$.

(4.7) LEMMA. *Let* $B(R) = \{\, u \in L^\infty(0,1) \,:\, \|\,u\,\|_\infty \leqslant R \,\}$. *There exists a constant* c, *dependent on* R, *such that for all* u, $v \in B(R)$,

$$c \,\|\, u - v \,\|_2^2 \leqslant \mathcal{D}(u,v) \leqslant c^{-1} \,\|\, u - v \,\|_2^2 \,.$$

(4.8) Corollary. *There exists a constant* c, *dependent on* f_o, *such that for all* $h \geqslant \lambda_n$,

$$c \,\|\, u_{nh} - w_h \,\|_2^2 \leqslant \mathcal{D}(u_{nh}, w_h) \leqslant c^{-1} \,\|\, u_{nh} - w_h \,\|_2^2 \,.$$

(4.9) EXERCISE. (a) Prove the elementary inequality for all real x, y,

$$\left(\mathrm{e}^x \wedge \mathrm{e}^y \right) |\, x - y \,|^2 \leqslant \left(\mathrm{e}^x - \mathrm{e}^y \right) \left(x - y \right) \leqslant \left(\mathrm{e}^x \vee \mathrm{e}^y \right) |\, x - y \,|^2 \,.$$

(b) Prove Lemma (4.7).

We are now ready to derive improved error estimates and begin with the asymptotic bias term.

(4.10) THEOREM. (a) *Under the assumption* (4.1),

$$\mathcal{D}(w_h, w_o) \leqslant 4 \, h^2 \,\|\, w_o^{(3)} \,\|_2^2 \,.$$

(b) *Under the additional assumption* (4.2), *there exists a constant* c *dependent on* f_o *such that*

$$\mathcal{D}(w_h, w_o) \leqslant c \, h^4 \,\|\, w_o^{(6)} \,\|_2^2 \,, \quad \|\, (w_h - w_o)^{(3)} \,\|_2 \leqslant c \, h \,\|\, w_o^{(6)} \,\|_2 \,.$$

PROOF. We only prove (b). The starting point is the inequality (3.19) translated to the present setting. The assumptions (4.2) guarantee that the boundary terms in the following integration by parts vanish, so that

$$\left\langle \, w_o^{(3)} \,,\, (w_o - w_h)^{(3)} \, \right\rangle = - \left\langle \, w_o^{(6)} \,,\, w_o - w_h \, \right\rangle$$
$$\leqslant \|\, w_o^{(6)} \,\|_2 \,\|\, w_o - w_h \,\|_2$$
$$\leqslant c \,\|\, w_o^{(6)} \,\|_2 \,\{\, \mathcal{D}(w_h, w_o) \,\}^{1/2} \,,$$

the last inequality by Lemma (4.7). Thus, from (3.20),

$$2 \, h^2 \,\|\, (w_o - w_h)^{(3)} \,\|_2^2 + \mathcal{D}(w_h, w_o) \leqslant c \,\|\, w_o^{(6)} \,\|_2 \,\{\, \mathcal{D}(w_h, w_o) \,\}^{1/2} \,,$$

and the conclusions follow.   Q.e.d.

(4.11) EXERCISE. Assuming (4.1), investigate whether the conditions (4.2) are in fact necessary for the conclusions of Theorem (4.10) to hold.

The next item on the agenda is the asymptotic variance term. For its treatment, we need an integration by parts trick similar to Lemma (2.34). To set it up, we first introduce the Green's function for a boundary value problem.

(4.12) LEMMA. *Let $\lambda > 0$ and define t_λ by*

$$t_\lambda(x) = \begin{cases} \dfrac{\lambda \exp(-\lambda(x-1))}{\exp(\lambda) - 1} \,, & x > 0 \,, \\[3mm] \dfrac{\lambda \exp(-\lambda x)}{\exp(\lambda) - 1} \,, & x < 0 \,. \end{cases}$$

Then, $t_\lambda(x - y)$ is the Green's function for the boundary value problem

$$\lambda^{-1} \varphi' + \varphi = \psi \,, \quad \text{in } (0,1) \,,$$
$$\varphi(1) = \varphi(0) \quad.$$

In particular, the solution is given by $\varphi = t_\lambda * \psi$, with

$$t_\lambda * \psi(x) = \int_0^1 t_\lambda(x - y)\, \psi(y)\, dy \,.$$

(4.13) EXERCISE. Prove Lemma (4.12).

(4.14) LEMMA. *If $v \in W^{2,1}(0,1)$ and Ψ is a distribution function, then*

$$\int_0^1 v \, d\Psi = \lambda^{-1}\,(\, v(1) - v(0)\,)\,[\,t_\lambda * d\Psi\,](1) + \int_0^1 (\,-\lambda^{-1}\,v' + v\,)\,t_\lambda * d\Psi \,.$$

(4.15) EXERCISE. Prove Lemma (4.14).

To proceed, we need the following technical lemma, an adaptation of the slightly simpler case of the half line, Theorem 5.3.1 in HILLE (1972).

(4.16) LEMMA. *There exists a constant $c > 0$ such that for all $v \in W^{3,2}(0,1)$,*

$$\| v' \|_2^2 \leqslant c \max(\, \| v \|_2^2, \, \| v \|_2 \, \| v'' \|_2 \,) \,,$$
$$\| v' \|_2^3 \leqslant c \max\left(\, \| v \|_2^3, \, \| v \|_2^2 \, \| v^{(3)} \|_2 \,\right) \,.$$

PROOF. Let $0 < s < 1$, $0 < t < 1 - s$. Then, from Taylor's theorem with exact remainder,

$$v'(t) = \frac{1}{s}\,(\, v(t+s) - v(t)\,) - \frac{1}{s}\,\text{rem} \,,$$

with

$$\text{rem} = \int_t^{t+s} (\,t + s - r\,)\,v''(r)\,dr \,.$$

Consequently,

$$\left\{ \int_0^{1-s} |v'(t)|^2\, dt \right\}^{1/2} \leqslant \frac{2}{s}\, \| v \|_2 + \frac{1}{s}\left\{ \int_0^{1-s} |\,\text{rem}\,|^2\, dt \right\}^{1/2} \,,$$

and the task at hand is to evaluate the last expression. With Cauchy-Schwarz,

$$
\int_0^{1-s} |\operatorname{rem}|^2 \, dt \leqslant \int_0^{1-s} \left\{ \int_t^{t+s} (t+s-r) \, dr \right\} \times
$$

$$
\int_t^{t+s} (t+s-r) \, | \, v''(r) \, |^2 \, dr \, dt
$$

$$
\leqslant \tfrac{1}{2} s^2 \int_0^1 | \, v''(r) \, |^2 \left\{ \int_{(r-s)\vee 0}^{r\wedge(1-s)} (t+s-r) \, dt \right\} dr \, ,
$$

and so

$$
\int_0^{1-s} |\operatorname{rem}|^2 \, dt \leqslant \tfrac{1}{4} s^4 \, \| \, v'' \, \|_2^2 \, .
$$

It follows that

$$
\int_0^{1-s} | \, v'(t) \, |^2 \, dt \leqslant \left\{ \frac{2}{s} \, \| \, v \, \|_2 + \frac{s}{2} \, \| \, v'' \, \|_2 \right\}^2 .
$$

By symmetry, we also get

$$
\int_s^1 | \, v'(t) \, |^2 \, dt \leqslant \left\{ \frac{2}{s} \, \| \, v \, \|_2 + \frac{s}{2} \, \| \, v'' \, \|_2 \right\}^2 ,
$$

and so by adding, for $0 < s < \tfrac{1}{2}$,

$$
\| \, v' \, \|_2 \leqslant \sqrt{2} \left(\frac{2}{s} \, \| \, v \, \|_2 + \frac{s}{2} \, \| \, v'' \, \|_2 \right) .
$$

The minimum of the right-hand side as a function of $s > 0$ is attained at $s = \sqrt{r}$ with $r = 4 \, \| \, v \, \|_2 / \| \, v'' \, \|_2$, so that in view of the restriction on s, for a suitable constant c,

$$
\| \, v' \, \|_2^2 \leqslant c \, (\, \| \, v \, \|_2^2 \,) \wedge (\, \| \, v \, \|_2 \, \| \, v'' \, \|_2) \, .
$$

This is the first inequality. The second one comes about by applying the first inequality to get bounds for $\| \, v'' \, \|_2$ in terms of $\| \, v' \, \|_2$ and $\| \, v^{(3)} \, \|_2$, and substituting back into the first inequality.   Q.e.d.

After these preliminaries, we are ready for the variance term proper.

(4.17) THEOREM. *Under the assumption* (4.1), *for all $s > 1$,*

$$
h^2 \, \| \, (u_{nh} - w_h)^{(3)} \, \|^2 + \mathcal{D}(u_{nh}, w_h) =_{\mathrm{as}} \mathcal{O}(\, h^{-1/3} \, n^{-1} (\log n)^s \,) \, ,
$$

provided $h \asymp n^{-\beta}$ for some $\beta < \tfrac{1}{2}$.

PROOF. The starting point is inequality (3.26). With Lemma (4.7), this may be written as

$$2\,h^2 \,\| \,(u_{nh} - w_h)^{(3)}\, \|_2^2 + c\, \| \, u_{nh} - w_h \,\|_2^2 \leqslant$$

(4.18)

$$\int_0^1 (u_{nh} - w_h)\,(dF_n - dF_o)\,.$$

We now apply Lemma (4.14), with $\Psi = F_n - F_o$, $v = u_{nh} - w_h$, and $\tau_h = t_\lambda$, with $\lambda = h^{-1/3}$. It then follows that

$$\int_0^1 (u_{nh} - w_h)\,(dF_n - dF_o) =$$

$$h^{1/3}\,(\,[u_{nh} - w_h](1) - [u_{nh} - w_h](0)\,)[\,\tau_h * (dF_n - dF_o)\,](1) +$$

$$\int_0^1 \left(-h^{1/3}\,(u_{nh} - w_h)' + u_{nh} - w_h\,\right) \tau_h * (dF_n - dF_o)\,.$$

The last integral may be bounded as

$$\left(\,h^{1/3}\,\| \,(u_{nh} - w_h)'\,\|_2 + \| \,u_{nh} - w_h\,\|_2\,\right) \| \,\tau_h * (dF_n - dF_o)\,\|_2\,.$$

The first term may be written and bounded as

$$h^{1/3}\,[\,\tau_h * (dF_n - dF_o)\,](1)\int_0^1 (u_{nh} - w_h)' \leqslant$$

$$h^{1/3}\,\big|\,[\,\tau_h * (dF_n - dF_o)\,](1)\,\big|\,\| \,(u_{nh} - w_h)'\,\|_2\,.$$

Finally, using a sloppy version of Lemma (4.16) yields

$$h^{1/3}\,\| \,(u_{nh} - w_h)'\,\|_2 \leqslant$$

(4.19)

$$\leqslant c\,h^{1/3}\,\| \,u_{nh} - w_h\,\|_2^{2/3}\,\| \,(u_{nh} - w_h)^{(3)}\,\|_2^{1/3}$$

$$\leqslant c\,\big(\| \,u_{nh} - w_h\,\|_2 + h\,\| \,(u_{nh} - w_h)^{(3)}\,\|_2\big)\,,$$

where the last line in (4.19) follows by the arithmetic-geometric means inequality. Putting everything together, we get

$$2\,h^2\,\| \,(u_{nh} - w_h)^{(3)}\,\|_2^2 + c\,\| \,u_{nh} - w_h\,\|_2^2 \leqslant$$

$$c\,\big(\| \,u_{nh} - w_h\,\|_2 + h\,\| \,(u_{nh} - w_h)^{(3)}\,\|_2\big) \times$$

$$\big(\,\big|\,[\,\tau_h * (dF_n - dF_o)\,](1)\,\big| + \| \,\tau_h * (dF_n - dF_o)\|_2\big)\,.$$

With the help of Lemma (4.21) below, the proof is done.   Q.e.d.

(4.20) EXERCISE. Correct the sloppiness of (4.19).

The following lemma is really an exercise in § 4.4.

(4.21) LEMMA. (a) *For $h \to 0$,*

$$\mathbb{E}\big[\,\|\,\tau_h * (dF_n - dF_o)\,\|_2^2\,\big] = \mathcal{O}(\,h^{-1/3}\,n^{-1}\,)\ .$$

(b) *If $h \asymp n^{-\beta}$ for some $0 < \beta < 1$, then for all $s > \frac{1}{2}$,*

$$\|\,\tau_h * (dF_n - dF)\,\|_2 =_{\mathrm{as}} \mathcal{O}(\,h^{-1/6}\,n^{-1/2}\,(\log n)^s\,)\ .$$

(c) *The same bounds hold for $[\,\tau_h * (dF_n - dF_o)\,](1)$.*

(4.22) EXERCISE. Prove Lemma (4.21).

We may now combine Theorems (4.10) and (4.17) to get the (presumably close to optimal) rates of convergence.

(4.23) THEOREM. (a) *Under the assumption (4.1), for $h \asymp n^{-3/7}$, and for all $s > \frac{1}{2}$,*

$$\|\,u_{nh} - w_o\,\|_2 =_{\mathrm{as}} \mathcal{O}(\,n^{-3/7}\,(\log n)^s\,)\ .$$

(b) *Under the assumptions (4.2), for $h \asymp n^{-3/13}$, and for all $s > \frac{1}{2}$,*

$$\|\,u_{nh} - w_o\,\|_2 =_{\mathrm{as}} \mathcal{O}(\,n^{-6/13}\,(\log n)^s\,)\ .$$

(4.24) EXERCISE. (a) Prove Theorem (4.23).
(b) Obtain bounds for $\|\,\exp(u_{nh}) - f_o\,\|_1$.
[Hint: For (b), keep Lemma (4.7) in mind, as well as the L^1 lower bound for the Kullback-Leibler distance.]

Note the minuscule improvement in the convergence rate resulting from the extra assumptions (4.2). As a final comment, it should be observed that the rates obtained are very close to the *parametric* rate. This could presumably also be achieved by (boundary) kernels of the appropriate high order, see § 4.7, but then the resulting kernel estimators are not (necessarily?) pdfs. In the context of bounded log-densities, the estimators are *always* pdfs. Finally, the authors do not know whether the above rates are optimal, but we venture to guess that they are.

EXERCISES: (4.5), (4.9), (4.11), (4.13), (4.15), (4.20), (4.22), (4.24).

5. Estimation under constraints

The maximum penalized likelihood approach to nonparametric density estimation lends itself well to using side information. Examples of interest include shape constraints on the unknown density f_o, such as monotonicity, convexity, unimodality, and log-concavity. It seems logical to require that the estimator satisfy the same constraints as f_o. Here, we study uni-modal and log-concave density estimation, with methods similar to those used in the previous sections. For log-concave densities, we get a concise treatment with just about optimal a.s. convergence rates, but for the case of unimodality, exponential tail conditions need to be imposed. In the next chapter, a more careful study of monotone and unimodal density estimation gets rid of these.

To get our ducks in a row, we begin with a definition.

(5.1) DEFINITION. (a) *A univariate density f is monotone on* (a, ∞) *if it is decreasing there, i.e.,*

$$f(x) \leqslant f(y) \quad \text{for all } x \geqslant y > a .$$

Likewise, a univariate density is monotone on $(-\infty, a)$ *if it is increasing there.*
(b) *A univariate density f is unimodal if there exists an* $m \in \mathbb{R}$ *(the mode) such that f is monotone on* $(-\infty, m)$ *and on* (m, ∞).
(c) *A density f is log-concave if* $\log f$ *is concave. Equivalently, if for all* $0 < \theta < 1$ *and for all* $x, y \in \mathbb{R}$,

$$f(\theta x + (1 - \theta)y) \geqslant [f(x)]^{\theta} [f(y)]^{1-\theta} .$$

Some comments are in order. If a density f satisfies any of the above properties, then so do its scaled versions $x \longmapsto h^{-1} f(h^{-1}x)$ for all $h > 0$. In the above, "decreasing" actually means "nonincreasing". Also, the mode of a unimodal density need not be unique, as witnessed by the uniform $(0,1)$ density.

(5.2) EXERCISE. Verify that the following densities are log-concave:
(a) the uniform;
(b) the standard normal;
(c) the two-sided standard exponential $x \longmapsto \frac{1}{2} \exp(-|x|)$;
(d) the Epanechnikov kernel $x \longmapsto \frac{3}{4}(1 - x^2)_+$;
(e) the triangular density $x \longmapsto (1 - |x|)_+$;
(f) but that the Cauchy distribution is not.

(5.3) EXERCISE. Show that a log-concave density is unimodal.

It should come as no surprise that in the ensuing development it would be helpful if the above-introduced shape constraints remained valid after smoothing with convolution kernels. The following beautiful result of IBRAGIMOV (1956) is just what the doctor ordered and provides a surprising connection between the notions of unimodality and log-concavity.

(5.4) LEMMA. [IBRAGIMOV (1956)] *Let* f *be a density on the line. Then, the following are equivalent.*
(a) f *is log-concave.*
(b) $f * g$ *is unimodal for every unimodal density* g.

(5.5) COROLLARY. *If* f *and* g *are log-concave densities on the line, then so is* $f * g$.

The lemma shall not be proved here, but the corollary is easy.

(5.6) EXERCISE. Prove Corollary (5.5).

We continue with some additional observations regarding log-concave densities. Since $\log f$ is unimodal, it is continuous and differentiable almost everywhere on $\{\, f > 0 \,\}$ and $\log f$ is the integral of its derivative. Then, Taylor's theorem gives, for any a with $f(a) > 0$,

$$\log f(x) = \log f(a) + \int_a^x \{\log f\}'(y) \, dy \ .$$

Since $\{\log f\}'$ is decreasing, it follows that

(5.7) $\log f(x) \leqslant \log f(a) + |\, x - a\,| \{\log f\}'(a) \ , \quad -\infty < x < \infty \ .$

Now, f is a log-concave density, so it must be strictly decreasing somewhere. That is, there exists an a such that $\{\log f\}'(a) < 0$. Then, (5.7) says that f decreases exponentially.

(5.8) LEMMA. *Let* X *be a real-valued random variable with density* f. *If* f *is log-concave, then there exist positive constants* c, k *such that*

$$f(x) \leqslant c \, e^{-k\,|x|} \ , \quad x \in \mathbb{R} \ .$$

Moreover, $\mathbb{E}\big[\, e^{r\,|X|} \,\big] < \infty$ *for all* $r < k$.

Now, back to constrained estimation. We begin with the success story of log-concave density estimation. Let $X_1, X_2, \cdots, X_n$ be iid random variables with unknown common log-concave density f_o. To estimate f_o, we consider the maximum smoothed likelihood estimation problem

(5.9)
$$\text{minimize} \quad L_{nh}(f) \overset{\text{def}}{=} -\int_{\mathbb{R}} A_h * dF_n(x) \log f(x) \, dx + \int_{\mathbb{R}} f(x) \, dx$$

$$\text{subject to} \quad f \text{ is log-concave} \ ,$$

where $A_h(x) = h^{-1} A(h^{-1}x)$, for some log-concave, symmetric pdf A. The solution of (5.9), denoted by f^{nh}, exists and is unique, see §11.5. The customary exercise shows that f^{nh} is a pdf, see Exercise (5.22). Note that, apart from terms independent of f, as usual, $L_{nh}(f) = \mathrm{KL}(A_h * dF_n, f)$. Thus, since $A_h * f_o$ is log-concave,

(5.10) $\mathrm{KL}(A_h * dF_n, f^{nh}) \leqslant \mathrm{KL}(A_h * dF_n, A_h * f_o)$.

Amazingly, this is about all there is to it!. Because f_o has a finite exponential moment, the material of §4.5 shows that for all $s > 1$,

(5.11) $\mathrm{KL}(A_h * dF_n, A_h * f_o) =_{\mathrm{as}} \mathcal{O}\big((nh)^{-1} \log(nh)^{-1} (\log n)^s\big)$,

provided $h \asymp n^{-\beta}$ for some $0 < \beta < 1$. It follows that for all $s > \frac{1}{2}$,

(5.12) $\| A_h * dF_n - A_h * f_o \|_1 =_{\mathrm{as}} \mathcal{O}\big(\{ (nh)^{-1} \log(nh)^{-1} \}^{1/2} (\log n)^s\big)$.

To treat the bias $\| f_o - A_h * f_o \|_1$, we assume that $f_o \in W^{2,1}(\mathbb{R})$, which can be slightly weakened to $f_o' \in \mathrm{BV}(\mathbb{R})$, see Exercise (4.2.33).

(5.13) EXERCISE. Prove or disprove: If $f \in W^{1,1}(\mathbb{R})$ is a log-concave density, then f' has bounded variation.

So, assuming $f \in W^{2,1}(\mathbb{R})$, then Theorem (4.2.9)(a) yields

(5.14) $\| A_h * f_o - f_o \|_1 = \mathcal{O}(h^2)$.

Combining (5.12) and (5.14) proves the following theorem.

(5.15) THEOREM. Assume that the density $f_o \in W^{2,1}(\mathbb{R})$ is log-concave. If the kernel A is symmetric about 0, and log-concave, then with $h \asymp n^{-1/5}$ for all $s > \frac{1}{2}$,

$$\| f^{nh} - f_o \|_1 =_{\mathrm{as}} \mathcal{O}(n^{-2/5}(\log n)^s) .$$

The above argument is generic and may be repeated for unimodal densities.

(5.16) THEOREM. Let the unimodal density f_o have a finite exponential moment, and assume $f_o \in W^{2,1}(\mathbb{R})$. Let the kernel A be log-concave and symmetric about 0. Then, with $h \asymp n^{-1/5}$ for all $s > \frac{1}{2}$,

$$\| f^{nh} - f_o \|_1 =_{\mathrm{as}} \mathcal{O}(n^{-2/5}(\log n)^s) .$$

(5.17) EXERCISE. Prove Theorem (5.16).

The key to the above arguments is the inequality (5.10). Thus, it is not so easy to get rid of the factor $(\log n)^s$ in Theorems (5.15) and (5.16), or to prove optimal rates in Theorem (5.16) under the (for density estimation) minimal tail conditions. There is of course an easy way to get estimators with optimal rates of convergence, viz. in (5.9), replace the Kullback-Leibler distance by the L^1 norm

(5.18)
$$\text{minimize} \quad \| f - A_h * dF_n \|_1$$
$$\text{subject to} \quad \text{the shape constraint on } f \ .$$

Problems of this kind go by the name of *minimum distance* problems.

It is obvious that any solution f^{nh} of (5.18), assuming it exists, satisfies

$$\| f^{nh} - A_h * dF_n \|_1 \leqslant \| f_o - A_h * dF_n \|_1 \ ,$$

and it follows that

(5.19)
$$\| f^{nh} - f_o \|_1 \leqslant 2 \| f_o - A_h * dF_n \|_1 \ .$$

So, in this instance, we get the exact same rate of convergence as in kernel density estimation. Of course, one would expect that the factor 2 is superfluous: Surely, the extra information about the shape of the density should not result in worse error bounds? This is another story, which will not be pursued here.

We finish this section with some (open) questions. The first one is whether smoothing in (5.9) is really necessary. Let $f_o \in W^{2,1}(\mathbb{R})$ be log-concave and consider the solution f_n of the maximum likelihood estimation problem

(5.20)
$$\text{minimize} \quad - \int_{\mathbb{R}} \log f(x) \, dF_n 9x) + \int_{\mathbb{R}} f(x) \, dx$$
$$\text{subject to} \quad f \text{ log-concave} \ .$$

Do we then have that for $n \to \infty$ and for all s big enough that

(5.21)
$$\| f_n - f_o \|_1 =_{\text{as}} \mathcal{O}(n^{-2/5} (\log n)^s) \quad ?$$

The same question can be asked for convex densities on $(0, \infty)$, with integrable second derivative. The authors have no clue, but venture that the answer to both questions is yes.

The second question concerns the computation of log-concave estimators. Obviously (5.9) and (5.20) must be discretized (how?) and then solved. The sequential quadratic programming approach comes to mind. This would be similar to Newton's method for the computation of roughness penalized log-densities, see O'SULLIVAN (1988). However, we shall not pursue this either.

Finally, the long-awaited problem.

(5.22) EXERCISE. Verify that any solution of (5.9) must be a pdf.

EXERCISES: (5.2), (5.3), (5.6), (5.13), (5.17), (5.22).

6. Additional notes and comments

Ad § 1 : The prehistory of maximum penalized likelihood estimation and indeed of nonparametric density estimation in general is described in some detail in THOMPSON and TAPIA (1990).

Ad § 2 : The roughness penalization for square root densities originated with GOOD (1971) and GOOD and GASKINS (1971), and was further investigated in the 1970's and early 1980's by DE MONTRICHER, TAPIA and THOMPSON (1975), see also THOMPSON and TAPIA (1990), and by KLONIAS (1982), KLONIAS (1984). It is not so surprising that close to 30 years after the fact a final solution has been reached, see EGGERMONT and LARICCIA (1999b). The probabilistic tools were available back then. Indeed, the difficulties are mostly on the (applied) mathematics side.

The analysis of the bias term in the Hellinger distance setup follows the standard variational treatment of differential equations. A nice reference is PRENTER (1975). For the variance term, one would like to do the same, but one has to fight the technical difficulties.

As an aside, quite a few problems in statistics lead to significant problems in applied mathematics, in particular, to infinite-dimensional minimization problems. A recent example in the spirit of this text is LANDSMAN (2000).

Ad § 3 : The observation of Exercise (3.8) is due to SILVERMAN (1982). We encountered variations of it in §§ 2 and 5, and we will encounter it in many more guises.

Ad § 4 : The roughness penalization of the log-density was analyzed by SILVERMAN (1982) and COX and O'SULLIVAN (1990), who cite a number of interesting applications, among which is the estimation of the log-hazard function in random censoring problems. There, establishing existence and rates of convergence of the estimators takes place on a finite interval. This is typical for exponential families, see CRAIN (1976b), SILVERMAN (1982), BARRON and SHEU (1991), STONE (1990), and COX and O'SULLIVAN (1990). The analysis in SILVERMAN (1982) and COX and O'SULLIVAN (1990) depends on linearization around the true solution and so it resembles the treatment of parametric estimation (§ 2.2). The method of proof presented here is similar in spirit to that of the treatment of the GOOD estimator in § 3.

Ad § 5 : Lemma (5.4) and Corollary (5.5) are due to IBRAGIMOV (1956). Proofs may be found in the readily available reference DHARMADHIKARI and JOAG-DEV (1988). The Corollary was shown directly (and independently) by LEKKERKERKER (1953). The source of the last reference is DEVROYE (1986), which contains a wealth of information in general.

Regarding minimum distance estimators, we quote YATRACOS (1985) and VAN DE GEER (1993), both of whom relate the convergence rates of the estimators to the compactness of the constraint set as measured in terms of the *Kolmogorov entropy.* See also VAN DE GEER (2000).

Regarding computational issues, we note that discretized versions of (5.18) with monotonicity or convexity constraints (and unimodality with fixed mode) are equivalent to *linear programming* problems, for which there are efficient algorithms, see BARRODALE and ROBERTS (1978), MADSEN and NIELSEN (1993), LI and SWETTITS (1998), and references therein.

6

Monotone and Unimodal Densities

1. Introduction

In this chapter, we undertake a detailed study of nonparametric maximum likelihood estimation of monotone densities on $(0, \infty)$ and unimodal densities on $(-\infty, \infty)$. The topic was lightly touched in § 5.5 as an instance of constrained density estimation, but the actual character of the constraints barely came into play. Here, the notion of monotonicity is dissected and put to good use in proving L^1 error bounds. The connection with unimodal density estimation is immediate, and in a way, this provides the motivation for the study of monotone densities.

Let $\mathcal{D}$ denote the set of decreasing densities on $(0, \infty)$. Actually, we do not enforce the pdf constraint, so

$$(1.1) \qquad \mathcal{D} \overset{\text{def}}{=} \{ f \in L^1(0, \infty) \, : \, f \geqslant 0 \, , \, f \text{ decreasing} \} \, .$$

Let $X_1, X_2, \cdots, X_n$ be nonnegative iid random variables with common density f_o, which is assumed to be decreasing ($\equiv$ nonincreasing) on $(0, \infty)$. The plain, unpenalized maximum likelihood estimation problem is

$$(1.2) \qquad \begin{aligned} \text{minimize} \quad & L_n(f) \overset{\text{def}}{=} -\frac{1}{n} \sum_{i=1}^{n} \log f(X_i) + \int_0^\infty f(y) \, dy \\ \text{subject to} \quad & f \in \mathcal{D} \, . \end{aligned}$$

(1.3) EXERCISE. Show that any solution of (1.2) must be a pdf.

We recall that without the restriction to decreasing densities, the maximum likelihood estimator does not exist. This is not exactly equivalent to saying that *with* the restriction in place the estimator would exist, but in fact it does. This realization is due to GRENANDER (1956). The first indication of happy days is that $L_n(f)$ restricted to $\mathcal{D}$ cannot go to $-\infty$, so the infimum *might* be attained.

(1.4) LEMMA. $L_n(f)$ *is a.s. bounded below on* $\mathcal{D}$.

PROOF. Let $X > 0$ and consider minimizing

$$L_n(f; X) = -\log f(X) + \int_0^\infty f(y)\, dy$$

over $\mathcal{D}$. By Exercise (1.3), we may restrict attention to decreasing pdfs f. Then,

$$1 \geqslant \int_0^X f(y)\, dy \geqslant X f(X) ,$$

whence $f(X) \leqslant 1/X$, and it follows that $L_n(f; X) \geqslant 1 + \log X$.

Now, let $X_1, X_2, \cdots, X_n$ be nonnegative iid random variables with common density f_o (decreasing or not). Then, each $X_i > 0$ almost surely, and

$$L_n(f) = \tfrac{1}{n} \sum_{i=1}^n L_n(f; X_i) \geqslant \tfrac{1}{n} \sum_{i=1}^n \{\, 1 + \log X_i \,\} > -\infty \quad \text{a.s.} \qquad \text{Q.e.d.}$$

In § 11.5, the existence of solutions to (1.2) for various shape restrictions is shown or indicated using compactness arguments, in which the boundedness below of the negative log-likelihood features prominently. In the next section, the existence of solutions to (1.2) is shown by "elementary" methods. The existence of the estimator f_n being under control, attention turns to its consistency. It is not hard to show that f_n converges if and only if f_o is decreasing, i.e.,

(1.5) $\| f_n - f_o \|_1 \longrightarrow_{\text{as}} 0$

if and only if f_o is monotone, see, e.g., DEVROYE (1987). As far as rates of convergence are concerned, this seems to be an open question, but the hope is that

(1.6) $\| f_n - f_o \|_1 =_{\text{as}} \mathcal{O}(n^{-1/3})$,

provided f_o is bounded and has compact support. BIRGÉ (1987b) constructs histogram estimators for which the analogue for expected values of (1.6) holds, uniformly over suitable classes of decreasing densities. (Here we also mention BIRGÉ (1989), in which precise bounds for the L^1 error are obtained.) It is also known that extra smoothness does not improve the rate of convergence, as stated in the following theorem.

(1.7) THEOREM. [GROENEBOOM (1985)] If f_o is strictly decreasing, has compact support in the interval $[0, B]$, and is twice continuously differentiable on its support, then

$$n^{1/6} \{\, n^{1/3} \| f_n - f_o \|_1 - \mu \,\} \longrightarrow_d Y \sim N(0, \sigma^2) ,$$

where, for known constants c_1 and σ, independent of f_o,

$$\mu = c_1 \int_0^B |\, \tfrac{1}{2} f(t)\, f'(t)\,|^{1/3}\, dt .$$

The theorem was finally proved in GROENEBOOM, HOOGHIEMSTRA and LOPUHAÄ (1999), but it should come as no surprise that we do not reproduce it here.

We note that the conclusion of the theorem fails for the uniform density, since then

$$(1.8) \qquad n^{1/2} \, \| \, f_n - f_o \, \|_1 \longrightarrow_d 2 \, Z \ ,$$

where Z is the maximum of a standard Brownian bridge, see GROENEBOOM (1985). We also note that under the stated assumptions one should be able to get estimators with the usual rate of convergence, viz. $n^{-2/5}$.

So, the question is whether there are estimators that give good convergence rates for smooth monotone densities. The interpretation of the kernel density estimator as the maximum smoothed likelihood estimator is now extremely useful because, as in the previous chapter, we may add the monotonicity constraint to the minimization problem. So, let $A_h(x, y)$ be a symmetric boundary kernel, and consider the density estimator (see § 4.8 for references)

$$(1.9) \qquad A_h dF_n(x) = \int_0^\infty A_h(x, y) \, dF_n(y) \ , \quad 0 < x < \infty \ .$$

A convenient, but not optimal, choice is

$$(1.10) \quad A_h(x, y) = h^{-1} A(\, h^{-1}(x - y) \,) + h^{-1} A(\, h^{-1}(x + y) \,) \ , \quad 0 < x, y < \infty \ ,$$

where A is a density symmetric about 0, with a finite exponential moment. The maximum smoothed likelihood estimation problem for monotone densities is then

$$(1.11) \qquad \text{minimize} \quad L_{nh}(f) \overset{\text{def}}{=} -\tfrac{1}{n} \sum_{i=1}^n [\, A_h\{ \log f \, \}](X_i) + \int_0^\infty f(y) \, dy$$

$$\text{subject to} \quad f \in \mathcal{D} \ .$$

As shown in § 1.2, without the constraint $f \in \mathcal{D}$, the solution to (1.11) is the kernel density estimator $f = A_h dF_n$. We show in the next three sections that the solution f^{nh} of (1.11) exists and is unique, and that the corresponding distribution F^{nh} is the least concave majorant of $A_h F_n$, the distribution with density $A_h dF_n$. At this point, it is useful to introduce the notations

$$(1.12) \qquad F^{nh} = \text{LCM}(\, A_h F_n \,) \ , \quad f^{nh} = \text{lcm}(\, A_h dF_n \,) \ .$$

We show in § 3 that for all realizations of $X_1, X_2, \cdots, X_n$,

$$(1.13) \qquad \| \, \text{lcm}(\, A_h dF_n \,) - f \, \|_1 \leqslant \| \, A_h dF_n - f \, \|_1 \ , \quad \text{for all } f \in \mathcal{D} \ .$$

Note that this is not a probabilistic or asymptotic statement. Consequently, if f_o is indeed monotone, then consistency and convergence rates for the L^1 error of f^{nh} follow from those for the boundary kernel density estimator. This holds even for random smoothing parameters.

(1.14) EXERCISE. Show that any solution of (1.11) must be a pdf.

(1.15) EXERCISE. Based on (1.13), derive rates of convergence for f^{nh} under appropriate conditions on f_o.

(1.16) EXERCISE. Compare (1.13) to the inequalities obtained for the minimum distance estimators in § 5.5.

We finish this section by pointing out the equivalence between the maximum likelihood approach and the least-squares approach for nonparametric monotone density estimation, but the details are left to the reader, see Exercise (2.27) in the next section. We also point out the connection with isotonic regression, which is plain nonparametric regression with the monotonicity constraint added. This also serves as an opportunity to elaborate on an unexpected property of the monotone estimator, viz. its *contractivity*. The continuous version of isotonic regression on $(0, \infty)$ reads as

(1.17)
$$\text{minimize} \quad \| f - \varphi \|_2^2$$
$$\text{subject to} \quad f \in \mathcal{D}_2 \, ,$$

where $\varphi \in L^2(0, \infty)$ is the data, with "distribution" $\Phi(x) = \int_0^x \varphi(t) \, dt$, and

$$\mathcal{D}_2 = \{ f \in L^2(0, \infty) : f \text{ decreasing} \} \, .$$

It is assumed that φ is close to the "true" $f_o \in \mathcal{D}_2$. It turns out that the solution f of (1.17) is the derivative of the least concave majorant of Φ, which we denoted earlier as $\text{lcm}(\varphi)$. Of course, the solution of (1.17) is (also) given by the (L^2) orthogonal projection of φ onto $\mathcal{D}_2$, and so

(1.18) $\quad \| \text{lcm}(\varphi) - \psi \|_2 \leqslant \| \varphi - \psi \|_2 \quad \text{for all } \psi \in \mathcal{D}_2 \, .$

For all of this and more, see the four B's, BARLOW, BARTHOLOMEW, BREMNER and BRUNK (1972). It follows that (1.13) holds in L^2 norm as well, and that (1.18) also holds for the L^1 norm. In fact, both inequalities follow from the contractivity of the mapping $\varphi \longmapsto \text{lcm}(\varphi)$ expressed by

(1.19) $\quad \| \text{lcm}(\varphi) - \text{lcm}(\psi) \|_1 \leqslant \| \varphi - \psi \|_1 \quad \text{for all } \varphi, \, \psi \in L^1(0, \infty) \, ,$

and its analogues for all L^p norms, the Kullback-Leibler, Hellinger, and Pearson's φ^2 distances. In particular, for any nonnegative convex function $J : \mathbb{R}^2 \longrightarrow \mathbb{R}$, with negative, mixed second-order partial derivative, and $J(x, x) = 0$ for all x, we have

(1.20) $\quad \displaystyle\int_0^\infty J\big([\text{lcm}(\varphi)](x), [\text{lcm}(\psi)](x) \big) \, dx \leqslant \int_0^\infty J\big(\varphi(x), \psi(x) \big) \, dx \, .$

We consider this in § 5.

The remainder of this chapter is put together as follows. In the next section, we give the characterization of the solutions to (1.2) as the density whose distribution is the least concave majorant of F_n. This is due to GRENANDER (1956). In §3, we do the same for the solution of the smoothed maximum likelihood problem (1.11). In §4, we prove the contractivity (in L^1 norm) of the mapping $\varphi \longmapsto \text{lcm}(\varphi)$ for densities φ that are step functions with finitely many steps. In §5, the general case is considered. In §6, we apply what we have learned about monotone estimation to the maximum smoothed likelihood estimation of unimodal densities. In §7, some alternative estimators of unimodal densities are surveyed. Finally, in §8, convex density estimation and the apparent lack of an analogy with the monotone case is briefly mentioned.

EXERCISES : (1.3), (1.14), (1.15), (1.16).

2. Monotone density estimation

In this section, we give the characterization of the solution to (1.2) as the density whose distribution is the least concave majorant of F_n. This is due to GRENANDER (1956). In §3, we do the same for the maximum smoothed likelihood solution. We also discuss, rather informally, geometric aspects of least convex majorization, which explains some "obvious" facts.

It turns out that a least concave majorant is exactly that, it is a majorant, it is concave, and it is the least one of all of those. Precisely:

(2.1) DEFINITION. Let F be a distribution on $(0, \infty)$. A function Φ is a concave majorant of F if
(a) Φ is a concave function, and
(b) Φ majorizes F, i.e., $\Phi(x) \geqslant F(x)$ for all $0 < x < \infty$.

Now, the least concave majorant (LCM) of a distribution F is defined for $x > 0$ by

$$(2.2) \qquad [\text{LCM}(F)](x) = \inf \{ \Phi(x) : \Phi \text{ is a concave majorant of } F \} ,$$

and is a concave majorant of F, see the following exercise. This is consistent with the usage in (1.12). Note that any concave distribution function Φ has a derivative almost everywhere, denoted by φ, and that $\varphi \in L^1(0, \infty)$. Moreover, Φ is the integral of its derivative. Another important observation is that the least concave majorant must be linear on intervals where it is strictly larger than the distribution it majorizes.

(2.3) EXERCISE. (a) Let $\mathcal{S}$ be a set of concave functions on an interval (a, b). Define the function Ψ by

$$\Psi(x) = \inf \{ \Phi(x) : \Phi \in \mathcal{S} \} , \quad x \in (a, b) .$$

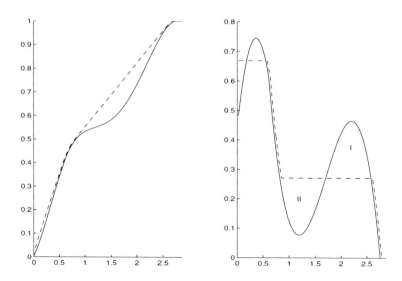

Figure 2.1. Illustration of the string theory for the least concave majorant of a distribution and the equal area rule for the corresponding densities. The solid lines represent the distribution F and corresponding density f. The dashed lines are for LCM(F) and lcm(f) and have been slightly shifted horizontally. Note the slight dip in F near the origin.

Show that Ψ is concave. (Note that the value $-\infty$ is allowed.)
(b) Let F be a distribution, and set $\Phi = $ LCM(F). If $\Phi(x) > F(x)$ for all $a < x < b$, and $\Phi(x) = F(x)$ for $x = a, b$, then

$$\Phi(x) = \frac{b - x}{b - a}\, F(a) + \frac{x - a}{b - a}\, F(b) , \quad a \leqslant x \leqslant b .$$

Thus, LCM(F) is linear wherever it is larger than F.

The least concave majorant of a distribution has some nice geometric properties. Let F be a probability distribution, and assume that the corresponding density f has compact support, say, on $[\,0\,,\,T\,]$. Let $\Phi = $ LCM(F), and let $\varphi = $ lcm(f). See Figure 2.1.

The first geometric observation is the *string theory*, due to BARLOW, BARTHOLOMEW, BREMNER and BRUNK (1972). In the xy-plane, fasten a string to the origin, and pull the string through a little ring at the point with coordinates $(T, 1)$. Now, pull the string taut, making sure the string lies above the graph of F. The taut string then represents the graph of LCM(F). It is also makes obvious the content of Exercise (2.3)(b).

The second geometric observation is that the *string theory* for distributions may be translated into the *equal area rule* for the corresponding densities. In Figure 2.1, observe that for $a \approx 0.8$, $b \approx 2.7$, $c \approx 0.7$,

$$(2.4) \qquad\qquad \Phi(x) > F(x) , \quad a < x < b ,$$

with equality for $c \leqslant x \leqslant a$ and $x \geqslant b$. By Exercise (2.3)(b), this implies that Φ is linear on $[a, b]$, and so φ is constant there. If F is smooth, then the derivatives f and φ of F and Φ coincide for $c \leqslant x \leqslant a$ and for $x \geqslant b$. The above implies that

$$(2.5) \qquad \int_a^b \varphi(x)\,dx = \int_a^b f(x)\,dx \ ,$$

and so the regions marked I and II in Figure 2.1 have equal areas. Also, note that if f is continuous at $x = 0$, then

$$(2.6) \qquad [\mathrm{lcm}(f)](0) \geqslant f(0) \ .$$

The reason for all of the above hoopla about least concave majorants is the following theorem. Note that this settles the existence question for maximum likelihood estimation of monotone densities.

(2.7) THEOREM. *The distribution corresponding to the solution of (1.2) is the least concave majorant of F_n.*

The first steps in the proof are to whittle down the set $\mathcal{D}$ of possible solutions of (1.2) to a finite-dimensional set. A minor observation is that in the minimization in (1.2), we may restrict attention to functions that are left-continuous ($\equiv$ continuous from the left) and decreasing. The following observation is more substantial.

(2.8) LEMMA. *If f solves (1.2), then f is piecewise constant, with jumps at (some of) the $X_1, X_2, \cdots, X_n$.*

PROOF. Let $X_{1,n} \leqslant X_{2,n} \leqslant \cdots \leqslant X_{n,n}$ be the order statistics of the observations $X_1, X_2, \cdots, X_n$, so that almost surely $X_{1,n} > 0$. Set $X_{0,n} = 0$. Let $f \in \mathcal{D}$ be a solution of (1.2). Construct the piecewise constant function

$$g(x) = f(X_{i,n}) \ , \qquad X_{i-1,n} < x \leqslant X_{i,n} \ , \ i = 1, 2, \cdots, n \ ,$$

and $g(x) = 0$ for $x > X_{n,n}$. Then, $g(x) \leqslant f(x)$ for all x, and $g(X_i) = f(X_i)$ for all i. So then

$$L_n(g) - L_n(f) = \int_0^\infty \{ g(x) - f(x) \}\,dx \ ,$$

and this is nonnegative, because f solves (1.2). Since $g(x) \leqslant f(x)$ for all x, then $f(x) = g(x)$ everywhere.   Q.e.d.

The lemma implies that (1.2) is a finite-dimensional problem : We need only look at densities of the form

$$(2.9) \qquad f(x) = \begin{cases} p_i \ , & x \in (X_{i-1,n}, X_{i,n}] \ , \quad i = 1, 2, \cdots, n \ , \\ 0 \ , & x > X_{n,n} \ , \end{cases}$$

with $p_1 \geqslant p_2 \geqslant \cdots \geqslant p_n \geqslant 0$. (Actually, $p_n > 0$ but never mind.) It is useful to introduce the set of all vectors satisfying this constraint

(2.10) $\mathcal{P}_{\downarrow} = \{ p \in \mathbb{R}^n : p_1 \geqslant p_2 \geqslant \cdots \geqslant p_n \geqslant 0 \}$

and the weights

(2.11) $d_i = X_{i,n} - X_{i-1,n}$, $i = 1, 2, \cdots, n$.

The problem (1.2) is then equivalent to

(2.12)
$$\text{minimize} \quad \Lambda_n(p) \stackrel{\text{def}}{=} -\frac{1}{n} \sum_{i=1}^{n} \log p_i + \sum_{i=1}^{n} d_i \, p_i$$

$$\text{subject to} \quad p \in \mathcal{P}_{\downarrow} .$$

(2.13) EXERCISE. Give an alternative proof of Lemma (1.4) by showing that apart from terms independent of p,

$$\Lambda_n(p) = \text{KL}(u, q) ,$$

where KL is the discrete Kullback-Leibler distance,

$$\text{KL}(a, b) = \sum_{i=1}^{n} a_i \log \frac{a_i}{b_i} + a_i - b_i ,$$

and $q_i = d_i p_i$ for all i, and $u \in \mathbb{R}^n$ is the vector all of whose components equal $1/n$.

Later, we need to discretize problems like (1.11), and to that end, it is useful to consider problems like (2.12), where the "weights", in this case $(1/n)$, are not necessarily equal. So we consider

(2.14)
$$\text{minimize} \quad \Lambda_n(p) \stackrel{\text{def}}{=} -\sum_{i=1}^{n} d_i \, w_i \log p_i + \sum_{i=1}^{n} d_i \, p_i$$

$$\text{subject to} \quad p \in \mathcal{P}_{\downarrow} ,$$

where the w_i are nonnegative and $\sum_{i=1}^{n} d_i \, w_i = 1$. Of course, (2.12) is a special case of this. Note that the d_i are positive (a.s.). The w_i define a density φ via

(2.15) $\varphi(x) = \begin{cases} w_i , & x \in (X_{i-1,n}, X_{i,n}] , \; i = 1, 2, \cdots, n , \\ 0 , & x > X_{n,n} , \end{cases}$

but of course there is no reason for $\varphi \in \mathcal{D}$.

(2.16) EXERCISE. Show that the solution of (2.14) satisfies

$$\sum_{i=1}^{n} d_i \, p_i = \sum_{i=1}^{n} d_i \, w_i .$$

At this point, it is important to observe that the objective function in (2.14) is convex, and that the constraint set is also convex. In other words, the problem is convex. One verifies that the objective function is differentiable, as long as $p_n > 0$, which can be assumed, since otherwise the objective function equals $+\infty$. We are now ready to show that any solution of (1.2) is a concave majorant of F_n, and actually a little bit more. Translated in terms of the p_i, this reads as follows.

(2.17) LEMMA. *The discrete density p solves (2.14) if and only if*

(a)
$$\sum_{i=1}^{k} d_i\, p_i \geqslant \sum_{i=1}^{k} d_i\, w_i\,, \quad k = 1, 2, \cdots, n\,,$$

and, for every $k = 1, 2, \cdots, n-1$,

(b)
$$p_k > p_{k+1} \implies \sum_{i=1}^{k} d_i\, p_i = \sum_{i=1}^{k} d_i\, w_i\,.$$

(2.18) EXERCISE. Let f be given by (2.9). Verify that its distribution function satisfies

$$F(X_{k,n}) = \sum_{i=1}^{k} d_i\, p_i\,, \quad k = 1, 2, \cdots, n\,.$$

PROOF OF LEMMA (2.17). $\implies$: Since (2.14) is a convex problem, the necessary and sufficient conditions for p to be a minimum are, see §9.2, Example (9.2.14) and Exercise (9.2.20),

(2.19)
$$\delta\Lambda_n(p, q - p) \geqslant 0\,, \quad \text{for all } q \in \mathcal{P}_{\downarrow}\,,$$

where $\delta\Lambda_n$ is the Gateaux variation of Λ_n,

$$\delta\Lambda_n(p, q - p) = \sum_{i=1}^{n} d_i\, (-w_i/p_i + 1)\,(q_i - p_i)\,.$$

This may be rewritten as

(2.20)
$$\delta\Lambda_n(p, q - p) = \sum_{i=1}^{n} d_i\, (p_i - w_i)\,(q_i/p_i - 1)\,.$$

By Exercise (2.16), we may drop the -1 in (2.20). Now, instead of considering all $q \in \mathcal{P}_{\downarrow}$, we only consider q of the form $q_i = \theta_i\, p_i$, $i = 1, 2, \cdots, n$, with $\theta \in \mathcal{P}_{\downarrow}$. Then, (2.20) implies that

(2.21)
$$\sum_{i=1}^{n} d_i\, (p_i - w_i)\,\theta_i \geqslant 0\,, \quad \text{for all } \theta \in \mathcal{P}_{\downarrow}\,.$$

We now do summation by parts. Let

$$P_i = \sum_{j=1}^{i} d_j\, p_j\,, \quad W_i = \sum_{j=1}^{i} d_j\, w_j\,, \quad i = 1, 2, \cdots, n\,,$$

with $P_o = W_o = 0$. For convenience, set $\theta_{n+1} = 0$. Then, (2.21) is equivalent to

(2.22) $$\sum_{i=1}^{n} (P_i - W_i)(\theta_i - \theta_{i+1}) \geqslant 0, \quad \text{for all } \theta \in \mathcal{P}_\downarrow,$$

and so $P_i - W_i \geqslant 0$ for all i. This is (a).

Now, let $1 \leqslant k \leqslant n-1$. If $p_k > p_{k+1}$, define θ by

$$\theta_i = \begin{cases} 1 & , & 1 \leqslant i \leqslant k, \\ 1 + \varepsilon & , & k+1 \leqslant i \leqslant n, \end{cases}$$

with $\varepsilon > 0$ small enough that q defined via $q_i = \theta_i\, p_i$ is still in $\mathcal{P}_\downarrow$, even though $\theta \notin \mathcal{P}_\downarrow$. All that is required is that $p_k \geqslant (1+\varepsilon)\,p_{k+1}$. Now, (2.22) gives that $(P_k - W_k)(-\varepsilon) \geqslant 0$, so that $P_k \leqslant W_k$. It follows that $P_k = W_k$. This is (b).

$\Longleftarrow$: For the converse, it suffices to show that (2.22) holds for $\theta = q/p$ with $q \in \mathcal{P}_\downarrow$ arbitrary. Let $I \subset \{1, 2, \cdots, n\}$ be those indices i for which

(2.23) $$P_i > W_i \quad \text{and} \quad \frac{q_i}{p_i} - \frac{q_{i+1}}{p_{i+1}} < 0.$$

The indices $i \in I$ are the only ones for which the terms in the sum in (2.22) are negative. All other ones are nonnegative. Now, for $i \in I$, we have $P_i > W_i$, so by (b) then $p_i = p_{i+1}$ and the second statement of (2.23) implies that $q_i < q_{i+1}$. This cannot be, since $q \in \mathcal{P}_\downarrow$. The conclusion is that $I = \varnothing$ and (2.22) holds.   Q.e.d.

(2.24) EXERCISE. (a) Verify that (2.21) is equivalent to (2.22).
(b) Verify that (2.22) indeed implies that $P_i \geqslant W_i$ for all i.

We are now able to make the leap to the least concave majorant aspects of the maximum likelihood estimator.

(2.25) LEMMA. *Let Φ be the distribution for the density φ defined in (2.15), and let the density f be given by (2.9). Then, $f = \mathrm{lcm}(\varphi)$ if and only if the following three conditions hold.*

(a) $p \in \mathcal{P}_\downarrow$.

(b) $\displaystyle\sum_{i=1}^{k} d_i\,(p_i - w_i) \geqslant 0, \quad k = 1, 2, \cdots, n.$

(c) $p_k > p_{k+1} \implies \displaystyle\sum_{i=1}^{k} d_i\,(p_i - w_i) = 0, \quad k = 1, 2, \cdots, n.$

(2.26) EXERCISE. Prove the Lemma.

All of the details are now in place for the proof of Theorem (2.7).

PROOF OF THEOREM (2.7). By Lemma (2.8), the minimization in (1.2) may be restricted to densities of the form (2.9), and by Lemma (2.17), we may further whittle down this set to densities of this form whose distribution function majorizes F_n (the concavity is built in). The density whose distribution is the least concave majorant of F_n belongs to this class and by Lemma (2.17), has the smallest value of $L_n(f)$, respectively $\Lambda_n(q)$, in this class, so this must be the solution. The uniqueness comes from the strict convexity of $\Lambda_n(q)$.   Q.e.d.

(2.27) EXERCISE. Show that the problem

$$\text{minimize} \quad \int_0^\infty f(x)\, dx - 2 \int_o^\infty f(x)\, dF_n(x)$$

$$\text{subject to} \quad f \text{ is monotone}$$

has a unique solution which is the right continuous derivative of $\mathrm{LCM}(F_n)$. Thus, the maximum likelihood and least-squares approaches for nonparametric monotone density estimation are equivalent, cf. § 1.1.

  This concludes our treatment of the GRENANDER estimator. The strong point of the estimator is that it is a.s. consistent in the L^1 norm, see (1.5). However, there is trouble in the L^∞ norm, in particular, at $x = 0$. For any monotone density, redefine $f(0)$ as $f(0) = \lim_{x \to 0+} f(x)$. It is clear that the least concave majorant of F_n on $(0, X_{1,n})$ exceeds (is not smaller than) $\Phi(x) = x/(n\, X_{1,n})$. Thus, the density on $(0, X_{1,n})$ exceeds $1/(n\, X_{1,n})$. Let $Y_n = n\, X_{1,n}$. One verifies that the distribution function Φ_n of Y_n satisfies

$$\Phi_n(y) = 1 - (1 - F_o(y/n))^n , \quad y > 0 ,$$

with density

(2.28) $\varphi_n(y) = (1 - F_o(y/n))^{n-1} f_o(y/n) , \quad y > 0 .$

So, if $f_o(0) < \infty$, then $Y_n \longrightarrow_d Y_o$, where Y_o has density

(2.29) $\varphi_o(y) = f_o(0) \exp(-y\, f_o(0)) , \quad y > 0 ,$

i.e., Y_o has an exponential distribution, with mean $1/f(0)$. So we certainly do not have that $Y_n \longrightarrow_{as} 1/f(0)$, or that $1/Y_n \longrightarrow_{as} f(0)$. As a matter of fact, $\mathbb{E}[1/(n\, X_{1,n})] = +\infty$, which is about as bad as it can get. See WOODROOFE and SUN (1993), who are interested in estimating $f(0)$, and mention applications.

(2.30) EXERCISE. Verify (2.28) and (2.29).

The remedy of WOODROOFE and SUN (1993) for the woes above is to consider the maximum penalized likelihood problem (!)

$$\text{minimize} \quad -\tfrac{1}{n} \sum_{i=1}^{n} \log f(X_i) + h^2 f(0)$$

(2.31)

$$\text{subject to} \quad f \in \mathcal{D} .$$

They show that for appropriate values of h, this indeed gives consistent estimators for $f(0)$. Since the penalization functional $R(f) = f(0) = \| f \|_{\mathrm{BV}}$ is not *strongly* convex on $\mathcal{D}$, we shall not consider the penalization aspects of (2.31). However, it is a nice exercise to show that the solution of (2.31) is still a step function with breaks only at (some of) the order statistics, and that it is still the derivative of the least concave majorant of "something".

(2.32) EXERCISE. (a) Show that any solution of (2.31) must be constant on each of the intervals $(X_{i-1,n}, X_{i,n}]$.
(b) Write the problem (2.31) in the form (2.12), keeping in mind that $f(0) = f(X_1)$, and derive the analogue of Lemma (2.17).

EXERCISES : (2.3), (2.13), (2.16), (2.18), (2.24), (2.26), (2.27), (2.30), (2.32).

3. Estimating smooth monotone densities

We now add the ingredient of smoothness to monotone density estimation, and consider the maximum smoothed likelihood estimation problem for monotone densities (1.11). Assuming that the kernel $A_h(x, y)$ is symmetric, then $L_{nh}(f)$ may be written as

$$(3.1) \qquad L_{nh}(f) = - \int_0^\infty [A_h dF_n](x) \, \log f(x) \, dx + \int_0^\infty f(x) \, dx ,$$

and we see that the estimator is based on the (boundary) kernel estimator $A_h dF_n$. However, the development works for any density φ on $(0, \infty)$, so we consider the estimation problem

$$\text{minimize} \quad L(f) \stackrel{\text{def}}{=} - \int_0^\infty \varphi(x) \, \log f(x) \, dx + \int_0^\infty f(x) \, dx$$

(3.2)

$$\text{subject to} \quad f \in \mathcal{D} ,$$

where $\mathcal{D}$ denotes the set of decreasing densities on $(0, \infty)$. The objective function $L(f)$ depends also on φ, but this is suppressed in the notation. Recall the operator "lcm" from (1.12). In this section, we prove the following theorem.

(3.3) THEOREM. *Let φ be a density. Then, the solution f of (3.2) exists and is given by $f = \mathrm{lcm}(\varphi)$.*

PROOF. The proof is analogous to the discrete case. The first observation is that the support of the solution f must contain the support of φ, since otherwise $L(f) = +\infty$. The next observation is that the objective function $L(f)$ is convex in f, and that the constraint set is convex. Thus, the necessary and sufficient conditions for a minimum of Theorem (10.2.12) apply. Specifically, f solves (3.2) if and only if

$$\delta L(f; \psi - f) \geqslant 0 \quad \text{for all } \psi \in \mathcal{D} ,$$

where δL is the Gateaux variation of L. One need consider only those ψ that are positive whenever φ is. Then,

$$\delta L(f; \psi - f) = \int_0^\infty \left(-\frac{\varphi}{f} + 1 \right) (\psi - f) = \int_0^\infty (f - \varphi)(\theta - 1) ,$$

where $\theta = \psi/f$. Since f and φ are pdfs, we may drop the -1 in the second factor of the integrand. Note that the integration may be restricted to the set where $f > 0$, so θ is finite there. Thus, the necessary and sufficient conditions for f being a minimum are

$$(3.4) \qquad \int_0^\infty (f - \varphi) \theta \geqslant 0 , \qquad \text{for } \theta = \psi/f, \ \text{for all } \psi \in \mathcal{D} .$$

Since $pq \in \mathcal{D}$ whenever $p \in \mathcal{D}$ and $q \in \mathcal{D}$, (3.4) must hold for all bounded smooth $\theta \in \mathcal{D}$, for which then $\theta' \in L^1(0, \infty)$. Integration by parts in (3.4) then gives

$$(3.5) \qquad -\int_0^\infty (F - \Phi) \theta' \geqslant 0 ,$$

for all bounded, smooth $\theta \in \mathcal{D}$. Here, F is the distribution corresponding to f. It follows that $F - \Phi \geqslant 0$, and so F is a concave majorant of Φ.

Now, suppose that f is strictly decreasing at $x = a$, i.e., $f(b) > f(c)$ for all $b < a < c$. Let $\varepsilon > 0$, and choose θ as

$$\theta(x) = \frac{1}{\text{median} \left(f(a - \varepsilon), f(x), f(a + \varepsilon) \right)} .$$

This is a needlessly clever way of saying that $\theta(x) = 1/f(x)$ on $|x - a| < \varepsilon$, is constant outside this interval, and continuous on $(0, \infty)$. Note that with this definition $\psi = \theta f \in \mathcal{D}$, and (3.5) becomes

$$-\int_{|x-a|<\varepsilon} (F(x) - \Phi(x)) \, d\{ 1/f(x) \} \geqslant 0 .$$

Since $F(x)$ and $\Phi(x)$ are continuous, the mean value theorem gives that

$$-(F(y) - \Phi(y)) \int_{|x-a|<\varepsilon} d\{ 1/f(x) \} =$$

$$-(F(y) - \Phi(y)) \left(\frac{1}{f(a + \varepsilon)} - \frac{1}{f(a - \varepsilon)} \right) \geqslant 0 ,$$

for some y in the interval $|x - a| < \varepsilon$. Since f is strictly decreasing at $x = a$, the last factor is strictly positive, and so letting $\varepsilon \to 0$ gives that $F(a) - \Phi(a) \leqslant 0$. Thus, $F(a) = \Phi(a)$.

This shows that for all $a > 0$,

$$f \text{ strictly decreasing at } x = a \implies F(a) = \Phi(a) .$$

Together with F being a concave majorant of Φ, this is the characterization of F as the *least* concave majorant of Φ.   Q.e.d.

We formulate the last statement of the proof as a lemma.

(3.6) LEMMA. *Let F and Φ be distributions on $(0, \infty)$ with densities f and φ. Then, $f = \mathrm{lcm}(\varphi)$ if and only if the following three conditions hold.*

(a) $f \in \mathcal{D}$.

(b) $F(x) \geqslant \Phi(x)$ *for all $x > 0$.*

(c) *f strictly decreasing at $x = a$* $\implies$ $F(a) = \Phi(a)$.

(3.7) EXERCISE. Prove Lemma (3.6).

The following useful result is left as an exercise.

(3.8) EXERCISE. Show that if φ is a continuous density, then $\mathrm{lcm}(\varphi)$ is continuous, and $[\mathrm{lcm}(\varphi)](0) \geqslant \varphi(0)$.

Finally, we mention the equivalence with least-squares estimation.

(3.9) EXERCISE. Let $\varphi \in L^2(0, \infty)$ be nonnegative. Show that the problem

$$\begin{aligned} \text{minimize} \quad & \| f - \varphi \|_2^2 \\ \text{subject to} \quad & f \in L^2(0, \infty) , \ f \text{ decreasing} \end{aligned}$$

has a unique solution, given by $f = \mathrm{lcm}(\varphi)$.

EXERCISES: (3.7), (3.8), (3.9).

4. Algorithms and contractivity

In this section, we take a closer look at the construction of the least concave majorant. We first consider the case of simple distributions, that is, absolutely continuous distributions whose derivatives are simple step functions (simple densities). This leads to the pool-adjacent-violators algorithm for computing the maximum likelihood estimator of a monotone density and, more importantly, provides a pathway to showing the *contractivity* of the least-concave-majorant mapping, viz. if φ and ψ are (simple) densities, then

(4.1) $$\| \operatorname{lcm}(\varphi) - \operatorname{lcm}(\psi) \|_1 \leqslant \| \varphi - \psi \|_1 \;.$$

The usual limiting argument proves it for arbitrary densities. All of this serves as preparation for a much more general contractivity result in §5. The algorithm is due to KRUSKAL (1964), see also BARLOW, BARTHOLO-MEW, BREMNER and BRUNK (1972) and references therein.

It is perhaps worthwhile to point out that the detour via simple densities and simple distributions is taken to avoid consideration of "weird" distributions, where there is little chance of explicitly constructing the least concave majorant. For simple distributions, everything is much simpler, in that one can actually draw diagrams illustrating what is going on. The disadvantage is that in the end one has to take limits.

We begin by precisely describing simple densities and simple distributions. Consider adjacent intervals $A_1, A_2, \cdots, A_n$ of the form

(4.2) $$A_j = (x_{j-1}, x_j] \;, \quad j = 1, 2, \cdots, n \;,$$

where $0 = x_0 < x_1 < \cdots < x_n < \infty$. For later reference, the length of the interval A_j is denoted by $| A_j |$. Functions of the form

(4.3) $$\varphi(x) = \begin{cases} \varphi_j \;, & x \in A_j \;, \\ 0 \;, & x > x_n \;, \end{cases}$$

or, equivalently,

(4.4) $$\varphi(x) = \sum_{j \in I} \varphi_j \, \mathbb{1}(x \in A_j) \;, \quad x \in (0, \infty) \;,$$

where $I = \{ 1, 2, \cdots, n \}$, are called simple step functions. The corresponding distributions are called simple distributions. We do not insist on the simple step functions being probability density functions (if they are, we call them *simple densities*), nor on the φ_j being distinct.

The immediate goal is now to construct an algorithm for computing $\operatorname{lcm}(\varphi)$, for a simple density φ. This algorithm goes by the name of pool-adjacent-violators algorithm (p-a-v-a). For its derivation, it is more than useful to first consider the simple case $n = 2$. So, let φ be given by (4.4), and denote its distribution by Φ. If $\varphi_1 \geqslant \varphi_2$, then $\varphi = \operatorname{lcm}(\varphi)$, so the interesting case is when $\varphi_1 < \varphi_2$. The situation is illustrated in Figure 4.1. The derivative of the least concave majorant of Φ is denoted by the dashed line. It corresponds to a density that is constant on $A_1 \cup A_2$. Since the total probability is unchanged, the areas of the two rectangles I and II are equal, and the new density is

$$\varphi^{\mathrm{new}}(x) = \frac{\varphi_1 | A_1 | + \varphi_2 | A_2 |}{| A_1 | + | A_2 |} \, \mathbb{1}(x \in A_1 \cup A_2) \;, \quad 0 < x < \infty \;.$$

Thus, $\varphi^{\mathrm{new}} = \operatorname{lcm}(\varphi)$ is the solution in the case $n = 2$.

What happens for $n > 2$? The surprising answer is that at most $n - 1$ steps of the above kind suffice to compute $\text{lcm}(\varphi)$. So, let $n > 2$, and let the density φ be given by (4.4), on adjacent intervals $A_1, A_2, \cdots, A_n$ as in (4.2). The "basic step" of the algorithm consists of locating a violation of the assumption that φ is decreasing, and fixing it. That is, find a pair of adjacent intervals A_j, A_{j+1} on which φ is not decreasing, i.e. $\varphi_j < \varphi_{j+1}$, and replace φ on $A_j \cup A_{j+1}$ by the constant function

$$(4.5) \qquad \varphi_j^{\text{new}} = \frac{\varphi_j |A_j| + \varphi_{j+1}|A_{j+1}|}{|A_j| + |A_{j+1}|} ,$$

and leave φ unchanged everywhere else. A bit of notation: The "new" function φ thus defined is denoted by $\varphi^{\text{new}} = \text{lcm}(\varphi, A_j \cup A_{j+1})$, and the corresponding distribution Φ^{new} by $\text{LCM}(\Phi, A_j \cup A_{j+1})$. Thus, in case $\varphi_j < \varphi_{j+1}$,

$$(4.6a) \qquad [\,\text{lcm}(\varphi, A_j \cup A_{j+1})\,](x) \overset{\text{def}}{=} \begin{cases} \varphi(x), & x \notin A_j \cup A_{j+1} , \\ \varphi_j^{\text{new}}, & x \in A_j \cup A_{j+1} , \end{cases}$$

and, if $\varphi_j \geqslant \varphi_{j+1}$,

$$(4.6b) \qquad [\,\text{lcm}(\varphi, A_j \cup A_{j+1})\,](x) = \varphi(x), \quad x > 0 .$$

Finally, it turns out to be important to construct the reduced representation of φ^{new} by joining A_j and A_{j+1}, so, with $A_j^{\text{new}} = A_j \cup A_{j+1}$, and $A_\ell^{\text{new}} = A_\ell$ for $\ell \neq j, j + 1$,

$$(4.7) \qquad \varphi^{\text{new}}(x) = \sum_{j \in I^{\text{new}}} \varphi_j^{\text{new}} \, \mathbb{1}(x \in A_j^{\text{new}}), \quad x \in (0, \infty) ,$$

where $I^{\text{new}} = I \setminus \{j + 1\}$. The crucial point is of course that the representation involves one less interval than before.

This concludes the description of the "basic step". The algorithm consists of REPEATING the basic step UNTIL $\varphi_j \geqslant \varphi_{j+1}$ for all j. In Algorithm 4.1, this is stated a bit more formally. As a final comment on the algorithm, we note that the particular rule for picking j such that $\varphi_j < \varphi_{j+1}$ is unimportant for our purposes, although it would seem to make sense to start at "$+\infty$".

There are at least two questions of interest. The easiest one to answer is whether Algorithm 4.1 terminates in a finite number of steps. The answer is yes: The number of intervals in the reduced representation can only be decreased $n - 1$ times. The second question is whether the algorithm computes $\text{lcm}(\varphi)$. To show this, it suffices to show that the densities φ before and after the "basic step" satisfy

$$\text{lcm}(\varphi^{\text{after}}) = \text{lcm}(\varphi^{\text{before}}) ,$$

in other words, that $\text{lcm}(\varphi)$ is an invariant of the algorithm. To phrase this formally, we first extend the definition of $\text{LCM}(\Phi, A_j \cup A_{j+1})$ to sets B

Algorithm 4.1. The pool-adjacent-violators-algorithm
for the computation of $\mathrm{lcm}(\varphi)$ for a simple density φ.

LET $\varphi(x) = \sum\limits_{j=1}^{n} \varphi_j \, \mathbb{1}(x \in A_j)$ BE A SIMPLE DENSITY;

REPEAT

    FIND j SUCH THAT $\varphi_j < \varphi_{j+1}$;

    COMPUTE $\varphi^{\mathrm{new}} = \mathrm{lcm}(\varphi, A_j \cup A_{j+1})$;

    CONSTRUCT THE REDUCED REPRESENTATION OF φ^{new}:

$$\varphi^{\mathrm{new}}(x) = \sum_{j \in I^{\mathrm{new}}} \varphi_j^{\mathrm{new}} \, \mathbb{1}(x \in A_j^{\mathrm{new}});$$

    REPLACE φ BY φ^{new};

UNTIL

$$\varphi_j \geqslant \varphi_{j+1} \text{ FOR ALL } j.$$

that are finite unions of intervals. Thus, if φ is a density with distribution Φ, then we define

$$(4.8) \qquad \mathrm{LCM}(\Phi, B) = \begin{cases} \text{the least majorant of } \Phi \text{ which is concave} \\ \text{on each interval contained in } B, \end{cases}$$

and

$$(4.9) \qquad \mathrm{lcm}(\varphi, B) = \begin{cases} \text{the function, continuous from the left,} \\ \text{which is a.e. equal to the derivative of} \\ \mathrm{LCM}(\Phi, B). \end{cases}$$

(4.10) LEMMA. *Let the function φ be a simple density on the adjacent intervals $A_1, A_2, \cdots, A_n$. Let B be the union of some of the A_j. Then,*

$$\mathrm{lcm}\big(\mathrm{lcm}(\varphi, B)\big) = \mathrm{lcm}(\varphi).$$

PROOF. It suffices to show that for any $1 \leqslant j \leqslant n - 1$,

$$\mathrm{lcm}\big(\mathrm{lcm}(\varphi, A_j \cup A_{j+1})\big) = \mathrm{lcm}(\varphi),$$

and to use induction. Q.e.d.

(4.11) EXERCISE. Fill in the details of the proof!

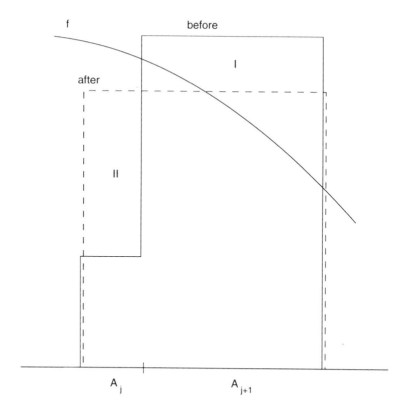

Figure 4.1. Illustration of the basic step of Algorithm 4.1, and of the inequality $\| \mathrm{lcm}(\varphi) - \psi \|_1 \leq \| \varphi - \psi \|_1$. The square regions I and II have the same area.

So we have proven the following result.

(4.12) THEOREM. *The Algorithm 4.1 computes* $\mathrm{lcm}(\varphi)$ *of the simple density* φ *in a finite number of steps.*

We are not quite ready for the contractivity property (4.1), but we are ready for an important special case.

(4.13) THEOREM. *Let* φ *be a simple step function on the adjacent intervals* $A_1, A_2, \cdots, A_n$, *and let* ψ *be a decreasing* L^1 *function. Then,*

$$\| \mathrm{lcm}(\varphi) - \psi \|_1 \leq \| \varphi - \psi \|_1 .$$

Using induction, the proof follows from the following lemma.

(4.14) LEMMA. *Under the same assumptions as in Theorem* (4.13), *for any* $1 \leqslant j \leqslant n-1$,

$$\| \operatorname{lcm}(\varphi, A_j \cup A_{j+1}) - \psi \|_1 \leqslant \| \varphi - \psi \|_1 .$$

PROOF. Let $\varphi^{\mathrm{new}} = \operatorname{lcm}(\varphi, A_j \cup A_{j+1})$. If $\varphi_j \geqslant \varphi_{j+1}$, then $\varphi^{\mathrm{new}} = \varphi$, and there is nothing to prove. So assume $\varphi_j < \varphi_{j+1}$. The situation is as in Figure 4.1. Let ψ be any decreasing L^1 function. Clearly, only its behavior on $A_j \cup A_{j+1}$ is important. If the graph of ψ lies above the regions I and II, then we have equality

$$\| \varphi^{\mathrm{new}} - \psi \|_1 = \| \varphi - \psi \|_1 .$$

The same is true if ψ lies below the regions I and II. There remains the case in which the graph of ψ intersects the regions I or II. Assume the graphs intersects region I. Then, region II lies below the graph of ψ and

$$\int_{A_j} | \varphi^{\mathrm{new}}(x) - \psi(x) | \, dx = \int_{A_j} | \varphi(x) - \psi(x) | \, dx - \text{area of region II} .$$

On A_{j+1}, we have $\varphi^{\mathrm{new}} = \varphi - h$, where h is the height of rectangle I, and so by the triangle inequality,

$$\int_{A_{j+1}} | \varphi^{\mathrm{new}}(x) - \psi(x) | \, dx \leqslant \int_{A_{j+1}} | \varphi(x) - \psi(x) | \, dx + \text{area of region I} .$$

Since the regions I and II have the same area, the result follows. The case in which the graph of ψ intersects region II is treated similarly.   Q.e.d.

(4.15) EXERCISE. Prove Theorem (4.13).

The next important problem is to extend Theorem (4.13) to arbitrary distributions. Going from simple distributions to arbitrary ones can be effected by a limiting process, but then we certainly need that the mapping $\varphi \longmapsto \operatorname{lcm}(\varphi)$ for simple step functions is continuous in the $L^1(0, \infty)$ topology. The continuity turns out to be contractivity.

(4.16) THEOREM. *Let* φ, ψ *be simple densities on* $(0, \infty)$. *Then,*

$$\| \operatorname{lcm}(\varphi) - \operatorname{lcm}(\psi) \|_1 \leqslant \| \varphi - \psi \|_1 .$$

To prove this, we may assume by Theorem (4.13) that neither φ nor ψ is decreasing on $(0, \infty)$. The proof then goes via the construction of an algorithm for the joint computation of $\operatorname{lcm}(\varphi)$ and $\operatorname{lcm}(\psi)$: Whenever we do a "basic step" for φ, we need to do the same step for ψ. Moreover, the algorithm has to terminate in a finite number of steps (an infinite number of steps would involve limits and might beg the question).

Let φ and ψ be simple step functions. A little thought, or reading ahead and backtracking, reveals that it is useful to represent φ on adjacent intervals $A_1, A_2, \cdots, A_n$, and to assume that ψ is a simple step function on the adjacent intervals $B_1, B_2, \cdots, B_m$, which constitute a refinement of $A_1, A_2, \cdots, A_n$, i.e.,

$$(4.17) \qquad \bigcup_{\ell=1}^{m} B_\ell = \bigcup_{j=1}^{n} A_j \ ,$$

and each B_ℓ lies in exactly one A_j. Now, let

$$(4.18) \qquad \begin{aligned} \varphi(x) &= \sum_{j=1}^{n} \varphi_j \, \mathbb{1}(x \in A_j) = \sum_{\ell=1}^{m} \theta_\ell \, \mathbb{1}(x \in B_\ell) \ , \\ \psi(x) &= \sum_{\ell=1}^{m} \psi_\ell \, \mathbb{1}(x \in B_\ell) \ . \end{aligned}$$

Note that the value $\varphi_n = 0$ is allowed.

The algorithm for simultaneously constructing $\mathrm{lcm}(\varphi)$ and $\mathrm{lcm}(\psi)$ is of course based on Algorithm 4.1. The first step is to find an index j for which $\varphi_j < \varphi_{j+1}$, but then what? If we do the $\varphi \longmapsto \mathrm{lcm}(\varphi, A_j \cup A_{j+1})$ step immediately, then we have no idea about the inequality sign in

$$\| \mathrm{lcm}(\varphi, A_j \cup A_{j+1}) - \psi \|_1 \gtreqless \| \varphi - \psi \|_1 \ ,$$

because ψ restricted to $A_j \cup A_{j+1}$ is arbitrary. However, having located the violating (for φ) intervals, we should instead apply Algorithm 4.1 to ψ (restricted to $A_j \cup A_{j+1}$), because the function φ is anything but arbitrary:

$$(4.19) \qquad \varphi \text{ restricted to } A_j \cup A_{j+1} \text{ is increasing!}$$

Now, on $A_j \cup A_{j+1}$, we use the representations on the adjacent intervals $B_1, B_2, \cdots, B_m$,

$$(4.20) \qquad \psi(x) = \sum_{\ell \in I} \psi_\ell \, \mathbb{1}(x \in B_\ell) \ , \quad \varphi(x) = \sum_{\ell \in I} \theta_\ell \, \mathbb{1}(x \in B_\ell) \ ,$$

where $I = \{ \ell : B_\ell \subset A_j \cup A_{j+1} \}$.

To run Algorithm 4.1 on ψ, restricted to $A_j \cup A_{j+1}$, we find a pair of adjacent intervals $B_\ell, B_{\ell+1}$ such that $\psi_\ell < \psi_{\ell+1}$, and perform the basic step on ψ,

$$(4.21) \qquad \psi^{\mathrm{new}} = \mathrm{lcm}(\psi, B_\ell \cup B_{\ell+1}) \ .$$

Now, what to do with φ? Note that $\theta_\ell \leqslant \theta_{\ell+1}$, since φ is increasing on $A_j \cup A_{j+1}$. Whether there is equality or not does not matter. In either case, we perform the "same" step for φ,

$$(4.22) \qquad \varphi^{\mathrm{new}} = \mathrm{lcm}(\varphi, B_\ell \cup B_{\ell+1}) \ ,$$

and note that

$$(4.23) \qquad \varphi^{\mathrm{new}} \text{ restricted to } A_j \cup A_{j+1} \text{ is again increasing} \ .$$

Algorithm 4.2. Simultaneously computing $\mathrm{lcm}(\varphi)$ and $\mathrm{lcm}(\psi)$ for simple distributions φ, ψ.

LET THE ADJACENT INTERVALS $B_1, B_2, \cdots, B_m$ BE A REFINEMENT OF $A_1, A_2, \cdots, A_n$.

LET $\varphi(x) = \sum\limits_{j=1}^{n} \varphi_j \, \mathbb{1}(x \in A_j) = \sum\limits_{\ell=1}^{m} \theta_\ell \, \mathbb{1}(x \in B_\ell)$,

$\psi(x) = \sum\limits_{\ell=1}^{m} \psi_\ell \, \mathbb{1}(x \in B_\ell)$.

REPEAT

    FIND j SUCH THAT $\varphi_j < \varphi_{j+1}$.

    RESTRICT φ, ψ to $A_j \cup A_{j+1}$:

        LET $I = \{ \ell : B_\ell \subset A_j \cup A_{j+1} \}$, AND SET

        $\varphi(x) = \sum\limits_{\ell \in I} \theta_\ell \, \mathbb{1}(x \in B_\ell)$, $\psi(x) = \sum\limits_{\ell \in I} \psi_\ell \, \mathbb{1}(x \in B_\ell)$.

    REPEAT

        FIND ℓ SUCH THAT $\psi_\ell < \psi_{\ell+1}$.

        COMPUTE $\psi^{\text{new}} = \mathrm{lcm}(\psi, B_\ell \cup B_{\ell+1})$,

                $\varphi^{\text{new}} = \mathrm{lcm}(\varphi, B_\ell \cup B_{\ell+1})$.

        CONSTRUCT THE REDUCED REPRESENTATIONS:

        $\psi^{\text{new}}(x) = \sum\limits_{\ell \in I^{\text{new}}} \psi_\ell^{\text{new}} \, \mathbb{1}(x \in B_\ell^{\text{new}})$,

        $\varphi^{\text{new}}(x) = \sum\limits_{\ell \in I^{\text{new}}} \theta_\ell^{\text{new}} \, \mathbb{1}(x \in B_\ell^{\text{new}})$.

    UNTIL

        $\psi_\ell \geqslant \psi_{\ell+1}$ FOR ALL $\ell \in I$.

    FINALLY, DO

        $\varphi^{\text{final}} = \mathrm{lcm}(\varphi, A_j \cup A_{j+1})$

UNTIL

    $\varphi_j \geqslant \varphi_{j+1}$ FOR ALL j .

TO FINISH, RUN ALGORITHM 3.1 ON $\psi : \psi = \mathrm{lcm}(\psi)$.

One verifies that

$$\int_{B_\ell \cup B_{\ell+1}} |\psi^{\text{new}}(x) - \varphi^{\text{new}}(x)| \, dx \leqslant \int_{B_\ell \cup B_{\ell+1}} |\psi(x) - \varphi(x)| \, dx \, ,$$

and so

(4.24)
$$\| \psi^{\text{new}} - \varphi^{\text{new}} \|_1 \leqslant \| \psi - \varphi \|_1 \, .$$

As before, the crucial last part of the basic step of Algorithm 4.2 is to join B_ℓ and $B_{\ell+1}$ in the representations of ψ as well as φ. This gives

(4.25)
$$\psi^{\text{new}}(x) = \sum_{\ell \in I^{\text{new}}} \psi_\ell^{\text{new}} \, \mathbb{1}(x \in B_\ell^{\text{new}}) \, ,$$

$$\varphi^{\text{new}}(x) = \sum_{\ell \in I^{\text{new}}} \theta_\ell^{\text{new}} \, \mathbb{1}(x \in B_\ell^{\text{new}}) \, ,$$

with

(4.26)
$$|I^{\text{new}}| = |I| - 1 \, .$$

We conclude that the algorithm for the computation of $\text{lcm}(\psi, A_j \cup A_{j+1})$ terminates after a finite number of steps, and if we do the same steps on φ, then upon termination, the terminal φ and ψ, denoted by φ^{term} and ψ^{term}, satisfy

$$\| \psi^{\text{term}} - \varphi^{\text{term}} \|_1 \leqslant \| \psi - \varphi \|_1 \, .$$

Finally, Algorithm 4.1 is applied to φ^{term}, restricted to the union of the intervals B_ℓ, $\ell \in I^{\text{term}}$, or, equivalently, restricted to $A_j \cup A_{j+1}$. Since ψ^{term} is decreasing on $A_j \cup A_{j+1}$, Theorem (4.13) implies that the final φ, denoted as φ^{final}, satisfies

(4.27)
$$\| \varphi^{\text{final}} - \psi^{\text{term}} \|_1 \leqslant \| \varphi^{\text{term}} - \psi^{\text{term}} \|_1 \leqslant \| \varphi - \psi \|_1 \, ,$$

and, by Lemma (4.10), $\varphi^{\text{final}} = \text{lcm}(\varphi, A_j \cup A_{j+1})$. This concludes the "grand step" of the algorithm for the simultaneous computation of $\text{lcm}(\varphi)$ and $\text{lcm}(\psi)$, which may be summarized as

(4.28)
$$\begin{pmatrix} \varphi \\ \psi \end{pmatrix} \longrightarrow \begin{pmatrix} \varphi^{\text{term}} \\ \psi^{\text{term}} \end{pmatrix} \longrightarrow \begin{pmatrix} \varphi^{\text{final}} \\ \psi^{\text{term}} \end{pmatrix} \, .$$

The next "grand step" begins with $\varphi = \varphi^{\text{final}}$, $\psi = \psi^{\text{term}}$. This procedure is continued until φ^{final} is monotone. This last φ^{final} is denoted by φ^{last}, and the corresponding ψ^{term} as ψ^{penult}.

After the computation of $\varphi^{\text{last}} = \text{lcm}(\varphi)$ has run its course, the really last step is to put the finishing touches on the computation of $\text{lcm}(\psi)$, viz.

(4.29)
$$\psi^{\text{last}} = \text{lcm}(\psi^{\text{penult}}) \, ,$$

and since φ^{last} is decreasing,

(4.30)
$$\| \psi^{\text{last}} - \varphi^{\text{last}} \|_1 \leqslant \| \psi^{\text{penult}} - \varphi^{\text{last}} \|_1 \leqslant \| \varphi - \psi \|_1 \, .$$

There are only a finite number of "grand steps" since the number of intervals $A_1, A_2, \cdots, A_n$ in the representation of φ decreases by one after each such step. So this algorithm is a finite algorithm. The resulting algorithm is formalized in Algorithm 4.2. We have proven the following theorem.

(4.31) THEOREM. *Let φ and ψ be simple densities. The Algorithm 4.2 terminates in a finite number of steps and upon termination returns $\mathrm{lcm}(\varphi)$ and $\mathrm{lcm}(\psi)$. Moreover,*

$$\| \mathrm{lcm}(\varphi) - \mathrm{lcm}(\psi) \|_1 \leqslant \| \varphi - \psi \|_1 .$$

(4.32) EXERCISE. Write a (more) formal proof of the Theorem.

We now turn to the proof of (1.19). It is a simple consequence of Theorem (4.31) and the continuity of the map $\varphi \longmapsto \mathrm{lcm}(\Phi)$, as expressed in the following lemma.

(4.33) LEMMA. *Let Φ_o be a distribution on $(0, \infty)$ with density φ_o, and let $\{\Phi_n\}_n$ be a sequence of simple distributions with densities φ_n, such that $\| \varphi_n - \varphi_o \|_1 \longrightarrow 0$. Then,*

$$\| \mathrm{lcm}(\varphi_n) - \mathrm{lcm}(\varphi_o) \|_1 \longrightarrow 0 .$$

PROOF. Let $\Psi_n = \mathrm{LCM}(\Phi_n)$ and $\psi_n = \mathrm{lcm}(\varphi_n)$. Now, Theorem (4.31) gives $\| \psi_n - \psi_m \|_1 \leqslant \| \varphi_n - \varphi_m \|_1$, so that $\{\psi_n\}_n$ is a Cauchy sequence. Thus, there exists a ψ_o with $\| \psi_n - \psi_o \|_1 \longrightarrow 0$, and all we need to show is that $\psi_o = \mathrm{lcm}(\varphi_o)$, or that $\Psi_o = \mathrm{LCM}(\Phi_o)$, where Ψ_o is the distribution with density ψ_o. Since

$$(4.34) \qquad \| \Psi_n - \Psi_o \|_\infty \leqslant \| \psi_n - \psi_o \|_1 \longrightarrow 0 ,$$

then Ψ_o is the pointwise limit of a sequence of concave distributions, and it follows that Ψ_o is concave. Also, since $\Psi_n \geqslant \Phi_n$ for all n, then $\Psi_o \geqslant \Phi_o$, and Ψ_o is a concave majorant of Φ_o.

To show that Ψ_o is the least concave majorant of Φ_o, let Θ be any concave majorant of Φ_o. Then, $\Theta \geqslant \Psi_n - \varepsilon_n$, where $\varepsilon_n = \| \Psi_n - \Psi_o \|_\infty$. This implies that $\Theta + \varepsilon_n$ is a majorant of Ψ_n, and it is concave. So, it dominates the least concave majorant, $\Theta + \varepsilon_n \geqslant \mathrm{LCM}(\Phi_n)$, and so $\Theta + \varepsilon_n \geqslant \Psi_n \geqslant \Psi_o - \varepsilon_n$, with ε_n as before. This implies that $\Theta \geqslant \Psi_o - 2\varepsilon_n$. Since $\varepsilon_n \to 0$, see (4.34), this says that $\Theta \geqslant \Psi_o$, and so $\Psi_o = \mathrm{LCM}(\Phi_o)$.   Q.e.d.

(4.35) EXERCISE. Prove (1.19).

EXERCISES : (4.11), (4.15), (4.32), (4.35).

5. Contractivity: the general case

Here, we consider the general contractivity result (1.20). The bulk of the material deals with simple densities; the limiting argument that proves it for arbitrary densities is short and technical. First, we give a precise formulation.

Let $\mathbb{R}_+ = [0, \infty)$. For densities φ, ψ on $(0, \infty)$, let

$$(5.1) \qquad \mathbb{J}(\varphi, \psi) = \int_0^\infty J\big(\varphi(x), \psi(x)\big)\, dx \;,$$

where $J : \mathbb{R}_+ \times \mathbb{R}_+ \longrightarrow \mathbb{R}_+ \cup \{+\infty\}$ satisfies

(5.2) $J(x, y)$ is continuous on $\mathbb{R}_+ \times (0, \infty)$,

(5.3) $J(x, y)$ is twice continuously differentiable on $(0, \infty) \times (0, \infty)$, and

$$(5.4) \qquad \frac{\partial^2 J}{\partial x^2}(x, y) \geqslant 0 \;, \qquad \frac{\partial^2 J}{\partial x\, \partial y}(x, y) \leqslant 0 \;, \qquad \text{for all } x, y > 0 \;.$$

In all that follows, conditions (5.3) and (5.4) may be replaced by the conditions of the following exercise.

(5.5) EXERCISE. Show that (5.4) implies that

$$\frac{J(b, p) - J(a, p)}{b - a} \leqslant \frac{J(c, q) - J(b, q)}{c - b}$$

for all $0 \leqslant a < b < c$ and all $p \geqslant q > 0$. [Hint: §9.1!]

We also require the condition expressed in Exercise (5.5) for the reverse function $(x, y) \longmapsto J(y, x)$, which under the circumstances amounts to

$$(5.6) \qquad \frac{\partial^2 J}{\partial y^2}(x, y) \geqslant 0 \;, \qquad \text{for all } x, y > 0 \;.$$

In addition, we need

$$(5.7) \qquad\qquad J(x, y) \text{ is convex in } x, y \text{ jointly} \;.$$

It is not clear to the authors whether all of these conditions are independent, but we shall not pursue the point. Abusing notation somewhat we say that $\mathbb{J}$ satisfies (5.2) if $\mathbb{J}$ is given by (5.1) and J satisfies (5.2), and so on.

The standard examples of functions J satisfying (5.2) through (5.7) are constructed as follows. Let f be a nonnegative, convex function on $\mathbb{R}$, with $f(0) = 0$, and define J by

$$(5.8) \qquad\qquad J(x, y) = f(x - y) \;, \qquad x \geqslant 0 \;, \; y \geqslant 0 \;.$$

The choice $f(x) = |x|^p$ $(p \geqslant 1)$ covers the L^p distances. Alternatively, for f a nonnegative, convex function on $\mathbb{R}_+$, with $f(1) = 0$, define J by

$$(5.9) \qquad J(x,y) = \begin{cases} y\, f(x/y) & , \quad x \geqslant 0 , \; y > 0 , \\ x \, \liminf\limits_{t \to \infty} t^{-1} f(t) , & x > 0 , \; y = 0 , \\ 0 & , \quad x = 0 , \; y = 0 . \end{cases}$$

Thus, $J(x,0) = \infty$ for all $x > 0$ or $J(x,0) < \infty$ for all x. Note that

$$x \liminf_{t \to \infty} t^{-1} f(t) = \liminf_{y \to 0+} y\, f(x/y) , \quad x > 0 .$$

The choices $f(x) = (\sqrt{x} - 1)^2$, $x \log x + 1 - x$, and $(x-1)^2$ cover the Hellinger, Kullback-Leibler, and Pearson's φ^2 distances.

A general class of functions J satisfying (5.2) through (5.4), but not necessarily (5.6) and (5.7), are functions of the form

$$(5.10) \qquad J(x,y) = f(x) - f(y) - f'(y)\,(x - y)$$

for (differentiable) convex functions f, e.g., if $f(x) = -\sqrt{x}$, $x \geqslant 0$, then (5.6) and (5.7) fail, since $J(x,y)$ is not convex in y.

We are now ready for the statement of the contractivity theorem.

(5.11) CONTRACTIVITY THEOREM FOR SIMPLE DENSITIES. (a) *Suppose that* $\mathbb{J}$ *satisfies* (5.2) *through* (5.4). *Then, for all simple densities* φ *and* ψ, *with* ψ *decreasing,*

$$\mathbb{J}\big(\operatorname{lcm}(\varphi), \psi\big) \leqslant \mathbb{J}\big(\varphi, \psi\big) .$$

(b) *If* $\mathbb{J}$ *satisfies* (5.2) *through* (5.7), *then for all simple densities* φ *and* ψ,

$$\mathbb{J}\big(\operatorname{lcm}(\varphi), \operatorname{lcm}(\psi)\big) \leqslant \mathbb{J}\big(\varphi, \psi\big) .$$

The proofs to follow are based on a judicious use of the pool-adjacent-violators-algorithm (p-a-v-a), as discussed in the previous section.

PROOF OF THEOREM (5.11)(a). Suppose φ and ψ are supported on $(0, T)$, for some $T > 0$, and write φ as

$$\varphi(x) = \sum_{j \in I} \varphi_j \, \mathbb{1}(x \in A_j) , \quad x \in (0, \infty) ,$$

for adjacent intervals $A_1, A_2, \cdots, A_n$, with $\bigcup_j A_j = (0, T)$. We may suppose that $\mathbb{J}(\varphi, \psi) < \infty$, since otherwise there is nothing to prove. This implies that $J(\varphi(x), \psi(x)) < \infty$ for all $x > 0$.

We compute $\operatorname{lcm}(\varphi)$ using the p-a-v-a. Thus, we find two adjacent intervals A_j and A_{j+1} such that $\varphi_j < \varphi_{j+1}$, and compute φ^{new} by (4.5). We show that

$$(5.12) \qquad \mathbb{J}(\varphi^{\text{new}}, \psi) \leqslant \mathbb{J}(\varphi, \psi) .$$

It suffices to consider the contributions of $A_j \cup A_{j+1}$ only, i.e., to show that

$$\int_{A_j \cup A_{j+1}} J(\varphi^{\text{new}}(x), \psi(x)) \, dx \leq$$

$$\int_{A_j} J(\varphi_j, \psi(x)) \, dx + \int_{A_{j+1}} J(\varphi_{j+1}, \psi(x)) \, dx ,$$

which may be rewritten as

$$(5.13) \quad \int_{A_j} \{ J(\varphi^{\text{new}}(x), \psi(x)) - J(\varphi_j, \psi(x)) \} \, dx \leq$$

$$\int_{A_{j+1}} \{ J(\varphi_{j+1}, \psi(x)) - J(\varphi^{\text{new}}(x), \psi(x)) \} \, dx .$$

It is annoying that the two integrals may be over intervals of different lengths. Using a simple change of variables to transform the intervals of intergration A_j and A_{j+1} into, respectively, $[-1, 0]$ and $[0, 1]$, one obtains

$$(5.14) \quad |A_j| \int_{-1}^{0} \{ J(\varphi^{\text{new}}, \psi_1(x)) - J(\varphi_j, \psi_1(x)) \} \, dx \leq$$

$$|A_{j+1}| \int_{0}^{1} \{ J(\varphi_{j+1}, \psi_2(x)) - J(\varphi^{\text{new}}, \psi_2(x)) \} \, dx ,$$

where $\psi_1(x)$ is the appropriate scaled version of $\psi(x)$ on A_j, and $\psi_2(x)$ the scaled version of $\psi(x)$ on A_{j+1}, and where we dropped the argument of φ^{new}, since it is constant on $A_j \cup A_{j+1}$. Now, note that

$$\varphi^{\text{new}} - \varphi_j = \frac{|A_{j+1}|(\varphi_{j+1} - \varphi_j)}{|A_j| + |A_{j+1}|} , \quad \varphi_{j+1} - \varphi^{\text{new}} = \frac{|A_j|(\varphi_{j+1} - \varphi_j)}{|A_j| + |A_{j+1}|} ,$$

and $\varphi_{j+1} - \varphi_j > 0$, so that (5.14) is equivalent to

$$(5.15) \quad \int_{-1}^{0} \frac{J(\varphi^{\text{new}}, \psi_1(x)) - J(\varphi_j, \psi_1(x))}{\varphi^{\text{new}} - \varphi_j} \, dx \leq$$

$$\int_{0}^{1} \frac{J(\varphi_{j+1}, \psi_2(x)) - J(\varphi^{\text{new}}, \psi_2(x))}{\varphi_{j+1} - \varphi^{\text{new}}} \, dx .$$

Thus, we are done if (5.15) holds. Since ψ is decreasing,

$$\psi_1(-x) \geq \psi_1(0) \geq \psi_2(0) \geq \psi_2(t) \quad \text{for all } x, t \in [0, 1] .$$

Since $\varphi_j \leq \varphi^{\text{new}} \leq \varphi_{j+1}$, then condition (5.4) via Exercise (5.5) implies that (5.15) holds, and so does (5.12).

The proof is concluded by running the p-a-v-a to completion. By (5.12), each step decreases $\mathbb{J}(\varphi, \psi)$, and after finitely many steps, we are done. Q.e.d.

We now turn our attention to proving Theorem (5.11)(b). Of course, Theorem (5.11)(a) covers the case in which ψ is decreasing. It also covers the reverse case, i.e.,

(5.16)   if φ is decreasing and J satisfies (5.2) through (5.6), then
$$\mathbb{J}(\varphi, \mathrm{lcm}(\psi)) \leqslant \mathbb{J}(\varphi, \psi) .$$

The other extreme case, in which ψ is increasing, is useful in establishing the general result.

(5.17) LEMMA. *Suppose $\mathbb{J}$ satisfies (5.2) through (5.7). Let φ and ψ be simple functions on the same partition, i.e., there exists adjacent intervals $A_1, A_2, \cdots, A_n$, such that $\bigcup_j A_j = (0, T)$ for some $T > 0$, and*
$$\varphi(x) = \sum_{j \in I} \varphi_j \,\mathbb{1}(x \in A_j) , \quad \psi(x) = \sum_{j \in I} \psi_j \,\mathbb{1}(x \in A_j) , \quad x \in (0, \infty) .$$

Let $0 \leqslant P < Q \leqslant T$, and assume that ψ is increasing on (P, Q). Then,
$$\mathbb{J}\big(\mathrm{lcm}(\varphi, (P, Q)) , \mathrm{lcm}(\psi, (P, Q))\big) \leqslant \mathbb{J}(\varphi , \psi) .$$

PROOF. We may assume that $\mathbb{J}(\varphi, \psi) < \infty$, so that $J(\varphi(x), \psi(x)) < \infty$ for all x. We may also assume that $(P, Q) = (0, T)$.

We apply the p-a-v-a to φ and ψ simultaneously. That is, we find intervals A_j, A_{j+1} such that $\varphi_j < \varphi_{j+1}$, and compute $\varphi^{\mathrm{new}} = \mathrm{lcm}(\varphi, A_j \cup A_{j+1})$. Since ψ is increasing, then $\psi_j \leqslant \psi_{j+1}$, and we do the "same" step on ψ, so we compute $\psi^{\mathrm{new}} = \mathrm{lcm}(\psi, A_j \cup A_{j+1})$. Note that after pooling the intervals A_j and A_{j+1}, both φ and ψ are simple functions on the new set of adjacent intervals.

Now, we show that

(5.18) $\mathbb{J}(\varphi^{\mathrm{new}} , \psi^{\mathrm{new}}) \leqslant \mathbb{J}(\varphi , \psi).$

Again, only the contributions of the intervals A_j and A_{j+1} need to be considered, and all functions in question are constant on these intervals. Thus, it suffices to show that
$$\big(|A_j| + |A_{j+1}|\big) J(\varphi^{\mathrm{new}} , \psi^{\mathrm{new}}) \leqslant |A_j| \, J(\varphi_j , \psi_j) + |A_{j+1}| \, J(\varphi_{j+1} , \psi_{j+1}),$$
and this holds by the convexity of $J(x, y)$ in x, y jointly.

This describes one step of p-a-v-a applied to φ and simultaneously to ψ. After finitely many step, we have thus computed $\mathrm{lcm}(\varphi)$. Denote the corresponding ψ by ψ^{last}. By induction, using (5.18), it follows that
$$\mathbb{J}\big(\mathrm{lcm}(\varphi) , \psi^{\mathrm{last}}\big) \leqslant \mathbb{J}(\varphi , \psi) .$$

Now, if $\psi^{\mathrm{last}} = \mathrm{lcm}(\psi)$, we are done. If $\psi^{\mathrm{last}} \neq \mathrm{lcm}(\psi)$, then we still have $\mathrm{lcm}(\psi) = \mathrm{lcm}(\psi^{\mathrm{last}})$, and since $\mathrm{lcm}(\varphi)$ is obviously decreasing, (5.16) implies
$$\mathbb{J}\big(\mathrm{lcm}(\varphi) , \mathrm{lcm}(\psi)\big) \leqslant \mathbb{J}\big(\mathrm{lcm}(\varphi) , \psi^{\mathrm{last}}\big) \leqslant \mathbb{J}(\varphi , \psi) .$$

This concludes the proof.   Q.e.d.

PROOF OF THEOREM (5.11)(b). Assume that $\mathbb{J}(\varphi, \psi) < \infty$. Suppose that ψ is a step function on the adjacent intervals A_j, $j = 1, 2, \cdots, n$. We apply a basic step of p-a-v-a to ψ, and find adjacent intervals A_j and A_{j+1} such that $\psi_j < \psi_{j+1}$. Thus, ψ is increasing on $A_j \cup A_{j+1}$. Since φ is a simple function, φ and ψ are simple functions on the adjacent intervals B_i, $i = 1, 2, \cdots, m$, with

$$\bigcup_{1 \leqslant i \leqslant m} B_i = A_j \cup A_{j+1} .$$

So Lemma (5.17) applies and gives

(5.19) $\mathbb{J}\bigl(\mathrm{lcm}(\varphi, A_j \cup A_{j+1}), \mathrm{lcm}(\psi, A_j \cup A_{j+1})\bigr) \leqslant \mathbb{J}(\varphi, \psi) .$

After finitely many steps of p-a-v-a applied to ψ, the algorithm terminates, and we have computed $\mathrm{lcm}(\psi)$ as well as $\mathrm{lcm}(\varphi, B)$, where B is the union of all violating subintervals A_j, A_{j+1}. By induction, using (5.19), we thus obtain

$$\mathbb{J}\bigl(\mathrm{lcm}(\varphi, B), \mathrm{lcm}(\psi)\bigr) \leqslant \mathbb{J}(\varphi, \psi) .$$

Since $\mathrm{lcm}(\psi)$ is obviously decreasing, Part (a) of the theorem applies, so that

$$\mathbb{J}\bigl(\mathrm{lcm}(\varphi), \mathrm{lcm}(\psi)\bigr) \leqslant \mathbb{J}\bigl(\mathrm{lcm}(\varphi, B), \mathrm{lcm}(\psi)\bigr) \leqslant \mathbb{J}(\varphi, \psi) . \qquad \text{Q.e.d.}$$

The next step is of course to prove that Theorem (5.11) holds for arbitrary densities φ, ψ, and not just for simple densities. This may be proved using Theorem (5.11) by approximating arbitrary densities by simple densities and taking limits. The special case for L^1 distances, already proved in Lemma (4.33), is useful here. We should warn the reader that the proof is an exercise in measure theory, see WHEEDEN and ZYGMUND (1977), but there seems to be no way around it.

(5.20) CONTRACTIVITY THEOREM FOR ARBITRARY DENSITIES.
(a) *Suppose that* $\mathbb{J}$ *satisfies* (5.2) *through* (5.4). *Then, for all densities* φ *and* ψ, *with* ψ *decreasing,*

$$\mathbb{J}\bigl(\mathrm{lcm}(\varphi), \psi\bigr) \leqslant \mathbb{J}(\varphi, \psi) .$$

(b) *If* $\mathbb{J}$ *satisfies* (5.2) *through* (5.7)*, then for all densities* φ *and* ψ,

$$\mathbb{J}\bigl(\mathrm{lcm}(\varphi), \mathrm{lcm}(\psi)\bigr) \leqslant \mathbb{J}(\varphi, \psi) .$$

PROOF OF THEOREM (5.20)(b). Assume that $\mathbb{J}(\varphi, \psi) < \infty$. We approximate φ_o and ψ_o by simple functions φ_n and ψ_n, and take limits in the inequality

$$\mathbb{J}\bigl(\mathrm{lcm}(\varphi_n), \mathrm{lcm}(\psi_n)\bigr) \leqslant \mathbb{J}(\varphi_n, \psi_n) .$$

Let $n \in \mathbb{N}$. Since φ_o, ψ_o, and $J_o \stackrel{\text{def}}{=} J\bigl(\varphi_o(\cdot), \psi_o(\cdot)\bigr)$ are nonnegative elements of $L^1(0, \infty)$, there exists a $T = T(n) > 0$ such that

$$\int_T^\infty \bigl(\varphi_o(x) + \psi_o(x) + J_o(x)\bigr)\, dx < \tfrac{1}{n} \ .$$

By Lusin's theorem, there exists a closed set $F \subset (0, T)$, with relative complement $G = \{\, x \in (0, T) \ : \ x \notin F \,\}$, such that φ_o, ψ_o and J_o are uniformly continuous relative to F, and

$$|\, G\,| < \tfrac{1}{n} \ , \quad \int_G \bigl(\varphi_o(x) + \psi_o(x) + J_o(x)\bigr)\, dx < \tfrac{1}{n} \ .$$

The uniform continuity relative to F means that

$$\forall\, \varepsilon > 0 \ \exists\, \delta > 0 \ \forall\, x, y \in F \ : \ |\,x - y\,| < \delta \implies |\,\varphi_o(x) - \varphi_o(y)\,| < \varepsilon \ .$$

Now, φ_o may be approximated by step functions in the following manner. For $k \in \mathbb{N}$, let

$$A_{ik} = \bigl(\, i/k \,, \, (i+1)/k \,\bigr] \ , \quad i = 0, 1, 2, \cdots \ ,$$

and let $I(k) = \{\, i \ : \ |\, A_{ik} \cap F\,| > 0 \,\}$. Then, for arbitrary $\theta_{ik} \in A_{ik} \cap F$,

$$\lim_{k \to \infty} \ \sum_{i \in I(k)} \ \int_{A_{ik} \cap F} \bigl|\, \varphi_o(\theta_{ik}) - \varphi_o(x)\,\bigr|\, dx = 0 \ .$$

Also, because φ_o is bounded on F, being a continuous function on a compact set, then

$$\limsup_{k \to \infty} \ \sum_{i \in I(k)} \bigl|\{\, x \in A_{ik} \ : \ x \notin F \,\}\bigr|\ \varphi_o(\theta_{ik}) \leqslant \tfrac{c}{n} \ ,$$

where $c = \max\{\, \varphi_o(x) \ : \ x \in F \,\}$. Now, choose $k = k(n)$ so large that

$$\sum_{i \in I(k)} \ \int_{A_{ik} \cap F} \bigl|\, \varphi_o(\theta_{ik}) - \varphi_o(x)\,\bigr|\, dx < \tfrac{1}{n} \ ,$$

$$\sum_{i \in I(k)} \bigl|\{\, x \in A_{ik} \ : \ x \notin F \,\}\bigr|\ \varphi_o(\theta_{ik}) < \tfrac{2c}{n} \ .$$

Arranging things so that the same inequalities hold for ψ_o and J_o, we define

$$\varphi_n(x) = \sum_{i \in I(k)} \varphi_o(\theta_{ik})\, \mathbb{1}(\, x \in A_{ik}\,) \ ,$$

$$\psi_n(x) = \sum_{i \in I(k)} \psi_o(\theta_{ik})\, \mathbb{1}(\, x \in A_{ik}\,) \ ,$$

and
$$J_n(x) = J(\varphi_n(x), \psi_n(x)) \ , \quad x > 0 \ .$$

This completes the construction of the approximating sequences. We now take limits. First,

$$\| \varphi_n - \varphi_o \|_1 \leqslant \int_T^\infty \varphi_o(x)\, dx + \int_G \varphi_o(x)\, dx + $$

$$\int_F | \varphi_k(x) - \varphi_o(x) |\, dx + \sum_{i \in I(k)} \big| \{ x \in A_{ik} : x \notin F \} \big| \, \varphi_o(\theta_{ik}) \ ,$$

and thus,

$$\| \varphi_n - \varphi_o \|_1 < \tfrac{1}{n} + \tfrac{1}{n} + \tfrac{1}{n} + \tfrac{2c}{n} \ ,$$

so $\varphi_n \longrightarrow \varphi_o$ in L^1. The same arguments give $\psi_n \longrightarrow \psi_o$ and $J_n \longrightarrow J_o$ in L^1. By Lemma (4.33), then also $\mathrm{lcm}(\varphi_n) \longrightarrow \mathrm{lcm}(\varphi_o)$ in L^1, and so, for a subsequence, $\mathrm{lcm}(\varphi_n) \longrightarrow \mathrm{lcm}(\varphi_o)$ almost everywhere. Likewise, along a subsequence of the subsequence,

$$\mathrm{lcm}(\psi_n) \longrightarrow \mathrm{lcm}(\psi_o) \ , \quad J_n \longrightarrow J_o \quad \text{almost everywhere} \ .$$

By Fatou's lemma along the appropriate subsequence,

$$\mathbb{J}\big(\mathrm{lcm}(\varphi_o), \mathrm{lcm}(\psi_o) \big) \leqslant \liminf_{n \to \infty} \mathbb{J}\big(\mathrm{lcm}(\varphi_n), \mathrm{lcm}(\psi_n) \big) \ .$$

Finally, since $J_n \longrightarrow J_o$ in L^1, then $\lim_{n \to \infty} \mathbb{J}(\varphi_n, \psi_n) = \mathbb{J}(\varphi_o, \psi_o)$. The theorem follows.
                                                                        Q.e.d.

(5.21) EXERCISE. Prove Theorem (5.20)(a).

EXERCISES : (5.5), (5.21).

6. Estimating smooth unimodal densities

In this section, we apply what we have learned about smooth monotone density estimation to the estimation of smooth unimodal densities. So the problem is to estimate the unimodal density f_o from an iid sample $X_1, X_2, \cdots, X_n$. The mode of f_o is denoted by m_o. If the mode were known, then unimodal density estimation can be achieved by monotone density estimation on (m_o, ∞) and on $(-\infty, m_o)$, and the results from §4 apply. So the interest is in estimating a unimodal density with the (unknown) mode being a nuisance parameter. Of course, the approach taken here is to estimate the density, and implicitly the mode, by solving the maximum smoothed likelihood problem with the unimodality constraint

(6.1) $\text{minimize} \quad L_{nh}(f) \overset{\mathrm{def}}{=} - \int_{\mathbb{R}} [\, A_h * dF_n \,](x) \log f(x)\, dx + \int_{\mathbb{R}} f(x)\, dx$

    $\text{subject to} \quad f \in L^1(\mathbb{R}) \ , \ f \geqslant 0 \ , \ f \text{ is unimodal} \ ,$

where $A_h * dF_n$ is the usual convolution kernel density estimator. As usual, the solution of (6.1) is denoted by f^{nh}. We assume as in § 5.5 that

(6.2)   the kernel A is a symmetric, log-concave density .

Thus, $A_h * f_o$ is unimodal whenever f_o is unimodal, by the IBRAGIMOV (1956) result of Lemma (5.5.4).

Here, the usual questions of existence (yes), uniqueness (no), and convergence rates of the density estimator are considered. Rates for the estimator of the mode are not foremost in our minds, but we need to establish some ludicrously slow ones. We require the usual nonparametric assumptions on the density f_o, viz.

(6.3) $f_o \in W^{1,2}(\mathbb{R})$,

(6.4) $\mathbb{E}[\,|\,X\,|^{\kappa}\,] < \infty$ for some $\kappa > 1$.

In addition, f_o must drop off from its mode or modes in a prescribed manner. For continuous unimodal f_o, let

(6.5) $M_o = \{\, m \in \mathbb{R} \,:\, f_o(m) = \max\limits_{x \in \mathbb{R}} f_o(x)\,\}$

be the interval of modes of f_o, and let

(6.6) $\mathrm{dist}(x, M_o) = \min\{\,|\,x - m\,| \,:\, m \in M_o\,\}$.

The assumptions are as follows.

(6.7) There exists an open neighborhood Ω of M_o, such that

$$f_o''(x) \leqslant 0 \quad \text{for all } x \in \Omega ,$$

and, with $|x - m_o| = \mathrm{dist}(x, M_o)$, for all $\varepsilon > 0$,

(6.8) $\liminf\limits_{|x - m_o| \to 0^+} \big(f_o(m_o) - f_o(x)\big)\exp\Big(\dfrac{\varepsilon}{|x - m_o|}\Big) > 0$.

(Actually, the limit then equals $+\infty$.)

Some simple sufficient conditions for (6.7) and (6.8) to hold are

$$f_o'' \text{ is continuous on } \Omega \text{ and } f_o''(m_o) < 0 ,$$

or, for some $c > 0$, $p > 0$,

$$f_o''(x) = -c\,|\,x - m_o\,|^p , \quad \text{for all } x \in \Omega$$

(and then the mode is unique). The convergence rates can now be stated.

(6.9) THEOREM. Let $0 < \beta < 1$ and $h = h_n \asymp n^{-\beta}$. Under the above conditions on A and f_o,

$$\|\,f^{nh} - f_o\,\|_1 \leqslant \|\,A_h * dF_n - f_o\,\|_1 + \delta_{nh} ,$$

with $\delta_{nh} =_{\text{as}} o\big((nh)^{-1/2} \big)$.

The theorem says that the unimodal estimator is asymptotically not worse than the plain kernel estimator.

Before continuing, note that the minimization in (6.1) is over unimodal, nonnegative L^1 functions. Therefore:

(6.10) EXERCISE. Show that any solution of (6.1) must be a pdf.

We now discuss the uniqueness, existence, and convergence rates of the density estimator, in that order.

The uniqueness of solutions to (6.1) is questionable, since the problem (6.1) is not convex (the set of all unimodal functions is not convex), although computational experience suggests that the solution is usually unique. The nonuniqueness is illustrated in the following exercise.

(6.11) EXERCISE. Let $n = 2$. For all h small enough, show that the maximum smoothed likelihood estimators of the mode are given by $m = X_1$ and $m = X_2$ [Hint: This problem is hard from scratch. It is really easy using Theorem (6.40).]

To show the existence of solutions to (6.1), it is useful to consider the estimation problem with an *a priori* fixed mode m, that is,

(6.12)   minimize $L_{nh}(f)$ over $f \in \mathcal{U}(m)$,

where

(6.13) $\mathcal{U}(m) = \{ f \in L^1(\mathbb{R}) : f \geqslant 0 , \ f \text{ is unimodal, with mode at } m \}$.

It is obvious that (6.12) has a unique solution, which we denote by $f(\cdot\,; m)$, and also by $f^{nh}(\cdot\,; m)$. Thus, the existence of solutions to (6.1) reduces to the existence of solutions to

(6.14)

$$\text{minimize} \quad L^*_{nh}(m) \stackrel{\text{def}}{=} \min \{ L_{nh}(f) : f \in \mathcal{U}(m) \}$$

$$\text{subject to} \quad -\infty < m < \infty .$$

Any solution of (6.14) will be denoted by m_{nh}. The function $L^*_{nh}(m)$ goes by the name of *profile* negative log-likelihood. Note that (6.14) is a *parametric* estimation problem, but unfortunately, no benefit seems to accrue from this observation.

(6.15) EXERCISE. Show that (6.14) has a solution if and only if (6.1) has a solution.

The following inequality regarding (6.14) is a nice little exercise. (Note that $L_{nh}(f) = \text{KL}(A_h * dF_n , f) +$ terms independent of f.)

(6.16) LEMMA. *Let φ, ψ be nonnegative $L^1(\mathbb{R})$ functions, and let $f_\varphi(\,\cdot\,;m)$ be the minimizer of $\mathrm{KL}(\varphi,f)$ over $f \in \mathcal{U}(m)$, and likewise for $f_\psi(\,\cdot\,;m)$. Then,*

$$\| f_\varphi(\,\cdot\,;m) - f_\psi(\,\cdot\,;m) \|_1 \leqslant \| \varphi - \psi \|_1 \; .$$

(6.17) EXERCISE. (a) Prove Lemma (6.16).

(b) Show that $\| f^{nh}(\,\cdot\,;m_o) - f_o \|_1 \leqslant \| A_h * dF_n - f_o \|_1$.

The existence of solutions to (6.14) follows from the following lemma.

(6.18) LEMMA. *For all realizations of $X_1, X_2, \cdots, X_n$,*

(a) $L^*_{nh}(m)$ *is lower semi continuous in m,*

(b) $L^*_{nh}(m)$ *is bounded below, and*

(c) $L^*_{nh}(m) \longrightarrow \infty$ *for $m \to \pm\infty$.*

PROOF. For part (b), observe as usual that $L_{nh}(f) = \mathrm{KL}(A_h * dF_n , f) +$ other (finite) terms independent of f. To prove (c), it is useful to write L_{nh} as

$$(6.19) \qquad\qquad L_{nh}(f) = \frac{1}{n} \sum_{i=1}^{n} L_h(f; X_i) \; ,$$

where

$$(6.20) \quad L_h(f; X_i) = -\int_{\mathbb{R}} A_h(x - X_i) \log f(x)\, dx + \int_{\mathbb{R}} f(x)\, dx \; .$$

Now, suppose that $m \to \infty$, so w.l.o.g. $m > X_{n,n}$, the largest order statistic. Let $1 \leqslant i \leqslant n$, and consider

$$(6.21) \qquad\qquad \text{minimize} \quad L_h(f; X_i) \quad \text{over } f \in \mathcal{U}(m) \; .$$

Since A is log-concave, then $A_h(x - X_i)$ is decreasing on (m, ∞), and hence the solution to (6.21) is given by

$$(6.22) \qquad\qquad f_{im}(x) = \begin{cases} A_h(x - X_i) \, , & x \notin [\lambda, m] \, , \\ A_h(\lambda - X_i) \, , & x \in [\lambda, m] \, , \end{cases}$$

where λ satisfies

$$(6.23) \qquad \int_{\lambda}^{m} A_h(x - X_i)\, dx = (m - \lambda)\, A_h(\lambda - X_i) \; .$$

See the illustration in Figure 6.1.

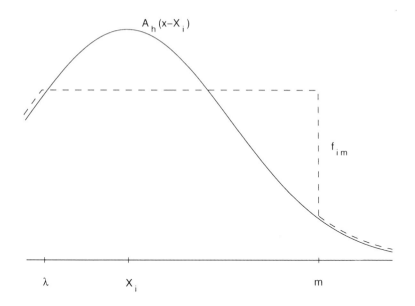

Figure 6.1. Illustration of the construction of f_{im}, see (6.22).

Now, consider $L_h(f_{im}; X_i)$. To begin, note that

$$(6.24) \quad \left| \int_{x \notin [\lambda, m]} A_h(x - X_i) \log f_{im}(x) \, dx \right| \leqslant$$

$$\int_{\mathbb{R}} A_h(x - X_i) \left| \log A_h(x - X_i) \right| dx ,$$

and this is independent of m (and X_i for that matter), and finite. (Exercise! Recall that A is continuous and log concave.)

Also,

$$\int_\lambda^m A_h(x - X_i) \log f_{im}(x) \, dx = \left(\log A_h(\lambda - X_i) \right) \int_\lambda^m A_h(x - X_i) \, dx .$$

To proceed, we need a lower bound for $\delta = A_h(\lambda - X_i)$. In view of the facts that $\lambda < X_i \leqslant X_{n,n}$ and

$$(m - \lambda)\delta = \int_\lambda^m A_h(x - X_i) \, dx \leqslant 1 ,$$

then $\delta \leqslant (m - \lambda)^{-1} < (m - X_{nn})^{-1}$, whence $-\log \delta > \log(m - X_{n,n})$. Also,

$$\int_\lambda^m A_h(x - X_i) \, dx \geqslant c(m) \stackrel{\text{def}}{=} \int_{X_i}^m A_h(x - X_i) \, dx \longrightarrow \tfrac{1}{2} \quad \text{as } m \to \infty .$$

Consequently,

$$-\int_\lambda^m A_h(x - X_i) \log f_{im}(x) \, dx \geqslant c(m) \log(m - X_{n,n}) .$$

Together with (6.24), this shows that $L_h(f_{im}; X_i) \longrightarrow \infty$ for $m \to \infty$. Finally, from (6.19), then

$$L_{nh}^*(m) \geqslant \frac{1}{n} \sum_{i=1}^{n} L_h(f_{im}; X_i) \longrightarrow \infty \quad \text{as } m \to \infty .$$

The case $m \to -\infty$ is treated similarly.

The lower semicontinuity of $L_{nh}^*(m)$ is implied by the lower semicontinuity of $L_{nh}(f)$ in the $L^1(\mathbb{R})$ topology, which follows from its strong convexity, see (11.5.7), Exercise (9.1.22), and the next lemma.   Q.e.d.

The following lemma was referred to in the proof above, but it is interesting and useful in its own right.

(6.25) LEMMA. *For all realizations of* $X_1, X_2, \cdots, X_n$, *and all real* m, μ,

$$\| f(\cdot ; m) - f(\cdot ; \mu) \|_1 \leqslant 2 \, | m - \mu | \, \| A_h * dF_n \|_\infty .$$

PROOF. Let $\varphi = A_h * dF_n$, and assume that $m < \mu$. First note that

$$f(x; \mu) = [\operatorname{lcm}(\varphi, (\mu, \infty))](x) , \quad x > \mu ,$$

so that not only

(6.26)
$$\int_\mu^\infty f(x; \mu) \, dx = \int_\mu^\infty \varphi(x) \, dx ,$$

but also by Lemma (4.10),

$$f(x; m) = [\operatorname{lcm}(\psi, (m, \infty))](x) , \quad x > m ,$$

where

$$\psi(x) = \begin{cases} f(x; \mu) , & x > \mu , \\ \varphi(x) , & m < x < \mu . \end{cases}$$

It then follows that

(6.27)
$$f(x; \mu) \geqslant f(x; m) , \quad \text{for } x > \mu .$$

Now, write $\| f(\cdot ; m) - f(\cdot ; \mu) \|_1$, as

$$\int_\mathbb{R} | f(x; m) - f(x; \mu) | \, dx = \int_{-\infty}^m \cdots + \int_m^\mu \cdots + \int_\mu^\infty \cdots ,$$

and consider each piece in turn. For the last integral, we obtain

$$\int_\mu^\infty |f(x;m) - f(x;\mu)|\, dx = \int_\mu^\infty (\, f(x;\mu) - f(x;m)\,)\, dx$$

$$= \int_\mu^\infty (\, \varphi(x) - f(x;m)\,)\, dx$$

(6.28)
$$= \int_m^\mu (\, f(x;m) - \varphi(x)\,)\, dx\ ,$$

where the first equality is by (6.27), the second one by (6.26), and the third one again by (6.26), with μ replaced by m.

Likewise, on $(-\infty, m)$,

(6.29) $$\int_{-\infty}^m |f(x;m) - f(x;\mu)|\, dx = \int_m^\mu (\, f(x;\mu) - \varphi(x)\,)\, dx\ .$$

Now, (6.28) and (6.29) imply

$$\| f(\,\cdot\,;m) - f(\,\cdot\,;\mu)\, \|_1 =$$

$$\int_m^\mu \{\, f(x;m) + f(x;\mu) - 2\,\varphi(x) + |f(x;m) - f(x;\mu)|\,\}\, dx =$$

$$2 \int_m^\mu \{\, f(x;m) \vee f(x;\mu) - \varphi(x)\,\}\, dx \leqslant$$

$$2 \int_m^\mu \{\, \|\,\varphi\,\|_\infty - \varphi(x)\,\}\, dx \leqslant 2\,(\mu - m)\, \|\,\varphi\,\|_\infty\ ,$$

where we used Exercise (3.8).   Q.e.d.

This concludes the proof that (6.1) has solutions.

On to convergence rates. Lemma (6.25) may be used to get error bounds for unimodal density estimation in terms of the mode selected.

(6.30) LEMMA. *With m_{nh} any solution of (6.14), and m_o a mode of f_o, for $nh \to \infty$, $h \to 0$,*

$$\| f^{nh} - f_o \|_1 \leqslant \| A_h * dF_n - f_o \|_1 + c_{nh}\, |\, m_{nh} - m_o\,|\, \| A_h * (dF_n - dF_o) \|_\infty\ ,$$

where $c_{nh} \longrightarrow_{as} \sqrt{32}$.

PROOF. Let $m_o \in M_o$ be such that $|\, m_{nh} - m_o\,| = \text{dist}(m_{nh}, M_o)$. Assume that $m_{nh} < m_o$. (The case $m_{nh} > m_o$ goes the same way.) As usual, we split the integral

$$\int_{\mathbb{R}} |f^{nh} - f_o| = \int_{-\infty}^{m_{nh}} \cdots + \int_{m_{nh}}^{m_o} \cdots + \int_{m_o}^\infty \cdots\ .$$

For the first integral, we have by Theorem (4.33),

$$\int_{-\infty}^{m_{nh}} |f^{nh} - f_o| \leqslant \int_{-\infty}^{m_{nh}} |A_h * dF_n - f_o|.$$

For the last integral, we have

$$\int_{m_o}^{\infty} |f^{nh} - f_o| \leqslant \int_{m_o}^{\infty} |f^{nh}(\cdot; m_{nh}) - f^{nh}(\cdot; m_o)| + \int_{m_o}^{\infty} |f^{nh}(\cdot; m_o) - f_o|,$$

and for the last integral of this, the L^1 contractivity result (1.19) gives

$$\int_{m_o}^{\infty} |f^{nh} - f_o| \leqslant \int_{m_o}^{\infty} |A_h * dF_n - f_o|.$$

For the first integral, we have similar to (6.28) and (6.29)

$$\int_{m_o}^{\infty} |f^{nh} - f^{nh}(\cdot; m_o)| = \int_{m_{nh}}^{m_o} f^{nh} - A_h * dF_n.$$

Finally, for the integral

$$\int_{m_{nh}}^{m_o} |f^{nh} - f_o|,$$

the triangle inequality suffices. Putting it all together gives the inequality

$$\| f^{nh} - f_o \|_1 \leqslant \| A_h * dF_n - f_o \|_1 + 2 \int_{m_{nh}}^{m_o} |f^{nh} - A_h * dF_n|.$$

The last remaining integral is duly bounded in the next rather technical lemma.   Q.e.d.

Note that in the above proof, no use was made of the fact that f^{nh} or m_{nh} were maximum smoothed likelihood solutions. In the next lemma, that changes.

(6.31) LEMMA. *Under the same conditions as Lemma* (6.30),

$$\left| \int_{m_{nh}}^{m_o} |f^{nh} - A_h * dF_n| \right| \leqslant c_{nh} |m_{nh} - m_o| \, \| A_h * (dF_n - dF_o) \|_\infty,$$

where $c_{nh} \longrightarrow_{as} \sqrt{8}$.

PROOF. Assume that $m_{nh} < m_o$, and construct a unimodal estimator as follows. Let

$$\varphi^{nh}(x) = \begin{cases} f^{nh}(x; m_{nh}), & x \notin [m_{nh}, m_o], \\ f_h(x) + \delta_{nh}, & x \in [m_{nh}, m_o], \end{cases}$$

where $f_h = A_h * dF_o$ and $\delta_{nh} = \| A_h * (dF_n - dF_o) \|_\infty$. Below, we prove that

(6.32) φ^{nh} is unimodal.

Consequently, φ^{nh} is an (unsuccessful) candidate for a solution of (6.1) and so

$$KL(A_h * dF_n, f^{nh}) \leqslant KL(A_h * dF_n, \varphi^{nh}) .$$

Both sides of this inequality are integrals over $(-\infty, \infty)$, but the integrands differ only on $[m_{nh}, m_o]$, so that

$$(6.33) \quad \int_{m_{nh}}^{m_o} A_h * dF_n \log \frac{A_h * dF_n}{f^{nh}} + f^{nh} - A_h * dF_n \leqslant$$

$$\int_{m_{nh}}^{m_o} A_h * dF_n \log \frac{A_h * dF_n}{\varphi^{nh}} + \varphi^{nh} - A_h * dF_n .$$

The integral on the right may be bounded by Pearson's φ^2 distance

$$\int_{m_{nh}}^{m_o} \frac{|A_h * dF_n - \varphi^{nh}|^2}{\varphi^{nh}} ,$$

the square root of which may be bounded as

$$\{ f_h(\mu^{nh}) \}^{-1/2} \left\{ \int_{m_{nh}}^{m_o} |A_h*(dF_n - dF_o)|^2 \right\}^{1/2} + \{ (m_o - m_{nh}) \}^{1/2} \delta_{nh} \leqslant$$

$$2 \{ f_h(m_{nh}) \}^{-1/2} |m_{nh} - m_o|^{1/2} \| A_h * (dF_n - dF_o) \|_\infty ,$$

where μ^{nh} is defined via $f_h(\mu^{nh}) = \min\{ f_h(x) : x \in [m_{nh}, m_o] \}$. The integral on the left of (6.33) may be bounded below by, see Exercise (9.1.22),

$$\left\{ \int_{m_{nh}}^{m_o} \tfrac{2}{3} A_h * dF_n + \tfrac{4}{3} f^{nh} \right\}^{-1} \left\{ \int_{m_{nh}}^{m_o} |f^{nh} - A_h * dF_n| \right\}^2 ,$$

and the first factor behaves as $\{2 f_o(m_o)(m_o - m_{nh}) \}^{-1}$ for $nh \to \infty$, $h \to 0$. [Here, we already use that $m_{nh} \longrightarrow_{as} m_o$, see Lemma (6.35).]

Since $f_h \longrightarrow f_o$ uniformly, as $nh \to \infty$, $h \to 0$, putting all this together proves the required bound.

We still must prove (6.32). Some of the crucial facts required are that

$$f^{nh}(m_{nh}; m_{nh}) = A_h * dF_n(m_{nh}), \quad f^{nh}(m_o; m_{nh}) \leqslant f^{nh}(m_o; m_o) ,$$

the first one of which is the content of Lemma (6.39) below. Let m_h be the mode of $A_h * f_o$. We distinguish between the cases $m_h > m_o$ and $m_h \leqslant m_o$. Suppose $m_h > m_o$. Then,

$$\varphi^{nh}(m_{nh}) = A_h * f_o(m_{nh}) + \delta_{nh} \geqslant A_h * dF_n(m_{nh}) = f^{nh}(m_{nh}; m_{nh}) .$$

Moreover, φ^{nh} is increasing on $[m_{nh}, m_o]$. Thus, φ^{nh} is increasing on $(-\infty, m_o)$. Since it is decreasing on (m_o, ∞), then φ^{nh} is unimodal with mode at m_o. This is (6.32) when $m_h > m_o$.

Suppose $m_h \leqslant m_o$. Then, $A_h * f_o$ is decreasing on (m_o, ∞). We now have that

$$f^{nh}(m_o; m_{nh}) \leqslant f^{nh}(m_o; m_o) \leqslant \max_{x \geqslant m_o} A_h * dF_n(x)$$

$$\leqslant \max_{x \geqslant m_o} A_h * dF_o(x) + \delta_{nh} = A_h * dF_o(m_o) + \delta_{nh} = \varphi^{nh}(m_o) .$$

At $x = m_{nh}$, we have

$$\varphi^{nh}(m_{nh}) = A_h * f_o(m_{nh}) + \delta_{nh} \geqslant A_h * dF_n(m_{nh}) = f^{nh}(m_{nh}; m_{nh}) .$$

Now, φ^{nh} is increasing on $(-\infty, m_{nh})$, decreasing on (m_o, ∞), and it is unimodal on $[m_{nh}, m_o]$. The conclusion is that φ^{nh} is unimodal. Thus, (6.32) holds when $m_h < m_o$ as well.   Q.e.d.

In the remainder of this section, we suitably bound $\| A_h * (dF_n - dF_o) \|_\infty$ and $m_{nh} - m_o$. We quote the following pertinent result, specialized to our needs. The reason we did not prove it in Chapter 4 and are not proving it here is that it requires probabilistic tools that fall outside of the scope of the text.

(6.34) LEMMA. [SILVERMAN (1978)] *Let A be a log-concave pdf, and assume that f_o is uniformly continuous. Let $0 < \beta < 1$ and $h \asymp n^{-\beta}$. Then,*

$$\| A_h * (dF_n - dF_o) \|_\infty =_{as} \mathcal{O}\big((nh)^{-1/2} \log h^{-1} \big) .$$

The above makes clear that to get the "correct" rates of convergence we need merely prove the following lemma.

(6.35) LEMMA. *Let $h \asymp n^{-\beta}$ with $0 < \beta < 1$. Under the assumptions (6.3) and (6.4),*

$$\mathrm{dist}(m_{nh}, M_o) \longrightarrow_{as} 0 .$$

Moreover, with the additional assumptions (6.7) and (6.8),

$$\mathrm{dist}(m_{nh}, M_o) =_{as} o\big((\log n)^{-1} \big) .$$

PROOF. Let m_o be the point in M_o closest to m_{nh}, and let us assume that $m_{nh} < m_o$. Let

$$\delta_{nh} = (1 + c_{nh}) \, | m_o - m_{nh} | \, \| A_h * (dF_n - dF_o) \|_\infty ,$$

where c_{nh} is given in Lemma (6.31). Then, the triangle inequality and Lemma (6.31) imply that

$$\int_{m_{nh}}^{m_o} | f_o(x) - f^{nh}(x) | \, dx \leqslant \delta_{nh} + \int_{m_{nh}}^{m_o} | f_o(x) - A_h * f_o(x) | \, dx .$$

For the last integral, we have the bound, for a suitable constant γ,

$$(6.36) \quad \int_{m_{nh}}^{m_o} |f_o(x) - A_h * f_o(x)| \, dx \leqslant (m_o - m_{nh}) \, \| f_o - A_h * f_o \|_\infty$$

$$\leqslant \gamma \, (m_o - m_{nh}) \, h \, \| (f_o)' \|_\infty \, .$$

Since $(f_o)'' \in L^1(\mathbb{R})$, then $(f_o)' \in L^1(\mathbb{R})$ and thus, for (another) suitable constant γ,

$$\int_{m_{nh}}^{m_o} |f_o(x) - f^{nh}(x)| \, dx \leqslant \delta_{nh} + \gamma \, h \, (m_o - m_{nh}) \, .$$

Now, on the interval $[m_{nh}, m_o]$, the function f_o is increasing and f^{nh} is decreasing. Thus, the increasing estimator of f^{nh} on this interval is a constant function, and it follows from the L^1 contractivity result (1.19) that

$$(6.37) \quad \min_c \int_{m_{nh}}^{m_o} |f_o(x) - c| \, dx \leqslant \delta_{nh} + c \, h \, (m_o - m_{nh}) \, .$$

Since f_o is continuous, it follows that $f_o(m_{nh}) - f_o(m_o) \longrightarrow_{as} 0$, and the conclusion follows.

Now, assume the additional conditions (6.7) and (6.8). Since $f_o'' \leqslant 0$, then f_o is increasing on $[m_{nh}, m_o]$ and so the optimal c in (6.37) equals $c = f_o(\mu)$, with $\mu = \frac{1}{2}(m_{nh} + m_o)$. Then,

$$\int_{m_{nh}}^{m_o} |f_o(x) - c| \, dx \geqslant \int_{\mu}^{m_o} (f_o(x) - f_o(\mu)) \, dx \, .$$

Since f_o is concave on $[\mu, m_o]$, geometric considerations show that the last integral equals at least one-half the area of the rectangle with opposite vertices $(\mu, f(\mu))$ and $(m_o, f_o(m_o))$, or

$$\delta_{nh} + \gamma \, h \, (m_o - m_{nh}) \geqslant$$

$$\min_c \int_{m_{nh}}^{m_o} |f_o(x) - c| \, dx \geqslant \tfrac{1}{2} \, (m_o - \mu) \, (f_o(m_o) - f_o(\mu)) \, .$$

Now, (6.37) and (6.8) imply that for all $\varepsilon > 0$, there exists a constant c such that

$$\exp\left(-\frac{\varepsilon}{m_o - \mu}\right) \leqslant c \, (\| A_h * (dF_n - dF_o) \|_\infty + \gamma \, h) \, ,$$

and in view of Lemma (6.34) for $h \asymp n^{-\beta}$, this implies asymptotically for a suitable constant c',

$$m_o - \mu \leqslant c' \, \varepsilon \, \log n \, .$$

Since $\varepsilon > 0$ is arbitrary, the conclusion follows.   Q.e.d.

(6.38) EXERCISE. Prove (6.36), cf. §4.2.

Finally, we establish some cute facts about the (implied) estimators of the mode, which are useful in computations and were used above. We recall that A_h is log-concave and continuous, so that $A_h * dF_n$ is continuous.

(6.39) LEMMA. *Let $A_h * dF_n$ be continuous, and let $f(\,\cdot\,;m)$ be the solution of (6.1). Then, (at least) one of the following three statements holds.*

(a) $f(\,\cdot\,;m)$ *is continuous, and* $f(m;m) = A_h * dF_n(m)$.

(b) $\exists \delta > 0 : f(\,\cdot\,;m)$ *is constant on* $(m - \delta, m)$ *and*
$$f(m - 0;m) \geqslant f(m + 0;m).$$

(c) $\exists \delta > 0 : f(\,\cdot\,;m)$ *is constant on* $(m, m + \delta)$ *and*
$$f(m + 0;m) \geqslant f(m - 0;m).$$

PROOF. Exercise (3.8) implies that $f(\,\cdot\,;m)$ is continuous everywhere, except possibly at $x = m$, and that $f(m \pm 0;m) \geqslant A_h * dF_n(m)$. If we now have $f(m + 0;m) = f(m - 0;m) = A_h * dF_n(m)$, then case (a) holds. If it does not, then $f(m + 0;m) > A_h * dF_n(m)$ or $f(m - 0;m) > A_h * dF_n(m)$. So suppose $f(m + 0;m) > A_h * dF_n(m)$. By the continuity of $f(\,\cdot\,;m)$ on (m, ∞), and the continuity of $A_h * dF_n$ there exists a $\delta > 0$ such that $f(x;m) > A_h * dF_n(x)$, for all $x \in (m, m + \delta)$. This can only happen if $f(\,\cdot\,;m)$ is constant on $(m, m + \delta)$. If now $f(m - 0;m) = A_h * dF_n(m)$, then case (c) holds. If, on the other hand, $f(m - 0;m) > A_h * dF_n(m)$, then by the same reasoning as above, $f(\,\cdot\,;m)$ is constant on some interval $(m - \varepsilon, m)$. Then, at least one of the cases (b) and (c) holds. The case $f(m - 0;m) > A_h * dF_n(m)$ goes the same way. Q.e.d.

The lemma immediately establishes the following useful (theoretically and practically) result regarding the maximum smoothed likelihood estimator of the mode.

(6.40) THEOREM. *The minimum of L_{nh}^* occurs at a local mode of $A_h * dF_n$.*

PROOF. Let $m \in \mathbb{R}$, and suppose that $f(\,\cdot\,;m)$ is constant on the interval $[m, m + \delta]$, with $\delta > 0$ chosen as large as possible, and $f(m + 0;m) \geqslant f(m - 0;m)$. Then, $f(\,\cdot\,;m) \in \mathcal{U}(\eta)$ for all $\eta \in [m, m + \delta]$. Consequently,

$$(6.41) \quad L_{nh}^*(m) = L_{nh}(f(\,\cdot\,;m)) \geqslant L_{nh}^*(\eta) , \quad \text{for all } \eta \in [m, m + \delta] .$$

Let $\eta_o \in [m, m + \delta]$ be the largest local mode of $A_h * dF_n$ on $[m, m + \delta]$. There are now two possibilities: Either $A_h * dF_n(\eta_o) = f(m + 0;m)$ or $A_h * dF_n(\eta_o) > f(m + 0;m)$. In the first case, $A_h * dF_n(x) = f(m + 0;m)$ for all $x \in [m, m + \delta]$, and m is a local mode of $A_h * dF_n$. In the second case, $f^{nh}(\,\cdot\,;\eta_o)$ is different from $f^{nh}(\,\cdot\,;m)$, since $f^{nh}(\eta_o;\eta_o) > f^{nh}(m + 0;m)$. Then, the uniqueness of $f^{nh}(\,\cdot\,;\eta_o)$ shows that $L_{nh}^*(m) > L_{nh}^*(\eta_o)$, and in view of (6.41), m is not a local minimum of L_{nh}^*. The same conclusion prevails if $f(\,\cdot\,;m)$ is constant on an interval $[m - \delta, m]$, and consequently,

$f(m - 0; m) \geqslant f(m + 0; m)$. Thus, if m is a local minimum of L^*_{nh}, then part (a) of Lemma (6.4) must hold, and m is a local mode of $A_h * dF_n$. Q.e.d.

(6.42) COROLLARY. If f^{nh} solves (6.1), then
$$f^{nh}(m_{nh}) = A_h * dF_n(m_{nh}) .$$

EXERCISES : (6.10), (6.11), (6.15), (6.17), (6.38).

7. Other unimodal density estimators

In this section, we discuss some alternative estimators of unimodal densities. Some experimental comparisons with the maximum smoothed likelihood estimator are briefly commented on, but are shown in Chapter 8.

(a) One can consider estimating the mode by the mode of the kernel density estimator and then estimate the unimodal density by the solution of (6.12) to the right, and similarly to the left of the mode. It appears that the resulting density estimator satisfies the same asymptotic and just about the same small sample behavior as the smoothed maximum likelihood estimator. It should be noted that our emphasis here is on the accuracy of the density estimator, not on the accuracy of the estimator of the mode. We should mention that the (optimal) estimation of the mode is an entirely different game. Estimating the mode of f_o by the mode of $A_h * dF_n$ goes back to PARZEN (1962). EDDY (1980) shows asymptotic normality of this estimator when the third derivative of f_o is absolutely continuous, and $f''_o(m_o) < 0$ (so the mode is unique). It should be noted that EDDY (1980) requires that the smoothing parameter h satisfy $h \asymp n^{-1/7}$, which is quite a bit larger than the optimal h for the L^1 error of the kernel density estimator under the usual nonparametric assumptions, viz. $h \asymp n^{-1/5}$. As a matter of fact, EDDY (1980) specifically excludes this rate. Bootstrapped versions are discussed in GRUND and HALL (1995).

(b) One can also consider various minimum distance estimators. One choice is to estimate the mode of the unimodal density by the solution to

(7.1)   minimize $\| f^{nh}(\,\cdot\,; m) - A_h * dF_n \|_1$ subject to $m \in \mathbb{R}$.

Here, $f^{nh}(\,\cdot\,; m)$ denotes the solution of (6.12) when the mode is a priori fixed at m, so (7.1) is not a plain minimum distance method. The solution, denoted by ψ^{nh}, exists because the local minima occur again at some of the modes of $A_h * dF_n$, and Lemma (6.30) and Theorem (6.9) hold for ψ^{nh} also, by the same arguments under the same assumptions as for the smoothed maximum likelihood estimator. The behavior in the small sample case of this estimator is also just about the same as for the maximum smoothed likelihood estimator.

(c) The approach of FOUGÈRES (1997) suggests replacing $f^{nh}(\,\cdot\,;m)$ by what may be called the unimodal rearrangement of $A_h * dF_n$, i.e., the decreasing rearrangements of $A_h * dF_n(x)$, $x \geqslant m$, on (m, ∞) and the increasing rearrangement of $A_h * dF_n(x)$, $x < m$, on $(-\infty, m)$. Here, the decreasing rearrangement of a nonnegative $f \in L^1(m, \infty)$ is defined to be that function $f^* \in L^1(m, \infty)$ that is decreasing on (m, ∞), and for which the sets

$$\{\, x \,:\, f^*(x) > \alpha \,\} \quad \text{and} \quad \{\, x \,:\, f(x) > \alpha \,\}$$

have the same (Lebesgue) measure for all $\alpha > 0$, see HARDY, LITTLEWOOD, and POLYA (1951) or LIEB and LOSS (1996). The increasing rearrangement of a nonnegative function $f \in L^1(-\infty, m)$ is defined analogously. The resulting unimodal estimator is denoted as $\varphi^{nh}(\,\cdot\,;m)$. It seems to be both a strength and a weakness of this estimator that it is extremely insensitive to outliers: In its simplest form, when A is symmetric, unimodal, and has compact support, for $n = 1$ and $X_1 \gg m$, the unimodal rearrangement with mode m of $A_h(x - X_1)$, $x \in \mathbb{R}$, is given by

$$\varphi^{1,h}(x\,;m) = \begin{cases} A_h\big(\tfrac{1}{2}(x - m)\big)\,, & x > m\,, \\[2mm] 0 & ,\quad x < m\,, \end{cases}$$

and the outlier has disappeared without a trace.

Two natural choices for the mode are the mode of $A_h * dF_n$ and the solution of

(7.2)   minimize $\|\, \varphi^{nh}(\,\cdot\,;m) - A_h * dF_n \,\|_1$ subject to $m \in \mathbb{R}\,.$

Both of these estimators are similar to each other, but behave quite differently compared with the smoothed maximum likelihood estimator.

(d) It should be noted that for unimodal estimation, the (unsmoothed) maximum likelihood approach does not work: For any j $(1 \leqslant j \leqslant n)$, the "density" corresponding to a point mass with mass $1/n$ at X_j and the Grenander estimator on each side of X_j (and omitting X_j from the data) makes the likelihood "equal" to $-\infty$. However, the approach of BIRGÉ (1997) says in effect that the point mass $1/n$ when the mode is at an observation can be ignored, considering the error bound $n^{-1/3}$ for monotone densities. Actually, BIRGÉ (1997) considers any estimator $f^{nh}(\,\cdot\,;\mu)$ (as per the above) with distribution $F^{nh}(\,\cdot\,;\mu)$ for $\mu \neq X_i$, $i = 1, 2, \cdots, n$, which satisfies

(7.3) $\|\, F^{nh}(\,\cdot\,;\mu) - F_n \,\|_\infty \leqslant \min_m \|\, F^{nh}(\,\cdot\,;m) - F_n \,\|_\infty + \eta\,,$

for $\eta = o(n^{-1/2})$, say. For this estimator BIRGÉ (1997) shows bounds of order $n^{-1/3}$ for the expected values.

(e) To avoid the point mass at one observation detailed in (d), BICKEL and FAN (1996) propose to eliminate X_j when doing maximum likelihood and solve

(7.4)

$$\text{minimize} \quad -\tfrac{1}{n-1} \sum_{\substack{i=1 \\ i \neq j}}^{n} \log f(X_i) + \int_{\mathbb{R}} f(x)\, dx$$

$$\text{subject to} \quad f \in L^1(\mathbb{R})\,,\ f \geqslant 0\,,\ f \text{ unimodal, with mode at } X_j\,,$$

where the minimization is also over $j = 1, 2, \cdots, n$. They prove the point-wise consistency of the estimator, except at the mode m_o, analogous to the monotone case, see WEGMAN (1970) and WOODROOFE and SUN (1993).

(f) BICKEL and FAN (1996) also consider "grouping" the observations followed by maximum likelihood estimation. This is close to considering (1.11) with the kernel estimator $A_h * dF_n$ (for small h) approximated by a step function on a fine partition of $\mathbb{R}$. The resulting estimator for smooth unimodal densities is only $\mathcal{O}(n^{-1/3})$ accurate, and in the small sample case, it is in fact not competitive for smooth unimodal densities. However, smoothing this estimator, e.g., using kernel smoothing, yields a competitive estimator for smooth densities. This is analogous to isotone regression, see MAMMEN (1991).

We finish this section with some more comments and exercises on the minimum distance methods discussed before.

(7.5) LEMMA. *The solution(s) of (7.1) occur at some of the local modes of $A_h * dF_n$.*

(7.6) EXERCISE. Prove Lemma (7.5).

(7.7) EXERCISE. Does the analogue of Lemma (7.5) hold for the minimum distance problem (7.2)?

Are the solutions of (7.1) and (6.1) in fact identical? The following example shows that they need not be.

(7.8) EXERCISE. Let $a > \tfrac{1}{2}$ and let $A_h * dF_n$ be a mixture of two uniforms

$$A_h * dF_n(x) = \tfrac{1}{2}\, \Pi_2(x + a) + \tfrac{1}{2}\, \Pi_2(x - a)\,,$$

where $\Pi(x) = \mathbb{1}(0 < x < 1)$ and $\Pi_t(x) = t^{-1} \Pi(t^{-1} x)$. Show that the solutions of (6.1) form the interval $[\,a, a+1\,]$ and that the solutions of (7.1) form the interval $[\,-a, -a + 2\,]$.

(7.9) EXERCISE. Determine a.s. convergence rates for $\| f^{nh}(\,\cdot\,; \nu^{nh}) - f_o \|_1$, where $m = \nu^{nh}$ is a solution of (7.1).

(7.10) EXERCISE. Likewise, for $\| \varphi^{nh}(\,\cdot\,; \lambda^{nh}) - f_o \|_1$, where $m = \lambda^{nh}$ is a solution of (7.2).

We finish this section with an exercise about modes.

(7.11) EXERCISE. Let f_o'' be uniformly continuous and $f_o''(m_o) < 0$. Let $nh \to \infty$, $h \to 0$. Show that the mode μ^{nh} of $A_h * dF_n$ lies in the interval $[\ell_{nh}, r_{nh}]$, where ℓ_{nh} and r_{nh} are the two (asymptotically unique) solutions to the equation

$$f_o(x) = f_o(m_o) - 2 \| A_h * dF_n - f_o \|_\infty \, ,$$

and that for a suitable constant C,

$$r_{nh} - \ell_{nh} \leqslant C \left(\| A_h * dF_n - f_o \|_\infty \right)^{1/2} .$$

EXERCISES: (7.6), (7.7), (7.8), (7.9), (7.10), (7.11).

8. Afterthoughts: convex densities

The authors had a rather nice section in mind on maximum likelihood estimation of convex densities, with application to the nonparametric estimation of densities with heavy convex tails. In particular, this section was to mimic the material on monotone density estimation. Things did not work out this way.

The problem is as follows. For $1 \leqslant p \leqslant \infty$, let C_p be the constraint set

(8.1) $$C_p \overset{\text{def}}{=} \{ f \in L^p(0, \infty) : f \geqslant 0 \, , \, f \text{ convex} \} .$$

Let F_n be the empirical distribution function of $X_1, X_2, \cdots, X_n$, an iid sample of the positive random variable X with density f_o, convex on $(0, \infty)$. Finally, let $A_h dF_n$ be a boundary kernel estimator of f_o. The maximum smoothed likelihood estimator of f_o is the solution f^{nh} of

(8.2)
$$\text{minimize} \quad -\int_0^\infty A_h dF_n(x) \log f(x) \, dx + \int_0^\infty f(x) \, dx$$
$$\text{subject to} \quad f \in C_1 .$$

Note that we are minimizing over a subset of $L^1(0, \infty)$. The solution exists and is unique, by a proof similar to the monotone case.

(8.3) EXERCISE. Verify that f^{nh} is a pdf.

The analogy with monotone estimation suggests that one should look for a pool-adjacent-violators algorithm and prove the inequality

(8.4) $$\| f^{nh} - \varphi \|_1 \leqslant \| A_h dF_n - \varphi \|_1 \quad \text{for all } \varphi \in C_1 .$$

Sadly enough, the following example dashes such hopes.

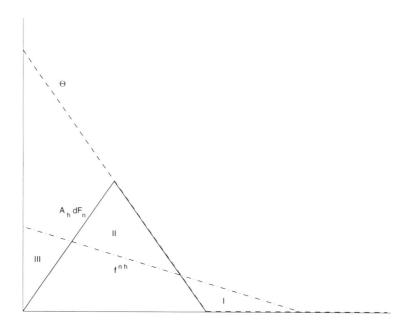

Figure 8.1. The failure of the inequality $\| f^{nh} - \theta \|_1 \leqslant \| \mathcal{A}_h dF_n - \theta \|_1$.

(8.5) COUNTEREXAMPLE. Consider the triangular pdf $\mathcal{A}_h dF_n$ in Figure 8.1. It is clear that the graph of the maximum smoothed likelihood estimator looks like the dot-dashed curve: in particular, that its relation to the vertices is as indicated. Since both the estimator and $\mathcal{A}_h dF_n$ are pdfs, we have that

$$\text{area}_I + \text{area}_{III} = \text{area}_{II} \ .$$

Now, let θ be convex and decreasing, with $\theta(x) \geqslant \mathcal{A}_h dF_n(x)$ for all x, given by the dashed curve in Figure 8.1. One verifies that

$$\| f^{nh} - \theta \|_1 = \| \mathcal{A}_h dF_n - \theta \|_1 + c \ ,$$

where $c = \text{area}_I + \text{area}_{II} - \text{area}_{III} = 2\,\text{area}_I$, so (8.4) fails.

There is another sense in which convex density estimation is different from the monotone case. Recall that for the monotone case, maximum likelihood and least-squares are equivalent, see Exercises (2.27) and (3.9).

(8.6) EXERCISE. Show that the problems

$$\text{minimize} \quad -\int_0^\infty \log f(x)\, dF_n(x) + \int_0^\infty f(x)\, dx$$

$$\text{subject to} \quad f \in \mathcal{C}_1$$

and

$$\text{minimize} \quad \int_0^\infty |f(x)|^2 \, dx - 2 \int_0^\infty f(x) \, dF_n(x)$$

$$\text{subject to} \quad f \in \mathcal{C}_1$$

are not equivalent. [Hint: Consider the case $n = 1$.]

In the absence of (8.4), one would still expect the maximum likelihood estimator f^{nh} to depend Lipschitz continuously on $A_h dF_n$. So, one would hope that there exists a universal constant C such that if φ^{nh} is the solution of (8.2) for the data $A_h dG_m$, where G_m is some (other) empirical distribution,

(8.7) $$\| f^{nh} - \varphi^{nh} \|_1 \leqslant C \, \| A_h dF_n - A_h dG_m \|_1 \, ,$$

but the authors do not know how to prove this. Thus, this section dies a peaceful death. However, empirical evidence suggests that the convex estimator is much better than the kernel estimator for small sample sizes.

For a thorough treatment of convex density estimation via least squares as well as maximum likelihood see GROENEBOOM, JONGBLOED and WELL-NER (2000).

EXERCISES: (8.3), (8.6).

9. Additional notes and comments

Ad §1: All of this chapter finds its origin with GRENANDER (1956). Regarding Theorem (1.7), we note that WANG (1995) has studied the case when $f'(x) = 0$ on subsets of its support.

An alternative to maximum smoothed likelihood estimation is to add a penalization functional, as in the GOOD (1971) approach. If only we knew how to do this.

Ad §3: That the solution of (3.2) is the derivative of the least concave majorant of Φ seems to be well known. MAMMEN (1991) and BICKEL and FAN (1996) mention it in passing.

Ad §4, 5: The contractivity results are from EGGERMONT and LARICCIA (2000). A forerunner of this result was obtained by BRUNK (1965), who in our setting, proves (1.20) for functions J of the form

(9.1) $$J(x, y) = j(x) - j(y) - j'(y) (x - y) \, , \quad x, y \in \text{domain} \, (D) \, ,$$

where j is a convex, differentiable function and $j'(y)$ is the derivative (or a subgradient) of j at y, cf. (5.10). In particular, this covers the squared L^2 norm, as well as the Kullback-Leibler distance, but other L^p norms and the Hellinger and Pearson's φ^2 distances are not included. ROBERTSON and WRIGHT (1974) prove a discrete version of (1.18) for the L^∞ norm, but wrongly interpret a remark of BRUNK (1965) as meaning that the L^p $(p > 2)$ errors are a special case of (9.1). Another forerunner of (1.19) was hinted at by MARSHALL (1970), who observed that for all concave distributions Φ,

$$\| F^{nh} - \Phi \|_\infty \leqslant \| A_h F_n - \Phi \|_\infty .$$

The general contractivity result of §5 came about after reading LIEB and LOSS (1996) regarding monotone (equimeasurable) rearrangements, being directed there by FOUGÈRES (1997). The map

$$\varphi \longmapsto \text{monotone rearrangement of } \varphi$$

has the same "localness" as the map $\varphi \longmapsto \text{lcm}(\varphi)$ expressed in Lemma (4.10), which helps explain the similar result. However, the authors misread $J(\varphi(x) - \psi(x))$ in LIEB and LOSS (1996) as $J(\varphi(x), \psi(x))$ and had to come up with the goods.

Ad §6: The present treatment follows EGGERMONT and LARICCIA (2000). In the nonsmooth setting, BIRGÉ (1987a) exhibits upper and lower bounds on the L^1 error.

Regarding Lemma (6.34) of SILVERMAN (1978) on uniform bounds for the error in the kernel estimator, we note that we only need a bound on

$$\sup \left\{ \, | \, [A_h * (dF_n - dF_o)](x) | \, : \, x \in [\, m_o - \varepsilon \, , \, m_o + \varepsilon \,] \, \right\} ,$$

for small ε (assuming m_o is unique for now). Since f_o is bounded away from 0 on the sets in question, many results in the literature apply. However, under very mild conditions on the kernel (compact support of the kernel not required) and the density, DEHEUVELS (2000a, 2000b) shows among other things that for $h \asymp n^{-\beta}$, with $0 < \beta < 1$,

$$(9.2) \qquad \lim_{n \to \infty} \frac{\displaystyle\sup_{a \leqslant x \leqslant b} \frac{nh}{2 \log(1/h)} \, \big| \, [A_h * (dF_n - dF_o)](x) \, \big|^2}{\displaystyle\sup_{a \leqslant x \leqslant b} f(x) \, \| A \|_2^2} =_{as} 1 \, ,$$

with $-\infty \leqslant a < b \leqslant +\infty$. For the whole line, the extra condition that f_o is uniformly continuous on $\mathbb{R}$ is needed. (Of course, a much more general dependence of h on n is allowed, and even weighted suprema are considered, but we shall not go into that). Recently, GINÉ and GUILLOU (2000) extended (9.2) to the multivariate case for suitable kernels with compact support.

Ad §7: In the context of isotone regression, MAMMEN (1991) shows the asymptotic equivalence of the following two procedures: (a) Construct the isotone estimator, and smooth it using kernel smoothing, say; and (b) smooth the data first and construct the isotone estimator using the smoothed data. He even shows under certain conditions that the estimator (a) is better. For density estimation, the authors would have thought the analogue of (b) to be better, but the simulations of §.8.6 do not necessarily support this. For the Kaplan-Meier estimator for censored data, the analogue of (b) seems to be better, see ACUSTA (1998).

7
Choosing the Smoothing Parameter

1. Introduction

We have now come to just about the most important aspect of nonparametric density estimation: choosing the smoothing parameter in kernel estimation that will give near-optimal results for large classes of densities. The same problem arises for maximum penalized likelihood estimation, or any other method for that matter. Actually, in kernel estimation, both the kernel and the smoothing parameter need to be chosen appropriately, whereas in maximum penalized likelihood estimation, the roughness penalization functional and the smoothing parameter are subject to choice. Choosing the roughness penalization is essentially uncharted territory. The authors do not even know whether it is an important question, and we shall not explore it. As discussed in § 4.7, in kernel estimation, the choice of the smoothing parameter is much more critical than the choice of the kernel. Why this should be so is not *a priori* clear, although deep statistical insights could be quoted.

In this chapter, we study some current methods for choosing the smoothing parameter h in the kernel estimator

$$(1.1) \qquad f^{nh}(x) = \tfrac{1}{n} \sum_{i=1}^{n} A_h(x - X_i) , \quad x \in \mathbb{R} ,$$

to wit, least-squares cross-validation and least-squares plug-in methods, the double kernel method, various L^1 plug-in methods, and a method based on a discrepancy principle. The L^1 plug-in methods require pilot estimators, of which we single out the double kernel method. We also discuss variational analogues of the plug-in methods, in which there is no need for pilot estimators. There are many more methods for smoothing parameter selection, of which methods based on the bootstrap and on spacings should be mentioned. However, these are not considered here. Instead, the reader is referred to the surveys and (simulation) comparisons in PARK and TURLACH (1992), BERLINET and DEVROYE (1994), CAO, CUEVAS and GONZÁLEZ-MANTEIGA (1994), and DEVROYE (1997). We also discuss a discrepancy principle for selecting the smoothing parameter for the GOOD estimator of § 5.2.

In the remainder of this introduction, we make some general observations regarding smoothing parameter selection procedures. Three questions are addressed. The first one addresses the issue of quantifying what the "optimal" smoothing parameter is supposed to achieve. In Chapter 8, this provides the basis for the objective comparison of various selection procedures with each other. The second question concerns the kind of densities one is likely to encounter in practice, and with the desired asymptotic behavior of the selected smoothing parameter under these circumstances. The third question deals with the fact that the smoothing parameter should be dependent on the data only, and not on the intuition or the deep statistical insight of the experimenter. One requirement is that the selected smoothing parameter should be scaling and translation invariant.

What is the purpose of selecting the smoothing parameter? From the L^1 point of view of this text, the "optimal" method would select h for each sample so as to

$$(1.2) \qquad \text{minimize} \quad \| f^{nh} - f_o \|_1 \quad \text{over } h > 0 .$$

This may be called *finite sample optimality*, which in practice must be deemed unattainable, even if we restrict f_o to "reasonable" classes of densities. A perhaps more accessible goal is to minimize the "risk", that is,

$$(1.3) \qquad \text{minimize} \quad \mathbb{E}\big[\, \| f^{nh} - f_o \|_1 \,\big] \quad \text{over } h > 0 ,$$

but even this is much too hard to achieve. The next choice is to strive for *asymptotic* optimality, that is, construct any kernel estimator $f_{n,\text{ANY}}$ that satisfies for *every* density f_o,

$$(1.4) \qquad \limsup_{n \to \infty} \frac{\| f_{n,\text{ANY}} - f_o \|_1}{\min_h \| f^{nh} - f_o \|_1} =_{\text{as}} 1 ,$$

but perhaps with expected values in both the numerator and the denominator. This is still hard to achieve. The best result until now is by DEVROYE and LUGOSI (1996), (1997): For every $\varepsilon > 0$, they can construct methods for which the limsup is $\leqslant_{\text{as}} 3 + \varepsilon$, for every density. For densities f_o satisfying the usual nonparametric assumptions, the kernel method satisfies the expected value version of (1.4), see DEVROYE (1989) and § 3 below. With L^2 norms and for bounded densities (1.4) is the famous result of STONE (1984) for least-squares cross-validation, see § 2. The practical significance of these *universal* asymptotically optimal selection procedures is limited, since typically the density to be estimated is known (assumed) to be smooth or to satisfy certain shape constraints. Morover, one is dealing with the small sample case, and small sample adjustments have to be made. This explains the plethora of techniques in the literature, of which we cover only a few.

What kind of densities are we likely to encounter in nonparametric density estimation? Naturally, we make the usual nonparametric assumptions regarding smoothness and light tails. The smoothness assumption appears to be controversial, but the authors find the following justification convincing. With small sample sizes, one can reasonably hope to recover only the global features of a density, and one must consider the small-scale features to be inaccessible. Alternatively, for small sample sizes, one cannot hope to distinguish between a very rough density and a smoothed version of it, cf. Exercise (8.1.1) in the next chapter. This is tantamount to saying that only a smoothed version of the unknown density can be estimated well. So, exaggerating a bit, in the small sample case, all densities are smooth. Regarding the tail conditions, we note that evidence regarding the (alleged) light tails is embodied in the sample and, thus, is open to inspection. However, the existence of a finite moment of order > 1 allows (roughly) the tail behavior

$$(1.5) \qquad f_o(x) = \mathcal{O}\big(|x|^{-\alpha}\big) , \quad |x| \to \infty ,$$

for some $\alpha > 2$. This should be contrasted with the two-sided exponential density, which in practice still has quite heavy tails. (Finite samples contain many outliers, see §2.5.) Thus, the nonparametric moment condition is not very stringent.

We next discuss the desired asymptotic behavior of the smoothing parameter, when considering densities that satisfy the usual nonparametric assumptions. Recall from Chapter 4 that for these densities, the asymptotically optimal smoothing parameter satisfies $h_{\mathrm{asymp}} \asymp n^{-1/5}$, with the corresponding L^1 error of order $n^{-2/5}$. So, as a minimal requirement, it is reasonable to insist that H_n, the smoothing parameter chosen, satisfies

$$(1.6) \qquad H_n \asymp_{\mathrm{as}} n^{-1/5} \quad \text{for } n \to \infty ,$$

by which we mean that

$$(1.7) \qquad 0 < \liminf_{n \to \infty} n^{1/5} H_n \leqslant \limsup_{n \to \infty} n^{1/5} H_n < \infty \quad \text{almost surely} ,$$

and that

$$(1.8) \qquad \| f^{nH_n} - f_o \|_1 =_{\mathrm{as}} \mathcal{O}\big(n^{-2/5}\big) .$$

In statements like (1.6), we usually drop the qualification $n \to \infty$, but it is intended nevertheless. Equation (1.8) is a case in point. As discussed before, one would like to achieve asymptotic or even small sample optimality, but that is outside of the scope of this text.

For a few selection procedures to be discussed later, we prove (1.6), under the usual assumptions. We also show, in §3, that (1.6) implies (1.8), provided A is the Gaussian or two-sided exponential kernel. For general kernels, the proof does not work. Then, our only alternative is to use fractional integration by parts, see §4.3, but this yields only the suboptimal rate of $n^{-2/5+\varepsilon}$, for arbitrary $\varepsilon > 0$. However, we venture to guess that here too (1.6) implies (1.8). For general kernels, the expected value version is treated in the next exercise.

(1.9) EXERCISE. Suppose H_n satisfies (1.6), i.e., for deterministic constants $0 < c < C < \infty$, assume that $H_n \in I_n$ almost surely, where

$$I_n = [h_n, h^n] \quad \text{with} \quad h_n = c n^{-1/5} \quad, \quad h^n = C n^{-1/5} .$$

Show that the bound (1.8) holds, under the usual nonparametric assumptions on f_o, and suitable conditions on A, as follows.
(a) Show that $\| A_h * (dF_n - dF_o) \|_1$ is a.e. differentiable with respect to h, and that

$$\left| \frac{d}{dh} \| A_h * (dF_n - dF_o) \|_1 \right| \leqslant h^{-1} \| B_h * (dF_n - dF_o) \|_1 ,$$

where $B(x) = -\dfrac{d}{dx} \left\{ x A(x) \right\}$ and, as usual, $B_h(x) = h^{-1} B(h^{-1} x)$.
(b) Show that

$$\| A_{H_n} * (dF_n - dF_o) \|_1 \leqslant \sup_{h \in I_n} \| A_h * (dF_n - dF_o) \|_1 ,$$

(c) and that

$$\sup_{h \in I_n} \| A_h * (dF_n - dF_o) \|_1 \leqslant$$

$$\| A_{h_n} * (dF_n - dF_o) \|_1 + \int_{h_n}^{h^n} h^{-1} \| B_h * (dF_n - dF_o) \|_1 \, dh .$$

(d) Now, take expectations in (b) and (c), and take care of the difference between $A_{H_n} * (dF_n - dF_o)$ and $A_{H_n} * dF_n - f_o$.

Finally, how should the selected h depend on the data? It goes almost without saying that the selected h should be a statistic. Equivalently, the procedure should be "rational", and not require input from the user, whatever that might mean exactly. However, how complicated a function of the data must it be? To partly answer this question, we investigate the scaling invariance of the smoothing parameter. By way of example, whether the Buffalo snowfall data are presented in inches or centimeters, one should insist that the selected h change accordingly, so that the two estimators based on the data in inches or in centimeters are "the same". A precise way of saying this is as follows. Suppose that the random variable X has density f_o. Then, for any (deterministic) $t > 0$, the random variable $t X$ has density f_t, with $f_t(x) = t^{-1} f_o(t^{-1} x)$, $t > 0$ (so $f_1 = f_o$). Let $X_1, X_2, \cdots, X_n$ be an iid sample with common pdf f. To simplify notation, let

(1.10) $$\mathbb{X}_n = (X_1, X_2, \cdots, X_n) \in \mathbb{R}^{1 \times n} ,$$

so then $t \mathbb{X}_n = (t X_1, t X_2, \cdots, t X_n)$ is an iid sample with common density f_t. Now, consider a kernel estimator of f based on the sample $\mathbb{X}_n$, written as

$$(1.11) \qquad f^{nh}(x\,;\,\mathbb{X}_n) = \tfrac{1}{nh} \sum_{i=1}^{n} A\big(h^{-1}(x - X_i)\big)\,, \qquad -\infty < x < \infty\,.$$

Then, the corresponding kernel estimator of f_t based on $t\,\mathbb{X}_n$ is

$$(1.12) \qquad f^{nh}(x\,;\,t\,\mathbb{X}_n) = \tfrac{1}{nht} \sum_{i=1}^{n} A\big(h^{-1}(x - t\,X_i)\big)\,,$$

and so,

$$(1.13) \qquad f^{nh}(x\,;\,t\,\mathbb{X}_n) = t^{-1} f^{n,ht}(t^{-1}x\,;\,\mathbb{X}_n)\,, \qquad -\infty < x < \infty\,.$$

Now, suppose a hypothetical selection procedure applied to $\mathbb{X}_n$ selects the smoothing parameter $H_{n,HYP} = H_{n,HYP}(\mathbb{X}_n)$. Denoting the corresponding kernel density estimator (1.11) as

$$(1.14) \qquad f^{n, H_{n,HYP}}(x\,;\,\mathbb{X}_n) = f_{n,HYP}(x\,;\,\mathbb{X}_n)\,,$$

the two estimators are "the same" if

$$(1.15) \qquad f_{n,HYP}(x\,;\,\mathbb{X}_n) = t^{-1} f_{n,HYP}(t^{-1}x\,;\,t\,\mathbb{X}_n)\,, \qquad -\infty < x < \infty\,.$$

This occurs if the selected $H_{n,HYP}$ is scaling invariant in the sense that

$$(1.16) \qquad H_{n,HYP}(\mathbb{X}_n) = t^{-1} H_{n,HYP}(t\,\mathbb{X}_n)\,.$$

(1.17) EXERCISE. (a) Verify (1.13) and that (1.16) implies (1.15).
(b) Show that for all $t > 0$,

$$\| f_t - f^{n,ht}(\,\cdot\,;\,t\,\mathbb{X}_n)\|_1 = \| f_o - f^{n,h}(\,\cdot\,;\,\mathbb{X}_n)\|_1\,.$$

It is easy to construct scaling-invariant smoothing parameters that even satisfy the asymptotic size information of (1.6), e.g., take H_n as

$$(1.18) \qquad H_n = c\,n^{-1/5} \big\{ \tfrac{1}{n} \sum_{i=1}^{n} |X_i - \overline{X}|^2 \big\}^{1/2}\,,$$

where c is a universal constant and $\overline{X}$ is the sample mean. Unfortunately, this hardly solves the problem. To illustrate this, consider the problem of estimating a normal density versus estimating a mixture of two normals, see Figure 1.1. The two densities in question are the normal $\phi_\sigma(x)$ with $\sigma = 0.714$ and the mixture

$$\tfrac{9}{10} \phi_{1/2}(x - 5) + \tfrac{1}{10} \phi_{1/2}(x - 7)\,,$$

which have the same standard deviations. Here, as usual, $\phi_\sigma = \sigma^{-1} \phi(\sigma^{-1}x)$. Graphs of these densities are shown in the left diagram of Figure 1.1. In the right diagram of Figure 1.1, we show the L^1 errors of the kernel estimators as functions of h, for two "typical" samples of size 100.

Two conclusions may be drawn. First, the mixture of normals being "rougher" than the normal, the (asymptotically) optimal values of the

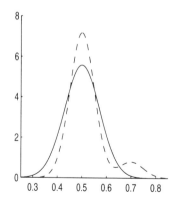 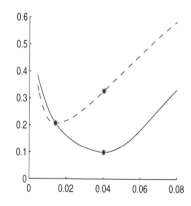

Figure 1.1. On the left, graphs of the normal density $\phi_\sigma(x-5)$ with $\sigma = 0.714$ and the mixture $\frac{9}{10}\phi_{1/2}(x-5) + \frac{1}{10}\phi_{1/2}(x-7)$, which have the same variance. On the right, graphs of the L_1 errors (as functions of h) of kernel estimators using the normal kernel, for two "typical" iid samples of size 100 from each density. The locations of both minima are indicated on both curves.

smoothing parameters are quite different in each case. Secondly, it is clear that a lot would be lost if a single value of h were used in both cases. The conclusion is that a data-driven smoothing parameter like (1.18) is not fully satisfactory in practice.

The remainder of this chapter is put together as follows. We discuss L^2 cross validation and L^2 plug-in methods in § 2. The remainder of the chapter deals with L^1 errors: For kernel estimation, the double kernel method and a compelling modification are discussed in § 3, and small sample and asymptotic plug-in methods in §§ 4 and 5. A discrepancy principle for kernel estimators and the GOOD estimator is discussed in §§ 6 and 7. Heuristic justifications of the various methods are given, as well as some proofs regarding asymptotic rates of convergence, but only when it can be done by the methods of Chapters 4 and 5.

EXERCISES : (1.9), (1.17).

2. Least-squares cross-validation and plug-in methods

In this section, we take the L^2 point of view, and study chosing h so as to

(2.1)   minimize $\| f^{nh} - f_o \|_2^2$ subject to $h > 0$.

The methods discussed are L^2 cross validation and plug-in methods. In the least-squares cross-validation method, one minimizes an *unbiased* estimator of $\| f^{nh} - f_o \|_2^2$. The idea was first considered by RUDEMO (1982) and BOWMAN (1984), and furthered by HALL (1983) and STONE (1984). See also WAHBA (1981). The second type of methods to be considered are plug-in methods, based on minimizing asymptotic expressions for the expected squared L^2 error. As discussed earlier, the drawback of these approaches is that they deal with the L^2 error, which has no obvious interpretation in the context of estimating densities.

In the cross-validation approach, one first derives an unbiased estimator of $\| f^{nh} - f_o \|_2^2$ by observing that

$$(2.2) \qquad \| f^{nh} - f_o \|_2^2 = \| f^{nh} \|_2^2 + \| f_o \|_2^2 - 2 \int_{\mathbb{R}} f^{nh}(x) \, dF_o(x) .$$

Now, the second term on the right is independent of h, and since ultimately we wish to minimize over h, only the last term needs to be estimated. It was written in a rather suggestive manner: A "natural" estimator for it is

$$(2.3) \qquad \int_{\mathbb{R}} f^{nh}(x) \, dF_o(x) \approx \int_{\mathbb{R}} f^{nh}(x) \, dF_n(x) = \frac{1}{n} \sum_{i=1}^{n} f^{nh}(X_i) .$$

However, this turns out to be a *biased* estimator of $\int_{\mathbb{R}} f^{nh} dF_o$. This may be traced to the fact that

$$(2.4) \qquad f^{nh}(X_i) = (nh)^{-1} A(0) + \frac{1}{n} \sum_{j \neq i} A_h(X_i - X_j) ,$$

and it is clear (?) that the first term does not "belong". Thus, the biasedness may be fixed by using the approximation

$$(2.5) \qquad \int_{\mathbb{R}} f^{nh}(x) \, dF_o(x) \approx \frac{1}{n} \sum_{i=1}^{n} f^{nh}_{(i)}(X_i) ,$$

where

$$(2.6) \qquad f^{nh}_{(i)}(x) = \frac{1}{n-1} \sum_{j \neq i} A_h(x - X_j) , \quad x \in \mathbb{R} .$$

One interpretation of (2.6) is that we are estimating $f_o(X_i)$ by a kernel estimator based on the data with X_i omitted. For this reason, this method goes by the name of the "leave-one-out method", but "cross validation method" is the standard designation.

To summarize, if we set

$$(2.7) \qquad CV(h) = \| f^{nh} \|_2^2 - \frac{2}{n} \sum_{i=1}^{n} f^{nh}_{(i)}(X_i) ,$$

then

$$(2.8) \qquad \mathbb{E}\big[CV(h) \big] = \| f^{nh} - f_o \|_2^2 - \| f_o \|_2^2 .$$

(2.9) EXERCISE. (a) Show that for $i \neq j$,

$$\mathbb{E}[A_h(X_i - X_j)] = \int_{\mathbb{R}} f_o(x) \, [A_h * f_o](x) \, dx .$$

(b) Verify that the estimator of (2.3) is a biased estimator of $\int_{\mathbb{R}} f^{nh} \, dF_o$, in general.

(c) Verify (2.8).

(2.10) EXERCISE. Verify that

$$CV(h) = (nh)^{-1} \| A \|_2^2 \; - \; [n(n-1)]^{-1} \sum_{i \neq j} B_h(X_i - X_j) \; -$$

$$[n^2(n-1)]^{-1} \sum_{i \neq j} [A_h * A_h](X_i - X_j) ,$$

where $B_h(x) = h^{-1} B(h^{-1} x)$ and $B = 2A - A * A$. Here, the summation over $i \neq j$ is over all i, j, with $i = 1, 2, \cdots, n$ and $j = 1, 2, \cdots, n$, but $i \neq j$.

We are thus lead to the least-squares cross-validation method.

(2.11) In the least-squares cross-validation method, the smoothing parameter is chosen so as to

minimize $CV(h)$ over $h > 0$.

The h so chosen is denoted by $H_{n,cv}$ and the corresponding kernel estimator by $f_{n,cv}$.

We shall not attempt to analyze this method and merely state its asymptotic optimality for L^2 errors.

(2.12) THEOREM. [STONE (1984)] *Let A be a symmetric, Hölder continuous kernel with compact support and integral equal to one. If f_o is a bounded density, then*

$$\limsup_{n \to \infty} \frac{\| f_{n,cv} - f_o \|_2}{\inf_h \| f^{nh} - f_o \|_2} =_{as} 1 .$$

An amazing feature of this theorem is the almost complete lack of conditions on the density f_o. It even holds in the multivariate case, if in addition all the one-dimensional marginals of f_o are bounded. Note also that higher order kernels are allowed. On the negative side, least-squares cross validation seems to have practical drawbacks. The smoothing parameter $H_{n,cv}$ selected seems to show too much variability and too large a negative correlation with the optimal smoothing parameter $h_{n,opt}$. The various fixes based on slight modifications of $CV(h)$ seem to have their own drawbacks, see SCOTT and TERRELL (1987), HALL, MARRON and PARK (1992), and HALL and JOHNSTONE (1992). Another issue is the scaling invariance of $H_{n,cv}$, prompted by the lack of scaling invariance of the L^2

distance. However, the above treatment can be repeated for

$$\frac{\|f^{nh} - f_o\|_2}{\|f_o\|_2} \, ,$$

and this expression is scaling invariant. Because the denominator is independent of h, minimizing the numerator over $h > 0$ is the same as minimizing the quotient. So the estimator is indeed scaling invariant.

(2.13) EXERCISE. Verify that $H_{n,cv}$ is scaling invariant in the sense of (1.16).

We now consider plug-in methods. In the asymptotic L^2 plug-in methods, the squared error $\|f^{nh} - f_o\|_2^2$ is estimated by first replacing it by the asymptotic expansion of its expected value

(2.14) $\mathbb{E}[\|f^{nh} - f_o\|_2^2] =$
$$\tfrac{1}{4} h^4 \sigma^4(A) \| (f_o)'' \|_2^2 + (nh)^{-1} \| A \|_2^2 + o(h^4 + (nh)^{-1}) \, ,$$

for $nh \to \infty$, $h \to 0$. Naturally, the assumptions required are that

(2.15) $$f_o \in W^{2,2}(\mathbb{R}) \, ,$$

i.e., f_o and its second derivative both belong to $L^2(\mathbb{R})$, and that

(2.16) the kernel A is a symmetric, square integrable pdf, with

$$\sigma(A) = \left\{ \int_{\mathbb{R}} x^2 A(x) \, dx \right\}^{1/2} < \infty \, .$$

Thus, a theoretically interesting choice for the smoothing parameter h is the one that minimizes

$$\tfrac{1}{4} h^4 \sigma^4(A) \| (f_o)'' \|_2^2 + (nh)^{-1} \| A \|_2^2$$

over $h > 0$, and this is given by

(2.17) $$h_{\text{asymp}} = n^{-1/5} \, r(A) \, \varrho(f_o) \, ,$$

where

(2.18) $$r(A) = \left\{ \frac{\| A \|_2}{\sigma^2(A)} \right\}^{2/5} \, , \qquad \varrho(f_o) = \left\{ \frac{1}{\| (f_o)'' \|_2} \right\}^{2/5} \, .$$

Of course, this hardly solves the problem: The factor $\varrho(f_o)$ or $\| (f_o)'' \|_2^2$ must be estimated from the data.

One of the first such estimators was proposed by DEHEUVELS (1977). His idea was to use a *parametric* estimator for f_o to estimate $\| (f_o)'' \|_2$ by pretending that f_o may be approximated well by some element from a specific parametric family. This is somewhat at odds with the nonparametric approach to density estimation, but never mind. Thus, consider a scaling family of densities $\gamma(\cdot \, ; \theta) = \theta^{-1} g(\theta^{-1} x)$ for some known density g, and let θ_n be an estimator of the "true" scale parameter θ_o. The estimator of h_{asymp} is then

$$(2.19) \qquad H_{n,Deh} \overset{\text{def}}{=} n^{-1/5}\, r(A)\, \varrho(\gamma(\,\cdot\,;\theta_n))\ ,$$

and it is an exercise to show that

$$(2.20) \qquad H_{n,Deh} = \theta_n\, r(A)\, \varrho(g)\, n^{-1/5}\ .$$

Moreover, if θ_n is scaling invariant, then so is $H_{n,Deh}$. Precisely, as usual, let $\mathbb{X}_n = (X_1, X_2, \cdots, X_n)$. Then, for all $t > 0$,

$$(2.21) \qquad \begin{aligned} &\text{If} \quad \theta_n(\mathbb{X}_n) = t^{-1}\,\theta_n(t\,\mathbb{X}_n), \quad \text{then} \\ &H_{n,Deh}(\mathbb{X}_n) = t^{-1}\,H_{n,Deh}(t\,\mathbb{X}_n)\ . \end{aligned}$$

(2.22) EXERCISE. Verify (2.20) and (2.21).

If θ_n is selected properly, then typically, something more can be said. That is, whether $\gamma(\,\cdot\,;\theta)$ is the correct parametric family or not, usually there exists a $\theta_o = \theta(f_o)$ such that

$$(2.23) \qquad \theta_n - \theta_o =_{\text{as}} \mathcal{O}\big((n^{-1}\log\log n)^{1/2}\big)\ ,$$

and hence,

$$(2.24) \qquad H_{n,Deh} =_{\text{as}} \theta_n\, r(A)\, \varrho(g)\, n^{-1/5} + \mathcal{O}\big(n^{-7/10}(\log\log n)^{1/2}\big)\ .$$

Thus, as long as (2.23) holds, $H_{n,Deh}$ passes the minimum requirements imposed on a smoothing parameter. DEHEUVELS (1977) proposed this with A the Epanechnikov kernel, g the standard normal, and $\theta_n = s_n$, the sample standard deviation. With these choices, (2.20)–(2.23) give the asymptotic plug-in-normal $H_{n,Deh}$ as

$$(2.25) \qquad H_{n,Deh} = 0.7443\, s_n\, n^{-1/5}\ .$$

(2.26) EXERCISE. Verify (2.25).

It is of course not surprising that we were able to pinpoint what is involved in establishing the a.s. asymptotic behavior of $H_{n,Deh}$, because everything is based on asymptotic considerations. But what about its small sample behavior? Then, everything depends on the appropriateness of the parametric family $\gamma(\,\cdot\,;\theta)$ and the scale estimator θ_n. Thus, in the present context, it seems more natural to use a nonparametric estimator for $\|(f_o)''\|_2^2$. Such a procedure was first proposed by WOODROOFE (1970), see also DEHEUVELS and HOMINAL (1980), and has resulted is a sizable literature. The development here is based on a long series of papers culminating in SHEATHER and JONES (1991). An obvious estimator for $\|(f_o)''\|_2^2$ is $\|(\varphi^{n\lambda})''\|_2^2$, where $\varphi^{n\lambda} = B_\lambda * dF_n$ is a kernel estimator for f_o based on a smooth symmetric kernel B, and we must select the new smoothing parameter λ! The precise assumptions on B are that it satisfies (2.16) and that $B'' \in L^1(\mathbb{R}) \cap L^2(\mathbb{R})$. Unfortunately, the above estimator is biased, as we now show. With

$$(2.27) \qquad \varphi^{n\lambda}(x) = \frac{1}{n} \sum_{i=1}^{n} B_\lambda(x - X_i) , \quad x \in \mathbb{R} ,$$

one has that

$$(2.28) \qquad \| (\varphi^{n\lambda})'' \|_2^2 = \frac{1}{n^2 \lambda^4} \sum_{i,j=1}^{n} C_\lambda(X_i - X_j) ,$$

in which $C = (B * B)^{(4)}$ (fourth derivative).

(2.29) EXERCISE. Derive (2.28).

One verifies that

$$(2.30) \qquad \mathbb{E}[\| (\varphi^{n\lambda})'' \|_2^2] = n^{-1}\lambda^{-5} \| B'' \|_2^2 + \frac{n-1}{n} \| B_\lambda * (f_o)'' \|_2^2 ,$$

provided $f_o \in W^{4,2}(\mathbb{R})$. Now, one could argue that the first term on the right does not belong and view $\| (\varphi^{n\lambda})'' \|_2^2 - n^{-1}\lambda^{-5} \| B'' \|_2^2$ as an estimator for $\| (f_o)'' \|_2^2$, but this is not what is done. Instead, one observes that

$$(2.31) \qquad \| B_\lambda * (f_o)'' \|_2^2 = \| (f_o)'' \|_2^2 - \tfrac{1}{2} \lambda^2 \sigma^2(B) \| (f_o)^{(3)} \|_2^2 + \mathcal{O}(\lambda^4) ,$$

and so

$$(2.32) \qquad \mathbb{E}[\frac{n}{n-1} \| (\varphi^{n\lambda})'' \|_2^2 - \| (f_o)'' \|_2^2] =$$
$$(n-1)^{-1} \lambda^{-5} \| B'' \|_2^2 - \tfrac{1}{2} \lambda^2 \sigma^2(B) \| (f_o)^{(3)} \|_2^2 + \mathcal{O}(\lambda^4) .$$

So, λ should/could be chosen so as to set the bias equal to 0. Omitting the $\mathcal{O}(\lambda^4)$ term in (2.32) and ignoring the difference between $n-1$ and n, the bias vanishes for $\lambda = \lambda_{\text{asymp}}$, given by

$$(2.33) \qquad \lambda_{\text{asymp}} = n^{-1/7} \left\{ \frac{2 \| B'' \|_2^2}{\sigma^2(B) \| (f_o)^{(3)} \|_2^2} \right\}^{1/7} .$$

Since the natural question now is how to estimate $\| (f_o)^{(3)} \|_2^2$, it seems that little progress has been made. However, at this point, SHEATHER and JONES (1991) observe that as functions of n, the smoothing parameters h_{asymp} of (2.17) and λ_{asymp} are asymptotically related by

$$\lambda_{\text{asymp}} \asymp \left(h_{\text{asymp}} \right)^{5/7} , \quad n \to \infty .$$

In particular, taking $A = B$ for simplicity,

$$(2.34) \qquad \lambda_{\text{asymp}} = c(B) \left\{ \frac{\| (f_o)'' \|_2}{\| (f_o)^{(3)} \|_2} \right\}^{2/7} \left(h_{\text{asymp}} \right)^{5/7} ,$$

with

$$(2.35) \qquad c(B) = \left\{ \frac{\sqrt{2}\, \sigma(B) \, \|\, B'' \,\|_2}{\|\, B \,\|_2} \right\}^{2/7}.$$

Now, to make (2.34) practicable, the expressions $\|\, (f_o)''\, \|_2^2$ and $\|\, (f_o)^{(3)}\, \|_2^2$ are estimated by the double sums

$$(2.36) \qquad S_4(\mu) = \frac{1}{n(n-1)\mu^5} \sum_{i,j=1}^{n} \phi^{(4)}\big(\mu^{-1}(X_i - X_j)\big)$$

and

$$(2.37) \qquad S_6(\nu) = \frac{1}{n(n-1)\nu^7} \sum_{i,j=1}^{n} \phi^{(6)}\big(\nu^{-1}(X_i - X_j)\big)\,,$$

with ϕ the normal kernel. The smoothing parameters μ and ν are chosen so as to obtain optimal estimators if f_o is a normal density. The hope is that the normal parametric model is sufficiently accurate for the purpose. The net result is that

$$(2.38) \qquad \mu = 0.920\ q_n\ n^{-1/7}\ ,\qquad \nu = 0.912\ q_n\ n^{-1/9}\ ,$$

where q_n is the sample interquartile range

$$(2.39)\qquad
\begin{aligned}
q_n &= \big(X_{[3n/4],n} - X_{[n/4],n} \big) \big/ \big(\Phi^{\text{inv}}(\tfrac{3}{4}) - \Phi^{\text{inv}}(\tfrac{1}{4}) \big)\\
&\doteq \big(X_{[3n/4],n} - X_{[n/4],n} \big) \big/ 1.35\ .
\end{aligned}$$

(Here, Φ is the distribution of the standard Gaussian density.)

So the asymptotic relationship (2.34) is replaced by the concrete one

$$(2.40) \qquad \lambda(h) = c(B) \left\{ \frac{S_4(\mu)}{S_6(\nu)} \right\}^{1/7} h^{5/7}\ .$$

Now, SHEATHER and JONES (1991) obtain their estimator $H_{n,SJ}$ of the smoothing parameter as the solution to

$$(2.41) \qquad h = n^{-1/5}\ r(B)\, \varrho\big(\varphi^{n,\lambda(h)}\big)\,,$$

with $\varphi^{n\lambda}$ as in (2.27). We refer to the solution $H_{n,SJ}$ of (2.40)–(2.41) as *the* SHEATHER-JONES estimator, although they have a number of estimators to their credit.

It is not at all obvious, but in practice the equations (2.40)–(2.41) have a unique solution, which is easily found using safe versions of the Secant method. It is possible to derive (quite amazing) bounds on $H_{n,SJ} - h_{\text{asymp}}$, but we shall not do so. WAND and JONES (1995) contains all of the details.

(2.42) EXERCISE. Verify (2.31). [Hint: Use

$$\int_{\mathbb{R}} f^{(2)}(x)\, f^{(4)}(x)\, dx = -\, \| (f_o)^{(3)} \|_2^2 \, .]$$

EXERCISES: (2.9), (2.10), (2.13), (2.22), (2.26), (2.29), (2.42).

3. The double kernel method

We now return to the L^1 point of view. It seems to make eminent sense to choose the smoothing parameter as the solution to

$$(3.1) \qquad\qquad \text{minimize} \quad \| f^{nh} - f_o \|_1 \quad \text{over } h > 0 ,$$

but, of course, the loss $\| f^{nh} - f_o \|_1$ must be estimated first. The niceties associated with the L^2 norm, especially (2.2), do not apply to the L^1 norm, so the goal of getting an *unbiased* estimator of $\| f^{nh} - f_o \|_1$ seems unattainable. There appears to be no other choice but to use another estimator of f_o. Thus, let B be some other kernel, and consider

$$(3.2) \qquad \varphi^{nh}(x) = B_h * dF_n(x) = \tfrac{1}{n} \sum_{i=1}^{n} B_h(x - X_i) , \qquad x \in \mathbb{R} ,$$

and suppose that φ^{nh} is much more accurate than f^{nh}. In view of § 4.7 on optimal kernels, if f_o is very smooth, one could take

$$(3.3) \qquad\qquad B = 2A - A * A .$$

So, assuming that

$$(3.4) \qquad\qquad \| \varphi^{nh} - f_o \|_1 \ll \| f^{nh} - f_o \|_1 ,$$

then

$$(3.5) \qquad\qquad \| f^{nh} - \varphi^{nh} \|_1 \approx \| f^{nh} - f_o \|_1 ,$$

and one would expect to do well by minimizing $\| f^{nh} - \varphi^{nh} \|_1$. This is the "double kernel method" for choosing the smoothing parameter, due to DEVROYE (1989).

So, for a suitable pair of kernels A and B,

(3.6)   in the double kernel method, the smoothing parameter
      is chosen so as to

$$\text{minimize} \quad DBL(h) \overset{\text{def}}{=} \| (A_h - B_h) * dF_n \|_1 \quad \text{over } h > 0 .$$

The resulting h is denoted as $H_{n,DBL}$ and the associated kernel estimator $f^{n,H_{n,DBL}}$ as $f_{n,DBL}$.

For the double kernel method to work, and to be able to say something about it, the kernels A and B must satisfy some minimal conditions. The various assumptions needed at one point or another are as follows.

(3.7) A and B are bounded, symmetric about 0,

have compact support, and

$$\int_{\mathbb{R}} A(x)\,dx = \int_{\mathbb{R}} B(x)\,dx = 1 \ .$$

Moreover, it is assumed that A and B are distinct, in the sense that there exists an $\omega_o > 0$ such that

(3.8) $\widehat{A}(\omega) \neq \widehat{B}(\omega)\ ,\quad 0 < |\omega| \leqslant \omega_o\ ,$

where $\widehat{A}$ and $\widehat{B}$ are the Fourier transforms of A and B. Finally, there should exist a constant c such that for all $h > 1$,

(3.9) $\| A - A_h \|_1 \leqslant c\,|\,1 - h\,|\ ,\quad \| B - B_h \|_1 \leqslant c\,|\,1 - h\,|\ .$

Some examples of pairs of kernels satisfying these conditions are A the uniform kernel or the Epanechnikov kernel, and $B = 2\,A - A * A$ (without proof).

What can we say about the double kernel method? The first concern is whether $H_{n,DBL}$ exists and is scaling invariant. Existence would be no problem, if we were to allow $H_{n,DBL} = 0$ or $= +\infty$. But, in fact, things are much nicer than that.

(3.10) EXERCISE. Let A and B satisfy (3.7). Show that for every realization of $X_1, X_2, \cdots, X_n$,
(a) $\| (A_h - B_h) * dF_n \|_1 \leqslant \| A - B \|_1\ ,$
(b) $\| (A_h - B_h) * dF_n \|_1 \longrightarrow \| A - B \|_1$ for $h \to 0$, as well as for $h \to \infty$,
(c) $\| (A_h - B_h) * dF_n \|_1$ is continuous in h, and
(d)   conclude that $H_{n,DBL}$ exists.

(3.11) EXERCISE. Show that $H_{n,DBL}$ is scale invariant, in the sense of (1.16).

The second worry is whether the double kernel method gives consistent estimators. Amazingly, it does so without any assumptions on the density.

(3.12) THEOREM. [DEVROYE (1989)] Under the assumptions (3.7), (3.8), and (3.9) on the kernels, for every density f_o,

$$\lim_{n \to \infty} \| f_{n,DBL} - f_o \|_1 =_{as} 0 \ .$$

The assumptions on the kernel can be relaxed somewhat. Moreover, the theorem holds for suitable higher order kernels. We shall not go into the details.

As the lack of smoothness and tail assumptions on f_o indicates, we are in no position to prove this. The next theorem states that under the usual smoothness and tail conditions on f_o, one gets the optimal rate of convergence, but even this is out of our reach.

(3.13) THEOREM. [DEVROYE (1989)] *Under the assumptions (3.7), (3.8), and (3.9) on the kernels, if $f_o \in W^{2,1}(\mathbb{R})$ and has a moment of order > 1, and*

$$\varepsilon \stackrel{\text{def}}{=} \{ 4 \| B \|_2 / \| A \|_2 \}^{1/2} < 1 \ ,$$

then

$$\limsup_{n \to \infty} \frac{\mathbb{E}[\, \| f_{n,DBL} - f_o \|_1 \,]}{\inf_h \mathbb{E}[\, \| f^{nh} - f_o \|_1 \,]} \leq_{\text{as}} \frac{1 + \varepsilon}{1 - \varepsilon} \ .$$

Note that one can make ε arbitrarily close to 0 by choosing the kernels A and B appropriately.

As lamented above, we cannot prove either theorem with the methods that were explored in Chapter 4. We are not even able to prove that

(3.14) $$H_{n,DBL} \asymp_{\text{as}} n^{-1/5} \ .$$

However, one side is easy, sort of.

(3.15) EXERCISE. Let $\varepsilon > 0$ be arbitrary. Show that if $f_o \in W^{2,1}(\mathbb{R})$ and $\sqrt{f_o} \in L^1(\mathbb{R})$, then $H_{n,DBL} =_{\text{as}} \mathcal{O}(\, n^{-1/5+\varepsilon}\,)$.

So, in view of Theorems (3.12) and (3.13), the double kernel method asymptotically is all peaches. Unfortunately, for small sample sizes, things are not as clear. By way of example, note that for the choice (3.3),

(3.16) $$f^{nh} - \varphi^{nh} = A_h * A_h * dF_n - A_h * dF_n \ ,$$

which is just the difference between "any" two kernel estimators, presumably of comparable accuracy: One would not expect either one to be much more accurate than the other. So the original motivation (3.4)–(3.5) may not be quite relevant. See also Exercise (3.48)(a) below. Actually, under the usual nonparametric conditions, asymptotically the optimal h for $A_h * dF_n$ satisfies $h \asymp n^{-1/5}$, and the optimal h for $B_h * dF_n$ satisfies $h \asymp n^{-1/9}$, provided $f_o \in W^{4,1}(\mathbb{R})$. Thus, it seems that drastically different scales are required for A and B. In fact, it suggests that B_h should be replaced by

(3.17) $$B_h = 2 A_\lambda - A_\lambda * A_\lambda \ ,$$

with $\lambda \asymp h^{5/9}$, if only one knew how to do this in a data-driven way. Although changing the scale of the second kernel just a little is enough to get the $(1+\varepsilon)/(1-\varepsilon)$ bound of Theorem (3.13), the above seems to explain the careful tweaking of the scales of A and B necessary to make the double kernel method work well in the small sample case, see Chapter 8.

Following DEVROYE (1989), in the simulations of Chapter 8, the following double kernel methods are considered. With A the Epanechnikov kernel, the second kernel is taken to be

(3.18) $B = L_s$, with $s \in \{ 2,\ 2.4,\ 2.88,\ 3 \}$ (the tweaking parameter) ,

where L is the Berlinet-Devroye kernel given by

$$(3.19) \qquad L(x) = \begin{cases} \frac{1}{4} \left(7 - 31\, x^2 \right) & , \quad |x| \leqslant \frac{1}{2} , \\ \frac{1}{4} \left(x^2 - 1 \right) & , \quad \frac{1}{2} < |x| \leqslant 1 , \end{cases}$$

and $= 0$ otherwise. There are good reasons for chosing this kernel L, which we shall not discuss. However, it is an exercise to show that L is a fourth-order kernel.

(3.20) EXERCISE. Investigate the theoretical and practical virtues, if any, of the following method for choosing h:

$$\text{minimize} \quad \| f^{nh} - \varphi^{n,h^{5/9}} \|_1 \quad \text{over } h > 0 .$$

Here, φ^{nh} is given by (3.2), and B by (3.3). [Hint: Use the methods below.]

It is perhaps worthwhile to pinpoint the difficulties in establishing (3.14). Following the martingale results of §4.4, we have

(3.21) $\| (A_h - B_h) * dF_n \|_1 =_{\text{as}}$
$$\mathbb{E}[\| (A_h - B_h) * dF_n \|_1] + \mathcal{O}\left((n^{-1} \log n)^{1/2} \right) .$$

The first difficulty is that the above holds for deterministic h only, but we pretend it holds for random h also. Now, by the triangle inequality,

(3.22) $\| (A_h - B_h) * dF_n \|_1 \geqslant$
$$\| (A_h - B_h) * f_o \|_1 - \| (A_h - B_h) * (dF_n - dF_o) \|_1$$
$$\geqslant_{\text{as}} c_1 h^2 - c_2 (nh)^{-1/2} .$$

This statement too holds for deterministic h only. Pretending that it holds for $h = H_{n,DBL}$ would give

$$c_1 (H_{n,DBL})^2 \leqslant_{\text{as}} c_3\, n^{-2/5} + c_2\, (nH_{n,DBL})^{-1/2} ,$$

which implies that there exists a constant c_4 such that

(3.23) $$H_{n,DBL} \leqslant_{\text{as}} c_4\, n^{-1/5} .$$

For the lower bound on $H_{n,DBL}$, we have similar to (3.22)

(3.24) $\| (A_h - B_h) * dF_n \|_1 \geqslant$
$$\| (A_h - B_h) * (dF_n - dF_o) \|_1 - \| (A_h - B_h) * f_o \|_1 ,$$

and we would be in great shape if

(3.25) $\| (A_h - B_h) * (dF_n - dF_o) \|_1 \geqslant_{as} c_5 (1 + nh)^{-1/2}$

for all $h \leqslant h_n \asymp n^{-1/5}$. For deterministically, smoothly varying h, this definitely holds, but it needs to hold for random h. Proceeding fearlessly, (3.25) gives the inequality

$$c_3 n^{-2/5} + c_1 (H_{n,DBL})^2 \leqslant_{as} c_5 (n H_{n,DBL})^{-1/2} ,$$

and this indeed implies that there exists a constant $c_6 > 0$ such that

(3.26) $H_{n,DBL} \geqslant_{as} c_6 n^{-1/5} .$

Thus, (3.14) would hold.

In the above, it is possible to finesse one's way around the randomness of $H_{n,DBL}$, but showing (3.25) is amazingly hard. In fact, it is not obvious that it is true. However, following the theme of this text (when in doubt, penalize), if we add a roughness penalization to the $DBL(h)$ function, then we ought to be able to show (the analogue) of (3.14). So,

(3.27) in the perverted double kernel method, the smoothing parameter
is taken to be the smallest solution to

minimize $PER(h) \overset{\text{def}}{=} \| (A_h - B_h) * dF_n \|_1 + (nh)^{-1/2} \mathrm{var}_n(B; h) ,$

where $\mathrm{var}_n(B; h)$ is defined as

$$\mathrm{var}_n(B; h) = \| \sqrt{(B^2)_h * dF_n - h (B_h * dF_n)^2} \|_1 .$$

The h so selected is denoted by $H_{n,PER}$ and the corresponding
kernel estimator by $f_{n,PER}$. Recall that $B = 2A - A * A$.

It should be noted that $(B_h)^2 = h^{-1} (B^2)_h$, so that $\mathrm{var}_n(B; h)$ is more like a (normalized) standard deviation than a variance. Notationally, we are safe by calling it a "variation" term.

A simple, but honest, motivation for this method is that it allows a proof of the analogue of (3.14) along the lines (3.21)–(3.26). In particular, there is no need to show (3.25), since there is already a term $(nh)^{-1/2}$ present. A more ambitious motivation is as follows. Note that

(3.28) $\| f^{nh} - f_o \|_1 \leqslant \| (A_h - B_h) * dF_n \|_1 +$
$$\| B_h * f_o - f_o \|_1 + \| B_h * (dF_n - dF_o) \|_1 .$$

Since B is a fourth-order kernel, if f_o is smooth enough, then

$$\| B_h * f_o - f_o \|_1 = \mathcal{O}(h^4) ,$$

which is much smaller than the bias in $A_h * dF_n$, and hence, we may ignore this term. Continuing, the third term on the right of (3.28) behaves pretty much like its expectation, see §4.4, and

(3.29) $\mathbb{E}[\| B_h * (dF_n - dF_o) \|_1] \leqslant (nh)^{-1/2} \mathrm{var}_o(B; h) ,$

with

(3.30) $\operatorname{var}_o(B; h) = \| \sqrt{(B^2)_h * dF_o - h\,(\,B_h * dF_o)^2}\,\|_1$.

Of course, $\operatorname{var}_n(B; h)$ is the obvious estimator of $\operatorname{var}_o(B; h)$. Now, if the above inequalities are just about equalities, then $PER(h)$ should be a good estimator of $\| f^{nh} - f_o \|_1$, and so minimizing $PER(h)$ should yield a good smoothing parameter.

So much for heuristics. We now prove that the modified kernel method works in the sense of (1.6) and (1.8).

(3.31) THEOREM. *Suppose* $f_o \in W^{2,1}(\mathbb{R})$ *and has a finite moment of order* $\lambda > 1$. *If* A *is the standard Gaussian density and* $B = 2A - A * A$, *then* $H_{n,PER} \asymp_{as} n^{-1/5}$ *and* $\| f_{n,PER} - f_o \|_1 =_{as} \mathcal{O}(n^{-2/5})$.

The crucial property to be used is that the normal density satisfies

(3.32) $A_h * A_\lambda = A_\sigma$,   with $\sigma^2 = h^2 + \lambda^2$,

which implies that expressions like

(3.33) $\| A_h * (dF_n - dF_o) \|_1$ and $\| \sqrt{(A^2)_h * dF_n}\,\|_1$

are monotone functions of h, see Exercises (3.48) and (3.51). For arbitrary kernels, this does not work, and we are up the creek without a paddle.

The first step in the study of $H_{n,PER}$ is to determine an upper bound for the minimum of $PER(h)$ over $h > 0$.

(3.34) LEMMA. *Under the assumptions of Theorem* (3.31), *there exists a constant* K *depending on* f_o *only such that*

$$\limsup_{n \to \infty} n^{2/5} \inf_h PER(h) \leqslant_{as} K .$$

Lower and upper bounds on $H_{n,PER}$ are then provided by the following two lemmas, which, combined, say that the infimum of $PER(h)$ occurs a.s. on an interval $\delta\,n^{-1/5} \leqslant h \leqslant \gamma\,n^{-1/5}$, for suitable (deterministic) δ and γ.

(3.35) LEMMA. *Under the assumptions of Theorem* (3.31), *for a large enough constant* γ, *depending on* f_o *only,*

$$\liminf_{n \to \infty} n^{2/5} \inf \big\{ PER(h) : h \geqslant \gamma n^{-1/5} \big\} >_{as} K .$$

(3.36) LEMMA. *Under the assumptions of Theorem* (3.31), *for a small enough positive constant* δ, *depending on* f_o *only,*

$$\liminf_{n \to \infty} n^{2/5} \inf \big\{ PER(h) : h \leqslant \delta n^{-1/5} \big\} >_{as} K .$$

As warning to the reader, in the proofs to follow, a careful distinction must be made between $(A_h)^2$ and $(A^2)_h$, which refer to the kernels

$$\left(h^{-1} A(h^{-1}x) \right)^2 \quad \text{and} \quad h^{-1}\left(A(h^{-1}x) \right)^2, \quad \text{respectively}.$$

PROOF OF LEMMA (3.34). Of the three, this proof is the simplest, since it suffices to stick to a deterministic choice of h. With $C = A - B$, we have

$$\| C_h * dF_n \|_1 \leqslant \| C_h * dF_o \|_1 + \| C_h * (dF_n - dF_o) \|_1$$

and

$$\| C_h*(dF_n - dF_o) \|_1 \leqslant$$
$$\leqslant \| A_h * A_h * (dF_n - dF_o) \|_1 + \| A_h * (dF_n - dF_o) \|_1$$
$$\leqslant 2 \| A_h * (dF_n - dF_o) \|_1 ,$$

where we used Young's inequality in the form

$$\| A_h * A_h * (dF_n - dF_o) \|_1 \leqslant \| A_h \|_1 \| A_h * (dF_n - dF_o) \|_1 .$$

It follows that

$$(3.37) \quad PER(h) \leqslant \| C_h*f_o \|_1 + 2 \| A_h*(dF_n - dF_o) \|_1 + (nh)^{-1/2} \, \mathrm{var}_n(B;h) .$$

An appeal to Exercise (3.49) gives the bound

$$\| C_h * f_o \|_1 \leqslant c h^2 ,$$

for a suitable constant c, depending on f_o.

Next, for suitable constants c_1 and c_2,

$$(3.38) \qquad \mathrm{var}_n(B;h) \leqslant \| \sqrt{(B^2)_h * dF_n} \|_1 \leqslant_{\mathrm{as}} c_1 + c_2 \, h^{1/(2\lambda)} ,$$

the last inequality by Exercise (4.2.26), since f_o has a finite moment of order $\lambda > 1$ and the density $B^2/\| B \|_2^2$ has finite *exponential* moments. Finally, the term $\| A_h * (dF_n - dF_o) \|_1$ has been adequately treated in §4.4: For $h \asymp n^{-\beta}$ (deterministic), with $0 < \beta < 1$,

$$(3.39) \quad \| A_h * (dF_n - dF_o) \|_1 =_{\mathrm{as}}$$
$$\mathbb{E}[\| A_h * (dF_n - dF_o) \|_1] + \mathcal{O}\left((n^{-1} \log n)^{1/2} \right) .$$

Moreover,

$$(3.40) \qquad \begin{aligned} \mathbb{E}[\| A_h * (dF_n - dF_o) \|_1] &\leqslant (nh)^{-1/2} \, \mathrm{var}_o(A;h) \\ &\leqslant c_4 \, (nh)^{-1/2} + c_5 \, n^{-1/2} . \end{aligned}$$

Putting everything together gives for $h \asymp n^{-\beta}$ and (other) suitable constants c_i,

$$PER(h) \leqslant_{\mathrm{as}} c_1 h^2 + c_2 (nh)^{-1/2}(1 + c_3 \, h^\mu) + c_4 n^{-1/2} , \quad n \to \infty .$$

Here, $\mu = \frac{1}{2} + \frac{1}{2\lambda}$. For $\beta = \frac{1}{5}$, this proves the lemma.   Q.e.d.

PROOF OF LEMMA (3.35). Let $h_n = \gamma n^{-1/5}$ for a suitable constant γ to be determined later. The starting point is the inequality

$$PER(h) \geqslant \| C_h * f_o \|_1 - \| A_h * (dF_n - dF_o) \|_1 ,$$

which implies that

$$(3.41) \quad \inf_{h \geqslant h_n} PER(h) \geqslant \inf_{h \geqslant h_n} \| C_h * f_o \|_1 - \sup_{h \geqslant h_n} \| A_h * (dF_n - dF_o) \|_1 .$$

By Exercise (3.49), we have the lower bound

$$\inf_{h \geqslant h_n} \| C_h * f_o \|_1 \geqslant c_1 \min((h_n)^2 , 1) , \quad h > 0 ,$$

for some positive constant c_1.

For the second term on the right of (3.41), we have by Exercise (3.48)(b) that $\| A_h * (dF_n - dF_o) \|_1$ is decreasing in h, which implies the bound

$$\sup_{h \geqslant h_n} \| A_h * (dF_n - dF_o) \|_1 = \| A_{h_n} * (dF_n - dF_o) \|_1 \leqslant c_2 (n h_n)^{-1/2} ,$$

the last bound by (3.39)–(3.40), for a suitable positive constant c_2. Combining these lower bounds gives

$$\inf_{h \geqslant h_n} PER(h) \geqslant_{as} c_1 \min((h_n)^2 , 1) - c_2 (n h_n)^{-1/2} .$$

It follows that

$$\liminf_{n \to \infty} n^{2/5} \inf_{h \geqslant h_n} PER(h) \geqslant_{as} c_1 \gamma^2 - c_2 \gamma^{-1/2} ,$$

and this dominates K for large enough γ.   Q.e.d.

PROOF OF LEMMA (3.36). Let $h_n = \delta n^{-1/5}$, for a small enough positive constant δ. The starting point is the inequality

$$PER(h) \geqslant (nh)^{-1/2} \mathrm{var}_n(B; h)$$
$$\geqslant (nh)^{-1/2} \| \sqrt{(B^2)_h} * dF_n \|_1 - n^{-1/2} \| B \|_1 ,$$

the last inequality by Exercise (3.50). Recall that

$$B = 2 A - A * A = 2 A - A_{\sqrt{2}} .$$

We work temporarily with $(B_h)^2$ rather than with $(B^2)_h$. The triangle inequality for the Euclidean norm on $\mathbb{R}^n$ gives

$$\sqrt{(B_h)^2 * dF_n(x)} \geqslant 2 \sqrt{(A_h)^2 * dF_n(x)} - \sqrt{(A_{h\sqrt{2}})^2 * dF_n(x)} .$$

Upon integration, we get

$$\| \sqrt{(B_h)^2 * dF_n} \|_1 \geqslant 2 \| \sqrt{(A_h)^2 * dF_n} \|_1 - \| \sqrt{(A_{h\sqrt{2}})^2 * dF_n} \|_1$$
$$\geqslant \| \sqrt{(A_h)^2 * dF_n} \|_1 ,$$

the last inequality by Exercise (3.51)(d). Thus, by translating back in terms of $(B^2)_h$ and $(A^2)_h$,

$$\| \sqrt{(B^2)_h * dF_n(x)} \|_1 \geqslant \| \sqrt{(A^2)_h * dF_n(x)} \|_1 .$$

Now, Exercise (3.51)(d) below implies that $\| \sqrt{(A_h)^2 * dF_n} \|_1$ is a decreasing function of h. Thus,

$$\inf_{h \leqslant h_n} PER(h) \geqslant \inf_{h \leqslant h_n} (nh)^{-1/2} \| \sqrt{(A^2)_h * dF_n} \|_1 - n^{-1/2} \| B \|_2$$

$$\geqslant (nh_n)^{-1/2} \| \sqrt{(A^2)_{h_n} * dF_n} \|_1 - n^{-1/2} \| B \|_2 .$$

In the next lemma, we show that there exists a positive constant c_3 such that for deterministic h, with $h \log n \to 0$,

$$\| \sqrt{(A^2)_h * dF_n} \|_1 \geqslant_{as} c_3 (nh)^{1/2} (1 + nh)^{-1/2} ,$$

and so

$$\liminf_{n \to \infty} n^{2/5} \inf_{h \leqslant h_n} PER(h) \geqslant_{as} c_3 \delta^{-1/2} .$$

For δ small enough, this dominates K.   Q.e.d.

To wrap up the above proof, we need to provide a.s. lower bounds on $\| \sqrt{(A^2)_h * dF_n} \|_1$. Since A^2 is nonnegative and integrable, it suffices to do this for an arbitrary nonnegative kernel K. Lucky for us, the material of §4.4 on martingales and exponential inequalities applies here as well and seems to give sharp bounds. We formulate it as a lemma.

(3.42) LEMMA. *Let K be a nonnegative kernel, with a finite exponential moment. There exists a positive constant c such that for deterministic h satisfying $h \log n \to 0$,*

$$(nh)^{-1/2} \| \sqrt{K_h * dF_n} \|_1 \geqslant_{as} c(1 + nh)^{-1/2} , \quad n \to \infty .$$

PROOF. This is an application of the DEVROYE (1991) approach to obtaining exponential inequalities for kernel estimators, see §4.4. Using the abbreviation $\mathbb{X}_n = (X_1, X_2, \cdots, X_n)$, let

$$(3.43) \qquad \psi_n(\mathbb{X}_n) = \| \sqrt{K_h * dF_n} \|_1 .$$

We leave it as an exercise to verify that, with $(x)_{n-1} = (x_1, x_2, \cdots, x_{n-1})$ and $((x)_{n-1}, u) = (x_1, x_2, \cdots, x_{n-1}, u)$,

$$(3.44) \qquad \sup_{u,w} | \psi_n((x)_{n-1}, u) - \psi_n((x)_{n-1}, w) | \leqslant 2 (h/n)^{1/2} .$$

Thus, Theorem (4.4.20) implies the exponential inequality

$$(3.45) \qquad \mathbb{P}\big[| \psi_n(\mathbb{X}_n) - \mathbb{E}[\psi_n(\mathbb{X}_n)] | > 2t\sqrt{h} \big] \leqslant 2 \exp(-2t^2) ,$$

which in turn implies the a.s. bound

(3.46) $\psi_n(\,\mathbb{X}_n\,) =_{\text{as}} \mathbb{E}[\,\psi_n(\,\mathbb{X}_n\,)] + \mathcal{O}\big(\,(\,h\,\log n\,)^{1/2}\,\big)\,.$

To obtain a lower bound on the expected value, note that Hölder's inequality implies for all x,

$$\mathbb{E}[\,K_h * dF_n(x)\,] \leqslant \big(\,\mathbb{E}[\,\sqrt{K_h * dF_n(x)}\,]\,\big)^{2/3}\,\big(\,\mathbb{E}[\,(\,K_h * dF_n(x)\,)^2\,]\,\big)^{1/3}\,,$$

and so

(3.47a) $\mathbb{E}[\,\|\,\sqrt{K_h * dF_n}\,\|_1\,] \geqslant \displaystyle\int_{\mathbb{R}} \frac{(\,K_h * f_o(x)\,)^{3/2}}{(\,Lf_o(x)\,)^{1/2}}\,dx\,,$

where $Lf_o(x) = \mathbb{E}[\,(\,K_h * dF_n(x)\,)^2\,]$, or

(3.47b) $Lf_o(x) = (nh)^{-1}\,[\,(K^2)_h * f_o\,](x) + (\,K_h * f_o(x)\,)^2\,.$

Now, once more using Hölder's inequality gives

$$\int_{\mathbb{R}} K_h * f_o(x)\,dx \leqslant \left\{\int_{\mathbb{R}} \frac{(\,K_h * f_o(x)\,)^{3/2}}{(\,Lf_o(x)\,)^{1/2}}\,dx\right\}^{2/3}\left\{\int_{\mathbb{R}} Lf_o(x)\,dx\right\}^{1/3}\,.$$

The integral on the left equals $\|\,K\,\|_1$, and obviously,

$$\int_{\mathbb{R}} Lf_o(x)\,dx = (nh)^{-1}\,\|\,K\,\|_2^2 + \|\,K\,\|_1\,.$$

It follows that

$$\int_{\mathbb{R}} \frac{(\,K_h * f_o(x)\,)^{3/2}}{(\,Lf_o(x)\,)^{1/2}}\,dx \geqslant \|\,K\,\|_1^{3/2}\,\big\{\,(nh)^{-1}\,\|\,K\,\|_2^2 + \|\,K\|_1\,\big\}^{-1/2}\,,$$

and (3.47a) clinches the argument.   Q.e.d.

In the above proofs, we referred to a number of results, which we formulate as exercises.

(3.48) EXERCISE. For the kernels of Theorem (3.31), show that
(a) $\|\,A_h * (dF_n - dF_o)\,\|_1 \leqslant \|\,B_h * (dF_n - dF_o)\,\|_1 \leqslant 3\,\|\,A_h * (dF_n - dF_o)\,\|_1\,,$
(b) $\|\,A_h * (dF_n - dF_o)\,\|_1$ is a decreasing function of h.
Likewise for the double exponential kernel. [Hint: See (3.32) for Part (b).]

(3.49) EXERCISE. Let the kernels A and B be as in Theorem (3.31), let $C = A - B$, and suppose the density f_o belongs to $W^{2,1}(\mathbb{R})$.
(a) Show that $\|\,C_h * f_o\,\|_1 = \frac{1}{2}\sigma^2(A)\,h^2\,\|\,f''\,\|_1 + o(h^2)\,,\quad h \to 0\,.$
(b) Show that $\displaystyle\lim_{h\to\infty} \|\,C_h * f_o\,\|_1 = \|\,C\,\|_1\,.$
(c) Show that $\|\,C_h * f_o\,\|_1$ is a continuous function of h, and that it is positive for all $h > 0$.

(d) Show that there exist constants $0 < c_1 \leqslant c_2 < \infty$ such that

$$c_1 \min(h^2, 1) \leqslant \| (A_h - B_h) * f_o \|_1 \leqslant c_2 \min(h^2, 1) \quad \text{for all } h > 0.$$

(e) Do the same for the double exponential kernel.

(3.50) EXERCISE. (a) Show that $\sqrt{x} - \sqrt{y} \leqslant \sqrt{x - y}$ for all $x \geqslant y \geqslant 0$.

(b) For any kernel K, show that

$$\left| \operatorname{var}_n(K; h) - \| \sqrt{(K^2)_h * dF_n} \|_1 \right| \leqslant \| K \|_1 \sqrt{h}.$$

(3.51) EXERCISE. Let A be the Gaussian kernel. For all $X_1, X_2, \cdots, X_n$, show that for every x,

(a) the map $\varphi \longmapsto \sqrt{[(\varphi^2) * dF_n](x)}$ is convex in φ,

(b) the map $\varphi \longmapsto \sqrt{[\varphi * dF_n](x)}$ is concave in φ (nonnegative),

(c) $\| \sqrt{(A^2)_h * dF_n} \|_1$ is an increasing function of h, and

(d) $\| \sqrt{(A_h)^2 * dF_n} \|_1$ is a decreasing function of h.

(3.52) EXERCISE. Complete the proof of Lemma (3.42) by verifying the statements (3.44)–(3.47).

(3.53) EXERCISE. Wrap up the proof of Theorem (3.31) by showing that

$$\| f_{n,PER} - f_o \|_1 =_{as} \mathcal{O}(n^{-2/5}),$$

using the fact that $H_{n,PER} \asymp_{as} n^{-1/5}$. [Hint: Use the monotonicity of $\| A_h * (dF_n - dF_o) \|_1$ as function of h.]

(3.54) EXERCISE. Use the integration by parts trick of §4.3 to prove the following (weakened) version of Theorem (3.31):

If A and B satisfy (3.7) and (3.8), and if $f \in W^{2,1}(\mathbb{R})$ and $\sqrt{f_o} \in L^1(\mathbb{R})$, then for all $\varepsilon > 0$ there exist positive constants c_1, c_2 such that

$$c_1 n^{-1/5-\varepsilon} \leqslant_{as} H_{n,PER} \leqslant_{as} c_2 n^{-1/5+\varepsilon}$$

and

$$\| f_{n,PER} - f_o \|_1 =_{as} \mathcal{O}(n^{-2/5+\varepsilon}).$$

(3.55) EXERCISE. For deterministic h, under the usual assumptions, provide a lower bound for

$$\mathbb{E}\left[\left| [A_h * (dF_n - dF_o)](x) \right| \right],$$

which will show that

$$\mathbb{E}[\| A_h * (dF_n - dF_o) \|_1] \geqslant c(1 + nh)^{-1/2}.$$

(3.56) EXERCISE. You thought we would forget about it, didn't you? Prove Theorem (4.1.53) for the normal and two-sided exponential kernel.

SOLUTION TO EXERCISE (3.56). We prove the "only if" part of Theorem (4.1.53). The proof is by way of contradiction.

So, suppose that the statement "$H \longrightarrow_{as} 0$, $nH \longrightarrow_{as} \infty$" is not true. First, assume that "$H \longrightarrow_{as} 0$" does not hold. Thus, pick a subsequence $\{H_{n_i}\}_i$ for which

$$\lim_{i \to \infty} H_{n_i} = 2\delta > 0,$$

and consider the sequence $\{\widetilde{H}_k\}_k$ with $\widetilde{H}_k = H_{n_i}$ for $n_i \leqslant k < n_{i+1}$. We denote $\widetilde{H}$ by just H. It now suffices to show that

$$\liminf_{n \to \infty} \| A_H * dF_n - f_o \|_1 > 0.$$

First, the triangle inequality gives

$$(3.57) \quad \| A_H * dF_n - f_o \|_1 \geqslant \| A_H * f_o - f_o \|_1 - \| A_H * (dF_n - dF_o) \|_1.$$

Since $H_n \geqslant \delta > 0$ for all n large enough (depending on the sample $X_1, X_2, \cdots, X_n$), then

$$\| A_H * (dF_n - dF_o) \|_1 \leqslant_{as} \| A_\delta * (dF_n - dF_o) \|_1 =_{as} \mathcal{O}\big((n^{-1} \log n)^{1/2} \big),$$

by Theorem (4.4.22). Also, because f_o is a density,

$$\| A_H * f_o - f_o \|_1 \geqslant_{as} \inf_{h > \delta} \| A_h * f_o - f_o \|_1 = \eta > 0.$$

From (3.57), it follows that $\liminf_n \| A_H * dF_n - f_o \|_1 >_{as} 0$, and the same holds for the original sequence $\{H_n\}_n$.

Now, suppose that $H \longrightarrow_{as} 0$, but $\limsup_n nH \leqslant_{as} C < \infty$. Take a subsequence for which $\liminf_n nH_n \leqslant_{as} C$, and replace the whole sequence by a sequence for which $\lim_n nH_n \leqslant_{as} C$, similar to the first part. The triangle inequality gives

$$\| A_H * dF_n - f_o \|_1 \geqslant \| A_H * (dF_n - dF_o) \|_1 - \| A_H * f_o - f_o \|_1.$$

The last term converges to 0 a.s., since $H \longrightarrow_{as} 0$. Now, since $nH \leqslant 2C$ for all n large enough (depending on $X_1, X_2, \cdots, X_n$), by monotonicity,

$$\| A_H * (dF_n - dF_o) \|_1 \geqslant_{as} \inf_{h \leqslant 2C/n} \| A_h * (dF_n - dF_o) \|_1$$

$$\geqslant_{as} \| A_\lambda * (dF_n - dF_o) \|_1,$$

with $\lambda = 2C n^{-1}$. Now, by Theorem (4.4.22),

$$\| A_\lambda * (dF_n - dF_o) \|_1 \geqslant_{as} \mathbb{E}[\| A_\lambda * (dF_n - dF_o) \|_1] - \mathcal{O}\big((n^{-1} \log n)^{1/2} \big).$$

By Exercise (3.55), the expected value exceeds $c(1 + n\lambda)^{-1/2}$, which is bounded away from 0. Thus, $\| A_\lambda * (dF_n - dF_o) \|_1$ will not tend to 0.

This proves the "only if" part of the Theorem. The "if" part goes along the same lines.   Q.e.d.

EXERCISES : (3.10), (3.11), (3.15), (3.20), (3.48), (3.49), (3.50), (3.51), (3.52), (3.53), (3.54), (3.55), (3.56).

4. Asymptotic plug-in methods

In the remainder of this chapter, we are interested in smooth densities for which kernel estimators can achieve the $n^{-2/5}$ rate of convergence for the L^1 error. With this in mind, the kernel A is assumed to be a symmetric, square integrable pdf, with finite variance. See (2.16).

We recall that the goal of smoothing parameter selection in kernel density estimation is to minimize $\| f^{nh} - f_o \|_1$. The starting point of plug-in methods is the realization of § 4.4 that there is not much of a difference between $\| f^{nh} - f_o \|_1$ and its expectation. This is followed by the decomposition of the expected L^1 error into bias and variance components,

$$(4.1) \qquad \mathbb{E}\big[\, \| f^{nh} - f_o \|_1 \, \big] \leqslant \mathrm{bias}(h) + (nh)^{-1/2} \, \mathrm{var}_o(A_h) \;,$$

where

$$(4.2) \qquad \mathrm{bias}(h) = \| A_h * f_o - f_o \|_1 \;,$$

$$(4.3) \qquad \mathrm{var}_o(A; h) = \| \sqrt{(A^2)_h * dF_o - h \, (A_h * dF_o)^2} \, \|_1 \;,$$

with the assumption that there is just about equality in (4.1). One additional step is taken, viz. the bias and variance terms in the bound (4.1) are replaced by the leading terms of their asymptotic expansions

$$(4.4) \quad \mathbb{E}\big[\, \| f^{nh} - f_o \|_1 \, \big] \leqslant \tfrac{1}{2} \sigma^2(A) \, \| f_o'' \|_1 \, h^2 +$$
$$(nh)^{-1/2} \, \| A \|_2 \, \| \sqrt{f_o} \|_1 + \cdots \; .$$

Thus, a theoretically interesting choice of h is the minimizer of the right-hand side of (4.4), that is,

$$(4.5) \qquad h_{n,AI'I} = r_1(A) \, \varrho_1(f_o) \, n^{-1/5} \;,$$

where

$$(4.6) \qquad r_1(A) = \big\{ \tfrac{1}{2} \| A \|_2 \, / \, \sigma^2(A) \big\}^{2/5} \;,$$

$$(4.7) \qquad \varrho_1(f_o) = \big\{ \| \sqrt{f_o} \|_1 \, / \, \| f_o'' \|_1 \big\}^{2/5} \; .$$

Note that $r_1(A)$ and $\varrho_1(f_o)$ differ from the corresponding expressions in the asymptotic L^2 error, see § 2.

In the above development, the inequalities in (4.1) and (4.4) must be considered blemishes. It may be corrected by the following interesting device connected with the Central Limit Theorem, see Chapter 5 of DEVROYE and GYÖRFI (1985) and also HALL and WAND (1988). The starting point is the *exact* decomposition

$$(4.8) \qquad f^{nh}(x) - f_o(x) = \mathrm{bias}(x; h) + [\, A_h * (dF_n - dF_o) \,](x) \;,$$

where

$$(4.9) \qquad \mathrm{bias}(x; h) = A_h * f_o(x) - f_o(x) \; .$$

Now, for fixed x and deterministically varying h,

$$n\,[\,A_h * (dF_n - dF_o)\,](x) = \sum_{i=1}^{n} Z_i$$

is the sum of the iid random variables

$$Z_i = A_h(x - X_i) - A_h * f_o(x) \,, \quad i = 1, 2, \cdots, n \,.$$

So by the Central Limit Theorem,

(4.10) $\qquad \sqrt{n}\,[\,A_h * (dF_n - dF_o)\,](x) \longrightarrow_d h^{-1/2}\,\mathrm{var}_o(A;h;x)\,Y \,,$

where $Y \sim N(0,1)$ and

(4.11) $\qquad \mathrm{var}_o(A;h;x) = \sqrt{(A^2)_h * f_o(x) - h\,(\,A_h * f_o(x)\,)^2} \,.$

It is then reasonable to conclude that

(4.12) $\qquad \mathbb{E}[\,|\,f^{nh}(x) - f_o(x)\,|\,] \longrightarrow [\,\Psi_{nh} f_o\,](x) \,,$

with

(4.13) $\qquad [\,\Psi_{nh} f_o\,](x) = \mathbb{E}[\,|\,\mathrm{bias}(x;h) + (nh)^{-1/2}\,\mathrm{var}_o(A;h;x)\,Y\,|\,] \,.$

Therefore, upon integration with respect to $x \in \mathbb{R}$, we find an asymptotic expression for $\mathbb{E}[\,\|\,f^{nh} - f_o\,\|_1\,]$, but its correctness is somewhat suspect.

The above may be made rigorous based on the following precise result, taken lock, stock, and barrel from DEVROYE and GYÖRFI (1985).

(4.14) LEMMA. [DEVROYE and GYÖRFI (1985)] *Let* $X_1, X_2, \cdots, X_n$ *be iid random variables with* $\mathbb{E}[\,X_1\,] = 0$, $\mathbb{E}[\,|\,X_1\,|^2\,] = 1$, *and* $\mathbb{E}[\,|\,X_1\,|^3\,] < \infty$. *Let*

$$S_n = n^{-1/2} \sum_{i=1}^{n} X_i \,,$$

and let $Y \sim N(0,1)$ *be independent of the* X_i. *Then,*

$$\sup_{a \in \mathbb{R}} \big|\,\mathbb{E}[\,|\,a + S_n\,|\,] - \mathbb{E}[\,|\,a + Y\,|\,]\,\big| \leqslant c\,n^{-1/2}\,\mathbb{E}[\,|\,X_1\,|^3\,] \,,$$

for some universal constant c.

PROOF. Let $P_n(x) = \mathbb{P}[\,S_n \leqslant x\,]$, and let $\Phi(x)$ be the standard normal distribution. Then,

$$\mathbb{E}[\,|\,a + S_n\,|\,] - \mathbb{E}[\,|\,a + Y\,|\,] = \int_{\mathbb{R}} |\,a + x\,|\,\{\,dP_n(x) - d\Phi(x)\,\}$$

$$= \int_{\mathbb{R}} \{\,P_n(x) - \Phi(x)\,\}\,\mathrm{sign}(\,a + x\,)\,dx \,,$$

and so

$$\big|\,\mathbb{E}[\,|\,a + S_n\,|\,] - \mathbb{E}[\,|\,a + Y\,|\,]\,\big| \leqslant \int_{\mathbb{R}} |\,P_n(x) - \Phi(x)\,|\,dx \,.$$

Now, using the bound of Berry-Esseen type,

$$| P_n(x) - \Phi(x) | \leqslant c\, n^{-1/2}\, \mathbb{E}[|X_1|^3]\,(1+|x|)^{-3}\,, \quad x \in \mathbb{R}\,,$$

proves the lemma.   Q.e.d.

The lemma implies that

$$\mathbb{E}[|f^{nh}(x) - f_o(x)|] = \mathbb{E}[|\operatorname{bias}(x;h) + (nh)^{-1/2}\operatorname{var}_o(A;h;x)\,Y|] + \varepsilon_{nh}(x)\,,$$

with

$$(4.15) \quad |\varepsilon_{nh}(x)| \leqslant c\,n^{-1}\,\frac{\mathbb{E}[|A_h(x-X_1) - A_h * f_o(x)|^3]}{\mathbb{E}[|A_h(x-X_1) - A_h * f_o(x)|^2]} \leqslant c_1\,(nh)^{-1}\,,$$

as $h \to 0$, for universal constants c, c_1.

Unfortunately, it is not permissible to integrate (4.15) with respect to $x \in \mathbb{R}$, but for any $T > 0$, we have

$$\mathbb{E}\left[\int_{|x|<T} |f^{nh}(x) - f_o(x)|\,dx\right] =$$

$$\int_{|x|<T} \mathbb{E}[|\operatorname{bias}(x;h) + (nh)^{-1/2}\operatorname{var}_o(A;h;x)\,Y|]\,dx \;+\; \delta_{nhT}\,,$$

where

$$|\delta_{nhT}| \leqslant 2\,c_1\,T\,(nh)^{-1}\,.$$

On $|x| > T$, we use that

$$\mathbb{E}[|f^{nh}(x) - f_o(x)|] \leqslant \operatorname{bias}(x;h) + (nh)^{-1/2}\operatorname{var}_o(A;h;x)\,,$$

which implies that

$$\mathbb{E}[\int_{|x|>T} |f^{nh}(x) - f_o(x)|\,dx] \leqslant C\,(h^2 + (nh)^{-1/2})\,\eta(T)\,,$$

for some function η, with $\eta(T) \longrightarrow 0$ for $T \to \infty$. Consequently,

$$\left|\mathbb{E}[\|f^{nh} - f_o\|_1] - \|\Psi_{nh}f_o\|_1\right| \leqslant 2\,c_1\,T\,(nh)^{-1} + C\,\eta(T)\,(h^2 + (nh)^{-1/2})\,.$$

Now, take $T = o(\{h^2 + (nh)^{-1/2}\}^{-1})$, and we have proven the following theorem.

(4.16) THEOREM. [DEVROYE and GYÖRFI (1985)] If f_o satisfies the usual nonparametric assumptions, as well as $\mathbb{E}[|X_1|^3] < \infty$, then for $h \to 0$, $nh \to \infty$,

$$\mathbb{E}[\|f^{nh} - f_o\|_1] = \|\Psi_{nh}f_o\|_1 + o(h^2 + (nh)^{-1/2})\,.$$

(4.17) COROLLARY. [DEVROYE and GYÖRFI (1985)]

$$\mathbb{E}[\|f^{nh} - f_o\|_1] \leqslant \operatorname{bias}(h) + \sqrt{2/(\pi nh)}\,\operatorname{var}_o(A;h) + o(h^2 + (nh)^{-1/2})\,.$$

Ignoring the $o(\cdots)$ terms in these two results then gives

(4.18) $\mathbb{E}[\,\|\,f^{nh} - f_o\,\|_1\,] =$

$$\int_{\mathbb{R}} \mathbb{E}[\,|\,\tfrac{1}{2}h^2\sigma^2(A)\,(f_o)''(x) + (nh)^{-1/2}\,\|\,A\,\|_2\,\sqrt{f_o(x)}\,Y\,|\,]\,dx$$

and

(4.19) $\mathbb{E}[\,\|\,f^{nh} - f_o\,\|_1\,] \leqslant \tfrac{1}{2}h^2\sigma^2(A)\,\|\,(f_o)''\,\|_1 +$

$$(2/\pi nh)^{1/2}\,\|\,A\,\|_2\,\|\,\sqrt{f_o}\,\|_1\;.$$

The first expression is the basis of L^1 plug-in methods, see HALL and WAND (1988) and the next section. Minimizing the second (asymptotic) upper bound gives the minimizer

(4.20) $$h_{n,API} = (2/\pi)^{1/5}\,r_1(A)\,\varrho_1(f_o)\,n^{-1/5}\;.$$

A useful upper bound on $h_{n,API}$ is given in the next exercise.

(4.21) EXERCISE. [DEVROYE and GYÖRFI (1985), p. 113] Show that for A the Epanechnikov kernel $h_{n,API} \leqslant h_{n,UP}$, where

$$h_{n,UP} = \sigma \left(\frac{98415}{65536} \frac{\pi^4}{n} \right)^{1/5} \doteq 2.71042\,\sigma\,n^{-1/5}\;,$$

in which σ is the standard deviation of X_1. Estimating σ, e.g., by the interquartile estimator q_n, see (2.39), then gives an estimator for this "upper bound", denoted by $H_{n,UP-E}$ for the Epanechnikov kernel. There is a similar bound for the upper bound $H_{n,UP-N}$ for the the normal kernel. [Hint: The framework of § 11.2 seems to fit.]

As in § 2, the expression (4.20) may be used as the starting point for estimation procedures. The factor $r_1(A)$ can be computed exactly, but $\varrho_1(f_o)$ is unknown and must be estimated. In the asymptotic plug-in methods, $\varrho_1(f_o)$ is estimated based on kernel estimators of f_o, with smoothing parameters chosen by some (other) method.

The literature on L^1 plug-in methods is not large. DEHEUVELS and HOMINAL (1980) suggest a normal pilot estimator, similar to DEHEUVELS (1977) in the L^2 context, see (2.25). The simulations of BERLINET and DE-VROYE (1994) and others suggest that the success of any plug-in method depends mainly on the performance of the pilot estimator. So the pilot estimator must be selected carefully. According to BERLINET and DEVROYE (1994) a carefully tweaked double kernel method works the best. Their L^1 plug-in method with the pilot double kernel method is as follows. Let A be the Epanechnikov kernel, let $B = L_3$, as in (3.18)–(3.19), so that B is a fourth-order kernel, and let $H_{n,DBL}$ be the corresponding double kernel smoothing parameter: It solves

$$\text{minimize}\quad \|\,(A_h - B_h) * dF_n\,\|_1\quad \text{over } h > 0\;.$$

Now, $\| \sqrt{f_o} \|_1$ is estimated by $\| \sqrt{f_{n,DBL}} \|_1$. For the term $\| f_o'' \|_1$, the starting point is (4.4), with the variance term omitted, thus,

$$\| (A_h - B_h) * dF_n \|_1 = \tfrac{1}{2} h^2 \sigma^2(A) \| f_o'' \|_1 + \cdots ,$$

and so $\| (f_o)'' \|_1$ is estimated by

$$(4.22) \qquad D_n = 2 \left\{ H_{n,DBL}\, \sigma(A) \right\}^{-2} \| (A_{H_{n,DBL}} - B_{H_{n,DBL}}) * dF_n \|_1 .$$

Actually, this is not it completely. To get the plug-in method to work for small sample sizes, a few more tweaks are required. First, to make sure that the smoothing parameter is large enough that the bias term in (4.4) is (much) larger than the variance term, in (4.22), the smoothing parameter $H_{n,DBL}$ is replaced by

$$(4.23) \qquad\qquad H_{n,DBL} \cdot \max(1, 10\,R),$$

where

$$(4.24) \qquad R = \frac{\| \sqrt{f_{n,DBL}} \|_1 \| A - B \|_2}{(n H_{n,DBL})^{1/2} \| (A_{H_{n,DBL}} - B_{H_{n,DBL}}) * dF_n \|_1} .$$

Here, R is an estimator for the ratio of the bias and the variance term. Secondly, the initial choice of the smoothing parameter is taken to be

$$(4.25) \qquad \widetilde{H}_{n,API-DBL} = \left(\| (f_{n,DBL})^{1/2} \|_1\, D_n^{-1} \right)^{-2/5} \left(15/2\pi n \right)^{-1/5} ,$$

with D_n as in (4.22). Finally, in view of Exercise (4.21), the final choice in the asymptotic plug-in method for the smoothing parameter is

$$(4.26) \qquad H_{n,API-DBL} = \min(\widetilde{H}_{n,API-DBL}, 2.71042\, q_n\, n^{-1/5}) ,$$

with q_n the inner quartile estimator of the scale, see (2.39).

As a closing comment, we note that this method is relatively simple and contains essentially only one tuning parameter, to wit, the stretch parameter in the underlying double kernel method. Therefore, its unparalleled practical performance, especially for "weird" densities, is quite remarkable. See BERLINET and DEVROYE (1994) and Chapter 8.

EXERCISE: (4.21).

5. Away with pilot estimators!?

There is something unsatisfying about the use of pilot estimators: If the pilot estimators are that good, then why bother plugging them into asymptotic formulas? The adjective "asymptotic" actually refers to two types of asymptotic behavior. On the one hand, the sample size n tends to ∞, and on the other, the window parameter h tends to 0. The former is stochastic in nature, the latter is mostly deterministic. In this section, we concentrate on the stochastic aspects, and let the window parameter be. It will transpire that this way one can eliminate the pilot estimators, at least for smooth densities. All of the methods turn out to be variational in nature, i.e., the window parameter is chosen by minimizing a certain estimator of the "error". We begin with a simple-minded approach suggested by the previous section, after which, we are fully armed to deal with the method of DEVROYE and GYÖRFI (1985) and HALL and WAND (1988).

As in § 4, we consider smooth densities, and assume that A is a symmetric, square integrable pdf with finite variance.

We are interested in estimating $\| f^{nh} - f_o \|_1$, but § 4.5 tells us we may just as well consider its expected value. One simple-minded message of the previous section is that

$$(5.1) \qquad \mathbb{E}[\, \| f^{nh} - f_o \|_1 \,] \approx \mathrm{bias}(h) + (2/\pi n h)^{-1/2} \, \mathrm{var}_o(A; h) \; ,$$

with $\mathrm{bias}(h)$ and $\mathrm{var}_o(A; h)$ as in (4.2)–(4.3). Now, the question arises of whether it is possible to estimate the right-hand side of (5.1).

As in § 3, we estimate $\mathrm{var}_o(A; h)$ by $\mathrm{var}_n(A; h)$,

$$(5.2) \qquad \mathrm{var}_n(A; h) = \| \sqrt{(A_h)^2 * dF_n - (A_h * dF_n)^2} \; \|_1 \; .$$

This should work reasonably well, for h not too small.

The bias term causes more difficulty. We do not know of a (good) method for estimating it over a wide range of h and, hence, must resort to small h asymptotics after all. For small h, we have the asymptotic expansion

$$(5.3) \qquad \mathbb{E}[\, \| A_h * f_o - f_o \|_1 \,] = \tfrac{1}{2} \sigma^2(A) \, h^2 \, \| (f_o)'' \|_1 + o(h^2) \; , \qquad h \to 0 \; ,$$

cf. Theorem (4.7.2), and this may be used to our advantage. A similar asymptotic behavior is obtained for $\| (A_h - B_h) * f_o \|_1$, provided that the kernel B is symmetric, with finite $\sigma^2(B)$, viz.

$$(5.4) \qquad \| (A_h - B_h) * f_o \|_1 = \tfrac{1}{2} h^2 \, |\sigma^2(A - B)| \, \| (f_o)'' \|_1 + o(h^2) \; .$$

To simplify the presentation, we henceforth assume that

$$(5.5) \qquad\qquad\qquad B \text{ is a fourth-order kernel} \; ,$$

beacuse then $\sigma^2(B) = 0$, and asymptotically, the right-hand sides of (5.3) and (5.4) are the same.

(5.6) EXERCISE. Prove (5.4).

It is useful to introduce the shorthand notations

$$(5.7) \qquad\qquad\qquad C = A - B$$

and $C_h = A_h - B_h$. Continuing, (5.3) and (5.4) imply that

$$\| C_h * dF_o \|_1 = \| A_h * f_o - f_o \|_1 + o(h^2) \; ,$$

and the left-hand side may be estimated by $\| C_h * dF_n \|_1$, except that this introduces another variance term. Employing the analogue of (5.1), we obtain

$$(5.8) \qquad \mathbb{E}[\, \| C_h * dF_n \|_1 \,] \approx \| C_h * f_o \| + (2/\pi nh)^{-1/2} \, \mathrm{var}_o(C; h) \,,$$

with $\mathrm{var}_o(C; h)$ as in (4.3). Again, we assume that there is just about equality in (5.8). Because we know how to estimate the variance term in (5.8), we are thus lead to estimating the error $\| A_h * dF_n - f_o \|_1$ by

$$(5.9) \quad SPI(h) \overset{\mathrm{def}}{=} \| (A_h - B_h) * dF_n \|_1 +$$
$$(2/\pi nh)^{-1/2} \{ \, \mathrm{var}_n(A; h) - \mathrm{var}_n(A - B; h) \, \} \,.$$

A good way to select the smoothing parameter h then ought to be:

(5.10) in the *SPI* method, the smoothing parameter
selected is the solution to

$$\text{minimize} \quad SPI(h) \quad \text{subject to} \quad h > 0 \,.$$

The h so selected is denoted by $H_{n,SPI}$ and the
corresponding kernel estimator by $f_{n,SPI}$.

Before continuing, some comments regarding the *SPI* method are in order. First, "*SPI*" stands for Small sample Plug-In, which seems apt enough. Second, there is a striking similarity with the perverted double kernel method. Third, the estimator of the bias term in the small sample plug-in method is copied from the double kernel method, although the motivation is somewhat different. Fourth, all of the considerations that went into the definition of $SPI(h)$ hinge on approximate equality in the inequalities (5.1) and (5.8). Although this is not true for all values of h (e.g., for $h \to \infty$), it seems to hold for all relevant h.

The above motivation of the *SPI* method is a very rough version of the asymptotic considerations of § 4. We may repeat those considerations in the present context, without (explicitly) letting $h \to 0$. Thus, from Theorem (4.16),

$$(5.11) \qquad \mathbb{E}[\, \| f^{nh} - f_o \|_1 \,] = \| \Psi_{nh} f_o \|_1 + o(\, (nh)^{-1/2} + h^2 \,) \,,$$

where, see (4.13),

$$(5.12) \qquad [\, \Psi_{nh} f_o \,](x) = \mathbb{E}[\, | \, \mathrm{bias}_o(x) + (nh)^{-1/2} \, \mathrm{var}_o(A; h; x) \, Y \, | \,] \,,$$

in which $Y \sim N(0, 1)$. As before, $\mathrm{var}_o(A; h; x)$ is estimated (accurately) by $\mathrm{var}_n(A; h; x)$. As usual, the bias term causes problems, but it is clear that its estimation should involve $C_h * dF_n = (A_h - B_h) * dF_n(x)$. Hindsight suggests one consider

$$\mathbb{E}_Z[\, | \, C_h * dF_n + (nh)^{-1/2} \, \text{stuff}\,(x) Z \, | \,] \,,$$

where $Z \sim N(0, 1)$ is independent of $X_1, X_2, \cdots, X_n$, and $\mathbb{E}_Z$ denotes expectation with respect to Z. Here, "stuff(x)" is independent of Z and is to be determined such that

(5.13) $\mathbb{E}_Z[\,|\,C_h * dF_n + (nh)^{-1/2}\,\text{stuff}\,(x)Z\,|\,] \approx [\,\Psi_{nh}f_o\,](x)$.

Now, similar to what went on in §4,

(5.14) $\mathbb{E}_Z[\,|\,C_h * dF_n + (nh)^{-1/2}\,\text{stuff}\,(x)Z\,|\,] \approx$

$\qquad \mathbb{E}_{Z,Y}[\,|\,C_h * dF_o + (nh)^{-1/2}\,\{\,\text{stuff}\,(x)Z + \text{var}_o(C;h;x)\,Y\,\}\,|\,]$,

where $Y \sim N(0,1)$ is independent of Z. Thus, since we know how to add independent normal random variables, the last expectation equals

$$\mathbb{E}_Z[\,|\,C_h * dF_o + (nh)^{-1/2}\,\sqrt{\text{stuff}\,(x)^2 + (\text{var}_o(C;h;x))^2}\ Z\,|\,] ,$$

where Z is another $N(0,1)$ random variable, and this expression equals $[\,\Psi_{nh}f_o\,](x)$ if

$$\left(\,\text{stuff}(x)\,\right)^2 = (\text{var}_o(A;h;x))^2 - (\text{var}_o(C;h;x))^2 .$$

Beacuse the right-hand side could well be negative, we choose

$$\text{stuff}\,(x) = \text{var}_o(A,C;h;x) ,$$

where

(5.15) $\text{var}_o(A,C;h;x) \stackrel{\text{def}}{=} \sqrt{0 \vee \{\,(\text{var}_o(A;h;x))^2 - (\text{var}_o(C;h;x))^2\,\}}$.

This may be estimated in the usual manner by $\text{var}_n(A,C;h;x)$, defined similarly to (5.15). Thus, we define the empirical functional, freely after DEVROYE and GYÖRFI (1985) and HALL and WAND (1988),

(5.16) $DGHW(h) \stackrel{\text{def}}{=} \big\|\,\mathbb{E}_Z[\,|\,C_h * dF_n + (nh)^{-1/2}\,\text{var}_n(A,C;h;\,\cdot\,)\,Z\,|\,]\,\big\|_1$,

where $C_h = A_h - B_h$, and $Y \sim N(0,1)$, which leads to the following method for choosing the smoothing parameter:

(5.17) in the $DGHW$ method, the smoothing parameter
is the solution to

$\qquad\qquad$ minimize $DGHW(h)$ subject to $h > 0$.

The h so selected is denoted by $H_{n,DGHW}$ and the
corresponding kernel estimator by $f_{n,DGHW}$.

Another method immediately suggests itself. The triangle inequality gives the upper bound for $DGHW(h)$

(5.18) $MOD(h) \stackrel{\text{def}}{=} \big\|\,(A_h - B_h) * dF_n\,\big\|_1 + \dfrac{2}{\sqrt{\pi nh}}\,\big\|\,\text{var}_n(A,C;h;\,\cdot\,)\,\big\|_1$,

and this gives yet another method for choosing the window parameter:

(5.19) in the *MOD* method, the smoothing parameter
is the solution to

$$\text{minimize} \quad MOD(h) \quad \text{subject to} \quad h > 0 \,.$$

The h so selected is denoted by $H_{n,MOD}$, and the
corresponding kernel estimator by $f_{n,MOD}$.

The rather limp designation *MOD* stands for MODified, as in modified
double kernel method.

We shall not prove that the methods *DGHW* and *MOD* work, either in
our sense or in the sense of asymptotic optimality. One would certainly
expect under the usual nonparametric smoothness and tail assumptions
that the *DGHW* method is asymptotically optimal, and that for the *MOD*
method, one can prove with "our" methods that

(5.20) $$H_{n,MOD} \asymp_{as} n^{-1/5} \,,$$

but we shall not do so.

At this point, something should be said about the choice of the kernels A
and B. We mention two possibilities based on A being the normal kernel.
The first choice for B is to take $B = 2\,A - A*A = 2\,A - A_{\sqrt{2}}$ or, admitting
a stretch (shrink) parameter,

(5.21) $$B = 2\,A_\lambda - A_{\lambda\sqrt{2}} \,,$$

for which $\sigma^2(B) = 0$. Simulations (unreported) suggest that $\lambda \approx 0.83$ gives
the best results. The other choice for the kernel B is

(5.22) $$B = A_\lambda \,,$$

with stretch parameter λ. Simulations (again unreported) seem to indicate
that $\lambda \approx 0.995$ is the most preferable choice. This also suggests the limiting
case $\lambda \to 1$, i.e., with the kernel C of (5.7) (recall A is the normal kernel)
given by

(5.23) $$C(x) = \frac{1}{\sqrt{8\pi}}\,(x^2 - 1)\,e^{-x^2/2} \,,$$

which performs just about the same as the previous method, except for
the uniform density where it is noticeably worse. This also indicates that
we are skating on thin ice with $\lambda = 0.995$. In Chapter 8, we show some
simulations for these methods.

We finish this section by proving that the *SPI* method works,
in the sense that

(5.24) $$H_{n,SPI} \asymp_{as} n^{-1/5} \,, \quad n \to \infty \,.$$

The conditions required are that A and B are as in (5.21) with $\lambda = 1$ (no
stretching), and that f_o satisfies the usual nonparametric assumptions, see

(4.1.47)–(4.1.48). The general approach to the proof of (5.24) is as in § 3, where most of the hard work was done. In particular, we leave as exercises the following three lemmas.

(5.25) LEMMA. *There exists a constant K depending on f_o only such that*

$$\limsup_{n\to\infty} n^{2/5} \inf_h SPI(h) \leqslant_{as} K .$$

(5.26) LEMMA. *For a large enough constant γ, depending on f_o only,*

$$\liminf_{n\to\infty} n^{2/5} \inf \left\{ SPI(h) : h \geqslant \gamma\, n^{-1/5} \right\} >_{as} K .$$

(5.27) LEMMA. *For a small enough positive constant δ, depending on f_o only,*

$$\liminf_{n\to\infty} n^{2/5} \inf \left\{ SPI(h) : h \leqslant \delta\, n^{-1/5} \right\} >_{as} K .$$

In comparison with § 3, an extra ingredient is required for the proof of Lemma (5.27). It would seem obvious that to prove this lemma, we need a suitable lower bound on the quantity $\mathrm{var}_n(A;h) - \mathrm{var}_n(A - B;h)$. The first step is to consider instead lower bounds for

$$\| \sqrt{(A^2)_h * dF_n} \|_1 - \| \sqrt{((A-B)^2)_h * dF_n} \|_1 .$$

A nice bound would follow if for some $0 < r < 1$,

(5.28) $$A(x) \geqslant r \,| A(x) - B(x) | , \quad x \in \mathbb{R} ,$$

but this fails for $x \to \pm\infty$. (Recall that A is the standard normal, and $B = 2A - A * A$.) Of course, where it matters, $| A - B |$ is much smaller than A. This may be expressed is as follows.

(5.29) EXERCISE. *For A the normal density and $B = 2A - A * A$ show that $\| A \|_2 = 0.531 \cdots$, and*

$$\| A \|_2 - \| A - B \|_2 = \sqrt{\frac{1}{\sqrt{4\pi}}} - \sqrt{\frac{1}{\sqrt{4\pi}} - \frac{2}{\sqrt{6\pi}} + \frac{1}{\sqrt{8\pi}}} = 0.38653 \cdots .$$

However, instead of (5.28), we have the following inequality, which we elevate to the status of lemma, although no formal proof is given.

(5.30) LEMMA. *For A the normal kernel,*

$$| A(x) - A_{\sqrt{2}}(x) | \leqslant r\, A_{\sqrt{3}}(x) , \quad x \in \mathbb{R} ,$$

where $r = 0.515 \cdots$.

PROOF BY GRAPHICS. Plot the ratio of the left- and right-hand side of the inequality, and read off the value of r.   Q.e.d.

(5.31) EXERCISE. As the graph of $|A - A_{\sqrt{2}}|/A_{\sqrt{3}}$ indicates, there are three local maxima, the middle one of which is a tad below the global maximum. This suggests that the comparison with $A_{\sqrt{3}}$ is not optimal. To get the optimal comparison, consider the problem

$$\text{minimize } \max_{x \in \mathbb{R}} \frac{|A(x) - A_{\sqrt{2}}(x)|}{A_\lambda(x)} \quad \text{subject to } \lambda > 0 .$$

($\lambda = \sqrt{3}$ is very close to the optimum.)

With the previous lemma in hand, we now prove the following lemma.

(5.32) LEMMA. *For all realizations of* $X_1, X_2, \cdots, X_n$,

$$\left\| \sqrt{(A_h)^2 * dF_n} \right\|_1 - \left\| \sqrt{(A_h - A_{h\sqrt{2}})^2 * dF_n} \right\|_1 \geq (1-r) \left\| \sqrt{(A_h)^2 * dF_n} \right\|_1 .$$

PROOF. By Lemma (5.30) and Exercise (3.51)(d), we have the inequalities

$$\left\| \sqrt{(A_h - A_{h\sqrt{2}})^2 * dF_n} \right\|_1 \leq r \left\| \sqrt{(A_{h\sqrt{3}})^2 * dF_n} \right\|_1$$

$$\leq r \left\| \sqrt{(A_h)^2 * dF_n} \right\|_1 .$$

The lemma follows.   Q.e.d.

(5.33) EXERCISE. (a) Prove Lemmas (5.25), (5.26), and (5.27), and that the *SPI* method works in the sense of (5.24).
(b) Prove that the *MOD* method works in the sense of (5.20).

(5.34) EXERCISE. Prove the asymptotic optimality of the *DGHW* method, and send the authors a copy: They are not actually sure they know how to do this!

(5.35) EXERCISE. For the actual computation of $DGHW(h)$, the following identity is useful (see DEVROYE and GYÖRFI (1985), Chapter 5). Let $Y \sim N(0, 1)$ and let $a, b \in \mathbb{R}$, with $b > 0$. Then,

$$\mathbb{E}[|a + bY|] = a \text{ erf}(x) + b (2/\pi)^{1/2} \exp(-x^2) ,$$

where $x = a/(b\sqrt{2})$ and $\text{erf}(x)$ is the error function

$$\text{erf}(x) = \frac{2}{\sqrt{\pi}} \int_0^x e^{-t^2} dt .$$

EXERCISES: (5.6), (5.29), (5.31), (5.33), (5.34), (5.35).

6. A discrepancy principle

Until now, the procedures for selecting the smoothing parameter consisted of minimizing some estimate of $\| f^{nh} - f_o \|_1$ over h. A different kind of method may be based on discrepancy principles. The notion was introduced by applied mathematicians: by REINSCH (1967) in the context of smoothing splines, and by MOROZOV (1966) and ARCANGELI (1966) for the selection of the regularization parameter in ill-posed least-squares problems. See Volume II. Generally speaking, statisticians shudder at the thought, but it is a notion worth considering. We describe here one method for selecting the smoothing parameter for kernel density estimation, based on a discrepancy principle.

The assumptions on the kernel A are the usual ones, i.e., A is a symmetric, square-integrable pdf with finite variance, see (2.16). Regarding the density f_o, the usual nonparametric assumptions suffice.

To begin, we recall the Kolmogorov-Smirnov statistic $\| F_n - \Psi \|_\infty$ for testing whether an iid sample with empirical distribution F_n has been drawn from a distribution Ψ. Also recall that $\| F_n - F_o \|_\infty$ is distribution free, i.e., its distribution does not depend on F_o,

$$(6.1) \qquad \| F_n - F_o \|_\infty =_{\mathrm{d}} \| U_n - U_o \|_\infty \,,$$

where $U_n(t)$ is the empirical distribution of an iid sample of size n from the uniform $(0,1)$ distribution and $U_o(t) = t$, $0 \leqslant t \leqslant 1$. Finally, recall the CHUNG (1949) law of the iterated logarithm

$$(6.2) \qquad \| F_n - F_o \|_\infty =_{\mathrm{as}} \mathcal{O}\big((n^{-1} \log \log n)^{1/2} \big) \,.$$

Putting two and two together, the following would seem to make sense. For arbitrary $h > 0$, let F^{nh} be the distribution associated with f^{nh}, the kernel estimator under discussion (or any other estimator depending on h). Now, choose h such that

$$(6.3) \qquad \| F_n - F^{nh} \|_\infty = c \, (n^{-1} \log \log n)^{1/2} \,,$$

for some (suitable) constant c. After all, we should not expect F^{nh} to be closer to F_n than the true F_o. Unfortunately, this does not quite work, apparently due to the ill-posedness of density estimation. As shown below, for all intents and purposes,

$$(6.4) \qquad \| F_n - F^{nh} \|_\infty \approx c \, h^2 \,,$$

so (6.3) would imply that $h \asymp n^{-1/4}$, give or take a power of $\log \log n$. However, asymptotically, the optimal h satisfies $h \asymp n^{-1/5}$, so in (6.3), we are demanding that F^{nh} is too close to F_n. From this point of view, replacing the right-hand side of (6.3) by $n^{-2/5}$ could be right:

(6.5)   In the *DP* method, the smoothing parameter h is selected
as the smallest solution of

$$\| F_n - F^{nh} \|_\infty = c_{DP}\, n^{-2/5} ,$$

where $c_{DP} = 0.35$. The h so selected is denoted by $H_{n,DP}$,
and the corresponding kernel estimator $f^{n,H_{n,DP}}$ by $f_{n,DP}$.

The numerical value of c_{DP}, is of course rather mysterious, but works remarkably well for smooth densities with light tails. This will be appreciated
more fully when we come to the simulations of Chapter 8.
   We now show that $H_{n,DP}$ is a reasonable smoothing parameter.

(6.6) THEOREM. *Let* $f_o \in W^{2,1}(\mathbb{R})$, $\sqrt{f_o} \in L^1(\mathbb{R})$. *Then,* $H_{n,DP}$, *exists and
satisfies*

$$H_{n,DP} =_{as} c\, n^{-1/5}\, (1 + o(1)) ,$$

where $c > 0$ *is given by*

$$c^2 = 2\, c_{DP} \left\{ \sigma^2(A) \| f_o' \|_\infty \right\}^{-1} .$$

(6.7) COROLLARY. *If* A *is the normal or two-sided exponential density,
then* $\| f_{n,DP} - f_o \|_1 =_{as} \mathcal{O}\big(n^{-2/5} \big)$.

(6.8) COROLLARY. $H_{n,DP}$ *is scale invariant in the sense of* (1.16).

The proof of Theorem (6.6) follows from a series of lemmas, which we
leave as exercises. The first series deals with the existence of $H_{n,DP}$.

(6.9) LEMMA. *Let* F_n *be the empirical distribution of* $X_1, X_2, \cdots, X_n$, *and
let* $X_{1,n} \leqslant X_{2,n} \leqslant \cdots \leqslant X_{n,n}$ *be the order statistics. If* Ψ *is a continuous
distribution, then*

$$\| F_n - \Psi \|_\infty = \max_{1 \leqslant i \leqslant n} \left\{ \left| \tfrac{i-1}{n} - \Psi(X_{i,n}) \right|, \left| \tfrac{i}{n} - \Psi(X_{i,n}) \right| \right\} .$$

It is useful to introduce the distribution $\mathbb{A}$ corresponding to the density A,

(6.10) $$\mathbb{A}(x) = \int_{-\infty}^x A(y)\, dy , \quad x \in \mathbb{R} .$$

Then, the distribution function for the density A_h is $\mathbb{A}_h(x) = \mathbb{A}(h^{-1}x)$ (note
the missing factor h^{-1}), and

(6.11) $$F^{nh}(x) = \tfrac{1}{n} \sum_{i=1}^n \mathbb{A}\big(h^{-1}(x - X_i) \big) , \quad x \in \mathbb{R} .$$

Now, observe that $\lim_{h \to 0} \mathbb{A}(h^{-1}z)$ equals 0 for $z < 0$ and equals 1 for
$z > 0$, and that $\lim_{h \to \infty} \mathbb{A}(h^{-1}z) = \mathbb{A}(0) = \tfrac{1}{2}$ for all z. This gives:

(6.12) LEMMA. *For almost all realizations of* $X_1, X_2, \cdots, X_n$, *and for* $1 \leqslant i \leqslant n$,

(a) $$F^{nh}(X_{i,n}) \longrightarrow \tfrac{i-1/2}{n} \quad \text{for } h \to 0 , \quad \text{and}$$

(b) $$F^{nh}(X_{i,n}) \longrightarrow \tfrac{1}{2} \quad \text{for } h \to \infty .$$

(6.13) LEMMA. *For almost all realizations of* $X_1, X_2, \cdots, X_n$,

(a) $$\lim_{h \to 0} \| F_n - F^{nh} \|_\infty = \tfrac{1}{2n} , \quad \text{and}$$

(b) $$\lim_{h \to \infty} \| F_n - F^{nh} \|_\infty = \tfrac{1}{2} .$$

Together with the continuity of $\| F_n - F^{nh} \|_\infty$ as a function of h, the above lemmas suffice to show the existence of $H_{n,\text{\tiny DP}}$.

To prove the asymptotic behavior, we need one last ingredient. Let F_h be the distribution for the density $A_h * f_o$, so

$$(6.14) \qquad\qquad F_h = A_h * F_o = \mathbb{A}_h * dF_o .$$

(6.15) LEMMA. *If* $f \in W^{2,1}(\mathbb{R})$, *then*

(a) $$\| F_o - F_h \|_\infty = \tfrac{1}{2} h^2 \sigma^2(A) \| f_o' \|_\infty + o(h^2) , \quad \text{and}$$

(b) $$\left| \| F_n - F^{nh} \|_\infty - \| F_o - F_h \|_\infty \right| \leqslant 2 \| F_o - F_n \|_\infty .$$

PROOF OF (a). Observe that

$$F_h(x) - F_o(x) = \int_{\mathbb{R}} A_h(x - y) \left(F_o(y) - F_o(x) \right) dy .$$

Now, Taylor's theorem with exact remainder gives

$$F_o(y) - F_o(x) = (y - x) f_o(x) +$$
$$\tfrac{1}{2} (x - y)^2 f_o'(x) - \int_y^x (y - z)^2 f_o''(z) \, dz .$$

It follows that

$$F_h(x) - F_o(x) = \tfrac{1}{2} h^2 \sigma^2(A) f_o'(x) + R ,$$

with

$$R = \int_{\mathbb{R}} A_h(x - y) \int_y^x (y - z)^2 f_o''(z) \, dz \, dy .$$

It thus suffices to show that $R = o(h^2)$, uniformly in x. Note that

$$(6.16) \qquad | R | \leqslant \int_{\mathbb{R}} (x - y)^2 A_h(x - y) \left| \int_y^x f_o''(z) \, dz \right| dy .$$

To bound the right-hand side of (6.16), we split the integral into two parts. Let

$$R_1 = \int_{|x-y|>\sqrt{h}} (x-y)^2 \, A_h(x-y) \left| \int_y^x f_o''(z)\,dz \right| dy \, ,$$

$$R_2 = \int_{|x-y|<\sqrt{h}} (x-y)^2 \, A_h(x-y) \left| \int_y^x f_o''(z)\,dz \right| dy \, .$$

First, uniformly in x,

$$R_1 \leqslant \| f_o'' \|_1 \int_{|y|>\sqrt{h}} y^2 \, A_h(y)\,dy$$

$$\leqslant h^2 \, \| f_o'' \|_1 \int_{|y|>1/\sqrt{h}} y^2 \, A(y)\,dy = o(h^2) \, .$$

Secondly, let

$$(6.17) \qquad \omega(\,f\,;\,h\,) = \sup_{|x-y|<\sqrt{h}} \left| \int_y^x (y-z)^2 \, f_o''(z)\,dz \right| .$$

Since $f_o'' \in L^1(\mathbb{R})$, then $\lim_{h\to 0} \omega(\,f\,;\,h\,) = 0$. Consequently, uniformly in x,

$$R_2 \leqslant \omega(\,f\,;\,h\,) \int_{|x-y|<\sqrt{h}} (x-y)^2 \, A_h(x-y)\,dy$$

$$\leqslant h^2 \, \sigma^2(A) \, \omega(\,f\,;\,h\,) = o(h^2) \, .$$

Part (a) follows.   Q.e.d.

(6.18) EXERCISE. Prove the remaining results of these sections.

EXERCISE: (6.18).

7. The Good estimator

The last item in this chapter is choosing the smoothing parameter for maximum penalized likelihood estimation. We consider only the roughness penalization of GOOD (1971) and GOOD and GASKINS (1971), as elaborated in § 5.2. We discuss an old and very interesting proposal of KLONIAS (1984) and a method based on the discrepancy principle of § 6.

We recall that the GOOD (1971) estimator f^{nh} is given by $f^{nh} = (u_{nh})^2$, where u_{nh} solves

$$(7.1) \qquad \begin{aligned} &\text{minimize} \quad -2 \int_{\mathbb{R}} \log u(x)\,dF_n(x) + \| u \|_2^2 + h^2 \, \| u' \|_2^2 \\ &\text{subject to} \quad u \in W^{2,1}(\mathbb{R}) \, , \ u \geqslant 0 \end{aligned}$$

($L^2(\mathbb{R})$ norms, and the prime denotes differentiation), and that

(7.2) $$\| u_{nh} \|_2^2 = 1 - h^2 \| (u_{nh})' \|_2^2 .$$

Also, recall that u_{nh} is implicitly given by

(7.3) $$u_{nh}(x) = \frac{1}{n} \sum_{i=1}^{n} \frac{\mathcal{B}_h(x - X_i)}{u_{nh}(X_i)} , \quad x \in \mathbb{R} ,$$

with $\mathcal{B}_h(x) = (2h)^{-1} \exp(-h^{-1} |x|)$, the scaled, two-sided exponential pdf.

The first method for choosing h is a very interesting proposal of KLONIAS (1984), viz.

(7.4) in the GK method, the smoothing parameter is chosen as the (smallest) solution of

$$\text{minimize} \quad h^2 \| (u_{nh})' \|_2^2 \quad \text{over } h > 0 .$$

The h so chosen is denoted by $H_{n,GK}$ and the resulting estimator by $f_{n,GK}$.

Actually, the proposal of KLONIAS (1984) was to minimize the Lagrange multiplier when the pdf constraint $\| u_{nh} \|_2 = 1$ is incorporated into (7.1), but this is equivalent to minimizing $h^2 \| (u_{nh})' \|_2$ in our context. See Exercise (11.3.21). Also, note that as a consequence of the Comparison Lemma (5.2.22), see (5.2.23), we may write

(7.5) $$h^2 \| (u_{nh})' \|_2^2 = \| f^{nh} - \mathcal{B}_{h/\sqrt{2}} * dF_n \|_1 ,$$

so that (7.4) may be thought of as a double kernel (or double estimator) method. It turns out that this method does not work, but it should be kept in mind that the proposal was made when little was known concerning rational ways of selecting the smoothing parameter. Be that as it may, in the course of its investigation, useful facts about the GOOD estimator have to be unearthed, which will be put to good use in the study of the discrepancy principle.

The first concern is the existence and scaling invariance of $H_{n,GK}$. The scaling invariance we leave to the end of this section. Establishing the existence involves some cute details, exhibited in the proofs of the following three lemmas.

(7.6) LEMMA. *For every n and almost every realization of $X_1, X_2, \cdots, X_n$,*
(a) $$0 \leqslant h^2 \| (u_{nh})' \|_2^2 \leqslant \tfrac{1}{2} ,$$
(b) $$h \| (u_{nh})' \|_2 \quad \text{is continuous for } h > 0 , \quad \text{and}$$
(c) $$h^2 \| (u_{nh})' \|_2^2 \longrightarrow \tfrac{1}{2} \quad \text{for } h \to 0 \quad \text{and for } h \to \infty .$$

PROOF. Part (b) follows from Exercise (5.2.54), and Part (a) from

$$h^2 \| (u_{nh})' \|_2^2 = 1 - \| u_{nh} \|_2^2 = 1 - \int_{\mathbb{R}} f^{nh}(x) \, dx ,$$

combined with the lower bound from the Comparison Lemma (5.2.22). In view of the equality above, to prove Part (c), it suffices to show that

$$\| u_{nh} \|_2^2 \longrightarrow \tfrac{1}{2} \ .$$

This we do in the following three lemmas.   Q.e.d.

(7.7) LEMMA. *For fixed n and almost all realizations of $X_1, X_2, \cdots, X_n$, for $i = 1, 2, \cdots, n$,*

$$(2\,nh)^{-1/2} \leqslant u_{nh}(X_i) = (2\,nh)^{-1/2}\,(1 + o(1))\ , \quad h \to 0.$$

PROOF. Let $u_i = u_{nh}(X_i)$, and write (7.3) as

$$(7.8) \qquad u_i = (2\,nh\,u_i)^{-1} + (2\,nh)^{-1} \sum_{j \neq i} [\,u_j\,]^{-1} \exp(-h^{-1}(X_i - X_j))\ ,$$

so that $u_i \geqslant (2\,nh\,u_i)^{-1}$, and the lower bound follows:

$$(7.9) \qquad\qquad u_i \geqslant (2\,nh)^{-1/2}\ , \quad i = 1, 2, \cdots, n\ .$$

(This holds for *all* $h > 0$, but never mind.) For the upper bound, consider $\delta = \min_{i \neq j} |X_i - X_j|$. Then, for fixed n, we have $\delta > 0$ almost surely. Consequently,

$$\varepsilon \overset{\text{def}}{=} \max_{i \neq j} (2h)^{-1} \exp(-h^{-1} |X_i - X_j|)$$

satisfies

$$\varepsilon = \mathcal{O}(\,(2h)^{-1} \exp(-h^{-1}\delta)\,) =_{\text{as}} o(\,1\,)\ .$$

So, from (7.8),

$$u_i \leqslant_{\text{as}} (2\,nh\,u_i)^{-1} + \frac{\varepsilon}{n} \sum_{j \neq i} [\,u_j\,]^{-1}\ ,$$

and then with (7.9),

$$u_i \leqslant_{\text{as}} (2\,nh)^{-1/2} + \varepsilon\,(2\,nh)^{1/2}\ ,$$

and the upper bound follows.   Q.e.d.

(7.10) LEMMA. *For fixed n and almost all realization of $X_1, X_2, \cdots, X_n$, for $i = 1, 2, \cdots, n$,*

$$u_{nh}(X_i) = (2\,h)^{-1/2}\,(1 + o(1))\ , \quad h \to \infty.$$

PROOF. Observe that for $h \to \infty$,

$$\mathcal{B}_h(X_i - X_j) = (2\,h)^{-1} \exp(-h^{-1} |X_i - X_j|) \overset{\text{def}}{=} (2\,h)^{-1}(1 + \varepsilon_{ij})\ .$$

Then, $\varepsilon_{ij} = \mathcal{O}(h^{-1}D)$, with

$$D = \max_{i \neq j} |X_i - X_j|\ .$$

So, for fixed n, we have $D < \infty$ almost surely. Then, from (7.3),

$$u_i = (2\,nh)^{-1} \sum_{j=1}^{n} [\,u_j\,]^{-1}(1 + \varepsilon_{ij})\,,$$

and so

(7.11) $(2\,nh)^{-1}\,(1 - \varepsilon) \sum_{j=1}^{n} [\,u_j\,]^{-1} \leqslant u_i \leqslant (2\,nh)^{-1}\,(1 + \varepsilon) \sum_{j=1}^{n} [\,u_j\,]^{-1}\,,$

with $\varepsilon = \max_{i \neq j} |\varepsilon_{ij}| = \mathcal{O}(\,h^{-1}\,D)$. Now, take reciprocals to obtain

$$[\,u_i\,]^{-1} \geqslant 2\,h\,(1 + \varepsilon)^{-1}\,\big(\tfrac{1}{n} \sum_{j=1}^{n} [\,u_j\,]^{-1}\,\big)^{-1}\,,$$

and sum over i. This yields

$$\tfrac{1}{n} \sum_{j=1}^{n} [\,u_j\,]^{-1} \geqslant (2\,h)^{1/2}\,(1 + \varepsilon)^{-1/2}\,.$$

With the lower bound of (7.11), this gives that

$$u_i \geqslant (2\,h)^{-1/2}\,(1 + \varepsilon)^{-1/2}\,(1 - \varepsilon)\,,$$

for each i. The upper bound follows similarly.   Q.e.d.

(7.12) LEMMA. *For fixed n, almost all realizations of $X_1, X_2, \cdots, X_n$, and all $h > 0$,*

$$\|\,u_{nh}\,\|_2^2 = n^{-2} \sum_{i,j=1}^{n} [\,u_{nh}(X_i)\,u_{nh}(X_j)\,]^{-1}\,a_{ij}(h)\,,$$

with $a_{ij}(h) = (4\,h)^{-1}\,(1 + h^{-1}\,|X_i - X_j|)\,\exp(-h^{-1}\,|X_i - X_j|)\,.$

In particular, with $\delta = \min_{i \neq j} |X_i - X_j| > 0$ almost surely,

$$a_{ij}(h) = (4\,h)^{-1}\,(1 + o(1))\qquad,\quad h \to \infty\,,\quad \text{all}\ \ i \neq j\,,$$
$$a_{ii}(h) = (4\,h)^{-1}\qquad\qquad\qquad,\quad \text{all}\ h,$$
$$a_{ij}(h) = \mathcal{O}(\,h^{-2}\,\exp(-h^{-1}\,\delta\,))\quad,\quad \text{for}\ \ h \to 0,\ i \neq j\,.$$

PROOF. From (7.3), one obtains

$$\|\,u_{nh}\,\|_2^2 = \frac{1}{n^2} \sum_{i,j} \frac{a_{ij}}{u_{nh}(X_i)\,u_{nh}(X_j)}\,,$$

with

(7.13) $a_{ij} = \displaystyle\int_{\mathbb{R}} \mathcal{B}_h(x)\,\mathcal{B}_h(x - X_i + X_j)\,dx = \mathcal{B}_h * \mathcal{B}_h(X_i - X_j)\,.$

Now, from the identity $\mathcal{B}_\lambda = (h/\lambda)^2 \mathcal{B}_h + (1 - (h/\lambda)^2) \mathcal{B}_\lambda * \mathcal{B}_h$, see § 4.6, it follows that

$$\mathcal{B}_\lambda * \mathcal{B}_h = \frac{\lambda^2 \mathcal{B}_\lambda - h^2 \mathcal{B}_h}{\lambda^2 - h^2} .$$

So, upon taking the limit as $\lambda \to h$,

$$\mathcal{B}_h * \mathcal{B}_h(x) = (2h)^{-1} \frac{\partial}{\partial h} \left[h^2 \mathcal{B}_h(x) \right] ,$$

and the lemma follows.   Q.e.d.

(7.14) EXERCISE. Prove Part (c) of Lemma (7.6).

The last concern is whether $H_{n,GK}$ has the correct asymptotic behavior for $n \to \infty$, under suitable conditions. First, we try to give sharp bounds on the minimum value of $h^2 \| (u_{nh})' \|_2^2$. Let w_h be the solution of the large sample asymptotic version of (7.1), and let $w_o = \sqrt{f_o}$. The triangle inequality implies that

(7.15) $h \| (u_{nh})' \|_2 = h \| w_o' \|_2 + \delta_{nh} ,$

where

(7.16) $| \delta_{nh} | \leqslant h \| (w_h - w_o)' \|_2 + h \| (u_{nh} - w_h)' \|_2 .$

From § 5.2, we have with $\lambda = h/\sqrt{2}$ that

(7.17) $h \| (u_{nh} - w_h)' \|_2 \leqslant c \| (T_\lambda * dF_n)^{1/2} - (T_\lambda * dF_o)^{1/2} \|_2 ,$

and so, if $f_o \in W^{2,1}(\mathbb{R})$ and f_o has a finite exponential moment, then for deterministic or random h, for all $\varepsilon > 0$, there exists a constant c such that

(7.18) $| \delta_{nh} | \leqslant_{as} c h^2 + c h^{-1/2-\varepsilon} n^{-1/2+\varepsilon} .$

The second term in (7.18) comes from the integration by parts tricks of § 4.3 and the bounds from § 4.4. With (7.15), this gives

$$h \| (u_{nh})' \|_2 \leqslant_{as} c h + c h^{-1/2-\varepsilon} n^{-1/2+\varepsilon} ,$$

and so, by taking $h \asymp n^{-1/3}$, for all $\varepsilon > 0$,

(7.19) $\min_{h>0} h \| (u_{nh})' \|_2 =_{as} \mathcal{O}(n^{-1/3+\varepsilon}) .$

For random h, similar arguments give

$$h \| (u_{nh})' \|_2 \geqslant h \| w_o' \|_2 - | \delta_{nh} | \geqslant_{as} h \| w_o' \|_2 - c h^{-1/2-\varepsilon} n^{-1/2+\varepsilon} ,$$

and it follows that for all $\varepsilon > 0$, there exists a constant c such that

(7.20) $H_{n,GK} \| w_o' \|_2 \leqslant_{as} c n^{-1/3+\varepsilon} + c (H_{n,GK})^{-1/2-\varepsilon} n^{-1/2+\varepsilon} .$

Consequently, for all $\varepsilon > 0$,

(7.21) $H_{n,GK} =_{as} \mathcal{O}(n^{-1/3+\varepsilon}) ,$

and we have proven the following result.

(7.22) LEMMA. *Let $\varepsilon > 0$ be arbitrary. If $f_o \in W^{2,1}(\mathbb{R})$, $\sqrt{f_o} \in W^{1,2}(\mathbb{R})$, and f_o has a finite exponential moment, then $H_{n,GK} =_{as} \mathcal{O}(n^{-1/3+\varepsilon})$.*

So, this choice of the smoothing parameter will not work asymptotically since $H_{n,GK} \ll n^{-1/5}$, the optimal rate for h.

(7.23) EXERCISE. Show that (7.20) implies (7.21).

Are there *good* ways to select h? Nothing seems to be known about this! Of all the methods discussed for kernel estimation, only the discrepancy principle of §6 is easily adapted to the present circumstances and easily analyzed. Thus, let F^{nh} be the distribution function with subdensity f^{nh}, the GOOD estimator. The discrepancy principle for selecting h may then be formulated as follows.

(7.24) In the *GDP* method, the smoothing parameter
is chosen as the smallest solution of

$$\| F_n - F^{nh} \|_\infty = c_{GDP}\, n^{-2/5} ,$$

where $c_{GDP} = 0.35$. The h so selected is denoted by $H_{n,GDP}$ and the corresponding GOOD estimator $f^{n,H_{n,GDP}}$ by $f_{n,GDP}$.

Note that $c_{GDP} = c_{DP}$, as in §6, but the authors have not experimented with other values of c_{GDP}.

We now address the usual concerns: existence, scaling invariance, and asymptotic behavior. The existence of $H_{n,GDP}$ is a consequence of the following lemma, whose proof we leave as an exercise.

(7.25) LEMMA. *For almost all realizations of $X_1, X_2, \cdots, X_n$,*

(a) $\| F_n - F^{nh} \|_\infty$ *is a continuous function of h,*

(b) $\lim_{h \to 0} \| F_n - F^{nh} \|_\infty = \frac{3}{4n}$, *and*

(c) $\lim_{h \to \infty} \| F_n - F^{nh} \|_\infty = \frac{1}{2}$.

(7.26) EXERCISE. Prove Lemma (7.25). [Hint: The asymptotic behavior of the $u_{nh}(X_i)$ for $h \to 0$ and $h \to \infty$ comes in handy here.]

We next consider the asymptotic behavior of $H_{n,GDP}$.

(7.27) THEOREM. *If $\sqrt{f_o} \in W^{1,2}(\mathbb{R})$ and f_o has a finite moment of order > 2, then*

$$H_{n,GDP} \asymp n^{-1/5} \quad almost\ surely .$$

PROOF. In the following, let $\lambda = h/\sqrt{2}$. We start with the decomposition

$$F_n - F^{nh} = \{ F_n - F_o - \mathbb{B}_\lambda * (F_n - F_o) \} + F_o - \mathbb{B}_\lambda * F_o +$$
$$h^2 \mathbb{B}_\lambda * |(u_{nh})'|^2 + \{ \mathbb{B}_\lambda * F_n - F^{nh} - h^2 \mathbb{B}_\lambda * |(u_{nh})'|^2 \} ,$$

where $\mathbb{B}_\lambda$ is the distribution corresponding to the density $\mathfrak{B}_\lambda$. The expression in curly brackets on the last line vanishes, cf. (5.2.18). The expression in curly brackets in the first line satisfies

$$\| F_n - F_o - \mathbb{B}_\lambda * (F_n - F_o) \|_\infty \leqslant 2 \| F_n - F_o \|_\infty =_{as} \mathcal{O}\big((n^{-1} \log \log n)^{1/2} \big) .$$

Finally, if $\sqrt{f_o} \in W^{1,2}(\mathbb{R})$, then

$$h^2 \mathbb{B}_\lambda * | (u_{nh})'|^2 = h^2 \mathbb{B}_\lambda * | (f_o^{1/2})'|^2 + \mathcal{O}(h^3) + \varepsilon_{nh}$$
$$= h^2 \mathfrak{B}_\lambda * \Phi_o + \mathcal{O}(h^3) + \varepsilon_{nh} ,$$

with

$$\Phi_o(x) = \int_{-\infty}^x | [(f_o^{1/2})'](y)|^2 \, dy$$

and

$$\varepsilon_{nh} = h^2 \mathbb{B}_\lambda * \big(| (u_{nh} - w_h)'|^2 \big) .$$

Here, w_h is the solution of the large sample asymptotic version of (7.1). With (7.17), we thus have

$$\| \varepsilon_{nh} \|_\infty \leqslant c h^2 \| \mathbb{B}_\lambda \|_\infty \| (u_{nh} - w_h)' \|_2^2$$
$$\leqslant c \| (T_\lambda * dF_n)^{1/2} - (T_\lambda * dF_o)^{1/2} \|_2^2$$
$$\leqslant c \, \mathrm{KL}(T_\lambda * dF_n , T_\lambda * dF_o) .$$

The bounds now follow from § 4.5. The randomness of λ causes no problem, by the monotonicity in λ of $\mathrm{KL}(T_\lambda * dF_n , T_\lambda * dF_o)$.

Putting everything together gives

$$\left| \| F_n - F^{nh} \|_\infty - \| F_o - \mathbb{B}_\lambda * dF_o + h^2 \mathfrak{B}_\lambda * \Phi_o \|_\infty \right| \leqslant$$
$$\| F_n - F_o - \mathbb{B}_\lambda * (F_n - F_o) \|_\infty =_{as} \mathcal{O}((n^{-1} \log \log n)^{1/2}) .$$

It follows that for a suitable constant c,

$$\| F_n - F^{nh} \|_\infty =_{as} c h^2 + \mathcal{O}(h^3) + \mathcal{O}((n^{-1} \log \log n)^{1/2}) + o((nh)^{-1/2}) .$$

Finally, the discrepancy principle says that this should equal $c_{GDP} \, n^{-2/5}$, and the conclusion follows.   Q.e.d.

The last concern of this section is the scaling invariance of $H_{n,GK}$ and $H_{n,GDP}$. To phrase this properly, we need to exhibit their dependence on the sample $\mathbb{X}_n = (X_1, X_2, \cdots , X_n)$, and show that they satisfy (1.16) and (1.17). So, let $f^{nh}(x ; \mathbb{X}_n)$ denote the GOOD estimator for a fixed value of

h and sample $\mathbb{X}_n$. Likewise, let $H_{n,GK}(\mathbb{X}_n)$ and $H_{n,GDP}(\mathbb{X}_n)$ denote $H_{n,GK}$ and $H_{n,GDP}$ for the given sample. The key to proving scaling invariance of the smoothing parameters selected is to see how scaling affects $f^{nh}(x;\mathbb{X}_n)$.

(7.28) LEMMA. *For all $h > 0$, $t > 0$ and all $X_1, X_2, \cdots, X_n$,*

$$t\,f^{n,\lambda}(t\,x;\ t\,\mathbb{X}_n) = f^{nh}(x;\,\mathbb{X}_n)\,, \quad x \in \mathbb{R}\,,$$

where $\lambda = t\,h$.

(7.29) EXERCISE. (a) For any smooth, nonnegative function f and $t > 0$, let f_t be defined by $f_t(x) = t^{-1}f(t^{-1}x)$. Show that

$$t^2\,\|\,\{\sqrt{f_t}\,\}'\,\|_2^2 = \|\,\{\sqrt{f}\,\}'\,\|_2^2\,.$$

(b) Prove Lemma (7.28).

The interpretation of Lemma(7.28) is that the parameter h in (7.1) acts like the smoothing parameter h in kernel density estimation. This was more or less predicted by the Comparison with kernel density estimation lemma (5.2.22).

(7.30) LEMMA. *For all realizations of $\mathbb{X}_n$,*
(a) $t^{-1}\,H_{n,GK}(t\,\mathbb{X}_n) = H_{n,GK}(\mathbb{X}_n)\,,$ *and*
(b) $t^{-1}\,H_{n,GDP}(t\,\mathbb{X}_n) = H_{n,GDP}(\mathbb{X}_n)\,.$

(7.31) EXERCISE. Prove the lemma.

EXERCISES: (7.14), (7.23), (7.26), (7.29), (7.31).

8. Additional notes and comments

Ad § 3: The idea of adding a penalization term to the double kernel estimator of the L^1 error is similar to the idea of BARRON, BIRGÉ and MASSART (1999) in the model selection context.

Ad § 6: The development here follows EGGERMONT and LARICCIA (1996).

Ad § 8: Had there been a real § 8, it would have dealt with the smoothing parameter selection for the roughness penalization of the log-density. Some useful references are STONE (1990), STONE and KOO (1986), KOOPERBERG and STONE (1991), GU and QIU (1993), and GU (1993).

Ad § 8: Current interest in nonparametric density estimation is in *adaptive* estimation methods. A method is called adaptive if it is asymptotically optimal over wide classes of densities. In a language similar to (1.4), one would like estimators f_n for which

$$(8.1) \qquad \lim_{n \to \infty} \sup_{f_o \in \mathcal{F}} \frac{\| f_n - f_o \|_1}{\inf_{\varphi_n \in \Phi_n} \| \varphi_n - f_o \|_1} =_{as} 1 \ ,$$

where Φ_n is the set of *all* estimators and $\mathcal{F}$ is the (large) class of densities one wishes to estimate. We have been concerned mostly with the class $PDF(C, C')$ of (1.5.1)–(1.5.2). In (8.1), the supremum should perhaps be outside the limit. Either way, shooting for (8.1) is quite ambitious, so typically one is satisfied if the above limit is finite. In the context of density estimation, one might want to estimate the (optimal) order of the kernel, as well as the smoothing parameter, see, e.g., DEVROYE, LUGOSI and UDINA (1998). In its own way, WATSON and LEADBETTER (1963) already investigated adaptive methods, by constructing optimal kernels based on a *priori* information on the unknown density. For more on adaptation in the density estimation and more general contexts, see BARRON, BIRGÉ and MASSART (1999) and references therein.

8

Nonparametric Density Estimation in Action

1. Introduction

We have come to the end of a long road, and it is time to put nonparametric density estimation to work, in particular, the various procedures for smoothing parameter selection. In the previous chapters, asymptotic properties were studied. Here, we emphasize the small sample behavior for sample sizes ranging from about 100 to 1000. We apply the various methods to our workhorse data sets, the Buffalo snow fall and the Old Faithful geyser data, and compare them in simulation experiments involving mostly smooth densities with light tails. One reason for this decision is that with small sample sizes, one can reasonably hope to recover only a smooth approximation to the unknown density. By way of example, with 100 observations, one can do a reasonable job of estimating a uniform density, but one cannot really hope to distinguish it from a sawtooth density with 20 peaks and valleys, say.

(1.1) Exercise. Verify by means of simulations whether the above statement is in fact true.

In the simulation experiments, the various methods are judged by their mean L^1 errors compared with the mean "optimal" L^1 errors. We recall how the "optimal" method is defined: With $f^{nh} = A_h * dF_n$,

(1.2)   in the *OPT* method, the smoothing parameter
        is the solution to

$$\text{minimize} \quad \| f^{nh} - f_o \|_1 \quad \text{subject to} \quad h > 0 .$$

The h so selected is denoted by $h_{n,OPT}$ and the corresponding kernel estimator by $f_{n,OPT}$.

Unfortunately, this is not a rational method, but the (L^1) methods of the previous chapter were designed to mimic it as well as possible.

The chapter is organized as follows. In § 2, we consider computational issues, in particular, the finite-dimensional approximations to kernel and maximum likelihood density estimators. The comparison of smoothing parameter selection procedures is carried out in § 3, with predictable results, sort of. The methods are then let loose on the Buffalo snow fall and Old Faithful geyser data in § 4. The irrelevance of higher order kernels in the context of small sample sizes is illustrated in § 5. As expected, for large sample sizes and smooth densities, the higher order kernels are at an advantage. We also compare some first-order kernels and observe that only the two-sided exponential density performs poorly, apparently due to its somewhat heavy tails. One would guess that this bodes ill for the maximum likelihood estimator with roughness penalization of the root density (the GOOD estimator), but one would be wrong! As a matter of fact, the GOOD estimator should be called the remarkably GOOD estimator. In § 6, unimodal density estimation is considered, with somewhat unexpected results.

EXERCISE : (1.1).

2. Finite-dimensional approximations

In this section, we discuss finite-dimensional approximations to kernel estimators, as well as to maximum likelihood estimators with smoothing or roughness penalization. An equivalent way of saying this is that we need to discretize the infinite-dimensional problem. In the authors' experience, this seems indispensable.

Is there really a need for finite-dimensional approximations to kernel estimators? Consider the computational aspects of kernel density estimation, in particular, the selection of the smoothing parameter, which typically involves calculating integrals such as $\| (A_h - B_h) * dF_n \|_1$. A standard way of computing these would be to calculate the integrand on a fine grid of points $x_j = j\,\delta\,, j = 0, \pm 1, \pm 2, \cdots, \pm N$, for some small $\delta > 0$, and compute the integral using Riemann sums or the trapezoidal or Simpson's rule. Of course, care must be taken that the functions in question do not vary wildly between grid points. For very small h, this means that the grid points must be spaced very close together, or one may encounter situations in which the computed value of $\| (A_h - B_h) * dF_n \|_1$ exceeds 2, even if A and B are nonnegative kernels. Thus, a lot of time and energy is spent on computations that are later discarded anyway: The "optimal" h is usually much larger than the smallest one considered. One way around this problem is to *meaningfully* approximate the set of densities

$$(2.1) \qquad\qquad \{ A_h * dF_n : h > 0 \}$$

by densities from a (fixed) finite-dimensional subspace of L^1. Here, this is done as follows. Define the partition $\{\, \omega_j \, : \, j \in \mathbb{Z} \,\}$ by

$$(2.2) \qquad \omega_j = (\, x_j \, , \, x_{j+1} \,] \, , \quad j \in \mathbb{Z} \, ,$$

where the x_j are the knots of the partitions, say, with $x_j = j\,\delta$ for some $\delta > 0$. Let f be a density on the line (more generally, a nonnegative $L^1(\mathbb{R})$ function), and approximate it by Tf, defined as

$$(2.3) \qquad Tf(x) = \frac{1}{|\,\omega_j\,|} \int_{\omega_j} f(y) \, dy \, , \quad x \in \omega_j \, .$$

On occasion, T is also denoted as T_δ. It is a fundamental result of the theory of Lebesgue integration that $\|\, f - T_\delta f \,\|_1 \longrightarrow 0$ for $\delta \to 0$. Moreover, if the density is smooth, then the convergence is reasonably fast.

The appeal of the above scheme is that $T(A_h * dF_n)$ may be computed explicitly in terms of the distribution $\mathbb{A}$ corresponding to the kernel A. In particular,

$$(2.4) \quad TA_h * dF_n(x) = \frac{\mathbb{A}_h * dF_n(x_{j+1}) - \mathbb{A}_h * dF_n(x_j)}{x_{j+1} - x_j} \, , \quad x \in \omega_j \, ,$$

where
$$\mathbb{A}_h * dF_n(x) = \int_{\mathbb{R}} \mathbb{A}\big(\, h^{-1}(x - y)\,\big) \, dF_n(y) \, .$$

Thus, $TA_h * dF_n$ is completely determined by its values at the points $\frac{1}{2}(\, x_j + x_{j+1}\,)$, $j \in \mathbb{Z}$. Now, $TA_h dF_n$ is still an infinite-dimensional object, but if A has compact support, then so does $TA_h * dF_n$, and a finite number of subintervals suffices. Even if A is the normal density, a finite number of subintervals suffices for most intents and purposes. So, $TA_h * dF_n$ is in effect a finite-dimensional object. Now, integrals like $\|\, \sqrt{A_h * dF_n} \,\|_1$ may be meaningfully approximated by the Riemann sums $\|\, \sqrt{TA_h * dF_n} \,\|_1$. In the simulation experiments, when we wish to compute the actual error in the kernel estimator, we approximate $\|\, f_o - f^{nh} \,\|_1$ by $\|\, Tf_o - Tf^{nh} \,\|_1$, which is again a Riemann sum.

The bin width δ in the intervals (2.2) should be such that for properly chosen h, the approximation error $\|\, A_h * dF_n - TA_h * dF_n \,\|_1$ is negligible compared with the expected variation in $A_h * dF_n$. For relatively smaller h, this approximation error will be large, but the computed estimator $TA_h * dF_n$ will still be a density. This lends a certain robustness to the computations.

We should note that this approximation scheme has its perils. By way of example, consider the double kernel method in which one must minimize $\|\, (A_h - B_h) * dF_n \,\|_1$. In the approximation scheme under discussion, this is replaced by

$$\text{minimize} \quad \|\, TA_h * dF_n - TB_h * dF_n \,\|_1 \quad \text{subject to} \quad h > 0 \, .$$

Unfortunately, we have
$$\lim_{h \to 0} \|\, (TA_h - TB_h) * dF_n \,\|_1 = 0 \, ,$$

rather than the "true" limit $\| A - B \|_1$. Thus, the smoothing parameter h must be restricted to $h \geqslant C\delta$ for a suitable constant C.

An alternative, used by DEVROYE (1989), is not to discretize at all and to compute the errors $\| f_o - f^{nh} \|_1$ and $\| (A_h - B_h) * dF_n \|_1$ exactly, as follows. First, determine all of the zeros (to machine precision, say) of the function $f_o(x) - f^{nh}(x)$, and write the L^1 norm as the sum of integrals over the subintervals on which $f_o - f^{nh}$ does not change sign. Each of these integrals may be evaluated exactly in terms of the corresponding distribution functions. The tricky point is to make sure that all sign changes of $f^{nh} - f_o$ have been determined. Of course, the same considerations apply to $\| (A_h - B_h) * dF_n \|_1$.

This concludes the discussion of finite-dimensional approximations to kernel estimators.

We next consider some of the maximum likelihood estimators of Chapters 5 and 6, beginning with the GOOD estimator of §5.2. As already explained in §5.2, the GOOD estimator may be treated similarly to kernel estimation. Let f^{nh} be the GOOD estimator, and recall that $u_{nh} = \sqrt{f^{nh}}$ satisfies

$$(2.5) \qquad u_{nh}(x) = \tfrac{1}{n} \sum_{i=1}^{n} [\, u_{nh}(X_i)\,]^{-1} \, \mathfrak{B}_h(x - X_i) \,, \quad -\infty < x < \infty \,,$$

where $\mathfrak{B}$ is the two-sided exponential density. Thus, once the $u_{nh}(X_i)$ have been computed, say, by Newton's method, then u_{nh} is a weighted kernel estimator, and the considerations regarding plain kernel estimators apply. So, u_{nh} may be approximated by $T u_{nh}$, and f^{nh} in turn by the step function

$$(2.6) \qquad\qquad\qquad \mathcal{T} f^{nh} = (T u_{nh})^2 \,.$$

On a fine enough grid, this works satisfactorily. An alternative would be to consider $T f^{nh}$. Since

$$(2.7) \quad f^{nh}(x) = \frac{1}{n^2} \sum_{i,j=1}^{n} [\, u_{nh}(X_i)\, u_{nh}(X_j)\,]^{-1} \, \mathfrak{B}_h(\cdot - X_i)\, \mathfrak{B}_h(\cdot - X_j) \,,$$

one must deal with expressions like $T\{\, \mathfrak{B}_h(\cdot - X_i)\, \mathfrak{B}_h(\cdot - X_j)\,\}$, but it is not clear that this is worth the trouble: The difference will be negligible for the relevant range of h values.

For unimodal estimation, the maximum smoothed likelihood problem itself is discretized. Thus, we approximate the maximum smoothed likelihood problem by

$$(2.8) \qquad \text{minimize} \quad -\int_{\mathbb{R}} T A_h * dF_n(x) \, \log f(x) \, dx + \int_{\mathbb{R}} f(x) \, dx$$

$$\text{subject to} \quad f \text{ nonnegative, unimodal} \,.$$

If the mode of the unimodal estimator is restricted to be one of the knots of the partition underlying T, then one verifies that the solution, again denoted by f^{nh}, is a step function on the partition, and that the pool-adjacent-violators algorithm of § 6.4 works as is. So the solution of (2.8) is a finite-dimensional object of the form $T\varphi$ for some density φ.

(2.9) EXERCISE. An alternative to the above scheme is approximation by continuous, piecewise linear functions on the grid (2.2). Work out the details, and run the experiments. In short, rewrite this chapter.

EXERCISE : (2.9).

3. Smoothing parameter selection

We now come to the experiments for comparing the various smoothing parameter selection procedures in kernel density estimation.

The first problem is to decide on a benchmark set of densities on which to try out the various methods. MARRON and WAND (1989) and BERLINET and DEVROYE (1994) have a close to exhaustive collection of densities, including many rough ones, but we are certainly not going to those lengths. So what should *our* benchmark family of densities consist of? It is a *sine qua non* that it should contain the normal as well as the uniform density. The normal density because it is the easiest of all, and every method should perform very well for it. The uniform should belong, because it is the simplest density that is not smooth (and it has no tails). We refrained from including the other extreme, say, the Cauchy density, which is infinitely smooth, but is all tail. Because estimating a normal density takes a large smoothing parameter compared to the scale of the density, we also incorporated a normal density with a small bump added, which is referred to as the nice mixture density. This density requires a relatively smaller smoothing parameter. Actually, it is the authors' experience that if a procedure works well for these three examples, the battle is half won (not the war, mind you). Another feature of the normal density is that it is symmetric, so a skewed (Beta) density was included, but it was made less smooth. To really test the limits of the various procedures, a density with two scales was added, implemented as a mixture of two normals with drastically different means and variances (the bad mixture). [This is an example for which a locally adaptive smoothing parameter would be very effective, but we shall not be concerned with this. See Exercise (3.2).] Finally, a bimodal density was added (courtesy of Paul Deheuvels in a successful attempt to expose the Achilles heel of the discrepancy principle of § 7.6). Although we are unabashed in making the usual nonparametric smoothness and tail assumptions, the benchmark family contains some significant, nonsmooth densities.

We now precisely define the set of densities. Because the normal density features so prominently, let $\phi(x) = (2\pi)^{-1/2} \exp(-x^2/2)$ denote the standard normal density with mean 0 and variance 1, and set $\phi_\sigma(x) = \sigma^{-1}\phi(\sigma^{-1}x)$. It is also useful to recall the Beta density with parameters α and β

$$\psi_{\alpha,\beta}(x) = B(\alpha,\beta)\,(\,x_+)^{\alpha-1}(\,(1-x)_+)^{\beta-1}\ .$$

To summarize, this select committee of judges consists of symmetric, skewed, unimodal, bimodal, and piecewise (non)smooth densities, and ought to provide a reasonably objective test of the procedures. Their graphs are shown in Figure 3.1. Thus, a precise description of the judges in the contest for the best all around smoothing parameter selection procedures are

nice mix $f_o(x) = \frac{9}{10}\phi_{1/2}(x-5) + \frac{1}{10}\phi_{1/2}(x-7)$,

normal $f_o(x) = \phi(x-5)$,

uniform $f_o(x) = \frac{1}{5}\,\mathbb{1}(\,3 < x < 8)$,

beta $f_o(x) = \frac{1}{5}\psi_{\alpha,\beta}\left(\frac{1}{5}(x-3)\right)$,   with $\alpha = 1.4$, $\beta = 2.6$,

bad mix $f_o(x) = \frac{1}{5}\phi_{9/5}(x-6) + \frac{4}{5}\phi_{1/10}(x-2)$,

bimodal $f_o(x) = \frac{1}{2}\phi_{1/2}(x-3.5) + \frac{1}{2}\phi_{1/2}(x-6.5)$,

As a practical observation, note that every distribution in the above list essentially lives on the interval $[0,10]$, and that a grid of 400 equally spaced subintervals is sufficient for the approximation scheme of § 2 to be effective. For the two-scale mixture of normals ("bad mix"), this has to be taken with a grain of salt, but it is unlikely to affect the conclusions (no one-scale method can estimate it very well on the basis of 100 observations, not even the optimal one-scale method).

For almost every smoothing parameter selection procedure, an empirical functional $\mathcal{L}(f^{nh})$ must be minimized over h, e.g., in the double kernel method $\| TA_h * dF_n - TB_h * dF_n \|_1$. Even the discrepancy principle of § 7.6 fits into this scheme. Rather than finding the exact minimum, we compute these functionals $\mathcal{L}(f^{nh})$ on a fine grid of h values and pick the smallest one among finitely many. This also permits us to graph $\mathcal{L}(f^{nh})$ as a function of h, which gives useful diagnostic information as to the validity of the process as a whole. Two examples are given in Figures 4.1 and 4.3. Inspection reveals or suggests that the functionals in question do not always have a useful minimum. To remedy this, for the methods *SPI*, *MOD*, and *DGHW*, we choose the local minimum of each functional *closest* to $H_{n,DP-N}$, see method II below. The range of h values was chosen in a somewhat unrational way, by first selecting a reasonable range for each density, and then (rationally) extending the range up to the upper bounds $H_{n,UP-E}$ and $H_{n,UP-N}$ of Exercise (7.4.21).

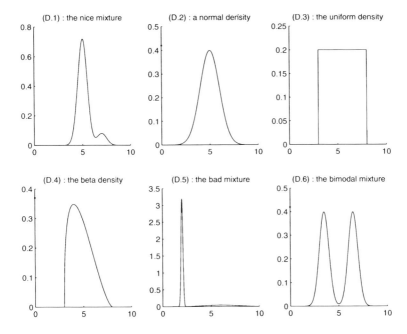

Figure 3.1. The judges : six densities used to judge the various smoothing parameter selection procedures. The various mixtures are mixtures of normal densities.

Now to the actual simulation setup. For each benchmark density, random samples of size 100 were generated, and the smoothing parameters determined according to the various criteria discussed in the previous chapter. In all cases, this was replicated 1000 times. Based on the 1000 replications, the (sample) means and standard deviations of $\| T f^{nH} - T f_o \|_1$ were computed for the various H. The kernels employed were either the Epanechnikov kernel or the normal density, with the Epanechnikov kernel scaled such that its variance equals 1. For L^1 errors, this is an invisible choice, but it does affect smoothing parameters in the obvious way.

In the comparisons below, an underlying theme is the comparison of the Epanechnikov kernel and the normal kernel. In the next section, we take a closer look at kernel selection, especially that of higher order kernels.

The smoothing parameter selection procedures to be compared are as follows. The acronyms serve mainly as entries into various tables.

   I. OP-E , OP-N : the optimal methods based on the Epanechnikov and
      normal kernels.
  II. DP-E , DP-N : the discrepancy principle of § 7.6, for the two kernels.
 III. DB-2.4, DB-4.2 : the double kernel method of § 7.3, with A the
      Epanechnikov kernel (with support $[-1, 1]$), and $B = L_\delta$, with

Table 3.1(a). Estimated means of $\| Tf^{nH} - Tf_o \|_1$ for various smoothing parameter selection procedures applied to various densities, for sample size 100, based on 1000 replications. The interval $[0,10]$ was divided into 400 subintervals.

	Nice mix	Nor-mal	Uni-form	Beta	Bad mix	Bi-modal
OP-E	0.158	0.127	0.217	0.162	0.335	0.187
OP-N	0.160	0.129	0.218	0.164	0.335	0.189
DP-E	0.180	0.151	0.243	0.178	0.360	0.237
DP-N	0.181	0.152	0.241	0.179	0.358	0.242
DB-2.4	0.223	0.193	0.239	0.203	0.406	0.256
DB-4.2	0.190	0.166	0.247	0.194	0.375	0.220
PER	0.176	0.148	0.250	0.182	0.370	0.204
SPI	0.175	0.146	0.247	0.183	0.361	0.201
MOD	0.174	0.144	0.258	0.186	0.352	0.199
DGHW	0.174	0.143	0.257	0.184	0.348	0.199
PI-DH	0.207	0.139	0.260	0.180	1.290	0.526
PI-DB	0.175	0.146	0.249	0.182	0.350	0.205
CV	0.227	0.185	0.242	0.203	0.406	0.259
PI-N	0.202	0.139	0.257	0.179	1.280	0.508
SJ	0.185	0.158	0.233	0.181	0.363	0.202

L given by (BD-I) of §4, and $\delta = 2.4$, respectively $\delta = 4.2$, all according to BERLINET and DEVROYE (1994).

IV. PER: the perverted double kernel method of §7.3, based on the double kernel method DB-4.2.

V. SPI, MOD, DGHW: The methods of §7.5, based on the kernel

$$C = \frac{A - A_\lambda}{1 - \lambda^2}, \quad \text{with} \quad \lambda = 0.995 .$$

VI. PI-DH: The L^1 asymptotic plug-in method with (parametric) normal pilot estimator of DEHEUVELS and HOMINAL (1980), using Epanechnikov kernels, see §7.4.

VII. PI-DB: The asymptotic plug-in method with (nonparametric) pilot estimator using the double kernel method, of BERLINET and DEVROYE (1994), as described in §7.4.

VIII. CV: the least-squares cross-validation method of RUDEMO (1982), based on the Epanechnikov kernel.

Table 3.1(b). Estimated standard deviations of $\| Tf^{nH} - Tf_o \|_1$ for various smoothing parameter selection procedures applied to various densities, for sample size 100, based on 1000 replications. The interval $[0,10]$ was divided into 400 subintervals.

	Nice mix	Normal	Uniform	Beta	Bad mix	Bimodal
OP-E	0.053	0.054	0.042	0.041	0.055	0.053
OP-N	0.053	0.054	0.042	0.041	0.055	0.053
DP-E	0.056	0.057	0.039	0.047	0.055	0.066
DP-N	0.056	0.057	0.039	0.047	0.055	0.066
DB-2.4	0.057	0.055	0.046	0.055	0.086	0.059
DB-4.2	0.064	0.066	0.055	0.058	0.056	0.063
PER	0.050	0.056	0.045	0.053	0.049	0.052
SPI	0.054	0.057	0.042	0.048	0.051	0.053
MOD	0.054	0.056	0.042	0.048	0.051	0.053
DGHW	0.054	0.057	0.038	0.047	0.051	0.054
PI-DH	0.061	0.054	0.034	0.045	0.039	0.029
PI-DB	0.055	0.056	0.045	0.048	0.053	0.057
CV	0.070	0.064	0.059	0.062	0.047	0.080
PI-N	0.060	0.054	0.034	0.045	0.030	0.030
SJ	0.055	0.058	0.041	0.047	0.053	0.055

IX. PI-N: The least-squares asymptotic plug-in method with (parametric) normal pilot estimator of DEHEUVELS (1977), using Epanechnikov kernels.

X. SJ: the SHEATHER and JONES (1991) method. For the last three methods, see § 7.2.

The results are given in Table 3.1. Some conclusions may be drawn. The asymptotic plug-in method with double kernel pilot estimator (PI-DB) is remarkably robust for all examples considered, even for the two-scale mixture of normal densities (the bad mix). However, for this density, even the optimal methods are not impressive (since they are one-scale estimators). It should be noted that for the uniform density, the PI-DB estimator is in fact provided by $H_{n,UP}$ in about one-quarter of the replications. We also draw attention to the methods MOD, SPI, and DGHW. On the smooth densities (nice mix, normal and bimodal), their performance is exemplary, but the rough densities cause problems. The SPI method is preferable to every method considered, if we ignore the bad mixture of normals. Considering all of the densities in our limited collection, the method DGHW is perhaps a little bit worse than PI-DB: On the smooth densities it gives just about the same results; for the bimodal density, it is a bit better; and on the uniform and Beta density, it is a bit worse. The overall message is that it is "easy" to get good methods for smooth densities, but quite difficult to get methods that also work well for rough densities. The behavior of the SJ method is somewhat odd, undoubtably due to the fact that we are judging it by the L^1 error rather than by the squared L^2 error.

(3.1) EXERCISE. This deals with the value $c_{pp} = 0.35$ used in the discrepancy principle, which seems arbitrary. Estimate

$$\mathbb{E}[\, \| \, f_{n,pp} - f_o \, \|_1 \,]$$

via simulation experiments, as a function of c_{pp}, for various densities f_o, and pick the best value of c_{pp}. This depends very much on the particular choices of f_o considered, but the collection of judges discussed earlier seems quite realistic for small sample sizes.

(3.2) EXERCISE. This exercise deals with locally adaptive smoothing parameters. The standard interpretation is to let h be a function of x, so that the kernel estimator looks like

$$\frac{1}{n} \sum_{i=1}^{n} (h(x))^{-1} A\big((h(x))^{-1}(x - X_i) \big) , \quad x \in \mathbb{R} ,$$

but this is not a density. The alternative is to let h depend on the data points, more specifically, on the k nearest order statistics. Thus, the estimator is

$$\frac{1}{n} \sum_{i=1}^{n} (h_{i,k})^{-1} A\big((h_{i,k})^{-1}(x - X_{i,n}) \big) , \quad x \in \mathbb{R} ,$$

where $h_{i,k} = H(X_{i-m,n}, \cdots, X_{i+m,n})$ for $k = 2\,m + 1$, with suitable corrections at the ends. See BREIMAN, MEISEL and PURCELL (1977). It seems reasonable to restrict attention to selectors based on

$$\overline{\Delta} = \frac{1}{2k} \sum_{j=i-m}^{i+m-1} \big\{ X_{j+1,n} - X_{j,n} \big\} ,$$

the average spacing of the $2\,m + 1$ neighboring observations. That is, to consider smoothing parameters of the form $h_{i,k} = \lambda \, \Psi(\overline{\Delta})$, where λ is a free parameter and Ψ is some function (which one?) of $\overline{\Delta}$. The parameter λ can then be chosen by the usual methods, but note that λ ought to be dimension free. Investigate this practically and theoretically. An alternative is to let

$$h_i = \lambda \, (\, f_o(X_i) \,)^{-1/2} ,$$

and to use a pilot estimator for f_o. See ABRAMSON (1982) and WAND and JONES (1995).

(3.3) EXERCISE. Investigate the methods MOD, SPI, and DGHW for the kernel

$$C = A - 2A_\lambda + A_{\lambda\sqrt{2}} \; ,$$

with A the normal kernel. What stretch parameter λ seems to work the best? Could another fourth-order kernel be used conveniently, so that the methods are easy to implement?

(3.4) EXERCISE. Connsider the methods MOD, SPI, and DGHW for the kernel

$$C = \frac{1}{\sqrt{8\pi}} \, (x^2 - 1) \, e^{-x^2/2}$$

and investigate the use of stretch parameters in C. (This is close to what BERLINET and DEVROYE (1994) do for the plug-in method with a double kernel pilot estimator.)

EXERCISES: (3.1), (3.2), (3.3), (3.4).

4. Two data sets

We apply some of the smoothing parameter selection procedures to the Buffalo snow fall data and the Old Faithful geyser data (see Appendix 1). Because this is about the only place where we report values of smoothing parameters, we emphasize that in this section the two kernels used are the Epanechnikov kernel and a scaled normal kernel, both with variance $1/5$.

We begin with the Buffalo snow fall data. Recall that in § 3.3 this data set was analyzed using mixtures of normals, with the conclusion that the trimodal mixtures of three normals seemed the most "natural". See Figure 3.3.1, set 4. Also, in § 1.1, graphs of the kernel estimators for a range of smoothing parameters were shown. There, we argued that the trimodal and unimodal estimators seemed the most natural, but that it was not so clear how to decide which of the two corresponds closer with the "truth". In Figure 4.2, the kernel estimators with smoothing parameter chosen by some of the selection procedures are shown. This confirms our earlier statement that the unimodal and trimodal estimators seem the most natural. But still: Which one corresponds closer to the truth?

To gain some more information regarding the "truth", and further insight into the inner workings of the various selection procedures, we now look at the graphs of the various functionals involved in the parameter selection procedures. See Figure 4.1. It seems clear that the two versions of the discrepancy principle (DP-N and DP-E, for the normal and Epanechnikov kernel respectively) provide "stable" estimators of the smoothing parameters: The curves are nicely increasing, and the location where they

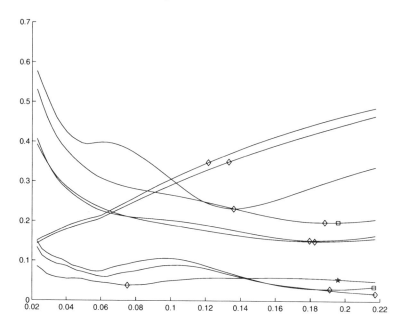

Figure 4.1. How the smoothing parameters get selected for the Buffalo snow fall data. The figure shows the graphs of the relevant functionals as functions of h for each method. The $\diamond$ indicates the minimum of each functional, but for the Discrepancy Principle methods (the two strictly increasing graphs), it indicates the point where the functional equals c_{DP}. The $\star$ indicates the position of $H_{n,PI-DB}$. The two squares ($\square$) on the right of the graphs indicate the upper bounds on the H, for the Epanechnikov and normal kernels. The graphs are identified as follows. The upper graph of the two Discrepancy Principle graphs is based on the Epanechnikov kernel, the lower one on the normal kernel. The other graphs from top to bottom at $h = 0.1$ are the graphs of $PER(h)$, $MOD(h)$, $SPI(h)$, and $DGHW(h)$. The bottom three graphs correspond to the double kernel methods (DB-E) with stretch parameters (from top to bottom, at $h = 0.1$) 3, 2.8, and 2.4. The acronyms are explained in § 7.3.

intersect the imaginary horizontal line at height $c_{DP} = 0.35$ will not change much when the curves move about a little. Likewise, the locations of the minima of the functionals for the methods PER, SPI, MOD, and DGHW seem fairly stable with respect to changes in the curves. The situation is much less clear for the double kernel methods. The (local) minima can be moved about by judiciously choosing the stretch parameter. However, remember that all of the other methods depend on double kernel methods (or double kernel functionals).

As a further comment, we note that estimators based on least-squares cross validation (CV) and on the Discrepancy Principle (DP-E) seem to

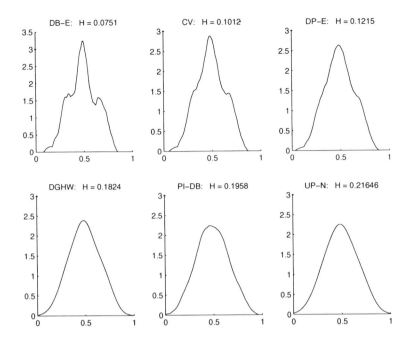

Figure 4.2. The kernel estimators for the Buffalo snow fall data, with smoothing parameters selected according to the various criteria. The acronyms are explained in § 3.

correspond the closest to the "truth", notwithstanding that one estimator is bimodal and the other unimodal.

For the Old Faithful geyser data, the situation is similar. There are two quite different candidates. The first set of kernel estimators corresponds to the mixture of two normals, of which the DGHW estimator is shown in Figure 4.4. However, compared with Figure 3.3.2, set 4, it seems oversmoothed. The second type is the essentially trimodal estimator provided, e.g., by the PI-DB method, but this estimator seems somewhat undersmoothed. It is interesting to note that again the smoothing parameters selected by each method appear quite stable, with the possible exception of the various double kernel methods, see Figure 4.3.

The simulation results of the previous section indicate that on average there is not much of a difference between the various procedures for selecting the smoothing parameters. On the other hand, this section illustrates the differences that can occur when these methods are applied to particular data sets. Which one of the resulting estimators comes closest to the truth? We do not know! However, with the sample sizes being as small as they are, it is doubtful that one can meaningfully distinguish between the accuracies of the various estimators.

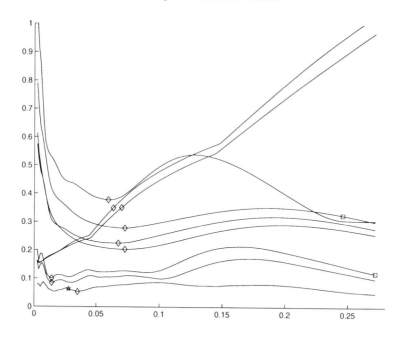

Figure 4.3. How the smoothing parameters get selected for the Old Faithful geyser data. The figure shows the graphs of the relevant functionals as
function of h for each method. The $\diamond$ indicates the local minimum closest
to $H_{n,DP-N}$ of each functional, but for the Discrepancy Principle methods,
it indicates the point where the functional equals c_{DP}. The $\star$ indicates
the position of the $H_{n,PI-DB}$. The two squares ($\square$) on the right of the
graphs indicate the upper bounds on the H, for the Epanechnikov and
normal kernels. The graphs are identified as follows. The upper graph of
the two Discrepancy Principle graphs is based on the Epanechnikov kernel,
the lower one on the normal kernel. The other graphs from top to bottom
at $h = 0.1$ are the graphs of $PER(h)$, $MOD(h)$, $SPI(h)$, and $DGHW(h)$.
The bottom three graphs correspond to the double kernel methods (DB-E)
with stretch parameters (from top to bottom, at $h = 0.1$) 3, 2.8, and 2.4.
The square ($\square$) on the graph for the double kernel method with stretch
parameter 2.4 indicates a local minimum of the function and is the minimum over the interval $[0, 0.1]$. The acronyms are explained in Section 3
and Chapter 7.

(4.1) EXERCISE. Apply the mixtures of normals approach of §3.3 to see
whether a trimodal estimator may be obtained, say, as for the PI-DB methods in Figure 4.4, and how the values of the negative log-likelihood function
compare.

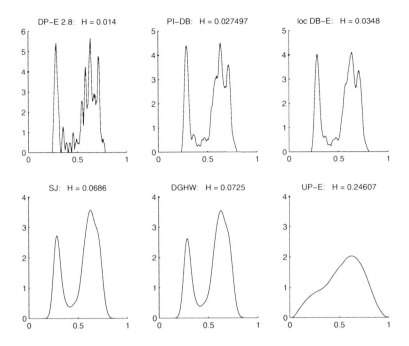

Figure 4.4. The kernel estimators for the Old Faithful geyser data with smoothing parameters selected according to the various criteria. The acronyms are explained in Section 3.

(4.2) EXERCISE. Repeat the experiments of this section for the Rubber Abrasion data (see Appendix 1).

EXERCISES : (4.1), (4.2).

5. Kernel selection

In this section, we offer selected results from simulation experiments regarding the choice of the kernel in kernel density estimation and for the GOOD estimator of § 5.2. The criterion by which the kernels/methods are judged is again the smallness of the L^1 error, but this time we choose the optimal smoothing parameter with respect to the L^1 error.

It belongs to the folklore of kernel density estimation that the choice of the kernel is not very important. There are actually two aspects to this folklore. Among nonnegative second-order kernels, the Epanechnikov kernel is asymptotically optimal. Moreover, this holds even for finite (small) sample sizes. However, in practice, the Epanechnikov and the normal kernel are used interchangeably, because the variability introduced by the smoothing parameter selection procedures more than hides the differences.

Formularium 5.1. The second- and fourth-order kernels to be compared. The second-order kernels

epan	$A(x) = \frac{3}{4}(1 - x^2)_+$	,	the Epanechnikov kernel ;
norm	$A(x) = \phi(x)$	,	the standard normal kernel ;
exp	$A(x) = \frac{1}{2}\exp(-|x|)$	,	the (two-sided) exponential ;
tria	$A(x) = (1 - |x|)_+$	,	the triangular kernel ;
sech	$A(x) = \pi^{-1}\operatorname{sech}(x)$	,	the hyperbolic secant kernel ;
unif	$A(x) = \mathbb{1}(|x| \leqslant \frac{1}{2})$	,	the uniform kernel ;
good	the GOOD estimator ;		

The fourth-order kernels

ZhFa	the ZHANG and FAN (2000) kernel, see § 11.2 ;
4nor	$A(x) = 2\phi(x) - \phi * \phi(x)$;
BD-I	$A(x) = \frac{150}{32}(1 - x^2)_+ - \frac{105}{32}(1 - x^4)_+$,

and

$$
\text{BD-II} \quad A(x) = \begin{cases} \frac{1}{4}(7 - 31x^2) & , \quad |x| \leqslant \frac{1}{2}, \\ \frac{1}{4}(x^2 - 1) & , \quad \frac{1}{2} < |x| \leqslant 1. \end{cases}
$$

It is also part of the folklore that the two-sided exponential kernel performs noticeably worse than the Epanechnikov kernel. If we may jump ahead to the conclusions, in view of the comparison with a two-sided exponential kernel density estimator, it is then rather surprising that the GOOD estimator is in fact a remarkably GOOD estimator.

The third aspect of the folklore regarding kernel selection is the irrelevance of higher order kernels for small sample sizes. Although they are a little bit better than the nonnegative second-order kernels for small sample sizes, this has no practical consequences in view of the (unsolved) problem of smoothing parameter selection for higher order kernels, and the variability introduced by them.

On to the simulation experiments. The judges of the contest are again the densities presented in the previous section, and we proceed to discuss the contestants. Among the nonnegative second-order kernels, the list obviously includes the Epanechnikov, normal, and two-sided exponential kernels. As stated earlier, the two-sided exponential is not very good. Inquiring minds wonder why: Is it the peak at the origin or the not-so-light tails? To find out, we also include the triangular kernel, which has the same peak, but no tails, and the hyperbolic secant kernel, which has the

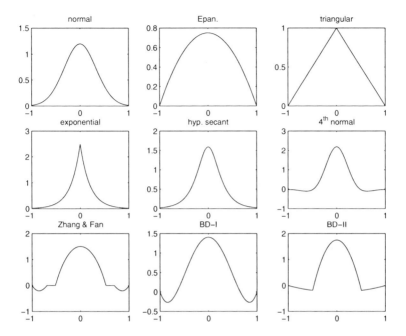

Figure 5.1. Nine kernels for kernel density estimation, suitably scaled.

same tails as the two-sided exponential, but is infinitely smooth. For the fun of it, we also include the uniform kernel. Finally, the GOOD estimator is included among the nonnegative second-order kernels: It is not a kernel estimator, but as a function of the smoothing parameter, it has the same bias and variance behavior as the second-order kernels.

As far as higher order kernels are concerned, we only look at fourth-order ones. In the list of contestants, we include the "optimal" kernel of ZHANG and FAN (2000), see § 4.7 and § 11.2. Also included are the two "optimal" kernels of BERLINET and DEVROYE (1994), as well as the fourth-order kernel based on the normal density. We refrain from including in the comparisons the maximum likelihood estimator with roughness penalization of the log-density as in § 5.3, but see the references in § 7.8.

The list of contestants is shown in Formularium 5.1. Note that the last kernel is in fact the fourth-order kernel of BERLINET and DEVROYE (1994), see (7.3.19). Their graphs are shown in Figure 5.1, but keep in mind that the two-sided exponential and hyperbolic secant kernels have infinite support.

The results of the simulations are tabulated in Tables 5.1 and 5.2. In Table 5.1, the results for sample size $n = 100$ are given, for all of the kernels. As expected, the Epanechnikov and normal kernels give just about indistinguishable optimal errors. The results for the triangular kernel are somewhat surprising. Based on simulation results not shown, the errors for

Table 5.1. Estimated means of $\| Tf_{n,OPT} - Tf_o \|_1$ for various kernels and the GOOD estimator applied to various densities, for sample size 100, based on 500 replications. The interval $[0,10]$ was divided into 400 subintervals.

	Nice mix	Normal	Uniform	Beta	Bad mix	Bimodal
Good	0.153	0.121	0.213	0.163	0.298	0.185
Epan	0.155	0.129	0.216	0.164	0.336	0.185
tria	0.156	0.129	0.215	0.164	0.334	0.185
norm	0.157	0.131	0.217	0.166	0.336	0.187
exp	0.172	0.145	0.226	0.179	0.345	0.203
sech	0.164	0.138	0.221	0.172	0.342	0.195
unif	0.163	0.136	0.224	0.171	0.345	0.193
ZhFa	0.144	0.115	0.218	0.161	0.314	0.172
4nor	0.145	0.116	0.215	0.159	0.317	0.173
BD-I	0.144	0.116	0.218	0.161	0.311	0.173
BD-II	0.144	0.116	0.219	0.161	0.313	0.173

the Epanechnikov and the triangular kernel are similar for each smoothing parameter h. Apparently, the two-sided exponential and hyperbolic secant kernels should be avoided. However, it should be noted that for larger h, the two-sided exponential has much smaller error than all of the other methods. This may have some advantages for mode estimation, see EDDY (1980), and for indirect estimation problems. The real surprise is that the GOOD estimator gives just about the same optimal L^1 errors as the Epanechnikov kernels. As illustrated in Figure 5.2, this is true not only on average, but also for each sample.

In Table 5.2, we report on larger sample sizes for a few methods and a few densities. Unreported simulation results suggest that the first kernel (BD-I) of BERLINET and DEVROYE (1994) and the ZHANG and FAN (2000) kernel are more or less indistinguishable, which leads to the conclusion that all reasonable fourth-order kernels have about the same average behavior. Again, the GOOD estimator and the Epanechnikov kernel estimator give just about the same results. Of course, for these sample sizes, the higher order kernels are at an advantage, modulo the problem of choosing the smoothing parameter. The difference between the GOOD estimator and the Epanechnikov kernel estimator when applied to normal data disappears for larger n, but the GOOD estimator is definitely holding its own.

(5.1) EXERCISE/PROJECT. Devise rational procedures for selecting the smoothing parameter for some fourth-order kernels.

Table 5.2. Estimated means of $\| Tf_{n,OPT} - Tf_o \|_1$ for various kernels and the GOOD estimator applied to the normal and Beta densities, for sample sizes 100 and 200 (500 replications), 400 (250 replications), and 800 (150 replications). The interval $[0,10]$ was divided into 400 subintervals.

| | | Normal | | | Beta | |
n	Good	Epan	ZhFa	Good	Epan	ZhFa
100	0.121	0.129	0.115	0.163	0.164	0.161
200	0.093	0.098	0.085	0.131	0.130	0.129
400	0.074	0.077	0.065	0.106	0.105	0.104
800	0.058	0.060	0.050	0.086	0.083	0.083

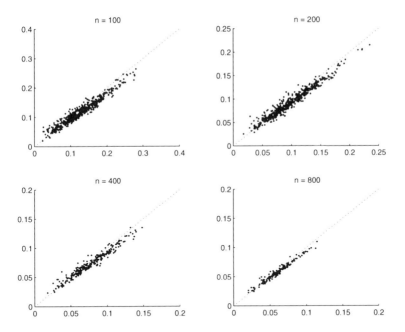

Figure 5.2. Scatter plots of the optimal L^1 errors of the GOOD estimator (the vertical axis) vs the Epanechnikov kernel estimator (the horizontal axis) as applied to the normal density, for sample sizes 100, 200, 400, and 800 (500, 500, 250, and 150 replications, respectively). The lines of equal errors are dotted in.

(5.2) EXERCISE/PROJECT. Compare various kernels for the purpose of the estimation of the mode of a density (unimodal, multimodal), see EDDY (1980), as detailed in §6.6.

(5.3) EXERCISE/PROJECT. The expensive part of computing the GOOD estimator is computing the values of $\{\, f^{nh}(X_i)\,\}^{1/2}$, $i = 1, 2, \cdots, n$. Investigate practically and theoretically whether the estimator $\varphi^{nh} = (w^{nh})^2$, with

$$ w^{nh}(x) = \tfrac{1}{n} \sum_{i=1}^{n} [\, \mathcal{B}_{h/\sqrt{2}} * dF_n(X_i)\,]^{-1/2} \, \mathcal{B}_h(x - X_i)\,, \quad -\infty < x < \infty\,, $$

is really "different" from the GOOD estimator, and if it is different, which one is to be preferred. Here, $\mathcal{B}(x)$ is the two-sided exponential kernel.

(5.4) EXERCISE/PROJECT. The results of this section just about scream out for replacing the two-sided exponential kernel in the GOOD estimator with other kernels,

$$ v^{nh}(x) = \tfrac{1}{n} \sum_{i=1}^{n} [\, v^{nh}(X_i)\,]^{-1} \, A_h(x - X_i)\,, \quad -\infty < x < \infty\,, $$

say, with A the Epanechnikov or the normal kernel, or maybe even fourth-order kernels. Investigate this (theoretically and) by simulations. Are there compelling choices for the new kernels? One direction is given by KLONIAS (1991).

EXERCISES: (5.1), (5.2), (5.3), (5.4).

6. Unimodal density estimation

In this section, we report on some simulation experiments for the nonparametric estimation of smooth unimodal densities.

Just about all estimators considered are based on kernel estimators, in which we exclusively use the Epanechnikov kernel. Of course, the discretization outlined in §2 is employed, with the discretized kernel estimator denoted by $TA_h * dF_n$. The estimators to be considered are built from two parts, viz. the selection of the mode of the unimodal estimator and the determination of the monotone estimators to the left and to the right of the mode. It is convenient to start with the monotone estimation part, given the mode. We consider two estimators, viz. the maximum smoothed likelihood estimator, which includes the GRENANDER estimator when smoothing is not incorporated, and the monotone (equimeasurable) rearrangement estimator of FOUGÈRES (1997), see §6.7. We recall from §§6.6 and 7 that the maximum smoothed likelihood estimator $f^{nh}(\cdot\,; m)$

is the solution to

(6.1)
$$\text{minimize} \quad -\int_{\mathbb{R}} [\, TA_h * dF_n \,](x) \, \log f(x) \, dx + \int_{\mathbb{R}} f(x) \, dx$$
$$\text{subject to} \quad f \in L^1(\mathbb{R}) \,, \ f \text{ unimodal, with mode at } m$$

and

(6.2)
$\varphi^{nh}(\,\cdot\,; m)$ is the unimodal (equimeasurable)
rearrangement of $TA_h * dF_n$ with the mode at m .

If the mode m coincides with one of the partition points x_j used to define T, then $\varphi^{nh}(\,\cdot\,; m)$ may be obtained by the appropriate sorting of the finitely many values of $TA_h * dF_n(y_j)$, where the y_j are the midpoints of the partition, say, and shuffling the partition accordingly.

For the determination of the mode, three methods suggest themselves: selection by maximum (smoothed) likelihood, by minimum L^1 distance, or as the mode of the kernel estimator. All of these were tried, that is, the mode m of the estimator $f^{nh}(\,\cdot\,; m)$ is selected by extending the minimization in (6.1) to include m as well; by taking m to be the mode of $TA_h * dF_n$; or by the solution to

(6.3)
$$\text{minimize} \quad \| \, f^{nh}(\,\cdot\,; m) - TA_h * dF_n \, \|_1$$
$$\text{subject to} \quad -\infty < m < \infty \,.$$

For the unimodal rearrangement estimator $\varphi^{nh}(\,\cdot\,; m)$, we select m as either the mode of $TA_h * dF_n$ or as the solution to

(6.4)
$$\text{minimize} \quad \| \, \varphi^{nh}(\,\cdot\,; m) - TA_h * dF_n \, \|_1$$
$$\text{subject to} \quad -\infty < m < \infty \,.$$

It was proven in §6.6 that to compute the maximum smoothed likelihood estimator of the mode, we need merely to inspect the local modes of the kernel estimator, and the same holds for the minimum L^1 distance method (6.3). For the unimodal rearrangement estimator, we assume that we may do the same, see Exercise (6.7.7). It should be noted that we may restrict the local modes of the discretized kernel estimator $TA_h * dF_n$ to the partition points x_j, see (2.2), $j = 1, 2, \cdots, m$, and then, as is readily seen, $f^{nh}(\,\cdot\,; x_j)$ and $\varphi^{nh}(\,\cdot\,; x_j)$ are step functions on the partition (2.2) as well.

In the experiments, we compare the various methods to each other. Each method is denoted by an acronym (O tempera, o mores!) of the from XXX-YYY, where YYY refers to the estimator, MLE for $f^{nh}(\,\cdot\,; m)$, and UR (unimodal rearrangement) for $\varphi^{nh}(\,\cdot\,; m)$. The XXX part refers to the way the mode m is chosen: by maximum likelihood (MLE), minimum L^1 distance (DIST), or as the mode of $TA_h dF_n$ (MODE). We emphasize that all estimators are based on $TA_h * dF_n$ rather than on $A_h * dF_n$. The specific methods considered are as follows.

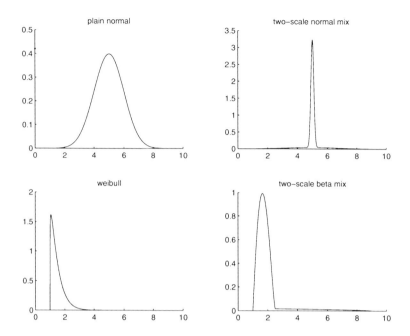

Figure 6.1. Four unimodal densities for nonparametric estimation.

   I. MLE-MLE: f^{nh}, the solution of (6.1), with minimization over m
      also.
  II. MODE-MLE: $f^{nh}(\,\cdot\,,\mu_{nh})$, with μ_{nh} the mode of $TA_h dF_n$.
 III. DIST-MLE: $f^{nh}(\,\cdot\,,\lambda_{nh})$, the solution of the minimum L^1 distance
      problem (6.3).
  IV. $A_h * $ MLE: the smoothed version of the maximum smoothed like-
      lihood estimator using grouped data, analogous to the regression
      case of MAMMEN (1991), see §6.7(f).
   V. MODE-UR: $\varphi^{nh}(\,\cdot\,,\mu_{nh})$, the unimodal rearrangement of $TA_h dF_n$
      with mode μ_{nh} as under (II).
  VI. DIST-UR: $\varphi^{nh}(\,\cdot\,;\eta_{nh})$, the solution of the minimum L^1 distance
      problem (6.4).

   In the experiments, the simulated data correspond to various unimodal
densities, restricted to the interval $[0\,,10]$. It is clear that the normal and
uniform densities have be considered, as well as a skewed unimodal density
with light tails. It turns out that unimodal estimation is especially useful
for heavy-tailed densities, so we include a symmetric and a skewed density
with heavy tails. These two are implemented by a mixture of two normals
with the same means but drastically different standard deviations and a
mixtures of two Beta densities. The precise choices are as follows. With
a slight change in notation, $\phi(\,\cdot\,;\mu,\sigma)$ denotes the Gaussian density with

Table 6.1. Estimated minimal L^1 errors, for various unimodal estimators of various densities, for sample size 100, based on 1000 replications. The minimality refers to the minimum over h. The interval $[0,10]$ was divided into 400 subintervals. The entry "cor. MLE-MLE" refers to the L^1 error of the MLE-MLE estimator for the value of h optimal for $TA_h * dF_n$. Likewise for "cor. DIST-UR".

	Nor-mal	Wei-bull	Uni-form	Beta mix	Normal mix
opt. $TA_h * dF_n$	0.126	0.250	0.216	0.200	0.319
cor. MLE-MLE	0.126	0.232	0.207	0.165	0.211
cor. DIST-UR	0.126	0.220	0.215	0.188	0.268
MLE-MLE	0.126	0.214	0.181	0.162	0.200
DIST-MLE	0.126	0.214	0.164	0.162	0.200
MODE-MLE	0.126	0.210	0.160	0.162	0.200
DIST-UR	0.122	0.175	0.215	0.186	0.267
MODE-UR	0.121	0.178	0.215	0.186	0.267
$A_h *$MLE	0.108	0.212	0.207	0.160	0.192

mean μ and standard deviation σ. We let $\text{wei}(x;\alpha)$ denote the density corresponding to the Weibull distribution $\text{Wei}(x;\alpha) = 1 - \exp(-x_+^\alpha)$, and $\text{beta}(x;\alpha,\beta) = B(\alpha,\beta)\, x_+^{\alpha-1}\,(1-x)_+^{\beta-1}$ the standard Beta density. The specific densities under consideration are as follows.

I. A normal density $\phi(\,\cdot\,;5,1)$.

II. The Weibull density $f_o(x) = \sigma^{-1}\text{wei}(x-\gamma;\alpha)$, with $\alpha = 1.1$, $\gamma = 1$ and $\sigma = 0.5$.

III. The uniform density on $[3,8]$.

IV. The mixture of two Beta densities

$$\tfrac{9}{10}\,\text{beta}_\sigma(x-\gamma;\alpha,\beta) + \tfrac{1}{10}\,\text{beta}_s(x-c;a,b)\,,$$

with $\alpha = 2.1$, $\beta = 2.45$, $\gamma = 1$, and $\sigma = 1.5$, and $a = 1.0$, $b = 1.5$, $c = 1$, and $s = 8$. Here, $\text{beta}_\sigma(x;\,\cdots) = \sigma^{-1}\text{beta}(\sigma^{-1}x;\,\cdots)$.

V. The mixture of two normals $\tfrac{4}{5}\phi(x;5,.1) + \tfrac{1}{5}\phi(x;5,1.8)$.

Graphs of the densities are shown in Figure 6.1.

In Table 6.1, we report on the estimated means

$$\mathbb{E}[\,\min_h \|\, f^{nh}(\,\cdot\,;m) - Tf_o\,\|_1\,]$$

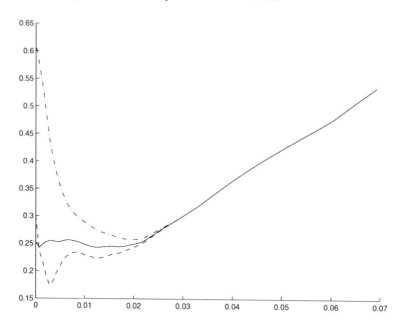

Figure 6.2. Graphs of the L^1 errors of the various density estimators as a function of the smoothing parameter, for a sample of size 100 drawn from the Weibull density: $TA_h * dF_n$ (dot-dashed), the MLE-MLE estimator (solid), and MODE-UR estimator (dashed) .

for the various choices of the mode m, and likewise for φ^{nh}. The sample standard deviations for each method were essentially the same for each density, but ranged from 0.04 to 0.07 over the various densities. The following conclusions may be drawn. For smooth densities with light tails, the unimodal rearrangement method à la FOUGÈRES (1997) works best, although we are definitely at a loss to explain the large improvement for the Weibull distribution. See Figure 6.2. For smooth densities with heavy tails, here simulated by the mixture of Beta densities and the mixture of normals, the smoothed maximum likelihood method (the $A_h * \text{MLE}$ method) works the best, with maximum smoothed likelihood a close second. The wonderful performance of the $A_h * \text{MLE}$ method for the normal density is noteworthy, as is its less than wonderful performance on the uniform density. Perhaps it should be recalled, see GROENEBOOM (1985), that when the Grenander (monotone) estimator is used to estimate a uniform density, then $\sqrt{n} \, \| f_n - f_o \|_1$ converges in distribution. The same is true for the unimodal (unsmoothed) maximum likelihood estimator, *but only for a deterministic choice of the mode.* The unimodal maximum likelihood estimator does not give good results. By way of eample, for the uniform density on $[\,3\,,8\,]$, with mode fixed at $m = 3$, the estimated expected L^1 error equals 0.117, as opposed to 0.210 when the mode is chosen by maximum likelihood.

A final word about smoothing parameter selection. In Table 6.1, we also report on the errors of the MLE-MLE and DIST-UR methods based on $TA_h dF_n$ with $h = h_{n,o \prime \prime}$ (the h that minimizes the error $\| TA_h dF_n - f_o \|_1$ in each replication). See the lines "cor. MLE-MLE" and "cor. DIST-UR" in Table 6.1. This shows that the smoothing parameter should be different when the unimodal estimator is used. In fact, the optimal h is substantially smaller for the various unimodal estimators, but it is not clear how this "optimal" h may be estimated in a rational way. In Figure 6.2, we show a typical (?) graph of the L^1 error for the MLE-MLE estimator and the MODE-UR estimator as function of the smoothing parameter. The sample size was $n = 100$, drawn from the Weibull density. The behavior of the MLE-MLE error seems predictable (but we do not know how), but less so for the MODE-UR estimator.

(6.5) EXERCISE. Develop data-based methods for selecting the smoothing parameter for both the MLE-MLE and MODE-UR estimators, and tell the authors about it.

EXERCISE : (6.5).

Part III:

Convexity and Optimization

9

Convex Optimization in Finite-Dimensional Spaces

1. Convex sets and convex functions

In the next three chapters, we consider convex optimization problems of the kind that arise in the maximum (penalized) likelihood approach to nonparametric estimation. We have seen some significant applications of convexity in the treatment and use of (sub)martingales (Chapter 4), in the discussion of the GOOD density estimator (§ 5.2), and in the estimation of log-concave (§ 5.5), and monotone densities (Chapter 6). Out of necessity, we will make considerable use of convexity in the treatment of indirect estimation problems in Volume II, such as in the Tikhonov regularization of ill-posed least-squares problems, and the nonparametric deconvolution problem. Additional applications arise in the form of "elementary inequalities" that have popped up at the least provocation. It seemed like a good idea to the authors to isolate convexity arguments and to collect them here in a form that can be used in other chapters. Under the motto that one must learn to walk before one can run, we begin with finite-dimensional convex optimization problems. (However, the definition of convexity does not depend on the dimension.)

Let X be a vector space. A subset C of X is *convex* if for every pair of elements of C, all elements on the line segment joining the two elements also lie in C. Equivalently, we have the following definition.

(1.1) DEFINITION. *Let X be a vector space. A set $C \subset X$ is convex if $\lambda x + (1 - \lambda)y \in C$ for all x, $y \in C$ and for all λ with $0 < \lambda < 1$.*

A typical example of a convex set in the plane ($\mathbb{R}^2$) is the unit disk (with or without some or all points on the circumference). The circumference of a circle in the plane is not a convex set.

Let $C \subset X$ be convex. A function $f : C \longrightarrow \mathbb{R}$ is said to be convex if the set $\mathrm{epi}(f)$ defined as

(1.2) $$\mathrm{epi}(f) = \{ (x, r) \in X \times \mathbb{R} : f(x) \geqslant r \}$$

is convex. The set $\mathrm{epi}(f)$ is called the *epigraph* of f. So, for a convex function f, the line segment joining any two points in $\mathrm{epi}(f)$ also lies in $\mathrm{epi}(f)$. In particular, if we take points on the boundary of $\mathrm{epi}(f)$, viz. points $(x, f(x))$ and $(y, f(y))$, then the line segment joining them lies above the corresponding portion of the graph, and this by itself is equivalent to $\mathrm{epi}(f)$ being convex. Formally, we have the following.

(1.3) DEFINITION. *Let $C \subset X$ be a convex set. The function $f : C \longrightarrow \mathbb{R}$ is convex if*

$$f\left(\lambda x + (1 - \lambda)y \right) \leqslant \lambda f(x) + (1 - \lambda) f(y)$$

for all x, $y \in C$ and for all λ with $0 < \lambda < 1$.

The above may be called the *convexity inequality*. An immediate consequence of the definition is the following.

(1.4) EXERCISE. JENSEN'S INEQUALITY. Let f be convex on C. Show that if $w_1, w_2, \cdots, w_n$ are positive numbers, with $\sum_i w_i = 1$, and $x_1, x_2, \cdots, x_n$ are points in C, then

$$f\left(\sum_i w_i x_i \right) \leqslant \sum_i w_i f(x_i) \ .$$

(1.5) EXERCISE. JENSEN'S INEQUALITY AGAIN. Let $C \subset \mathbb{R}^d$ be convex, and let $\Omega \subset \mathbb{R}^p$ be (Lebesgue) measurable. Suppose $g : \Omega \longrightarrow C$ is a measurable function, and $f : C \longrightarrow \mathbb{R}$ is convex. For w a pdf on C, show that

$$f\left(\int_\Omega w(x)\, g(x)\, dx \right) \leqslant \int_\Omega w(x)\, f(g(x))\, dx \ .$$

[The proof of this is an exercise in Lebesgue measure theory. The measurability of convex sets follows from Theorem (1.40) below.]

(1.6) EXERCISE. Let $f : C \longrightarrow \mathbb{R}$ be convex. Suppose x, $y \in C$ are such that $z = \lambda\, x + (1 - \lambda)\, y \in C$ for some real number $\lambda \notin [\, 0, 1\,]$. Show that

$$f(z) \geqslant \lambda\, f(x) + (1 - \lambda)\, f(y) \ .$$

Draw some diagrams for $C = \mathbb{R}$ to illustrate what it means.

Of course, the question arises of how one can recognize whether a function is convex. For functions of one variable, we know from the calculus that a twice differentiable function is convex if the second derivative is positive (nonnegative, actually). We show that properly interpreted, this holds in general. Recall the notations, for any natural number d,

$$\langle x, y \rangle = \sum_{i=1}^{d} x_i y_i$$

for the inner product on $\mathbb{R}^d$, as well as ∇f and $\nabla^2 f$ for the gradient and Hessian of a function f defined on (a subset of) $\mathbb{R}^d$, see $\S 2.2$.

(1.7) THEOREM. Let $C \subset \mathbb{R}^d$ be convex, and let $f : C \longrightarrow \mathbb{R}$ be a differentiable function. Then, the following conditions are equivalent.
(a) f is convex.
(b) $f(x) - f(y) - \langle \nabla f(y), x - y \rangle \geqslant 0$ for all $x, y \in C$.
(c) $\langle \nabla f(x) - \nabla f(y), x - y \rangle \geqslant 0$ for all $x, y \in C$.
(d)   If, in addition, f is twice differentiable, then

$$\langle \nabla^2 f(x)(x - y), x - y \rangle \geqslant 0 \text{ for all } x, y \in C .$$

PROOF. (a) $\Longrightarrow$ (b): Suppose that f is convex. Let $x, y \in C$. Then, for $0 < \lambda < 1$,

$$\frac{f(y + \lambda(x - y)) - f(y)}{\lambda} \leqslant f(x) - f(y) .$$

Because f is differentiable, letting $\lambda \longrightarrow 0$ gives

$$\langle \nabla f(y), x - y \rangle \leqslant f(x) - f(y) .$$

This is (b).
(b) $\Longrightarrow$ (c): Reversing the roles of x and y in the above inequality gives

$$\langle \nabla f(x), y - x \rangle \leqslant f(y) - f(x) .$$

Adding these two inequalities results in

$$\langle \nabla f(y) - \nabla f(x), x - y \rangle \leqslant 0 ,$$

which is (c).
(c) $\Longrightarrow$ (a): In (c), replace x by $\lambda x + (1 - \lambda)y$ with $0 < \lambda < 1$ to obtain

(1.8) $\langle \nabla f(\lambda x + (1 - \lambda)y) - \nabla f(y), x - y \rangle \geqslant 0$.

Now, observe that the expression on the left is equal to

$$\frac{d}{d\lambda} \{ f(\lambda x + (1 - \lambda)y) - \lambda \langle \nabla f(y), x - y \rangle \} ,$$

so that integrating (1.8) with respect to λ from 0 to 1 gives

(1.9) $f(x) - f(y) - \langle \nabla f(y), x - y \rangle \geqslant 0$.

(This shows (c) $\Longrightarrow$ (b), actually, but never mind.) Now, let $0 < \lambda < 1$, and set $z = \lambda x + (1 - \lambda) y$. This may be rewritten as $y = \theta z + (1 - \theta) x$, with $\theta = 1/(1-\lambda) > 1$. Now, (1.9) gives $f(y) - f(z) - \langle \nabla f(z), y - z \rangle \geqslant 0$, or

(1.10) $\qquad f(y) - f(z) - (\theta - 1)\langle \nabla f(z), z - x \rangle \geqslant 0$.

From (1.9), we also obtain $f(x) - f(z) - \langle \nabla f(z), x - z \rangle \geqslant 0$. Multiplying this by $\theta - 1$, which is positive, and adding this to (1.10) yields

$$f(y) + (\theta - 1) f(x) - \theta f(z) \geqslant 0 ,$$

but this is equivalent to $f(z) \leqslant \lambda f(x) + (1 - \lambda) f(y)$, so f is convex.
(c) $\implies$ (d): In (c), replace x by $\lambda x + (1 - \lambda)y$, with $0 < \lambda < 1$. After division by λ, we obtain

$$\langle \lambda^{-1} \{ \nabla f(\lambda x + (1 - \lambda)y) - \nabla f(y) \}, x - y \rangle \geqslant 0 .$$

Letting $\lambda \longrightarrow 0$ gives $\langle \nabla^2 f(y) (x - y), x - y \rangle \geqslant 0$. This is (d).
(d) $\implies$ (c): Since

$$0 \leqslant \langle \nabla^2 f(\lambda x + (1 - \lambda)y) (x - y), x - y \rangle$$
$$= \frac{d}{d\lambda}\langle \nabla f(\lambda x + (1 - \lambda)y), x - y \rangle ,$$

integration between 0 and 1 yields (c). $\qquad\qquad\qquad\qquad$ Q.e.d.

The definitions of convexity work in any vector space, but we now list some examples in finite-dimensional spaces that are relevant to our endeavor. We begin with some examples on (part of) the real line.

(1.11) EXAMPLES/EXERCISES. Show that the following functions are convex.

(a) $f(t) = t^2$. $\qquad\qquad\qquad\qquad\qquad$ (b) $f(t) = t \log t + 1 - t$.

(c) $f(t) = |t|^p$, $1 \leqslant p < \infty$. $\qquad\qquad$ (d) $f(t) = -\log t$.

(e) $f(t) = e^t$. $\qquad\qquad\qquad\qquad\qquad$ (f) $f(t) = \log \cosh t$.

(g) $f(t) = (\frac{2}{3} t + \frac{4}{3})(t \log t + 1 - t) - (t - 1)^2$.

Examples (b), (d), and (g) are on $\mathbb{R}_+$ only.

(1.12) EXAMPLES/EXERCISES. Let $A \in \mathbb{R}^{d \times d}$ and $b \in \mathbb{R}^d$, with suitable conditions added if necessary. Show that the following functions are convex. [You may want the look at Theorem (1.14) first.]

(a) $\| x \|$ for any norm $\| \cdot \|$ on $\mathbb{R}^d$.

(b) $\| Ax - b \|^2 = \sum_i | (Ax)_i - b_i |^2$ on $\mathbb{R}^d$.

(c) $\mathrm{PHI}(b, Ax) = \sum_i \dfrac{|b_i - (Ax)_i|^2}{(Ax)_i}$ on $\mathbb{R}^d_+$.

(d) $-\sum_i b_i \log(Ax)_i$ on $\mathbb{R}^d_+$.

(e) $KL(b, Ax) = \sum_i b_i \log(b_i/(Ax)_i) + (Ax)_i - b_i$ on $\mathbb{R}^d_+$.

(f) $H(b, Ax) = \sum_i |\sqrt{b_i} - \sqrt{(Ax)_i}|^2$ on $\mathbb{R}^d_+$.

In examples (c) through (f), A and b should be nonnegative component wise.

How would we verify that these are indeed convex functionals? Note that the examples in (1.12) are of the form $f(Ax)$, where $f(x) = \sum_i \varphi(x_i)$, with φ a convex function of one variable.

(1.13) EXERCISE. Suppose $\varphi(t)$ is a convex function of $t \in \mathbb{R}$. Show that $f(x) = \sum_{i=1}^d \varphi(x_i)$ is a convex function on $\mathbb{R}^d$.

From this exercise and the following theorem, we can then conclude that the examples in (1.12) are indeed convex functions.

(1.14) THEOREM. Let $A \in \mathbb{R}^{\ell \times d}$, and let $C \subset \mathbb{R}^d$ be a convex set. If $f : A(C) \longrightarrow \mathbb{R}$ is convex, then $f \circ A : C \longrightarrow \mathbb{R}$ is also convex.

PROOF. Let $x, y \in C$, and $0 < \lambda < 1$. Then, $\lambda A x + (1 - \lambda) A y \in A(C)$, and by convexity of f, then

$$f(\lambda A x + (1 - \lambda) A y) \leqslant \lambda f(A x) + (1 - \lambda) f(A y) .$$

Writing the left-hand side as $f \circ A (\lambda x + (1 - \lambda) y)$ shows that $f \circ A$ is convex.   Q.e.d.

A related result is given in the next exercise.

(1.15) EXERCISE. (a) Show that if $f : C \longrightarrow \mathbb{R}$ and $\varphi : f(C) \longrightarrow \mathbb{R}$ are convex and φ is increasing, then $\varphi \circ f$ is convex.
(b) Likewise, if $f : C \longrightarrow \mathbb{R}$ is concave and $\varphi : f(C) \longrightarrow \mathbb{R}$ is convex and decreasing, then $\varphi \circ f$ is convex.

Convexity and convexity arguments are pervasive in applied mathematics and analysis. In fact, some well-known inequalities are due to convexity, as we now discuss.

(1.16) THE GEOMETRIC-ARITHMETIC MEAN INEQUALITY. If $w_1, \cdots, w_n$ are nonnegative and add up to one, then for arbitrary nonnegative numbers $a_1, a_2, \cdots, a_n$, we have

$$\prod_{i=1}^n a_i^{w_i} \leqslant \sum_{i=1}^n w_i a_i .$$

This is an instance of Exercise (1.4). Specifically, write the left-hand side as $\exp\left(\sum_i w_i \log a_i\right)$, and note that the function $t \longmapsto e^t$ is convex.

(1.17) MINKOWSKI'S INEQUALITY. Let $1 \leqslant p < \infty$. For (measurable) functions f on the open set $\Omega \subset \mathbb{R}$, define

$$(1.18) \qquad \| f \|_p = \left\{ \int_\Omega | f(x) |^p \, dx \right\}^{1/p} .$$

The set $L^p(\Omega)$ is the collection of all f with $\| f \|_p < \infty$. Then, for all functions f and $g \in L^p(\Omega)$,

$$(1.19) \qquad \| f + g \|_p \leqslant \| f \|_p + \| g \|_p .$$

The proof of this depends on the convexity of the function $t \longmapsto |t|^p$ on the real line. Assuming that both f and g are nonzero (since otherwise there is nothing to prove), define

$$w = \frac{\| f \|_p}{\| f \|_p + \| g \|_p} .$$

Since $0 < w < 1$,

$$| f(x) + g(x) |^p = \{\| f \|_p + \| g \|_p\}^p \left| w \frac{f(x)}{\| f \|_p} + (1 - w) \frac{g(x)}{\| g \|_p} \right|^p$$

$$\leqslant \{\| f \|_p + \| g \|_p\}^p \left\{ w \frac{|f(x)|^p}{\| f \|_p^p} + (1 - w) \frac{|g(x)|^p}{\| g \|_p^p} \right\} ,$$

with the inequality due to convexity. The expression in curly brackets integrates to 1, so integration over Ω of the above inequality gives

$$\| f + g \|_p^p \leqslant \{\| f \|_p + \| g \|_p\}^p ,$$

and (1.19) follows.

(1.20) HÖLDER'S INEQUALITY. Let $1 < p < \infty$, and let q be the "dual" exponent defined by $(1/p) + (1/q) = 1$. So $1 < q < \infty$. If $\| f \|_p$ and $\| g \|_q$ are finite, then

$$\int_\Omega \left| f(x) \, g(x) \right| dx \leqslant \| f \|_p \| g \|_q .$$

(This also holds for $p = 1$, $q = \infty$.)

The proof again depends on the convexity of the function $t \longmapsto |t|^p$, for $p \geqslant 1$. However, in this case, the first step is to make the simple look complex. We write

$$\left\{ \int_\Omega | f(x) | \, | g(x) | \, dx \right\}^q = \| f \|_p^{pq} \left\{ \int_\Omega \frac{|f(x)|^p}{\| f \|_p^p} \, | f(x) |^{1-p} | g(x) | \, dx \right\}^q ,$$

and note that $|f(x)|^p/\| f \|_p^p$ is a pdf. By the convexity of $t \longmapsto |t|^q$, then, cf. Exercise (1.5),

$$\left\{ \int_\Omega | f(x) | \, | g(x) | \, dx \right\}^q \leqslant \| f \|_p^{pq} \int_\Omega \frac{| f(x) |^p}{\| f \|_p^p} \, | f(x) |^{q-pq} \, | g(x) |^q \, dx$$

$$\leqslant \| f \|_p^{pq-p} \int_\Omega | f(x) |^{p+q-pq} \, | g(x) |^q \, dx \; .$$

From $(1/p) + (1/q) = 1$, it follows that $p + q - pq = 0$ and $pq - p = q$. Thus, the above inequality reads as

$$\left\{ \int_\Omega | f(x) | \, | g(x) | \, dx \right\}^q \leqslant \| f \|_p^q \, \| g \|_q^q \; ,$$

and HÖLDER's inequality follows.

(1.21) EXERCISE. (a) Prove the discrete analogues of the MINKOWSKI and HÖLDER inequalities. Let $1 < p < \infty$, and let q be the dual exponent of p. For $x \in \mathbb{R}^n$, define

$$\| x \|_p = \left\{ \sum_{i=1}^n | x_i |^p \right\}^{1/p} \; .$$

Then, the MINKOWSKI and HÖLDER inequalities read as : For all $x, y \in \mathbb{R}^n$,

$$\| x + y \|_p \leqslant \| x \|_p + \| y \|_p \; , \quad \text{and} \quad | \langle x, y \rangle | \leqslant \| x \|_p \| y \|_q \; .$$

(b) The ∞-norm on $\mathbb{R}^n$ is defined as

$$\| x \|_\infty = \max_{1 \leqslant i \leqslant n} | x_i | \; .$$

Show that for all $x, y \in \mathbb{R}^n$,

$$\| x + y \|_\infty \leqslant \| x \|_\infty + \| y \|_\infty \quad \text{and} \quad | \langle x, y \rangle | \leqslant \| x \|_1 \| y \|_\infty \; .$$

(MINKOWSKI's inequality implies that $\| \cdot \|_p$ is a norm on $\mathbb{R}^n$.)

We leave some nice inequalities as exercises.

(1.22) EXERCISE. (a) Prove the following elementary inequalities:

$$(\tfrac{2}{3}x + \tfrac{4}{3}) \, (x \log x + 1 - x) \geqslant (x - 1)^2 \, , \quad x \geqslant 0 \; ,$$

$$x \log x + 1 - x \geqslant (1 - \sqrt{x})^2 \, , \quad x \geqslant 0 \; .$$

(b) Let f and g be pdfs on the real line, and let $\mathrm{KL}(f, g)$, $\mathrm{H}(f, g)$, and $\mathrm{PHI}(f, g)$ denote, respectively, the Kullback-Leibler distance, the Hellinger distance, and the Pearson's φ^2 distance between f and g, see § 1.3. Show that

$$\tfrac{1}{4} \| f - g \|_1^2 \leqslant \mathrm{H}(f, g) \leqslant \mathrm{KL}(f, g) \leqslant \mathrm{PHI}(f, g) \; .$$

(c) Something is lost in part (b), though. Show that for pdfs f and g,

$$\tfrac{1}{2} \| f - g \|_1^2 \leqslant \mathrm{KL}(f, g) \; , \quad \| f - g \|_1^2 \leqslant \mathrm{PHI}(f, g) \; .$$

[Hint for part (a): See (1.11)(g). Parts (a) and (b) are due to KEMPERMAN (1967).]

(1.23) EXERCISE. Let $F(u, v) = u^2/v$, defined for $(u, v) \in \mathbb{R} \times \mathbb{R}_{++}$. Here, $\mathbb{R}_{++} = (0, \infty)$. Show that for all (u, v), $(a, b) \in \mathbb{R} \times \mathbb{R}_{++}$,

$$F(u, v) - F(a, b) - (u - a)\frac{\partial F}{\partial u}(a, b) - (v - b)\frac{\partial F}{\partial v}(a, b) = v\left(\frac{u}{v} - \frac{a}{b}\right)^2 ,$$

and conclude that $F(u, v)$ is convex.

(1.24) EXERCISE. (a) Let $m \geqslant 1$, and set $F(u, v) = |u|^{m+1}/v^m$, defined for $(u, v) \in \mathbb{R} \times \mathbb{R}_{++}$. Show that F is convex.
(b) Let $\Phi(u, v) = |u - v|^2/v$, defined for $(u, v) \in \mathbb{R} \times \mathbb{R}_{++}$. Show that $\Phi(u, v)$ is convex.

(1.25) EXERCISE. (a) Let $S \in \mathbb{R}^{n \times m}$ be a nonnegative matrix, with

$$\sum_{i=1}^{n} [Sx]_i = \sum_{j=1}^{m} x_j \quad \text{for all } x \in \mathbb{R}^m .$$

Show that $\mathrm{KL}(Sx, Sy) \leqslant \mathrm{KL}(x, y)$, for all nonnegative $x, y \in \mathbb{R}^m$.
(b) Let $\Omega \subset \mathbb{R}^d$ be an open region, and let $\mathcal{S} : L^1(\Omega) \longmapsto L^1(\Omega)$ be an integral operator, for $f \in L^1(\Omega)$ defined by

$$\mathcal{S}f(x) = \int_{\Omega} s(x, y) f(y) \, dy , \quad x \in \Omega ,$$

where s is nonnegative and measurable, and integration is with respect to Lebesgue measure. Assuming that

$$\int_{\Omega} \mathcal{S}f(x) \, dx = \int_{\Omega} f(y) \, dy , \quad \text{for all } f \in L^1(\Omega) ,$$

show that $\mathrm{KL}(\mathcal{S}f, \mathcal{S}g) \leqslant \mathrm{KL}(f, g)$ for all nonnegative $f, g \in L^1(\Omega)$.

Next, we take a closer look at the differentiability of convex functions. The starting point is Theorem (1.7), parts (a) and (b). It is an easy exercise to show directly that part (b) implies part (a). Although the gradient $\nabla f(y)$ makes a cameo appearance, all that is required is that $\gamma = \nabla f(y)$ satisfies $f(x) - f(y) - \langle \gamma, x - y \rangle \geqslant 0$, for all $x \in C$. This indicates that there is no need for the differentiability of f, or for the uniqueness of γ, and leads to the definition of the subgradient of f.

(1.26) DEFINITION. Let $C \subset \mathbb{R}^d$ be convex, and let f be a real-valued function defined on C. Let $y \in C$. The subgradient of f at y is defined as

$$\partial f(y) = \{\gamma \in \mathbb{R}^d : f(x) - f(y) - \langle \gamma, x - y \rangle \geqslant 0 \text{ for all } x \in C\} .$$

To show that this definition makes sense, we formulate a modification of Theorem (1.7).

(1.27) THEOREM. *Let $C \subset \mathbb{R}^d$ be open and convex, and let $f : C \longrightarrow \mathbb{R}$ be a real-valued function. Then the following are equivalent.*
(a)   *f is convex on C.*
(b)   *$\partial f(y)$ is nonempty for all $y \in C$.*
(c)   *For all x, $y \in C$ and all $\eta \in \partial f(x)$, $\gamma \in \partial f(y)$,*

$$\langle\, \eta - \gamma \,,\, x - y \,\rangle \geqslant 0 \,.$$

(1.28) EXERCISE. Prove Theorem (1.27). [This is not as easy as it seems.]

What has been gained by the introduction of subgradients? For differentiable functions, absolutely nothing.

(1.29) EXERCISE. (a) Under the conditions of Theorem (1.27), if f is differentiable on C, then $\partial f(x) = \{\, \nabla f(x) \,\}$ for all $x \in C$.
(b) Suppose that x lies in the interior of C, and that $\partial f(x)$ is a singleton. Show that f is differentiable at x. [Do the one-dimensional case first.]

So the gain must be with respect to nondifferentiable convex functions. The following exercises are the standard easy examples.

(1.30) EXERCISE. Let $f(x) = |\,x\,|$ on the real line. Show that f is convex, and that $\partial f(0) = [-1\,,\,1]$.

(1.31) EXERCISE. Let $f : [\,0\,,\,1\,] \longrightarrow \mathbb{R}$ be a convex function. Then, the minimum of f occurs at x_o in the interior of $[\,0\,,\,1\,]$ if and only if $0 \in \partial f(x_o)$. The minimum occurs at 0 if and only if $\partial f(0) \subset \mathbb{R}_+$, and at 1 if and only if $-\partial f(1) \subset \mathbb{R}_+$.

The moral of all this is that the subgradient plays the role of the gradient, so the usual differentiability condition can be relaxed. We shall not be overly concerned with this, though.

We finish this section with some observations regarding the "representation" of convex sets by means of convex functions. The first step is to allow functions that take on the value $+\infty$, but then it is useful to define the effective domain, or simply the domain, of a function f by

(1.32) $\operatorname{domain}(\,f\,) = \{\, x \in X \,:\, f(x) < \infty \,\} \,.$

We emphasize that the value $-\infty$ is never allowed. The definition of convexity may then be modified as follows.

(1.33) DEFINITION. *Let X be a vector space. A function*

$$f : X \longrightarrow \mathbb{R} \cup \{+\infty\}$$

is convex if and only if its domain is convex and for all $x, y \in$ domain (f), *and for all* $0 < \lambda < 1$,

$$f(\lambda x + (1 - \lambda) y) \leqslant \lambda f(x) + (1 - \lambda) f(y) .$$

The first approach to representing convex sets by functions is quite simple. Specifically, for a (convex) set C, let $\mathbb{1}_C$ be the "reciprocal" of the indicator function of C, that is,

(1.34)
$$\mathbb{1}_C(x) = \begin{cases} 1 & , \quad x \in C , \\ +\infty , & x \notin C . \end{cases}$$

(1.35) EXERCISE. Verify that $\mathbb{1}_C$ is a convex function if and only if C is convex.

However, this is only a game: It is not clear that anything has been gained by (1.34). A more interesting representation is possible, which lets us represent a convex set by means of inequalities. However, we must still allow $+\infty$ as function values.

(1.36) DEFINITION. *Let* $C \subset \mathbb{R}^d$ *be a convex set and let* $x_o \in C$ *be a fixed point. The functional* $p_C : \mathbb{R}^d \longrightarrow \mathbb{R} \cup \{+\infty\}$ *defined by* $p_C(x_o) = 0$ *and for* $x \neq x_o$ *by*

$$p_C(x) = \inf\{t : t > 0 , \quad x_o + t^{-1}(x - x_o) \in C\}$$

is called the MINKOWSKI *functional of* C. *Here,* $\inf(\varnothing) = +\infty$.

(1.37) EXERCISE. Show that the MINKOWSKI functional p_C of a convex set $C \subset \mathbb{R}^d$ is convex on all of $\mathbb{R}^d$. Moreover, show that if C is closed, then $C = \{x \in \mathbb{R}^d : p_C(x) \leqslant 1\}$, whereas $C = \{x \in \mathbb{R}^d : p_C(x) < 1\}$ if C is open.

The final representation of a convex set $C \subset \mathbb{R}^d$ is by means of the support function $h_C : \mathbb{R}^d \longrightarrow \mathbb{R} \cup \{+\infty\}$. For any set $C \subset \mathbb{R}^d$, the support function h_C is defined by

(1.38)
$$h_C(v) = \sup\{\langle v, x \rangle : x \in C\} .$$

(1.39) EXERCISE. Let $C \subset \mathbb{R}^d$ be arbitrary. Prove the following.

(a) $h_C(v)$ is finite for every $v \in \mathbb{R}^d$ if and only if C is bounded.

(b) h_C is positively homogeneous: $h_C(\lambda v) = \lambda h_C(v)$ for every $\lambda \geqslant 0$.

(c) h_C is subadditive: $h_C(v + \eta) \leqslant h_C(v) + h_C(\eta)$ for all v, η.

(d) h_C is convex.

The next theorem says that every closed convex set is the intersection of all half spaces containing it. One part of the equality is easy and requires neither convexity nor closedness. The other half requires the Separation Lemma of § 2. We will come back to it then.

(1.40) THEOREM. *If* $C \subset \mathbb{R}^d$ *is closed and convex, then*

$$C = \{ x \in \mathbb{R}^d : \forall \nu \, (\langle \nu , x \rangle \leqslant h_C(\nu)) \} \, .$$

We really finish with the notion of log-convexity.

(1.41) DEFINITION. *Let* $C \subset \mathbb{R}^d$ *be convex, and let* $f : C \longrightarrow \mathbb{R}$ *be non-negative. We say that* f *is log-convex if* $\log f$ *is convex.*

(1.42) EXERCISE. (a) With f as in Definition (1.41), show that f is log-convex if and only if for all x, $y \in C$, and for all $0 < \theta < 1$,

$$f(\theta \, x + (1 - \theta) \, y) \leqslant [f(x)]^\theta \, [f(y)]^{1-\theta} \, .$$

(b) Show that if f is log-convex, then it is convex.

EXERCISES : (1.4), (1.5), (1.6), (1.11), (1.12), (1.13), (1.15), (1.22), (1.23), (1.24), (1.25), (1.28), (1.29), (1.30), (1.31), (1.35), (1.37), (1.39), (1.42).

2. Convex minimization problems

Let X be a finite-dimensional vector space with $C \subset X$ convex, and let $L : C \longrightarrow \mathbb{R}$ be a convex function. Consider the convex minimization problem

(2.1)
$$\begin{aligned} &\text{minimize} \quad L(x) \\ &\text{subject to} \quad x \in C \, . \end{aligned}$$

In this section and the next, we determine necessary and sufficient conditions for (2.1) to have a solution. In particular, the goal is to prove the Lagrange Multiplier Theorem. We assume throughout that L is differentiable. However, everything can be done using *subgradients* only.

We begin with a very simple necessary and sufficient condition for a minimum.

(2.2) THEOREM. *The element* $x_o \in C$ *solves* (2.1) *if and only if*

$$\langle \nabla L(x_o) , x - x_o \rangle \geqslant 0 \quad \text{for all} \quad x \in C \, .$$

PROOF. $\Longleftarrow$: By the convexity of L, we have for all $x \in C$ that

$$L(x) - L(x_o) \geqslant \langle \nabla L(x_o) , x - x_o \rangle \, ,$$

which is nonnegative by assumption. Thus, x_o is a minimum.

$\Longrightarrow$: Since x_o solves (2.1), then $L(x) - L(x_o) \geqslant 0$ for all $x \in C$. Fix $x \in C$, and consider the function $\ell(t) = L(x_o + t(x - x_o))$, $0 \leqslant t \leqslant 1$. Then, $\ell(t)$ is differentiable, with

$$\ell'(t) = \big\langle \nabla L(x_o + t(x - x_o)) , x - x_o \big\rangle .$$

Obviously, if x_o solves (2.1), then $\ell(t)$ is minimal at $t = 0$, and so $\ell'(0) \geqslant 0$. This gives the required result.   Q.e.d.

(2.3) COROLLARY. *If the solution of* (2.1) *lies in the interior of* C, *then* $\nabla L(x_o) = 0$.

PROOF. If x_o lies in the interior of C, then there exists a positive (but possibly small) t such that $x = x_o - t\nabla L(x_o)$ lies in C. From Theorem 2.2, then

$$\big\langle \nabla L(x_o) , x - x_o \big\rangle = -t \,\|\nabla L(x_o)\|^2 \geqslant 0 .$$

Since $t > 0$, this implies $\nabla L(x_o) = 0$.   Q.e.d.

(2.4) EXERCISE. The solution of (2.1) need not be unique nor need it exist. Either way, show that the set S of solutions of (2.1) is convex.

Theorem (2.2) has another consequence that features prominently in the theory of convex minimization problems, viz. the separation of a closed convex set C and a point $x \notin C$ by a hyperplane.

(2.5) SEPARATION LEMMA. *Let* $C \subset \mathbb{R}^d$ *be closed and convex, and let* $y \notin C$. *Then, there exist vectors* $\nu \in \mathbb{R}^d$ *and* $y_o \in C$ *such that*

$$\big\langle \nu , y - y_o \big\rangle > 0 \quad , \quad \text{and} \quad \forall \, x \in C : \big\langle \nu , x - y_o \big\rangle \leqslant 0 .$$

PROOF. Consider the minimization problem

(2.6)   minimize $\|y - x\|^2$ subject to $x \in C$.

This problem has a solution: If C is bounded, then C is in fact compact, and the continuous function $x \longmapsto \|y - x\|$ attains its infimum; if C is unbounded, then C may be replaced by its intersection with the ball $\{\,x : \|y - x\| \leqslant \|y - z\|\,\}$, where z is an arbitrary element in C.

Denote the solution of (2.6) by y_o. Since $y \notin C$, and $y_o \in C$, then surely $y \neq y_o$. By Theorem (2.2), now

$$\big\langle y - y_o , x - y_o \big\rangle \leqslant 0$$

for all $x \in C$. Finally, choose $\nu = y - y_o$, and we are done.   Q.e.d.

(2.7) EXERCISE. Prove Theorem (1.40): Every closed convex set is equal to the intersection of all half spaces containing it.

We finish this section with three minimization problems that are close to our hearts.

(2.8) EXAMPLE. Let $A \in \mathbb{R}^{n \times m}$, $b \in \mathbb{R}^n$, and consider the least-squares problem

$$\begin{aligned} \text{minimize} \quad & L(x) = \| Ax - b \|^2 \\ \text{subject to} \quad & x \in \mathbb{R}^m \ . \end{aligned} \tag{2.9}$$

As we have seen, L is convex. If a minimum occurs, then it surely happens in the interior of $\mathbb{R}^m$, and so the minimum occurs for those x for which $\nabla L(x) = 2 A^T(Ax - b) = 0$. These equations are known as the normal equations, usually written as $A^T A x = A^T b$. It is an exercise in elementary linear algebra to show that the normal equations always have a solution, and that all solutions of the normal equations solve (2.9).

(2.10) EXAMPLE. Let L be as in the first example, and consider the constrained least-squares problem

$$\begin{aligned} \text{minimize} \quad & L(x) = \| Ax - b \|^2 \\ \text{subject to} \quad & x \in \mathbb{R}^m_+ \ . \end{aligned} \tag{2.11}$$

Thus, the solution $x_o \in \mathbb{R}^m$ must be nonnegative component wise. The necessary and sufficient conditions for x_o to solve (2.11) are that

$$\langle A^T(Ax_o - b) , x - x_o \rangle \geqslant 0 , \quad \text{for all } x \geqslant 0 \ . \tag{2.12}$$

We now work out in detail what this means. In particular, we show that x_o solves the complimentarity problem

$$x_o \geqslant 0 , \quad A^T(Ax_o - b) \geqslant 0 , \quad \langle A^T(Ax_o - b) , x_o \rangle = 0 \ . \tag{2.13}$$

To prove this, consider the j-th component of x_o. If $x_{oj} > 0$, then define $x = x_o + t e_j$, where e_j is the j-th element of the standard basis of $\mathbb{R}^m$. So $x \geqslant 0$ for $t \geqslant -x_{oj}$, and the condition (2.12) for a minimum implies that

$$t \, (A^T(Ax_o - b))_j \geqslant 0 \ .$$

Since this must hold for some positive as well as some negative values of t, the conclusion is that

$$(A^T(Ax_o - b))_j = 0 \ .$$

This shows that x_{oj} and $(A^T(Ax_o - b))_j$ cannot both be (strictly) positive, so that $\langle A^T(Ax_o - b) , x_o \rangle = 0$. This is the third part of (2.13). This now implies that the condition (2.12) reduces to $\langle A^T(Ax_o - b) , x \rangle \geqslant 0$, for all $x \geqslant 0$. Taking $x = e_j$ then shows that $(A^T(Ax_o - b))_j \geqslant 0 , j = 1, 2, \cdots , m$. Thus, if x_o solves (2.11), then it solves (2.13). The converse follows likewise, because (2.13) implies (2.12). To summarize, x_o solves (2.11) if and only if it solves (2.13), and hence, the minimization problem (2.11) and the complimentarity problem (2.13) are equivalent.

(2.14) EXAMPLE. Let $A \in \mathbb{R}^{n \times m}$ and $b \in \mathbb{R}^n$ be nonnegative component wise, and assume further that A has column sums equal to 1, i.e.,

$$(2.15) \qquad \sum_{i=1}^{n} a_{ij} = 1 , \quad j = 1, 2, \cdots, m ,$$

and $\sum_i b_i = 1$. A nonnegative vector $x \in \mathbb{R}^m$ with $\sum_i x_i = 1$ may be considered a discrete probability distribution (dpd). The above condition on A implies that if x is a dpd, then so is Ax. We now consider the minimization problem

$$(2.16) \qquad \text{minimize} \quad L(x) = \sum_{i=1}^{m} b_i \log \frac{b_i}{(Ax)_i} + (Ax)_i - b_i$$

$$\text{subject to} \quad x \geqslant 0 .$$

Note that $L(x) = \text{KL}(b, Ax)$, the Kullback-Leibler distance between b and Ax. As shown in Exercise (1.12)(e), the function L is convex. Its gradient is given by

$$\nabla L(x) = A^T(-r + 1) ,$$

in which $r = b/(Ax)$ (component-wise division). So the conditions for a minimum are, similar to the constrained least-squares case,

$$(2.17) \qquad x_o \geqslant 0 , \quad A^T(-r_o + 1) \geqslant 0 , \quad \langle A^T(-r_o + 1), x_o \rangle = 0 ,$$

with $r_o = b/Ax_o$.

(2.18) EXERCISE. Show that the last equation of (2.17) is equivalent to

$$\sum_{i=1}^{n} b_i = \sum_{j=1}^{m} x_{oj} .$$

(2.19) EXERCISE. With L defined by (2.16), show that for any dpd x for which $L(x)$ is finite,

$$L(tx) \geqslant L(x) , \quad t > 0 ,$$

with equality if and only if $t = 1$. [This is another way of establishing the identity of Exercise (2.18).]

(2.20) EXERCISE. Work out the consequence of Theorem (2.2) for the problem (2.16) when the the requirement (2.15) is dropped.

(2.21) EXERCISE. Let A and b be as in Example (2.14). Work out the consequences of Theorem (2.2) for the problems

(a) $\qquad$ minimize $\text{PHI}(b, Ax)$ subject to $x \in \mathbb{R}_+^m$

and

(b)   minimize $H(b, Ax)$ subject to $x \in \mathbb{R}^m_+$.

See Example (1.12).

(2.22) EXERCISE. Suppose that $L(p, q)$ is convex in (p, q) on its domain, and that $L(p, q)$ attains its minimum for each q in the projected domain, i.e., for each q for which there exists a p such that $(p, q) \in$ domain(L). Show that

$$L^*(q) \overset{\text{def}}{=} \min_p L(p, q)$$

is convex.

EXERCISES : (2.4), (2.7), (2.17), (2.18), (2.19), (2.20), (2.21), (2.22).

3. Lagrange multipliers

In this section, we develop necessary and sufficient conditions for the solution of a convex minimization problem in the form of the Lagrange Multiplier Theorem. The problem and the setting are as in the previous section, but we take a more concrete problem formulation. Again, let $L : \mathbb{R}^d \longrightarrow \mathbb{R}$ be a convex function, and consider the convex minimization problem

(3.1)
$$\begin{aligned} \text{minimize} \quad & L(x) \\ \text{subject to} \quad & x \in C , \end{aligned}$$

but now the constraint set C is specialized to

(3.2)
$$C = \{ \bigcap_{j=1}^m C_j \} \cap \{ \bigcap_{j=1}^k D_j \} ,$$

with

(3.3)
$$\begin{aligned} C_j &= \{ x \in \mathbb{R}^d \ : \ p_j(x) \leqslant 0 \} , \quad j = 1, 2, \cdots, m , \\ D_j &= \{ x \in \mathbb{R}^d \ : \ q_j(x) = 0 \} , \quad j = 1, 2, \cdots, k , \end{aligned}$$

in which the p_j are differentiable convex functions defined on $\mathbb{R}^d$, and the q_j are affine, $q_j(x) = \langle r_j , x \rangle - t_j, j = 1, 2, \cdots, k$, for nonzero elements r_j of $\mathbb{R}^d$, and scalars t_j. We assume that L is differentiable as well. Everything can be done without differentiability of the objective and the constraint functions using subgradients, but the details are definitely more involved.

We make the following regularity assumption regarding C, which goes by the name of constraint qualification.

(3.4)   There exists $x^* \in C$ such that $p_j(x^*) < 0$ for all $j = 1, 2, \cdots, m$.

The purpose of the constraint qualification is to make sure that the constraint set C is "big" enough. By way of example, we definitely want to avoid the situation in which C is a singleton, since then problem (3.1) may look like a minimization problem, but in fact it would not be one. Unfortunately, there are more pathological situations that must be avoided as well. See Figure 3.1. Condition (3.4) does just that, although we do not want to create the impression that it is the "optimal" way of doing so.

(3.5) LAGRANGE MULTIPLIER THEOREM. *Necessary and sufficient conditions for x_o to solve (3.1)–(3.2) are that there exists Lagrange multipliers $\lambda_o = (\lambda_{o,1}, \cdots, \lambda_{o,m})^T$ and $\mu_o = (\mu_{o,1}, \cdots, \mu_{o,k})^T$ with the $\lambda_{o,j} \geqslant 0$ for all j such that*

(a) $x_o \in C$,

(b) $\lambda_{o,j}\, p_j(x_o) = 0$, $j = 1, 2, \cdots, m$,   *and*

(c) $\nabla L(x_o) + \sum\limits_{j=1}^{m} \lambda_{o,j} \nabla p_j(x_o) + \sum\limits_{j=1}^{k} \mu_{o,j} \nabla q_j(x_o) = 0$.

(Of course $\nabla q_j(x) = r_j$ for all x, but never mind.)

Before proving this, we note an important (though not for us) consequence. The Lagrangian associated with the problem (3.1)–(3.2) is

$$(3.6) \qquad L(x, \lambda, \mu) = L(x) + \sum_{j=1}^{m} \lambda_j\, p_j(x) + \sum_{j=1}^{k} \mu_j\, q_j(x) ,$$

defined for all $x \in C$, all $\lambda_j \geqslant 0$, and all μ_j.

(3.7) COROLLARY. *If x_o, λ_o and μ_o satisfy the conditions of Theorem (3.5), then (x_o, λ_o, μ_o) is a saddle point of $L(x, \lambda, \mu)$, i.e., for all $x \in C$, $\lambda \in \mathbb{R}^m_+$ and all $\mu \in \mathbb{R}^k$,*

$$L(x_o, \lambda, \mu) \leqslant L(x_o, \lambda_o, \mu_o) \leqslant L(x, \lambda_o, \mu_o) .$$

PROOF. By convexity of L, we have

$$L(x_o, \lambda_o, \mu_o) - L(x, \lambda_o, \mu_o) \leqslant \langle\, \nabla L(x_o)\,,\ x_o - x \,\rangle \leqslant 0.$$

Even easier,

$$L(x_o, \lambda, \mu) - L(x_o, \lambda_o, \mu_o) = \textstyle\sum_j \lambda_j\, p_j(x_o) \leqslant 0,$$

since $\lambda_j \geqslant 0$, $p_j(x_o) \leqslant 0$, and $q_j(x_o) = 0$ for all j.   Q.e.d.

To prove the Lagrange Multiplier Theorem, we need some geometric observations regarding convexity. A set $K \subset \mathbb{R}^d$ is a *convex cone* if

$$(3.8) \qquad \forall\, x,\, y \in K \quad \forall \lambda \geqslant 0 \quad \forall\, \mu \geqslant 0 \quad : \quad \lambda\, x + \mu\, y \in K .$$

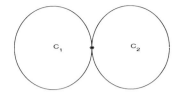 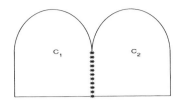

Figure 3.1. Two cases in the plane for which the constraint qualification (3.4) does not hold. In the example on the left, $C_1 \cap C_2$ is a singleton, and on the right, it is a line segment.

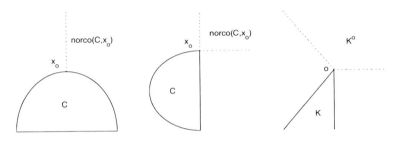

Figure 3.2. The normal cones for a half disk and a convex cone in the plane.

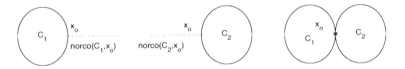

Figure 3.3. The normal cone of an intersection may be larger than the sum of the normal cones: $\mathrm{norco}(C_i, x_o)$, $i = 1, 2$, are half lines of the same line, whereas $\mathrm{norco}(C_1 \cap C_2, x_o) = \mathbb{R}^2$.

If $V \subset \mathbb{R}^d$, then the *closed positive convex span* of V is defined as the closure of the set of all positive combinations of elements in C

$$(3.9) \quad \mathrm{poco}\,(V) = \text{closure} \left\{ \sum_{j=1}^m \lambda_j\, x_j \ : \ m \in \mathbb{N}\,, \ \begin{matrix} x_1, x_2, \ \cdots\, , x_m \in V \\ \lambda_1, \lambda_2, \ \cdots\, , \lambda_m \geqslant 0 \end{matrix} \right\}.$$

Let $C \subset \mathbb{R}^d$ be convex, and let $x_o \in C$. If $y \in \mathbb{R}^d$, $y \neq 0$ is such that $\langle\, y\,, x - x_o\,\rangle \leqslant 0$ for all $x \in C$, then $\{\, x \in \mathbb{R}^d \ : \ \langle\, y\,, x - x_o\,\rangle = 0\,\}$ is called a *supporting hyperplane* of C at x_o. The set of "normals" of supporting hyperplanes is called (here) the *normal cone* of C at x_o

$$(3.10) \qquad \mathrm{norco}(C\,, x_o) = \{\, y \in \mathbb{R}^d \ : \ \forall\, x \in C\, (\langle\, y\,, x - x_o\,\rangle \leqslant 0)\,\}\,.$$

For convenience, we let $0 \in \text{norco}(C, x_o)$, but there is no hyperplane associated with it, of course. When K is a convex cone, then $\text{norco}(K, 0)$ is called the *polar cone* or simply the *polar* of K, and it is denoted by K°, so

$$(3.11) \qquad K^\circ = \text{norco}(K, 0) \quad (K \text{ a convex cone}) .$$

In Figure 3.2, some of these notions are illustrated.

(3.12) EXERCISE. If $V \subset \mathbb{R}^d$, show that $\text{poco}(V)$ is a convex cone.

(3.13) EXERCISE. If C is convex, and $x_o \in C$, show that $\text{norco}(C, x_o)$ is a closed convex cone.

(3.14) EXERCISE. If K is a convex cone, show that $\{\text{closure}(K)\}^\circ = K^\circ$.

Now consider the following easy reformulation of Theorem (2.2).

(3.15) THEOREM. *The element x_o solves (3.1)–(3.2) if and only if*

$$-\nabla L(x_o) \in \text{norco}(C, x_o) .$$

PROOF. The element x_o solves (3.1) if and only if $\langle \nabla L(x_o), x - x_o \rangle \leqslant 0$ for all $x \in C$, but this is equivalent to $-\nabla L(x_o) \in \text{norco}(C, x_o)$.   Q.e.d.

The above theorem says that in order to prove the Lagrange Multiplier Theorem, we need "merely" compute $\text{norco}(C, x_o)$ with the C from (3.2)–(3.3), under the constraint qualification (3.4). For single constraints, this is not a problem, see Exercises (3.16) and (3.17), but the difficulty occurs when several are combined, see Exercise (3.18).

(3.16) EXERCISE. Let $C = \{x \in \mathbb{R}^d : p(x) \leqslant 0\}$, where p is a convex differentiable function. Suppose that there exists an x^* such that $p(x^*) < 0$.
(a) If $p(x_o) = 0$, show that $\text{norco}(C, x_o) = \text{poco}(\nabla p(x_o))$.
(b) If $p(x_o) < 0$, show that $\text{norco}(C, x_o) = \{0\}$.

(3.17) EXERCISE. Let $r \in \mathbb{R}^d$, $t \in \mathbb{R}$, and define $D = \{x : \langle r, x \rangle = t\}$. For $x_o \in D$, show that $\text{norco}(D, x_o) = \{\lambda r : \lambda \in \mathbb{R}\}$.

(3.18) EXERCISE. Let $A, B \subset \mathbb{R}^d$ be closed convex sets, with $x_o \in A \cap B$. Show that

$$\text{norco}(A, x_o) + \text{norco}(B, x_o) \subset \text{norco}(A \cap B, x_o) .$$

Exercise (3.18) is the crux of the matter. If we had equality of the two sets, then the computation of $\text{norco}(C, x_o)$ would have been easy, and without further ado, the proof of the Lagrange Multiplier Theorem would follow. However, there need not be equality in Exercise (3.18), as illustrated in Figure 3.3. Some sort of constraint qualification is necessary. So let us start the business of computing $\mathrm{norco}(C, x_o)$. The first step is to linearize the constraints.

(3.19) LEMMA. Let p_j, $j = 1, 2, \cdots, m$ be convex and differentiable on $\mathbb{R}^d$. Let $p_j(x_o) = 0$ for all j. Then, for each $x \in \mathbb{R}^d$ the following two statements are equivalent.

(a) $\forall\, 1 \leqslant j \leqslant m \;:\; \langle \nabla p_j(x_o), x - x_o \rangle < 0$.

(b) $(\forall\, 1 \leqslant j \leqslant m)\, (\exists\, t_j > 0) \;:\; p_j(x_o + t_j(x - x_0)) < p_j(x_o)$.

PROOF. (b) $\Longrightarrow$ (a): By the hypothesis and convexity for all j,

$$0 > p_j(x_o + t_j(x - x_0)) - p(x_o) \geqslant t_j \langle \nabla p_j(x_o), x - x_o \rangle,$$

and the conclusion follows.
$\neg$(b) $\Longrightarrow$ $\neg$(a): If (b) does not hold, then there exists a j such that for all $t > 0$,

$$p_j(x_o + t(x - x_0)) \geqslant p_j(x_o) .$$

So x_o minimizes $p_j(x)$ over the half line $L = \{x_o + t(x - x_o) : t \geqslant 0\}$. By Theorem (2.2), then $\langle \nabla p_j(x_o), y - x_o \rangle \geqslant 0$ for all $y \in L$. Since $x \in L$, then (a) does not hold.   Q.e.d.

We note that the above lemma is not true if the strict inequalities are replaced by $\leqslant$, as the case of the circle in the plane makes clear: Tangent lines touch the circle in one point only. In fact, if the lemma had been true with $\leqslant$ everywhere, then the constraint qualification would be unnecessary.

(3.20) EXERCISE. Under the same conditions as Lemma (3.19), show that

$$(\forall\, 1 \leqslant j \leqslant m)\, (\exists\, t_j > 0) \;:\; p_j(x_o + t(x - x_0)) \leqslant p_j(x_o)$$

implies

$$\forall\, 1 \leqslant j \leqslant m \;:\; \langle \nabla p_j(x_o), x - x_o \rangle \leqslant 0 ,$$

but that the converse fails in general.

The lemma and exercise allow us to linearize the sets C_j about x_o. If $p_j(x_o) = 0$, then the tangent half space of C_j at x_o is defined as

(3.21) $TC_j(x_o) = \{x \in \mathbb{R}^d : \langle \nabla p_j(x_o), x - x_o \rangle \leqslant 0\}$.

(3.22) LEMMA. *Let $x_o \in C$ satisfy $p_j(x_o) = 0$ for all j. Assuming the constraint qualification (3.4) holds,*

$$\text{norco}(C, x_o) = \text{norco}\left(\left\{ \bigcap_{j=1}^{m} TC_j(x_o) \right\} \cap \left\{ \bigcap_{j=1}^{k} D_j \right\}, x_o \right).$$

PROOF. It is useful to introduce the sets

$$C_o = \left\{ x : \begin{array}{ll} p_j(x) < 0, & j = 1, 2, \cdots, m \\ q_j(x) = 0, & j = 1, 2, \cdots, k \end{array} \right\},$$

$$L_o = \left\{ x : \begin{array}{ll} \langle \nabla p_j(x_o), x - x_o \rangle < 0, & j = 1, 2, \cdots, m \\ q_j(x) = 0, & j = 1, 2, \cdots, k \end{array} \right\},$$

and

$$L = \left\{ x : \begin{array}{ll} \langle \nabla p_j(x_o), x - x_o \rangle \leqslant 0, & j = 1, 2, \cdots, m \\ q_j(x) = 0, & j = 1, 2, \cdots, k \end{array} \right\}.$$

So

$$L = \left\{ \bigcap_{j=1}^{m} TC_j(x_o) \right\} \cap \left\{ \bigcap_{j=1}^{k} D_j \right\}.$$

By the constraint qualification, then C_o is not empty, so that $C = \text{closure}(C_o)$. Likewise, $L = \text{closure}(L_o)$.

Now to the actual proof. We must prove the equality of sets. The $\subset$ part is easy and is left as an exercise.

Now for the $\supset$ part. Let $\nu \in \text{norco}(L, x_o)$. Then, surely, $\nu \in \text{norco}(L_o, x_o)$. By Lemma (3.19), then $\nu \in \text{norco}(C_o, x_o)$.

Now, consider a point $z \in C$, but $z \notin C_o$. Since C is the closure of C_o, there exists a sequence $\{ y_p \}_p \subset C_o$ which converges to z. Then,

$$\langle \nu, z - x_o \rangle = \lim_{p \to \infty} \langle \nu, y_p - x_o \rangle \leqslant 0.$$

So $\langle \nu, x - x_o \rangle \leqslant 0$ for all $x \in C$, thus, $\nu \in \text{norco}(C, x_o)$.   Q.e.d.

(3.23) EXERCISE. (a) Let $A \subset B \subset \mathbb{R}^d$, and let $x_o \in A \cap B$. Show that $\text{norco}(A, x_o) \supset \text{norco}(B, x_o)$.

(b) Prove the easy part of Lemma (3.22), using Exercise (3.20).

Finally, we need to compute the normal cone to convex cones of the form

$$A_b C_d = \{ x : Ax \leqslant b, Cx = d \},$$

where $A \in \mathbb{R}^{n \times m}$, $b \in \mathbb{R}^m$, $C \in \mathbb{R}^{k \times m}$, and $d \in \mathbb{R}^k$. This result is known as (a variation of) the Lemma of Farkas.

(3.24) FARKAS' LEMMA. *Suppose x_o satisfies $Ax_o = b$, $Cx_o = d$. Then,*

$$\operatorname{norco}(A_b C_d, x_o) = \{A^{\mathsf{T}} y + C^{\mathsf{T}} z : y \geqslant 0, z \text{ arbitrary}\}.$$

The following development to prove Lemma (3.24) could be condensed considerably, but in the process, we learn some nice things regarding convex cones. We begin with an exercise.

(3.25) EXERCISE. *Let $A \in \mathbb{R}^{n \times m}$, $C \in \mathbb{R}^{k \times m}$. Show that*

$$\operatorname{norco}(\{A^{\mathsf{T}} y + C^{\mathsf{T}} z : y \geqslant 0, z \in \mathbb{R}^k\}, 0) = \{x : Ax \leqslant 0, Cx = 0\}.$$

Again, this exercise would do it if the norco operation was on the other side of the equality.

(3.26) SEPARATION LEMMA. *Let K be a closed convex cone, and let $y \notin K$. Then there exists a $\nu \in \mathbb{R}^d$ and a $y_o \in K$ such that $\langle \nu, y \rangle > 0$, and*

$$\forall x \in K \; (\langle \nu, x \rangle \leqslant 0).$$

PROOF. Let y_o be the projection of y onto K. By Theorem (2.2), then

$$\forall x \in K \; \langle y - y_o, x - y_o \rangle \leqslant 0.$$

Now take $\nu = y - y_o$. If $y_o = 0$, then we are done. If $y_o \neq 0$, then consider the minimization of $\|x - y\|$ over all $x \in \{t y_o : t \in \mathbb{R}\}$. A simple geometric diagram shows that this projection is y_o, and by Theorem (2.2), then $\langle y - y_o, y_o \rangle = 0$. So $\langle \nu, y_o \rangle = 0$. Then, for all $x \in K$,

$$\langle \nu, x \rangle = \langle \nu, x - y_o \rangle \leqslant 0,$$

and also $\langle \nu, y - y_o \rangle = \langle \nu, y \rangle > 0$, and we are indeed done.   Q.e.d.

(3.27) LEMMA. *Let K be a closed convex cone. Then, $K^{\circ\circ} = K$.*

PROOF. Let $y \in K$. Then, $\langle \nu, y \rangle \leqslant 0$ for all $\nu \in K^\circ$. But this says that $y \in K^{\circ\circ}$.

Suppose $y \notin K$. By the Separation Lemma for Convex Cones, there exists a $\nu \in \mathbb{R}^d$ such that $\langle \nu, y \rangle > 0$, and

$$\forall x \in K \; \langle \nu, x \rangle \leqslant 0.$$

So this says that $\nu \in K^\circ$ and $\langle \nu, y \rangle > 0$. The conclusion is that $y \notin K^{\circ\circ}$. Q.e.d.

(3.28) LEMMA. *Let $A \in \mathbb{R}^{n \times m}$, $C \in \mathbb{R}^{k \times m}$. Then, the cone*

$$K = \{A^{\mathsf{T}} y + C^{\mathsf{T}} z : y \geqslant 0\}$$

is closed.

PROOF. Let $\{x_p\}_{p\geqslant 1} \subset K$ and suppose $x_p \longrightarrow x_o \in \mathbb{R}^m$. We must show that $x_o \in K$. We have

$$(3.29) \qquad x_p = A^T y_p + C^T z_p \;,$$

with $y_p \geqslant 0$. If the matrix $(A^T \,|\, C^T)$ has independent columns, then (3.29) implies that there exists a constant c such that

$$\| y_p \| + \| z_p \| \leqslant c \| x_p \| \;.$$

(Augment $(A^T \,|\, C^T)$ with a matrix E such that $(A^T \,|\, C^T \,|\, E)$ is nonsingular. The constant is the norm of the inverse.) But then also for all p, q,

$$\| y_p - y_q \| + \| z_p - z_q \| \leqslant c \| x_p - x_q \| \;.$$

Since $\{x_p\}_p$ converges, it is Cauchy, and the above inequality shows that $\{y_p\}_p$ and $\{z_p\}_p$ are also Cauchy. Hence, they converge, say, to y_o and z_o, and $x_o = A^T y_o + C^T z_o$. Surely $y_o \geqslant 0$, so $x_o \in K$.

What if the columns of $(A^T \,|\, C^T)$ are linearly dependent ? The trick is to write K as a finite union of closed cones K_j,

$$(3.30) \qquad K = \bigcup_j K_j \;.$$

So then K would be closed.

To prove (3.30), denote the columns of A and C by a_j and c_j. Let $x \in K$, so we may write x as

$$x = A^T y + C^T z = \sum_{j=1}^m y_j \, a_j + \sum_{j=1}^k z_j \, c_j$$

for $y \geqslant 0$ and (arbitrary) z. Now, the vectors a_j and c_j are (assumed) linearly dependent, so there exist numbers t_j, s_j not all equal to zero such that

$$\sum_{j=1}^m t_j \, a_j + \sum_{j=1}^k s_j \, c_j = 0.$$

So for arbitrary scalar λ,

$$(3.31) \qquad x = \sum_{j=1}^m (\, y_j - \lambda \, t_j)\, a_j + \sum_{j=1}^k (\, z_j - \lambda \, s_j)\, c_j \;.$$

Now, choose λ in such a way that (at least) one of the $m + k$ terms of $y_j - \lambda \, t_j$, $j = 1, 2, \cdots, m$, and $z_j - \lambda \, s_j$, $j = 1, 2, \cdots, k$. vanishes, while keeping

$$y_j - \lambda \, t_j \geqslant 0 \;, \quad j = 1, 2, \cdots, m \;.$$

Then, one term in the two sums of (3.31) vanishes. So we have eliminated one of the columns of $(A^T \,|\, C^T)$. By repeating this process if necessary, we arrive at a representation of x as a linear combination

$$x = \sum_{j \in \mathcal{I}} \tilde{y}_j \, a_j + \sum_{j \in \mathcal{J}} \tilde{z}_j \, c_j$$

for suitable index sets $\mathcal{I}$ and $\mathcal{J}$ (dependent on x), with $\tilde{y}_j \geqslant 0$ for all $j \in \mathcal{I}$, and such that the vectors a_j, $j \in \mathcal{I}$, c_j, $j \in \mathcal{J}$ are linearly independent. So x lies in the convex cone K_x,

$$K_x = \Big\{ \sum_{j \in \mathcal{I}} \lambda_j \, a_j + \sum_{j \in \mathcal{J}} \mu_j \, c_j \, : \, \lambda_j \geqslant 0 \, , \, \mu_j \text{ arbitrary} \Big\} \, ,$$

which is closed by the first part of the proof. Since obviously $K_x \subset K$, then

$$K = \bigcup_{x \in K} K_x \, .$$

Now, the collection of all cones K_x is finite: There are at most as many cones as there are ways of choosing independent collections of columns from the matrix $(A^T \,|\, C^T)$. So the above union is in fact a finite union. Since the K_x are closed, this shows that K is closed. Q.e.d.

PROOF OF FARKAS' LEMMA (3.24). Apply Lemma (3.27) to

$$\mathrm{norco}\big(\{ A^T y + C^T z \, : \, y \geqslant 0, \, z \in \mathbb{R}^k \} \, , 0 \big) = \{ x \, : \, A x \leqslant 0 \, , \, C x = 0 \} \, ,$$

see Exercise (3.25). Since $\{ A^T y + C^T z \, : \, y \geqslant 0, \, z \in \mathbb{R}^k \}$ is in fact a closed convex cone, this is allowed. Q.e.d.

The proof of the Lagrange Multiplier Theorem is now complete.

(3.32) EXERCISE. Fill in the details of the proof of the Lagrange Multiplier Theorem (3.5).

(3.33) EXERCISE. Apply the Lagrange Multiplier Theorem to the problems

(a) minimize $\| A x - b \|^2$

  subject to $C x \leqslant d \, , \ E x = f \, , \ \| G x - h \| \leqslant i \, ,$

for matrices A, C, E, and G, and vectors b, d, f, h, and i, of suitable dimensions;

(b) minimize $L(x)$

  subject to $b \leqslant A x \leqslant c \, ,$

for convex differentiable L on $\mathbb{R}^d$, and matrix A and vectors b, c.

EXERCISES: (3.12), (3.13), (3.14), (3.16), (3.17), (3.18), (3.20), (3.23), (3.25), (3.32), (3.33).

4. Strict and strong convexity

In this and the next section, we explore conditions for the existence and uniqueness of solutions of (2.1). Mere convexity of the objective function not being enough for existence and uniqueness of the solution of the min-imization problem, we investigate strengthened versions of convexity. In the next section, compactness arguments are briefly investigated.

We consider the following setting. Let $C \subset \mathbb{R}^d$ be a convex set, sometimes assumed to be closed. Let

$$L : C \longrightarrow \mathbb{R} \text{ be a convex differentiable function}.$$

Again, the differentiability assumption may be avoided, but we shall not belabor the point.

We begin with two strengthened notions of convexity.

(4.1) DEFINITION. (a) L is strictly convex if for all $x \neq y \in C$,

$$L(x) - L(y) - \langle \nabla L(y), x - y \rangle > 0 .$$

(b) L is strongly convex if there exists an $\alpha > 0$ such that for all $x, y \in C$,

$$L(x) - L(y) - \langle \nabla L(y), x - y \rangle \geqslant \alpha \| x - y \|^2 .$$

We note that the choice of norm in part (b) is unimportant, since all norms on $\mathbb{R}^d$ are equivalent. However, the constant α will depend on the particular norm used. We also note the implications

$$\text{strong convexity} \implies \text{strict convexity} \implies \text{convexity} .$$

Strict convexity is more or less designed for showing the uniqueness of solutions, but it is not helpful in establishing their existence or for determining any nice properties a minimizer might have. We emphasize that our usage of the phrase "the solution of (2.1) is unique" does not imply that a solution exists. It only means that there is at most one solution.

(4.2) THEOREM. If L is strictly convex, then the solution of (2.1) is unique.

PROOF. Suppose x_o solves (2.1), and let $y \in C$, $y \neq x_o$. Strict convexity gives $L(y) - L(x_o) - \langle \nabla L(x_o), y - x_o \rangle > 0$. Now, from Theorem (2.2), it follows that $L(y) - L(x_o) > \langle \nabla L(x_o), y - x_o \rangle \geqslant 0$, and so y does not solve (2.1).   Q.e.d.

How can one tell whether a convex function is strictly or strongly convex ? Fortunately, there are analogues of Theorem (1.7) for strict convexity and strong convexity. We only state the strong convexity result.

(4.3) THEOREM. Let $C \subset \mathbb{R}^d$ be convex, and let $f : C \longrightarrow \mathbb{R}$ be a differentiable function. Then, the following conditions are equivalent.

(a) $f(x) - f(y) - \langle \nabla f(y) , x - y \rangle \geqslant \alpha \| x - y \|^2$ for all x, $y \in C$.

(b) $\langle \nabla f(x) - \nabla f(y) , x - y \rangle \geqslant \alpha \| x - y \|^2$ for all x, $y \in C$.

(c)   If, in addition, f is twice differentiable, then for all x, $y \in C$,

$$\langle \nabla^2 f(x) (x - y) , x - y \rangle \geqslant \alpha \| x - y \|^2 .$$

(4.4) EXERCISE. Prove Theorem (4.3). Compare your proof with that of Theorem (1.7).

(4.5) EXERCISE. State and prove the theorem corresponding to Theorem (4.3) for strict convexity.

(4.6) EXERCISE. Determine which convex functions from Example (1.11) are strictly, respectively strongly, convex. For the strongly convex functions, exhibit the constant α.

(4.7) EXERCISE. Show that the functional L in Example (2.8) is strongly convex if and only if A has full column rank.

(4.8) EXERCISE. Show that a necessary condition for the functional L in Example (2.14) to be strongly convex is that A has full column rank. Is it also sufficient ?

(4.9) COROLLARY. If L is strongly convex and x_o solves (2.1), then for all $x \in C$,

$$L(x) - L(x_o) \geqslant \alpha \| x - x_o \|^2 .$$

PROOF. From Theorem (4.3) (a) and Theorem (2.2).   Q.e.d.

We are now ready for the first taste of existence deliberations. The following formal definition is of fundamental importance, as simple a concept as it is.

(4.10) DEFINITION. Let $S \subset \mathbb{R}^d$, and suppose that $f : S \longrightarrow \mathbb{R}$. A minimizing sequence for f is a sequence $\{ x_n \}_n \subset S$ for which

$$\lim_{n \to \infty} f(x_n) = \inf \{ f(x) : x \in S \} .$$

Note that in this definition, the infimum need not be attained, and that the infimum may in fact equal $-\infty$.

In order to proceed, we need to introduce the *dual norm* $\| \cdot \|_*$ of $\| \cdot \|$. This is defined as follows.

(4.11) DEFINITION. *If* $\| \cdot \|$ *is a norm on* $\mathbb{R}^d$, *then its dual norm* $\| \cdot \|_*$ *is defined as*

$$\| x \|_* = \max \left\{ \langle x, y \rangle : \| y \| = 1 \right\} .$$

(4.12) EXERCISE. Let $\| \cdot \|$ and $\| \cdot \|_*$ be dual norms. Show that
(a) $\| \cdot \|_*$ is indeed a norm on $\mathbb{R}^d$;
(b) $\| \cdot \|$ itself is the dual norm of $\| \cdot \|_*$;
(c) $\left| \langle x, y \rangle \right| \leqslant \| x \| \| y \|_*$ for all $x, y \in \mathbb{R}^d$.

(4.13) EXERCISE. Let $1 \leqslant p \leqslant \infty$, and let q be the dual exponent, defined formally via $(1/p) + (1/q) = 1$. Show that the norms $\| \cdot \|_p$ and $\| \cdot \|_q$ are dual to each other.

(4.14) THEOREM. *Let* $C \subset \mathbb{R}^d$ *be closed and convex. If* $L : C \longrightarrow \mathbb{R}$ *is strongly convex, then the solution of (2.1) exists and is unique.*

PROOF. The uniqueness follows from Theorem (4.2). As for the existence, we first show that L is bounded from below on C, then that any minimizing sequence is bounded, and finally that the limit of any convergent subsequence of the minimizing sequence is a solution of (2.1).

To show that L is bounded from below, fix an element $y \in C$. Then, for all $x \in C$, we have

$$L(x) \geqslant L(y) + \langle \nabla L(y), x - y \rangle + \alpha \| x - y \|^2$$
$$\geqslant L(y) - \| \nabla L(y) \|_* \| x - y \| + \alpha \| x - y \|^2$$
$$\geqslant L(y) + \alpha \left\{ \| x - y \| - (2\alpha)^{-1} \| \nabla L(y) \|_* \right\}^2 - (4\alpha)^{-1} \| \nabla L(y) \|_*^2$$
$$\geqslant L(y) - (4\alpha)^{-1} \| \nabla L(y) \|_*^2 .$$

This implies that L is bounded from below on C.

Now, L being bounded from below on C implies that $\inf\{ L(x) : x \in C \}$ is finite. Let $\{x_n\}_n \subset C$ be a minimizing sequence, i.e.,

$$\lim_{n \to \infty} L(x_n) = \inf_{x \in C} L(x) .$$

Without loss of generality, assume that $L(x_n) \leqslant L(x_1)$ for all n. The strong convexity implies that

$$\alpha \| x_n - x_1 \|^2 \leqslant L(x_n) - L(x_1) - \langle \nabla L(x_1), x_n - x_1 \rangle$$
$$\leqslant \| \nabla L(x_1) \|_* \| x_n - x_1 \| ,$$

which implies that

$$\alpha \| x_n - x_1 \| \leqslant \| \nabla L(x_1) \|_* .$$

In other words, the sequence $\{x_n\}_n$ is bounded.

Since $\{x_n\}_n \subset \mathbb{R}^d$ is bounded, it has a convergent subsequence, and since C is closed, its limit belongs to C. Without loss of generality, assume that the whole sequence $\{x_n\}_n \subset C$ converges to some element $x_o \in C$. Finally, since L is continuous on C, then

$$\lim_{n \to \infty} L(x_n) = L(x_o) \, ,$$

and because $\{x_n\}_n$ is a minimizing sequence, then x_o solves (2.1).   Q.e.d.

(4.15) REMARK. In infinite-dimensional spaces, the above argument does not work as is. The argument that L is bounded from below on C would go through essentially unchanged, as does the boundedness of any minimizing sequence. At that point, the argument comes to a grinding halt, because in infinite-dimensional spaces, bounded sequences do not necessarily have norm convergent subsequences. In some infinite-dimensional spaces, bounded sequences would have subsequences that converge in some weak sense, and that is sometimes/usually sufficient. In our favorite space $L^1(\mathbb{R})$, this does not work either (a sequence of pdfs might converge to a point mass, say). However, we can still use strong convexity to our advantage, see the next chapter.

EXERCISES: (4.4), (4.5), (4.6), (4.7), (4.8), (4.12), (4.13).

5. Compactness arguments

If the objective function in a convex minimization problem is not strongly convex, then other conditions are required to establish existence of solutions. In this section, we explore existence results based on compactness arguments. Such arguments have already made an appearance in the proof of the existence of solutions to minimization problems with strongly convex objective functions. In finite-dimensional spaces, compactness arguments are straightforward. The corresponding section in the infinite-dimensional case is much more interesting, see § 10.4. Compactness arguments are variations on the theme that continuous functions on compact sets attain their minimum. Another class of conditions are coerciveness conditions. These are variations on the theme $L(x) \longrightarrow \infty$ for $\|x\| \longrightarrow \infty$, which guarantee that solutions of (2.1) are a priori bounded.

Throughout this section, $C \subset \mathbb{R}^d$ is assumed to be a closed set, but not necessarily bounded, and $L : C \longrightarrow \mathbb{R}$ is not necessarily convex. The following two notions are useful for establishing the existence of a minimum.

(5.1) DEFINITION. The level sets of L (on C) are defined by

$$\{\, x \in C \, : \, L(x) \leqslant \ell \,\} \, ,$$

for $\ell \in \mathbb{R}$. The value of ℓ is referred to as the level of the level set.

(5.2) DEFINITION. *L is lower semicontinuous on C if for every sequence* $\{x_n\}_n \subset C$, *which converges to* $x_o \in C$,

$$L(x_o) \leqslant \liminf_{n \to \infty} L(x_n) .$$

(5.3) EXERCISE. (a) Show that the function $f : \mathbb{R}^2 \longrightarrow \mathbb{R}$ defined by

$$f(x,y) = \frac{x^2 - y^2}{x^2 + y^2} , \quad (x,y) \neq (0,0) ,$$

and $f(0,0) = -1$ is lower semicontinuous,
(b) but that the function g, defined by $g(x,y) = f(x,y)$, $(x,y) \neq (0,0)$, and $g(0,0) = 1$, is not.

(5.4) EXERCISE. The interaction between lower semicontinuity and convexity is interesting. Let C be convex, and let $L : C \longrightarrow \mathbb{R}$ be a convex function. Show that L is lower semicontinuous if and only if epi(L) is closed. [Note that if you are not too careful you will actually prove this for convex functions on topological vector spaces.]

A very useful assumption may now be expressed as follows:

(5.5)   *L has at least one (nonempty) compact level set* .

We remind the reader that a subset of $\mathbb{R}^d$ is compact if and only if it is closed and bounded.
    We now concern ourselves with the problem

(5.6)
$$\begin{aligned} \text{minimize} \quad & L(x) \\ \text{subject to} \quad & x \in C . \end{aligned}$$

(5.7) THEOREM. *If L is lower semicontinuous, bounded below, and satisfies* (5.5), *then the problem* (5.6) *has a solution.*

PROOF. Let $y \in C$ be such that the level set $\{x \in C : L(x) \leqslant L(y)\}$ is compact. (By (5.5), such a y exists.) Let $\{x_n\}_n \subset C$ be a minimizing sequence. Without loss of generality, assume that $L(x_n) \leqslant L(y)$ for all n. So the sequence lies in a compact level set, and hence, it has a convergent subsequence, with limit $x_o \in C$. By the lower semicontinuity of L, then

$$L(x_o) \leqslant \liminf_{n \to \infty} L(x_n) = \inf_{x \in C} L(x) .$$

It follows that x_o solves the minimization problem.   Q.e.d.

An alternative, useful condition is that of coerciveness, i.e.,

(5.8) $$L(x) \longrightarrow \infty \text{ for } \|x\| \to \infty \ (x \in C) .$$

This implies that minimizing sequences must be bounded.

(5.9) THEOREM. *If L is coercive, lower semicontinuous, and bounded from below, then (5.6) has a solution.*

(5.10) EXERCISE. Prove Theorem (5.9).

EXERCISES: (5.3), (5.4), (5.10).

6. Additional notes and comments

The bible of convex analysis on finite-dimensional spaces is ROCKAFELLAR (1970). If you wish to know more on convex optimization problems, you will find it in there. A recent rendition is the two-volume work of HIRIART-URRUTY and LEMARÉCHAL (1993). Some classic books on optimization theory and methods are LUENBERGER (1968), FIACCO and McCORMICK (1968), and MANGASARIAN (1969), and a new one is CENSOR and ZENIOS (1997).

10

Convex Optimization in Infinite-Dimensional Spaces

1. Convex functions

Having learned how to walk in finite-dimensional spaces, we now attempt to run in infinite-dimensional spaces. We begin by considering some abstract aspects of convex minimization problems. In the next section, these are specialized to convex functions defined by means of *convex integrals*, and derive suitable analogues of the results of Chapter 9. In this chapter, there is a definite paucity of "little" examples. The convexity aspects of the "big" examples (penalized least-squares problems and penalized maximum likelihood estimation) are treated in the next chapter.

First, some notations. Throughout this chapter, X denotes a Banach space and X^* denotes its dual space (the space of bounded linear functionals on X), with $\langle \cdot , \cdot \rangle_X$ the duality pairing on $X^* \times X$. That is, if $f \in X^*$, $x \in X$, then $f(x)$ is denoted as

$$(1.1) \qquad\qquad f(x) = \langle f , x \rangle_X .$$

It is assumed that X is an infinite-dimensional vector space, by which we mean a not necessarily finite-dimensional space. (For a review of the notions of Banach spaces and dual spaces, see Appendix 3.)

The definitions of §10.1 of convex sets, convex functions, and epigraphs of functions are independent of the dimension of the space, so they work in infinite-dimensional spaces as well. Let $C \subset X$ be a closed convex set, and let $L : C \longrightarrow \mathbb{R} \cup \{\infty\}$ be a not necessarily convex function. The effective domain of L is defined as domain $(L) = \{ x \in C : L(x) < \infty \}$. If L is convex, then domain (L) is obviously a convex set.

As in the previous chapter, we usually assume that the convex functions to be considered are differentiable in a reasonable sense. Thus, the notion of differentiability in infinite-dimensional spaces needs to be made precise. There are many different kinds of differentiability, but the Gateaux variation, or directional derivative, suits our purposes.

(1.2) DEFINITION. *Let $L : \mathcal{C} \longrightarrow \mathbb{R} \cup \{\infty\}$, and let $x \in$ domain (L). A vector $v \in X$ is said to be an "admissible direction" at x if there exists a scalar $t_o > 0$ with $x + t\,v \in$ domain (L) for all $0 < t < t_o$. The Gateaux variation of L at x in the (admissible) direction v is defined as the one-sided limit*

$$\delta L(x, v) = \lim_{t \to 0^+} t^{-1} \left\{ L(x + t\,v) - L(x) \right\} .$$

If there exists an $f \in X^*$ such that $\delta L(x, v) = \langle\, f , v\, \rangle_X$ for all admissible v, then f is called the Gateaux derivative of L at x, and we write $f = L'(x)$.

The notion of Gateaux variations blends in nicely with convexity. First, an easy exercise and some examples.

(1.3) EXERCISE. *Let $L : \mathcal{C} \longrightarrow \mathbb{R} \cup \{\infty\}$, $x \in \mathcal{C}$, and let $v \in X$ be an admissible direction at x.* (a) *Show that $\delta L(x, \lambda\,v) = \lambda\,\delta L(x, v)$ for all $\lambda > 0$. So $\delta L(x, v)$ is positively homogeneous in v.*
(b) *If $-v$ is an admissible direction as well, show that*

$$-\delta L(x, -v) \leqslant \delta L(x, v).$$

(c) *Assume that L is convex. Let $x \in \mathcal{C}$, and suppose that v and $w \in X$ are admissible directions at x. Show that $\delta L(x, v + w) \leqslant \delta L(x, v) + \delta L(x, w)$. So $\delta L(x, v)$ is sublinear in v.*

(1.4) EXERCISE. *For each of the following functionals, determine the set of admissible directions and the Gateaux variation at each point in its effective domain.*
(a) $L(f) = \| f \|_2^2$ on $L^2(\mathbb{R})$.
(b) $L(f) = \| f \|_2^2$ on $L^1(\mathbb{R})$. [Effective domain $L^2(\mathbb{R}) \cap L^1(\mathbb{R})$.]
(c) $L(f) = \| f \|_1$ on $L^1(\mathbb{R})$.
(d) $L(f) = \mathrm{H}(f, g)$ on the set $\mathcal{P} = \{ f \in L^1(\mathbb{R}) : f \geqslant 0 \}$, with $g \in \mathcal{P}$ fixed. Here, H is Hellinger distance.
(e) $L(f) = \mathrm{KL}(f, g)$, on the set $\mathcal{P}$, with $g \in \mathcal{P}$ fixed. Here, KL is Kullback-Leibler distance.

We now show that convex functionals possess Gateaux variations. It is clear that the Gateaux derivative does not necessarily exist, as the one-dimensional example $L(x) = |x|$, $x \in \mathbb{R}$, shows. The proof uses the fact that $\varphi(\lambda) = L(\lambda\,x + (1 - \lambda)\,y)$ is a convex function on $[\,0, 1\,]$, so it is helpful to first consider the slopes of secant lines for a convex function.

(1.5) LEMMA. *Let $\varphi : [\,0, 1\,] \longrightarrow \mathbb{R}$ be convex. Consider the "chords" $0 \leqslant p < q < 1$, and $0 < s < t \leqslant 1$, with $p \leqslant s$ and $q \leqslant t$. Then,*

$$\frac{\varphi(q) - \varphi(p)}{q - p} \leqslant \frac{\varphi(t) - \varphi(s)}{t - s} .$$

PROOF. Since $p < q \leqslant t$, then

$$q = \frac{t - q}{t - p} \, p + \frac{q - p}{t - p} \, t$$

is a convex combination of p and t, so the convexity of φ gives

$$\varphi(q) \leqslant \frac{t - q}{t - p} \, \varphi(p) + \frac{q - p}{t - p} \, \varphi(t) \, .$$

This is equivalent to

(1.6) $$\frac{\varphi(q) - \varphi(p)}{q - p} \leqslant \frac{\varphi(t) - \varphi(p)}{t - p} \, .$$

Now, apply this inequality to the function $\psi(\lambda) = \varphi(1 - \lambda)$ and the points $1 - t < 1 - s \leqslant 1 - p$. The result may be (re)written as

$$\frac{\varphi(t) - \varphi(p)}{t - p} \leqslant \frac{\varphi(t) - \varphi(s)}{t - s} \, .$$

Together with (1.6), this is the required inequality.   Q.e.d.

(1.7) LEMMA. *Let $L : \mathcal{C} \longrightarrow \mathbb{R} \cup \{ \infty \}$ be convex, and let x, $y \in$ domain (L). Then, for every $z = \lambda x + (1 - \lambda) y$, $0 < \lambda < 1$, the Gateaux variation $\delta L(z, x - y)$ exists and is an increasing function of λ. If $\delta L(x, y - x)$ and $\delta L(y, x - y)$ are both finite, then $L(z, x - y)$ is Lipschitz continuous in λ, and*

$$L(x) - L(y) = \int_0^1 \delta L(\lambda \, x + (1 - \lambda) \, y \, , \, x - y) \, d\lambda \, ,$$

where the integral is in the Lebesgue sense.

PROOF. Let $\varphi(\lambda) = L(\lambda \, x + (1 - \lambda) \, y)$, and apply Lemma (1.5) for the points $p = s = \lambda < q \leqslant t$ to obtain

$$\frac{\varphi(q) - \varphi(\lambda)}{q - \lambda} \leqslant \frac{\varphi(t) - \varphi(\lambda)}{t - \lambda} \, .$$

It follows that $\psi(q) = (\varphi(q) - \varphi(\lambda))/(q - \lambda)$ is an increasing function of q, $q > \lambda$. Note that

$$\psi(q) = t^{-1} \left\{ L(z + t(x - y)) - L(z) \right\} \, , \quad t = q - \lambda \, .$$

Applying Lemma (1.5) to the points $p < s = q = \lambda < t$ shows that

$$\psi(p) = \frac{\varphi(\lambda) - \varphi(p)}{\lambda - p} \leqslant \frac{\varphi(t) - \varphi(\lambda)}{t - \lambda} = \psi(t) \, .$$

Hence, $\psi(t)$ is increasing and bounded below for $t > \lambda$, and thus,

$$\lim_{t \to \lambda^+} \psi(t)$$

exists. Equivalently, the Gateaux variation $\delta L(\lambda x + (1-\lambda) y , x - y)$ exists for $0 < \lambda < 1$. By taking limits in the inequality of Lemma (1.5), viz. $q \to p^+$ and $t \to s^+$, we see that $\varphi'(p) \leqslant \varphi'(s)$ (one-sided derivatives) for a.e. p and s, with $p \leqslant s$. So $\delta L(\lambda x + (1 - \lambda) y , x - y)$ is an increasing function of λ.

Suppose $\delta L(x, y - x)$ and $\delta L(y, x - y)$ are finite. Let $p, s, t,$ and q satisfy $0 < p < s < t < q < 1$. Then, by Lemma (1.5),

$$\frac{\varphi(p) - \varphi(0)}{p} \leqslant \frac{\varphi(t) - \varphi(s)}{t - s} \leqslant \frac{\varphi(1) - \varphi(q)}{1 - q} .$$

Now, let $p \to 0^+$, $q \to 1^-$ $(1 - q \to 0^+)$, to obtain that

$$\delta L(x, y - x) \leqslant \frac{\varphi(t) - \varphi(s)}{t - s} \leqslant \delta L(y, x - y) ,$$

and hence, for an appropriate constant M,

$$| \varphi(t) - \varphi(s) | \leqslant M | t - s | .$$

In other words, $\varphi(\lambda)$ is Lipschitz continuous, so its derivative exists and is bounded (Lebesgue) almost everywhere. Moreover, φ is then the anti-derivative of its derivative. In particular,

$$\varphi(1) - \varphi(0) = \int_0^1 \varphi'(\lambda) \, d\lambda .$$

Since $\varphi'(\lambda) = \delta L(\lambda x + (1 - \lambda) y , x - y)$ a.e., the last statement of the lemma follows.   Q.e.d.

(1.8) COROLLARY. *Let $L : C \longrightarrow \mathbb{R} \cup \{ \infty \}$ be convex. Then $\delta L(x, y - x)$ exists and is $< \infty$ (but could equal $-\infty$) for all $x, y \in$ domain (L).*

To recapitulate, the lemma and corollary say that convex functions L satisfy the condition

(1.9) $\delta L(\lambda x + (1 - \lambda) y, x - y)$ is a measurable function of λ
          and is $< \infty$ for all $x, y \in$ domain (L).

We now get the same characterization of convex functions as in the finite-dimensional case.

(1.10) THEOREM. *Let $L : C \longrightarrow \mathbb{R} \cup \{ \infty \}$ satisfy (1.9). Then, the following are equivalent.*
(a) *L is convex.*
(b) *$L(x) - L(y) - \delta L(y, x - y) \geqslant 0$ for all $x, y \in$ domain (L).*
(c) *$-\delta L(y, x - y) - \delta L(x, y - x) \geqslant 0$ for all $x, y \in$ domain (L).*

PROOF. (a) $\Longrightarrow$ (b): By Lemma (1.7),

$$L(x) - L(y) = \int_0^1 \delta L(\lambda\, x + (1 - \lambda)\, y\, ,\, x - y)\, d\lambda \geqslant \delta L(y, x - y)\ .$$

(b) $\Longrightarrow$ (c) and (b) $\Longrightarrow$ (a): These are as in Theorem (9.1.7).
(c) $\Longrightarrow$ (b): Let x, $y \in C$. Now, (c) says that $\delta L(y, x - y) \leqslant \delta L(x; x - y)$, and so we may assume that

$$-\infty < \delta L(y, x - y) \leqslant \delta L(x, x - y) < +\infty\, ,$$

since otherwise there is nothing to prove.
   Now, consider the following. For $0 \leqslant \mu < \lambda \leqslant 1$, let

$$\tilde{x} = \lambda\, x + (1 - \lambda)\, y\ ,\quad \tilde{y} = \mu\, x + (1 - \mu)\, y\ .$$

Then, $\tilde{x} - \tilde{y} = (\lambda - \mu)\, (x - y)$, and since $\lambda - \mu > 0$ condition (c) implies that $\delta L(\tilde{x}, x - y) \geqslant \delta L(\tilde{y}, x - y)$. So $\delta L(\lambda\, x + (1 - \lambda)\, y, x - y)$ is an increasing function of $0 \leqslant \lambda \leqslant 1$. By Lemma (1.7), then

$$L(x) - L(y) = \int_0^1 \delta L(\lambda\, x + (1 - \lambda)\, y, x - y)\, d\lambda \geqslant \delta L(y, x - y)\ .$$

Thus, L is convex.   Q.e.d.

   The analogue of Theorem (9.2.2) now follows almost immediately. Consider the minimization problem

(1.11)
$$\begin{aligned} &\text{minimize}\quad L(x) \\ &\text{subject to}\quad x \in C\ . \end{aligned}$$

(1.12) THEOREM. Let $L : C \longrightarrow \mathbb{R} \cup \{\infty\}$ be convex. Then, x_o is a solution of (1.11) if and only if

$$\delta L(x_o, x - x_o) \geqslant 0 \quad \text{for all } x \in \text{domain}\,(L)\ .$$

(1.13) EXERCISE. Prove Theorem (1.12).

   It would be inappropriate to finish this section without remarking on other notions of differentiability. We mention the following two.
   Let X be a Banach space. A mapping $L : X \longrightarrow \mathbb{R}$ is Fréchet differentiable at x if there exists a $g \in X^*$, denoted as $g = L'(x)$, such that

(1.14)
$$\lim_{t \to 0} \frac{|\, L(x + t\, a) - L(x) - t\, \langle\, L'(x)\, ,\, a\, \rangle_X\, |}{|\, t\, |} = 0\ ,$$

uniformly in a, with $\|\, a\, \|_X = 1$. In Gateaux derivative language, the limit converges uniformly in all directions.

A mapping $L : X \longrightarrow \mathbb{R}$ is Hadamard differentiable at x_o if there exists a $g \in X^*$, denoted as $g = L'(x_o)$, such that for every compact $A \subset X$,

$$(1.15) \qquad \lim_{t \to 0} \frac{|L(x_o + t\,a) - L(x_o) - t\langle L'(x_o), a \rangle_X|}{t} = 0 \ ,$$

uniformly in $a \in A$. If L is Fréchet differentiable at every point of an open set, then we say it is Fréchet differentiable on that set. We shall have the same usage for a function being Gateaux or Hadamard differentiable on an open set. In finite-dimensional spaces, the notions of Gateaux, Hadamard, and Fréchet differentiability on open sets coincide. In infinite-dimensional spaces, it is a different story. The implications (point wise, or on sets)

$$\text{Fréchet} \implies \text{Hadamard} \implies \text{Gateaux differentiability}$$

are obvious. To illustrate the differences, we consider some examples.

(1.16) EXAMPLE. Let $L : \operatorname{domain}(L) \subset L^1(\mathbb{R}) \longrightarrow \mathbb{R}$ be defined by

$$L(f) = \int_{\mathbb{R}} x^2\, f(x)\, dx \ ,$$

with $\operatorname{domain}(L) = \{\, f \in L^1(\mathbb{R}) : \int_{\mathbb{R}} x^2\,|f(x)|\,dx < \infty \,\}$. It is apparent that for $f, g \in \operatorname{domain}(L)$,

$$\delta L(f, g - f) = \int_{\mathbb{R}} x^2\,\{\, g(x) - f(x) \,\}\, dx$$

is well defined and finite. So the Gateaux variation exists at every point in $\operatorname{domain}(L)$, but L is not Gateaux differentiable, since $\delta L(f, \cdot)$ is not a bounded linear functional on $L^1(\mathbb{R})$. Note that the linearity of L implies that it is convex.

(1.17) EXERCISE. Let Ω be an open and bounded subset of $\mathbb{R}$. Show that $L(f) = \int_{\Omega} x^2\, f(x)\, dx$ is Fréchet differentiable on $L^1(\Omega)$.

(1.18) EXAMPLE. This example is crucial to the endeavor of this text. Let φ be a pdf on the line, and define $L : \operatorname{domain}(L) \subset L^1(\mathbb{R}) \longrightarrow \mathbb{R}$ by

$$L(f) = \mathrm{KL}(f, \varphi) = \int_{\mathbb{R}} \left\{\, f(x) \log \frac{f(x)}{\varphi(x)} + \varphi(x) - f(x) \,\right\} dx \ .$$

Obviously, $\operatorname{domain}(L)$ is a subset of all nonnegative $L^1(\mathbb{R})$ functions. It seems clear that if the Gateaux variation exists, it must be

$$(1.19) \qquad \delta L(f, g - f) = \int_{\mathbb{R}} (g - f) \log \frac{f}{\varphi} \ ,$$

and although conditions under which this holds are obvious, at the same time, they beg the question. See Exercise (1.20). Of course, Lemma (1.7) applies and says that the Gateaux variation exists if f lies in an open line segment

$$\{ \lambda g + (1 - \lambda) h \; : \; 0 < \lambda < 1 \} \subset \text{domain}(L) \; ,$$

but it does not tell us that (1.19) holds. In the next section, we study convex integrals, and show that (1.19) does indeed hold, as a simple example of a general theory.

(1.20) EXERCISE. Show that (1.19) holds if $(g - f) \log(f/\varphi) \in L^1(\mathbb{R})$.

We finish this section with the standard examples of Fréchet differentiable functions on infinite-dimensional spaces.

(1.21) EXERCISE. Let X be a Hilbert space with norm denoted by $\| \cdot \|$. Show that $L(x) = \| x \|^2$ is Fréchet differentiable everywhere on X.

(1.22) EXERCISE. Let $1 < p < \infty$. Let $\Omega \subset \mathbb{R}^d$ be an open set, and let $X = L^p(\Omega)$, with norm $\| \cdot \|_p$. Show that $L(f) = \| f \|_p^p$ is Fréchet differentiable everywhere on X.

EXERCISES : (1.3), (1.4), (1.13), (1.17), (1.20), (1.21), (1.22).

2. Convex integrals

Motivated by Example (1.18), we now concretize our treatment of convexity and convex minimization problems in infinite-dimensional spaces. The infinite-dimensional spaces we have in mind are the Lebesgue spaces $L^p(\Omega)$ and the Sobolev spaces $W^{m,p}(\Omega)$, where $\Omega \subset \mathbb{R}^d$ is a nice region, e.g., a generalized rectangle. The convex functions to be studied are integrals of the form

$$(2.1) \qquad L(f) = \int_\Omega \ell(x, f(x), \nabla f(x), \cdots, \nabla^m f(x)) \, dx \; ,$$

where

$$(2.2) \qquad \begin{aligned} &\ell : \Omega \times \mathbb{R} \times \mathbb{R}^d \times \mathbb{R}^{d \times d} \times \cdots \longrightarrow \mathbb{R} \text{ is measurable, and} \\ &\ell(x, \cdot) \text{ is convex and differentiable for a.e. } x \in \Omega \; . \end{aligned}$$

We refer to such functionals as *convex integrals*. Convex functionals of the form

$$(2.3) \qquad L_n(f) = \tfrac{1}{n} \sum_{i=1}^n \ell(x_i, f(x_i), \nabla f(x_i), \cdots, \nabla^m f(x_i)) \; ,$$

with ℓ continuous and satisfying (2.2) also fall under the heading of (2.1) *in the proper setting*. A few special cases previously encountered in Chapters 5 and 6 are discussed later in Chapter 11 and in Volume II.

Despite appearences, the treatment here will not be in the generality implied above. We shall always have $d = 1$ or $= 2$ ($d = 1$ mostly), $p = 1$ or $= 2$ (both), and $m = 0$ or $= 1$. The case $m = 2$ will be treated only for a few specific examples. Also, typically, the derivatives will appear separately in the form of $\| \nabla f \|^2$ or $\| \nabla^2 f \|^2$ in a suitable Hilbert space norm.

In this section, we study the continuity and the differentiability of convex integrals and establish elementary facts about minimization problems for convex integrals. Not surprisingly, this involves basic aspects of Lebesgue integration, see, e.g., ROYDEN (1968) or WHEEDEN and ZYGMUND (1977).

Let $X = W^{1,p}(\Omega)$, and define the operator $\mathcal{L} : \Omega \times X \longrightarrow \mathbb{R}$ by

$$(2.4) \qquad \mathcal{L}f(x) = \ell(x, f(x), \nabla f(x)) \quad \text{a.e. } x \in \Omega \ ,$$

with ℓ as in (2.2). [If $\ell(x, u, w)$ is independent of w, then the proper choice for X is $X = L^p(\Omega)$, and the obvious changes must be made. We shall not belabor the point.] Note that (2.2) implies that $\mathcal{L}f(x)$ is convex in f for fixed x.

The first task at hand is to determine when L is well defined. Recall that in the context of convex functions, values of $+\infty$ are admissible, but values of $-\infty$ must be avoided. The following one-sided growth condition on ℓ,

$$(2.5) \qquad \exists c \geqslant 0 \ \exists m \in L^p(\Omega) \ : \ \ell(x, u, w) \geqslant m(x) - c\,|\,u\,|^p - c\,|\,w\,|^p \ ,$$

is precisely what is needed to avoid this problem. Here, $|\cdot|$ denotes both the absolute value on $\mathbb{R}$, and the (Euclidean) norm on $\mathbb{R}^d$.

(2.6) LEMMA. *If ℓ satisfies (2.2) and (2.5), then*

$$L : W^{1,p}(\Omega) \longrightarrow \mathbb{R} \cup \{+\infty\}$$

is well defined and convex.

PROOF. Obviously, $\mathcal{L}f$ is measurable on Ω, and (2.5) implies that for a suitable constant $c \geqslant 0$,

$$\mathcal{L}f(x) \geqslant m(x) - c\,|\,f(x)\,|^p - c\,|\,\nabla f(x)\,|^p \ .$$

It follows that $\mathcal{L}f$ is integrable and that $L(f) > -\infty$. The convexity of L is easy. Let $f, g \in \text{domain}\,(L)$ and let $0 < \lambda < 1$. For ease of notation, let

$$i(\lambda, x) = \big[\, \mathcal{L}\{\lambda\,f + (1 - \lambda)\,g\}\,\big](x).$$

The assumption (2.5) implies that $i(\lambda, x)$ is bounded below by an $L^1(\Omega)$ function. The convexity of $\mathcal{L}f(x)$ in f gives for a.e. $x \in \Omega$,

$$i(x, \lambda) \leqslant \lambda\,\mathcal{L}f(x) + (1 - \lambda)\,\mathcal{L}g(x) \ .$$

Now, since $L(f)$ and $L(g)$ are finite, then $\mathcal{L}f, \mathcal{L}g \in L^1(\Omega)$, so that $i(\cdot, \lambda)$ is bounded above by an $L^1(\Omega)$ function as well. It follows that the functional $L(\lambda\,f + (1 - \lambda)\,g)$ is finite and that the effective domain of L is convex.

Moreover,

$$L(\lambda f + (1 - \lambda) g) \leqslant \lambda L(f) + (1 - \lambda) L(g) .$$

Thus, L is convex.   Q.e.d.

If $L : Y \longrightarrow \mathbb{R} \cup \{\infty\}$ for a Banach space Y, then the (effective) domain of L is defined as

(2.7) $$\text{domain}(L) = \{ f \in Y : L(f) < \infty \} .$$

The next item on the agenda is the differentiability of L. One would assume that (ordinary) differentiability of $\ell(x, u, w)$ with respect to u and w is sufficient for L to be Gateaux differentiable, but it is not so simple. In particular, it is certainly possible that $\delta L(f, g - f) = -\infty$. But if the Gateaux variation or the Gateaux derivative exists, then it may be represented the way one expects. To state the result, define $\mathcal{L}_1 : \Omega \times X \longrightarrow \mathbb{R}$ and $\mathcal{L}_2 : \Omega \times X \longrightarrow \mathbb{R}^d$ by

(2.8)
$$\mathcal{L}_1 f(x) = \frac{\partial}{\partial u} \ell(x, u, w) \Big|_{(u,w) = \left(f(x), \nabla f(x) \right)} ,$$
$$\mathcal{L}_2 f(x) = \nabla_w \ell(x, u, w) \Big|_{(u,w) = \left(f(x), \nabla f(x) \right)} ,$$

where ∇_w denotes the gradient with respect to w.

(2.9) LEMMA. Let $f, g \in \text{domain}(L)$. If $\delta L(f, g - f)$ is finite, then the function ψ defined by

$$\psi(x) = \mathcal{L}_1 f(x) \{ g(x) - f(x) \} + \langle \mathcal{L}_2 f(x) , \nabla g(x) - \nabla f(x) \rangle , \quad x \in \Omega ,$$

belongs to $L^1(\Omega)$ and $\delta L(f, g - f) = \int_\Omega \psi(x) \, dx$.

PROOF. Note that

(2.10) $$\lim_{t \to 0^+} t^{-1} \{ L(t g + (1 - t) f) - L(f) \} =$$

$$\lim_{t \to 0^+} \int_\Omega t^{-1} \{ [\mathcal{L}\{t g + (1 - t) f\}](x) - [\mathcal{L}f](x) \} \, dx ,$$

so the problem is "merely" to pass the limit under the integral sign. But note that the integrand is an increasing function of $t > 0$, by Lemma (1.7). So the integrand in (2.10) is bounded above by an $L^1(\Omega)$ function, and the Monotone Convergence Theorem says that we may indeed pass the limit under the integration sign.   Q.e.d.

We note the following translation of Theorem (1.12) regarding the minimization problem

(2.11)
$$\text{minimize} \quad L(f) = \int_{\Omega} \ell\big(x, f(x), \nabla f(x)\big)\, dx$$
$$\text{subject to} \quad f \in C\,,$$

where $C \subset X$ is closed an convex.

(2.12) THEOREM. *If f_o solves (2.11), then for every $g \in \text{domain}\,(L) \cap C$,*

$$\delta L(f_o, g - f_o) \geq 0\,.$$

In particular, $\delta L(f_o, g - f_o)$ is finite for every $g \in \text{domain}\,(L) \cap C$.

With differentiability out of the way, we turn to lower semicontinuity. The following assumption is very useful and in our applications usually easy to verify.

(2.13)
There exists a $g_o \in X$ such that
$$\mathcal{L}_1\, g_o \in L^q(\Omega) \quad, \quad \mathcal{L}_2\, g_o \in L^q(\Omega)\,,$$
where $(1/p) + (1/q) = 1\,.$

Recalling that $X = W^{1,p}(\Omega)$, this implies

(2.14)
$$|\,\delta L(g_o, f - g_o)\,| \leq c\,\| f - g_o \|_X$$

for every $f \in \text{domain}\,(L)$, and so $\delta L(g_o, f - g_o)$ defines (can be extended to) a bounded linear functional on X.

(2.15) THEOREM. *With assumption (2.13), L is lower semicontinuous on its domain.*

PROOF. Let $\{ f_n \}_{n \geq 1} \subset \text{domain}\,(L)$ and $f_o \in \text{domain}\,(L)$, and suppose that $\| f_n - f_o \|_X \longrightarrow 0$. Consider a subsequence of $\{ f_n \}_n$ for which

$$\liminf_{n \to \infty} L(f_n)$$

is realized. Without loss of generality, we may assume that $\{ L(f_n) \}_n$ is decreasing. The subsequence of $\{ f_n \}_n$ in question has a subsequence, say, $\{ \varphi_n \}_n$, which converges to f_o almost everywhere, in the sense that

(2.16)
$$\varphi_n(x) \longrightarrow f_o(x) \quad \text{and} \quad \nabla\varphi_n(x) \longrightarrow \nabla f_o(x) \quad \text{a.e.}\,.$$

Now, the convexity of $\mathcal{L}$ gives us that

$$\mathcal{L}\,\varphi_n(x) - \mathcal{L}\, g_o(x) - \mathcal{L}_1\, g_o(x)\{\, \varphi_n(x) - g_o(x)\, \}$$
$$- \big\langle\, \mathcal{L}_2\, g_o(x)\,,\, \nabla\varphi_n(x) - \nabla g_o(x)\,\big\rangle \geq 0\,,$$

and so by Fatou's Lemma and Lemma (2.9),

$$\liminf_{n\to\infty} L(\varphi_n) - L(g_o) - \delta L(g_o, \varphi_n - g_o) \geqslant$$

$$L(f_o) - L(g_o) - \delta L(g_o, f_o - g_o) .$$

But $\delta L(g_o, \psi)$ is a bounded linear functional of $\psi \in X$, so that then $\delta L(g_o, \varphi_n - g_o) \longrightarrow \delta L(g_o, f_o - g_o)$, and the conclusion follows.   Q.e.d.

We finish with stating a more general lower semicontinuity result, in which the function $\ell(x, u, w)$ need not be convex in u, w jointly and need only satisfy

$$\ell : \Omega \times \mathbb{R} \times \mathbb{R}^d \longrightarrow \mathbb{R} \text{ is measurable, and}$$
(2.17) $\ell(x, \cdot, \cdot)$ is differentiable for a.e. $x \in \Omega$, and

$$\ell(x, u, \cdot) \text{ is convex for a.e. } x \in \Omega \text{ and all } u \in \mathbb{R} ,$$

with assumption (2.13) relaxed to

There exists $a \in L^1(\Omega)$, $b \in \left(L^q(\Omega)\right)^d$, such that
(2.18) $\ell(x, u, w) \geqslant a(x) + \langle b(x), w \rangle_{\mathbb{R}^d}$,

for a.e. $x \in \Omega$, , and all $u \in \mathbb{R}$, $w \in \mathbb{R}^d$.

The following theorem is a somewhat weaker version of a well-known theorem, see, e.g., DACOROGNA (1989), Chapter 3, Theorem 3.4.

(2.19) THEOREM. *Under the assumptions* (2.17) *and* (2.18), *L is lower semicontinuous on its domain.*

3. Strong convexity

In this section, we investigate the existence of solutions to convex minimization problems under the assumption of lower semicontinuity and strong convexity of the objective function. The treatment is a mixture of the general and the special. Of course, we have convex integral minimization problems in mind. The advantage of the concrete setting is that there are natural criteria that guarantee lower semicontinuity and strong convexity of convex integrals. In the general setting, we just have to assume it.

The definition of strong convexity remains the same as before, but for convenience, we repeat it. Throughout, we let X be a Banach space, with norm $\| \cdot \|_X$, and duality pairing $\langle \cdot, \cdot \rangle_X$.

(3.1) DEFINITION. *Let $C \subset X$ be convex, and let $L : C \longrightarrow \mathbb{R} \cup \{\infty\}$ be convex. We say L is strongly convex if there exists an $\alpha > 0$ such that for* all $f, g \in \text{domain}(L)$,

$$L(f) - L(g) - \delta L(g, f - g) \geqslant \alpha \| f - g \|_X^2 .$$

The constant α is referred to as the (strong) convexity constant of L.

Note that it suffices for the inequality to hold for all $f, g \in \text{domain}(L)$ for which $\delta L(g, f - g)$ exists and is finite.

(3.2) EXERCISE. Formulate and prove the analogue of Theorem (9.4.3).

The following corollary of Theorem (1.12) is immediate.

(3.3) COROLLARY. If L is strongly convex with convexity constant α, and f_o solves the minimization problem (1.11), then

$$L(f) - L(f_o) \geqslant \alpha \| f - f_o \|_X^2 \quad \text{for all } f \in C .$$

Strongly convex functionals need not be lower semicontinuous, as the example

$$(3.4) \qquad\qquad L(x) = \begin{cases} x^2 , & 0 < x \leqslant 1 , \\ 1 , & x = 0 , \end{cases}$$

shows, and so strongly convex minimization problems need not have solutions. However, combined with lower semicontinuity, the required result follows. Note that for convex integral functions L, the condition (2.13) implies lower semicontinuity. Together with strong convexity, it also implies that L is bounded below.

(3.5) THEOREM. If $L : C \longrightarrow \mathbb{R} \cup \{\infty\}$ is strongly convex, lower semicontinuous, and bounded below, then the problem (1.11) has a unique solution.

PROOF. Since L is bounded below, there exists a minimizing sequence $\{f_n\}_n \subset C$, i.e., $L(f_n) \longrightarrow \inf \{ L(f) : f \in C \} > -\infty$. For $n = 1, 2, \cdots$, define the finite-dimensional subspaces $V_n = \text{span}\{ f_1, f_2, \cdots, f_n \}$, and consider the minimization problem

$$(3.6) \qquad\qquad \begin{aligned} &\text{minimize} \quad L(f) \\ &\text{subject to} \quad f \in C \cap V_n . \end{aligned}$$

Since this problem is finite-dimensional, a solution exists. Denote it by φ_n. Then, $L(\varphi_n) \leqslant L(f_n)$, so that $\{\varphi_n\}_n$ is also a minimizing sequence. Next, Corollary (3.3) implies $L(f) - L(\varphi_n) \geqslant \alpha \| f - \varphi_n \|_X^2$ for all $f \in C \cap V_n$. So taking $f = \varphi_m$ $(m < n)$ gives

$$\| \varphi_m - \varphi_n \|_X^2 \leqslant \alpha^{-1} \left\{ L(\varphi_m) - L(\varphi_n) \right\} .$$

Since $\{L(\varphi_n)\}_n$ converges, this implies that $\{\varphi_n\}_n$ is a Cauchy sequence and, hence, converges to some $\varphi_o \in C$ (since C is closed). The lower semi-continuity implies that

$$L(\varphi_o) \leqslant \liminf_{n \to \infty} L(\varphi_n) \,,$$

and since $\{\varphi_n\}_n$ is a minimizing sequence, then φ_o minimizes $L(f)$ over C. Q.e.d.

(3.7) REMARK. The proof of Theorem (3.5) is satisfying for more than one reason. If we are to compute solutions to actual infinite-dimensional problems, then finite-dimensional approximations are mandatory. The approximating problems (3.6) or variants thereof suggest themselves. Of course, we may get in a vicious circle: The computation of (the first few terms of) a minimizing sequence requires a choice for the subspaces V_n. However, for separable Banach spaces, see Appendix 3, §4, there exist subspaces $V_1 \subset V_2 \subset \cdots$ whose union is dense in X, i.e.,

(3.8) $$\lim_{n \to \infty} \inf \{ \, \| \, x - v \, \|_X \, : \, v \in V_n \, \} = 0 \quad \text{for all } x \in X \,.$$

Of course, the construction of good subspaces V_n for the standard Banach spaces is a nontrivial task and is another job for Approximation Theory. See also §1.6 on sieves.

The above is the general spiel, and preciously little needs specializing to convex integrals. Let $\Omega \subset \mathbb{R}^d$ be a generalized rectangle and $X = W^{1,p}(\Omega)$. As discussed before, the boundedness below and lower semicontinuity of convex integrals are implied by the reasonable condition (2.13). For strong convexity, the situation is not so clear. How can we recognize strongly convex integral functions when we meet them? A good guess would seem to be: If $\ell(x, u, w)$ is strongly convex in u, w jointly (and uniformly in x), then L is strongly convex. Unfortunately, it is not so straightforward. To be specific, $\ell(x, \cdot, \cdot)$ is strongly convex, uniformly in $x \in \Omega$, if there exists a constant $\alpha > 0$ such that for all $x \in \Omega$, $u, u_o \in \mathbb{R}$, and $w, w_o \in \mathbb{R}^d$,

$$\ell(x, u, w) - \ell(x, u_o, w_o) -$$
$$\ell_u(x, u_o, w_o) (u - u_o) - \langle \, \ell_w(x, u_o, w_o) \, , \, w - w_o \, \rangle \geqslant$$
(3.9) $$\alpha \, | \, u - u_o \, |^2 - \alpha \, | \, w - w_o \, |^2 \,.$$

(3.10) LEMMA. *Let* $p = 2$. *If* ℓ *is strongly convex in the sense of* (3.9), *then* L *is strongly convex.*

(3.11) EXERCISE. Prove Lemma (3.10).

It seems clear that (3.9) is no good for proving strong convexity on $W^{1,p}(\Omega)$ when $p \neq 2$. Indeed, in the next chapter, this is discussed in detail for specific examples. But some simple examples should be instructive.

(3.12) EXERCISE. Let $2 < p < \infty$ and let $\Omega \subset \mathbb{R}^d$ open.
(a) Show that there exists a constant $c_p > 0$ such that for all $x \neq y \in \mathbb{R}$,

$$\frac{|x|^p - |y|^p - p|y|^{p-1}\operatorname{sign}(y)(x-y)}{|x-y|^p} \geq c_p \;.$$

(b) Define the functional $L : L^p(\Omega) \longrightarrow \mathbb{R}$ by $L(f) = \int_\Omega |f(x)|^p \, dx$. Show that it is strongly convex, with strong convexity constant c_p.
(c) Determine the optimal value of c_p.

But such examples are the exception. By way of example, things are less nice for $1 < p < 2$.

(3.13) EXERCISE. Let $1 < p < 2$ and let $\Omega \subset \mathbb{R}^d$ open.
(a) Show that there exists a constant $\tilde{c}_p > 0$ such that for all $x \neq y \in \mathbb{R}$,

$$\frac{\{\,|x|^p - |y|^p - p|y|^{p-1}\operatorname{sign}(y)(x-y)\,\}\{\,|x|^{2-p} + |y|^{2-p}\,\}}{|x-y|^2} \geq \tilde{c}_p \;.$$

(b) Define the functional $L : L^p(\Omega) \longrightarrow \mathbb{R}$ by $L(f) = \int_\Omega |f(x)|^p \, dx$. Show that it satisfies, for some (other) constant $\tilde{c}_p > 0$,

$$L(f) - L(g) - \delta L(g, f - g) \geq \tilde{c}_p \, \frac{\|f - g\|_p^2}{\|f\|_p^{2-p} + \|g\|_p^{2-p}} \;.$$

(c) Determine the optimal value of $\tilde{c}_p$.
Part (b) shows that L is locally strongly convex. What we call "strong convexity" is actually "global strong convexity". Is there in fact strong convexity in (b)?

EXERCISES: (3.2), (3.11), (3.12), (3.13).

4. Compactness arguments

We continue the study of the existence of solutions to the minimization problem

(4.1)
$$\begin{array}{ll}
\text{minimize} & L(x) \\
\text{subject to} & x \in \mathcal{C} \;,
\end{array}$$

where $\mathcal{C}$ is a subset of a Banach space X. The focus in this section is compactness arguments. All results are variations on the following theme.

(4.2) THEOREM. *If with respect to some topology on X, the set $\mathcal{C} \subset X$ is sequentially compact, and $L : \mathcal{C} \longrightarrow \mathbb{R}$ is lower semicontinuous, then (4.1) has a solution.*

The interest of the theorem is that the topology on X need not be the norm topology (but of course it could be!). In fact, the interesting topologies are the ones associated with weak convergence and weak$-*$ convergence. The convexity of C and L plays a very helpful role in this. We begin with the compactness of C.

The key observation in the study of convex sets is the Separation Theorem, also called the geometric version of the Hahn-Banach theorem.

(4.3) SEPARATION THEOREM. *Let* $C \subset X$ *be a closed convex set, and let* $y \notin C$. *Then, there exists an* $f \in X^*$ *and an* $\alpha \in \mathbb{R}$ *such that*

$$\langle f, y \rangle_X > \alpha, \quad \text{and} \quad \langle f, x \rangle_X \leqslant \alpha \quad \text{for all } x \in C .$$

We shall not prove this. In general it is quite involved, and can be found in any text on functional analysis. In finite-dimensional spaces, or Hilbert spaces, it is easy to prove, see Chapter 9, but one observes that this proof does not extend to general Banach spaces.

As an aside, we note that there are several analytic versions of the Hahn-Banach theorem. One reads as follows.

(4.4) HAHN-BANACH THEOREM. *Let* X *be a Banach space, with norm* $\| \cdot \|_X$, *and let* V *be a closed subspace of* X. *Suppose* $p : V \longrightarrow \mathbb{R}$ *is a bounded linear functional, i.e., there exists a constant* c *such that*

$$|p(v)| \leqslant c \| v \|_X , \quad \text{for all } v \in V .$$

Then, there exists an $f \in X^*$ *such that*

$$\langle f, v \rangle_X = p(v) , \quad \text{for all } v \in V .$$

Since f agrees with p on V, we say that f is an extension of p to all of X. For another version of the Hahn-Banach theorem, see Appendix 3, §4.

Note that the above theorem makes no claim regarding the uniqueness of such an extension, nor in characterizing what this extension is. In this regard, the following example is illuminating.

(4.5) EXAMPLE. Consider $X = L^\infty(0,1)$, the set of all bounded (Lebesgue) measurable functions on $(0,1)$, with the essential-supremum norm. Let $V = C(0,1)$ be the space of all continuous functions on the closed interval $[0,1]$. Then, V is a closed subspace of X. Let $x_o \in [0,1]$ be fixed and define the functional $p : X \longrightarrow V$ by

(4.6) $$p(f) = f(x_o) , \quad \text{for all } f \in V .$$

Then, p is a bounded linear functional on V. By Theorem (4.4), it has an extension $q \in X^*$. The intriguing part is to specify even one such extension.

Returning to the Separation Theorem, an important consequence is that a closed convex set is equal to the intersection of all half spaces containing it. Here, a half space is defined as

(4.7) $$\mathcal{H}(f,\alpha) = \left\{\, x \in X \,:\, \langle\, f\,,\, x\,\rangle_X \leqslant \alpha \,\right\},$$

where $f \in X^*$, $f \neq 0$, and $\alpha \in \mathbb{R}$.

(4.8) COROLLARY. *Let* $C \subset X$ *be closed and convex. Then,*

$$C = \bigcap \left\{\, \mathcal{H}(f,\alpha) \,:\, C \subset \mathcal{H}(f,\alpha) \,\right\}.$$

This in turn implies the following corollary.

(4.9) COROLLARY. *If* $C \subset X$ *is closed and convex, then it is weakly closed.*

This may be phrased equivalently in terms of convex functions.

(4.10) COROLLARY. *Let* $L : X \longrightarrow \mathbb{R} \cup \{\infty\}$ *be a convex function. If* L *is lower semicontinuous, then it is weakly lower semicontinuous.*

(4.11) EXERCISE. Prove these three corollaries.

We next address the question of when closed convex sets are weakly sequentially compact. If $X = Y^*$ for some separable Banach space Y, then Theorem (4.9) of Appendix 3 says that a convex set C is weak$-*$ sequentially compact if it is closed and bounded, and implies that a convex set is weak$-*$ closed if it is closed. The proof of the theorem also shows that if X^* is separable, then bounded closed sets are weakly sequentially compact. We phrase this as a theorem.

(4.12) THEOREM. (a) *If* X *is the dual of a separable Banach space, then closed convex sets are weak$-*$ closed, and closed, bounded, convex sets are weak$-*$ sequentially compact.*
(b) *If* X^* *is separable, then bounded, closed, convex sets are weakly sequentially compact.*

We may again translate this in terms of convex functions.

(4.13) COROLLARY. *Let* X *be the dual of a separable Banach space. If* $L : X \longrightarrow \mathbb{R} \cup \{\infty\}$ *is lower semicontinuous, then it is weak$-*$ lower semicontinuous.*

As in the previous section, proofs of the existence of solutions depend on the compactness of level sets of L. As a starting point, we need that level sets are bounded, but this is implied by, e.g., the coercivity of L, i.e.,

$L(x) \longrightarrow \infty$ for $\| x \|_X \to \infty$, see Definition (9.5.8). The closedness is the subject of the next exercise.

(4.14) EXERCISE. A function $J : X \longrightarrow \mathbb{R} \cup \infty$ has closed level sets if and only if it is lower semicontinuous.

(4.15) COROLLARY. *Let* $L : X \longrightarrow \mathbb{R} \cup \{\infty\}$ *be lower semicontinuous and coercive.*
(a) *If* X *is the dual of a separable Banach space, then the level sets of* L *are weak−∗ sequentially compact.*
(b) *If* X^* *is separable, then the level sets of* L *are weakly sequentially compact.*

It is now possible to state various existence theorems. We restrict ourselves to the following.

(4.16) EXISTENCE THEOREM. *Let* $\mathcal{C} \subset X$ *be closed and convex, and suppose* $L : X \longrightarrow \mathbb{R} \cup \{\infty\}$ *is lower semicontinuous and coercive. The minimization problem* (4.1) *has a solution if* X *is the dual of a separable Banach space, or if* X^* *is separable.*

We now specialize to the convex integral setting. We shall state various results regarding subsets of $L^p(\Omega)$ and $W^{1,p}(\Omega)$, but will not pursue the consequences in extreme generality. Some simple examples must suffice. More realistic examples are discussed in the next chapter. The very first compactness result, in the usual sense, belongs to the theory of Sobolev spaces. It is part of the Rellich-Sobolev Embedding Theorem, see ADAMS (1975) or MAZ'JA (1985).

(4.17) THEOREM. *Let* $\Omega \subset \mathbb{R}^d$ *be a bounded generalized rectangle and let* $1 \leqslant p \leqslant \infty$. *Then, closed, bounded subsets of* $W^{1,p}(\Omega)$ *are compact subsets of* $L^p(\Omega)$. *Precisely, if* $\mathcal{S} \subset W^{1,p}(\Omega)$ *is closed and bounded, i.e., there exists a constant* K *such that*

$$\| f \|_p + \| \nabla f \|_p \leqslant K \quad \text{for all } f \in \mathcal{S} \,,$$

then $\mathcal{S}$ *is a compact subset of* $L^p(\Omega)$. *(The norms are* $L^p(\Omega)$ *norms.)*

It is also useful to characterize sequentially compact and weakly sequentially compact subsets of $L^p(\Omega)$.

(4.18) THEOREM. (FRÉCHET-KOLMOGOROV) *Let* $\Omega \subset \mathbb{R}^d$ *be open and let* $1 \leqslant p < \infty$. *A set* $\mathcal{S} \subset L^p(\Omega)$ *is sequentially compact if it is closed and bounded, and the following two conditions hold.*

(a) $$\sup_{f \in S} \int_{\Omega} |f(x + t) - f(x)|^p \, dx \longrightarrow 0 \quad (t \to 0) \,.$$

(b) $$\sup_{f \in S} \int_{|x| > R} |f(x)|^p \, dx \longrightarrow 0 \quad (R \to \infty) \,.$$

In (a), if necessary, the functions f are extended by 0 to all of $\mathbb{R}^d$. In (b), the integration is actually over $\{\, x \in \Omega : |x| > R \,\}$.

Note that condition (a) is an equicontinuity condition, viz. that the translation operator $f \longmapsto f(\cdot + t)$ is uniformly continuous on S. The proof is nowadays a standard exercise in measure theory. The theorem for weak sequential compactness is somewhat similar.

(4.19) THEOREM. (DUNFORD-PETTIS, DE LA VALLÉE POUSSIN)
Let $\Omega \subset \mathbb{R}^d$ be open and let $S \subset L^1(\Omega)$ be closed and bounded. Then, the following are equivalent.

(a)   *S is weakly sequentially compact.*

(b) $\sup\limits_{f, B} \left\{ \int_{B \cap \Omega} |f(x)| \, dx \, : \, |B| \leqslant \varepsilon \right\} \longrightarrow 0 \quad \text{for } \varepsilon \to 0 \,.$

(c)   *There exists a positive measurable function V with*

$$\lim_{t \to \infty} t^{-1} V(t) = \infty \,, \quad \sup_{f \in S} \int_{\Omega} V(f(x)) \, dx < \infty \,.$$

In (b), the supremum is over all measurable subsets $B \subset \mathbb{R}^d$, as well as over all $f \in S$. (Here, $|B|$ denotes the Lebesgue measure of B.)

Condition (b) is known as an equi-integrability condition. The corresponding theorem for $1 < p < \infty$ follows from general principles: The dual space of $L^p(\Omega)$ is $L^q(\Omega)$ with $(1/p) + (1/q) = 1$, so that $1 < q < \infty$ as well. Thus, $L^p(\Omega)$ has a separable dual, and so the following theorem holds.

(4.20) THEOREM. *Bounded, closed subsets of $L^p(\Omega)$ ($1 < p < \infty$) are weakly sequentially compact.*

What can be said about bounded subsets of our favorite space $L^1(\Omega)$? Sadly, not much: Any set of pdfs is obviously bounded in $L^1(\Omega)$, but it may contain sequences that "converge" to a point mass, say. This may be interpreted as follows. Obviously, $L^1(\Omega)$ generates bounded linear functionals on $C(\Omega)$, but unfortunately, not all of them. To be specific, let $\Omega \subset \mathbb{R}^d$ be compact, and let $\mathfrak{M}(\Omega)$ be the vector space of regular Borel measures on Ω, with norm

(4.21) $\| \mu \|_{\mathfrak{M}} = \sup \left\{ \left| \int_{\Omega} f(x) \, d\mu(x) \right| \, : \, f \in C(\Omega) \,, \ \| f \|_{C(\Omega)} = 1 \right\} \,.$

So an element $\mu \in \mathfrak{M}(\Omega)$ is viewed as a bounded linear functional on $C(\Omega)$, and indeed $(C(\Omega))^*$ may be identified with $\mathfrak{M}(\Omega)$. Since $C(\Omega)$ is separable, we have the ensuing theorem.

(4.22) THEOREM. Let $\Omega \subset \mathbb{R}^d$ be compact. Bounded subsets of $L^1(\Omega)$ are weak$-*$ sequentially compact, i.e., every bounded sequence contains a subsequence $\{\, f_n\,\}_n \subset L^1(\Omega)$ for which there exists an $\mu \in \mathfrak{M}(\Omega)$ such that

$$\int_\Omega \varphi(x)\, f_n(x)\, dx \longrightarrow \int_\Omega \varphi(x)\, d\mu(x) \quad \text{for all } \varphi \in C(\Omega) \ .$$

What happens when Ω is an unbounded rectangle? Some of the above results are valid for this case, but not others, e.g., the compact embedding of $W^{1,p}(\Omega)$ into $L^p(\Omega)$ is not valid for Ω unbounded [Theorem (4.17)], and neither is the characterization of $(C(\Omega))^*$ of Theorem (4.22). But these results still hold on compact subsets of Ω, and lucky for us, this is usually good enough. We shall give the reader fair warning when we apply this trick.

EXERCISES : (4.11), (4.14).

5. Euler equations

In this section, we consider necessary (and sufficient) conditions for a function to be a minimum of integral minimization problems with convex roughness penalization functionals, but without constraints. These conditions go by the name of Euler equations. For the one-dimensional case, there are many elegant expositions of the theory under minimal conditions, see TROUTMAN (1983) and references therein. In the multidimensional case, life is not so simple, see DACOROGNA (1989) and its references.

To make life easier, we study only one case of special interest in this text: For $\Omega \subset \mathbb{R}^d$ a nice open region,

(5.1)
$$\begin{aligned} \text{minimize} \quad & L(f) \\ \text{subject to} \quad & f \in W^{1,2}(\Omega) \ , \end{aligned}$$

where $L(f)$ is of the form

(5.2) $$L(f) = \int_\Omega m(x, f(x))\, dx + \tfrac{1}{2}\, h^2 \, \| \, \nabla f \, \|^2$$

($L^2(\Omega)$ norm), and we assume that

(5.3)
$$\begin{aligned} & m(\, \cdot \,, f(\, \cdot \,)) \in L^1(\Omega) \text{ for all } f \in W^{1,2}(\Omega) \ , \\ & m(x, \, \cdot \,) \text{ is differentiable on } \mathbb{R} \text{ for all } x \in \Omega \ , \text{ and} \\ & m_u(\, \cdot \,, f(\, \cdot \,)) \in L^2(\Omega) \text{ for all } f \in W^{1,2}(\Omega) \ . \end{aligned}$$

Here, $m_u(x, u)$ is the derivative of $m(x, u)$ with respect to u. So $m(x, \cdot)$ is not necessarily convex, and neither is L, but L is convex in ∇f. One verifies that L is finite everywhere on $W^{1,2}(\Omega)$.

If L were convex, the necessary and sufficient condition for f_o to solve (5.1) would be

$$\delta L(f_o, \varphi - f_o) \geqslant 0 \text{ for all } \varphi \in \text{domain}(L) .$$

Since domain$(L) = W^{1,2}(\Omega)$, this is equivalent to

$$(5.4) \qquad \delta L(f_o, \varphi) = 0 \quad \text{for all } \varphi \in W^{1,2}(\Omega) .$$

If L is not convex, then this condition is only necessary for f_o to solve (5.1).

The condition (5.4) is not very transparent, but we show that it is equivalent to the condition that f_o solves the boundary value problem

$$(5.5) \qquad \begin{aligned} -h^2 \Delta f(x) + m_u(x, f(x)) &= 0 , & x \in \Omega , \\ \frac{\partial f}{\partial n} &= 0 & , & x \in \text{bndry}(\Omega) , \end{aligned}$$

where Δ is the Laplacian, and n is the outward normal on bndry(Ω), the boundary of Ω. These equations are known as the Euler equations for (5.1). The fact that L is quadratic in the (highest) derivative allows for a simple proof.

But first, why would one suspect that the solution of (5.1) satisfies this equation? Note that the Gateaux variation of L is given by

$$\delta L(f, \varphi - f) = \int_\Omega m_u(x, f(x))(\varphi(x) - f(x)) +$$

$$h^2 \left\langle \nabla f(x) , \nabla(\varphi(x) - f(x)) \right\rangle dx .$$

Here and elsewhere in this section $\langle \cdot , \cdot \rangle$ denotes the inner product on $\mathbb{R}^d$. Now, assuming $f \in W^{2,2}(\Omega)$ and $\dfrac{\partial f}{\partial n} = 0$ on bndry(Ω), an application of Green's formula, see, e.g., COURANT and HILBERT (1953), yields

$$(5.6) \quad \delta L(f, \varphi - f) = \int_\Omega \{ m_u(x, f(x)) - h^2 \Delta f(x) \}(\varphi(x) - f(x)) dx .$$

By (5.4), this vanishes for all φ, so (5.5) must hold.

There are a couple of big "ifs" in the above. Why should the normal derivative vanish on the boundary, and why should the solution be smooth? Note that if f_o solves (5.1), then $\| \nabla f_o \|$ is finite, but it does not say that f_o should be twice differentiable. To make progress, let $s_h(x, y)$ be the Green's function for the boundary value problem

$$(5.7) \qquad \begin{aligned} -h^2 \Delta f(x) + f(x) &= g(x) , & \text{a.e. } x \in \Omega , \\ \frac{\partial f}{\partial n} &= 0 & , & \text{a.e. } x \in \text{bndry}(\Omega) , \end{aligned}$$

and let $S_h : L^2(\Omega) \longrightarrow L^2(\Omega)$ be the operator defined by

$$(5.8) \qquad S_h f(x) = \int_\Omega s_h(x,y)\, f(y)\, dy \;.$$

Actually, we even have $S_h : L^2(\Omega) \longrightarrow W^{2,2}(\Omega)$ is bounded. Note that the one-dimensional version on the line is the two-sided exponential kernel $\mathfrak{B}_h$.

So if $g \in L^2(\Omega)$, then the solution f of (5.7) is given by $f = S_h g$, and $f \in W^{2,2}(\Omega)$. Now, write the Euler equations (5.5) in the form of (5.7) with

$$(5.9) \qquad g(x) = -m_u(x, f_o(x)) + f_o(x) \;.$$

The assumptions on m guarantee that $g \in L^2(\Omega)$, and so if f_o solves the Euler equations, then $f_o = \varphi_o$, where

$$(5.10) \qquad \varphi_o = S_h\, g \;.$$

Here comes the trick: Note that $\varphi_o = S_h g$ is smooth and that its normal derivative vanishes on bndry(Ω). So we can apply Green's identity:

$$\int_\Omega g(x)(\varphi_o(x) - f_o(x))\, dx$$

$$= \int_\Omega \left(-h^2\, \Delta\varphi_o(x) + \varphi_o(x) \right) (\varphi_o(x) - f_o(x))\, dx$$

$$= h^2 \int_\Omega \left\langle \nabla\varphi_o(x)\,,\, \nabla\varphi_o(x) - \nabla f_o(x) \right\rangle dx \,+$$

$$\int_\Omega \varphi_o(x) \left(\varphi_o(x) - f_o(x) \right) dx \;.$$

It follows that

$$\int_\Omega m_u(x, f_o(x)) \left(\varphi_o(x) - f_o(x) \right) dx = \int_\Omega \left(-g + f_o \right) \left(\varphi_o - f_o \right)$$

$$= \int_\Omega \left(h^2\, \Delta\varphi_o - \varphi_o + f_o \right) \left(\varphi_o - f_o \right)$$

$$= -h^2 \int_\Omega \left\langle \nabla\varphi_o(x)\,,\, \nabla\varphi_o(x) - \nabla f_o(x) \right\rangle dx - \| \varphi_o - f_o \|^2 \;,$$

and so

$$(5.11) \quad \delta L(f_o, \varphi_o - f_o)$$

$$= \int_\Omega m_u(\cdot, f_o(\cdot))(\varphi_o - f_o) + h^2 \left\langle \nabla f_o\,,\, \nabla\varphi_o - \nabla f_o \right\rangle$$

$$= -h^2\, \| \nabla\varphi_o - \nabla f_o \|^2 - \| \varphi_o - f_o \|^2 \;.$$

Since a necessary condition for a minimum is that the Gateaux variation is nonnegative, it follows that $f_o = \varphi_o$, and so f_o satisfies the Euler equations. We have proven a theorem.

(5.12) THEOREM. *Suppose m satisfies (5.3). If the function $f_o \in W^{1,2}(\Omega)$ solves the problem (5.1), then $f_o \in W^{2,2}(\Omega)$, and $f = f_o$ satisfies the Euler equations (5.5). If in addition $m(x, u)$ is convex in u, then the converse holds.*

(5.13) EXERCISE. Prove the converse part of Theorem (5.12).

(5.14) EXERCISE. State and prove the analogous theorem for the problem

$$\text{minimize} \quad \int_\Omega m\big(x, f(x), \nabla f(x), \cdots, \nabla^m f(x)\big) + h^2 \, \| \, \nabla^{m+1} f \, \|^2$$

$$\text{subject to} \quad f \in W^{m+1,2}(\Omega) \ .$$

EXERCISES : (5.13), (5.14).

6. Finitely many constraints

We now extend the convex minimization problems of the previous section to include finitely many linear constraints. So the problem is

(6.1)
$$\text{minimize} \quad L(f) \overset{\text{def}}{=} \int_\Omega m(x, f(x)) \, dx + \tfrac{1}{2} \, h^2 \, \| \, \nabla f \, \|^2$$

$$\text{subject to} \quad f \in W^{1,2}(\Omega) \ , \ A f = c \ , \ B f \leqslant d \ ,$$

where $m(x, u)$ satisfies the usual assumptions (5.3), and

$$A : L^2(\Omega) \longrightarrow \mathbb{R}^p \quad \text{and} \quad B : L^2(\Omega) \longrightarrow \mathbb{R}^q$$

are bounded linear operators. Using the Riesz-Fischer theorem, they may be formally represented as follows. For A, we have

(6.2)
$$[\, A f \,]_i = \langle \, A_i \, , \, f \, \rangle_{L^2(\Omega)} \ , \quad i = 1, 2, \cdots, p \ ,$$

where $A_i \in L^2(\Omega)$ for each i. The adjoint $A^* : \mathbb{R}^p \longrightarrow L^2(\Omega)$ of A may then be represented as

(6.3)
$$A^* b(x) = \sum_{i=1}^{p} b_i \, A_i(x) \ , \quad x \in \Omega \ ,$$

for all $b \in \mathbb{R}^p$. The same observations apply to the operator B. As in the finite-dimensional case, we need a constraint qualification and, hence, assume that

(6.4)
$$A_c B_d \overset{\text{def}}{=} \{ \, f \in L^2(\Omega) : A f = c \ , \ B f < d \, \} \neq \varnothing \ .$$

However, we do not wish to convey the impression that (6.4) is the "best" way to accomplish this.

We now formulate the Lagrange multiplier theorem for the problem (6.1).

(6.5) LAGRANGE MULTIPLIER THEOREM. *The function $f_o \in W^{1,2}(\Omega)$ solves (6.1) if and only if $f_o \in W^{2,2}(\Omega)$, and there exist $\lambda_o \in \mathbb{R}^p$ and $\mu_o \in \mathbb{R}^q$ with $\mu_o \geqslant 0$ component wise, such that (f_o, λ_o, μ_o) solves the boundary value problem with constraints*

$$-h^2 \Delta f(x) + m_u(x, f(x)) + A^*\lambda(x) + B^*\mu(x) = 0 , \quad x \in \Omega ,$$

$$\frac{\partial f}{\partial n} = 0 \qquad\qquad\qquad , \quad x \in \text{bndry}(\Omega) ,$$

$$Af = c , \quad Bf \leqslant d ,$$

where n is the outward normal on bndry(Ω).

(6.6) EXERCISE. Show the if part of Theorem (6.5).

PROOF. We prove the only if part of the theorem and begin with the case involving only linear equality constraints. Then, the proof goes essentially as in the previous section. So let f_o be a solution of (6.1), with the constraint $Bf \leqslant d$ omitted, and let $f = \varphi_o$, $\lambda = \lambda_o$ solve

(6.7)
$$-h^2 \Delta f(x) + f(x) = f_o(x) - m_u(x, f_o(x)) - A^*\lambda(x) , \quad x \in \Omega ,$$

$$\frac{\partial f}{\partial n} = 0 \qquad\qquad\qquad , \quad x \in \text{bndry}(\Omega) ,$$

$$Af = c .$$

Do φ_o and λ_o indeed exist ? It is clear that for each λ, a solution φ_o to the boundary value problem exists, but it is not clear that λ can be chosen such that the constraint $A\varphi_o = c$ is satisfied. Recalling the Green's function operator $\mathcal{S}_h$ from § 5, we get from the first two equations of (6.7) that for every λ,

$$\varphi_o = \mathcal{S}_h \{ f_o - m_u(\cdot, f_o(\cdot)) - A^*\lambda \} ,$$

and so it boils down to whether λ can be chosen such that

$$A\mathcal{S}_h A^*\lambda = -c + A\mathcal{S}_h \{ f_o - m_u(\cdot, f_o(\cdot)) \} .$$

Despite the complications, this basically looks like the normal equations for a least-squares problem, so a solution ought to exist. In Lemma (6.13), we show that this is indeed the case. So (6.7) does have a solution.

We next wish to show that the solution of (6.1) satisfies the Euler equations (6.7). As before, we compute the Gateaux variation of L at f_o. In preparation, note that

$$\langle A^*\lambda, \varphi_o - f_o \rangle_{L^2(\Omega)} = \langle \lambda, A(\varphi_o - f_o) \rangle_{\mathbb{R}^d} = 0 ,$$

and so

$$\left\langle\, m_u(\,\cdot\,, f_o(\,\cdot\,))\,,\ \varphi_o - f_o \,\right\rangle_{L^2(\Omega)} =$$
$$= \left\langle\, -A^*\lambda - \varphi_o + f_o + h^2 \Delta\varphi_o\,,\ \varphi_o - f_o \,\right\rangle_{L^2(\Omega)}$$
$$= -\|\varphi_o - f_o\|^2 - h^2 \left\langle\, \nabla\varphi_o\,,\ \nabla\varphi_o - \nabla f_o \,\right\rangle .$$

The last inner product is on $(L^2(\Omega))^d$. This yields

$$\delta L(f_o, \varphi_o - f_o) = \left\langle\, m_u(\,\cdot\,, f_o(\,\cdot\,))\,,\ \varphi_o - f_o \,\right\rangle + h^2 \left\langle\, \nabla f_o\,,\ \nabla\varphi_o - \nabla f_o \,\right\rangle$$
$$= -\|\varphi_o - f_o\|^2 - h^2 \|\nabla\varphi_o - \nabla f_o\|^2 .$$

Since the Gateaux variation at a minimum must be nonnegative, it follows that $\varphi_o = f_o$. Consequently, f_o satisfies (6.7), and the theorem is proven for the case of equality constraints.

Now for the case of inequality constraints. With f_o the solution of (6.1), consider the inequality constraints $Bf_o \leqslant d$. We may partition these equations into the active and inactive constraints

$$F f_o = e\,, \quad E f_o < g\,.$$

Now consider (6.1) with the equality constraints

$$A f = c\,, \quad F f = e\,,$$

and no inequality constraints. Then, f_o is the solution of this problem as well, and it follows that the Euler equations are satisfied, say, with λ, μ. Then, as in §5, the necessary condition for a minimum is

$$(6.8) \qquad \delta L(f_o, \varphi - f_o) \geqslant 0 \quad \text{for all } \varphi \in L^2(\Omega) \cap \{AF\}_{ce}\,,$$

with $\{AF\}_{ce} = \{\, f \in L^2(\Omega)\ :\ Af = c\,,\ Ff = e\,\}$. It is useful to rephrase this as

$$(6.9) \quad h^2 \Delta f_o - m_u(\,\cdot\,, f_o(\,\cdot\,)) \in \mathrm{norco}(\,\{AF\}_{ce}\,,\ f_o\,) \subset \mathrm{norco}(\,A_b C_d\,,\ f_o\,)\,,$$

with the normal cone defined as in the finite-dimensional case: For $\psi \in C$,

$$(6.10) \quad \mathrm{norco}(\,C\,,\ \psi\,) = \{\, \nu \in L^2(\Omega)\ :\ \forall \varphi \in C\,(\,\langle\, \nu\,,\ \varphi - \psi \,\rangle \leqslant 0)\,\}\,.$$

The problem looks like it is infinite-dimensional, but it is similar to the material in §5. In fact, the proof is now finished by computing the normal cone, which we leave as an exercise.

(6.11) LEMMA. *With A and B as above, and $f_o \in A_c B_d$,*

$$\mathrm{norco}(\,A_c B_d\,,\ f_o\,) = \{\, A^*\lambda + B^*\mu\ :\ \lambda \in \mathbb{R}^p\,,\ \mu \in \mathbb{R}^q\,,\mu \geqslant 0\,\}\,.$$

(6.12) Exercise. Prove the lemma, as follows. [See §9.4 for inspiration.]
(a) Show that

$$A_c B_d = \mathrm{norco}(\{\, A^*\lambda + B^*\mu\ :\ \lambda \in \mathbb{R}^p\,,\ \mu \in \mathbb{R}^q\,,\mu \geqslant 0\,\}, f_o\,)\,.$$

(b) Formulate and prove the analogue of the Separation Lemma for Convex Cones (9.3.26) in Hilbert space.

(c) Show that norco(norco(K , 0)) $=$ K, when $K \subset L^2(\Omega)$ is a closed convex cone.

(d) Show that $A_c B_d$ is a closed convex cone.

(e) Put the above together, and prove Lemma (6.11).

To finish the proof, we must still prove the following lemma.

(6.13) LEMMA. *The system of equations* $A \mathcal{S}_h A^* \lambda = A\nu$ *has a solution* λ *for every* $\nu \in L^2(\Omega)$.

PROOF. The range and null space of a linear operator A are denoted by $R(A)$ and $N(A)$, respectively. It suffices to show that $R(A \mathcal{S}_h A^*) = R(A)$, or, equivalently, since the ranges are finite-dimensional and, hence, closed,

$$N(A \mathcal{S}_h A^*) = N(A^*) .$$

Obviously, $N(A^*) \subset N(A \mathcal{S}_h A^*)$. For the converse, suppose that $x \in \mathbb{R}^p$ satisfies $A \mathcal{S}_h A^* x = 0$. After premultiplying by x^T, we see that

$$\langle w , \mathcal{S}_h w \rangle_{L^2(\Omega)} = 0 ,$$

where $w = A^* x$. With Green's identity, then

$$h^2 \| \nabla w \|^2 + \| w \|^2 = 0 ,$$

so that the above shows that $w = A^* x = 0$, or $x \in N(A^*)$.   Q.e.d.

The above proof can easily be modified to prove the Lagrange Multiplier Theorem for the following problem. Let X be a Banach space, with X^* its dual space, and let $J : X \longrightarrow \mathbb{R} \cup \{+\infty\}$ be convex and Gateaux differentiable. Denote the Gateaux derivative by $J'(f)$. So $J'(f) \in X^*$. Let $A : X \longrightarrow \mathbb{R}^p$ and $B : X \longrightarrow \mathbb{R}^q$ be bounded linear operators. Then, the adjoints $A^* : \mathbb{R}^p \longrightarrow X^*$ and $B^* : \mathbb{R}^q \longrightarrow X^*$ are bounded linear operators also. Now, let $c \in \mathbb{R}^p$, $d \in \mathbb{R}^q$, and consider the minimization problem

(6.14)
$$\begin{aligned} \text{minimize} \quad & J(f) \\ \text{subject to} \quad & f \in X , \ Af = c , \ Bf \leqslant d . \end{aligned}$$

We make the usual constraint qualification, viz.

(6.15) $A_c B_d = \{ x \in X : Ax = c , \ Bx < d \} \neq \varnothing .$

(6.16) LAGRANGE MULTIPLIER THEOREM. *Let* $J : \text{domain}\, (J) \subset X \longrightarrow \mathbb{R}$ *be convex and Gateaux differentiable on its domain. If the constraint qualification* (6.15) *holds, then* f_o *solves* (6.14) *if and only if there exists*

$\lambda \in \mathbb{R}^p$ and $\mu \in \mathbb{R}^q$, with $\mu \geqslant 0$ *component wise, such that*

$$J'(f_o) + A^*\lambda + B^*\mu = 0 ,$$
$$Af_o = c , \ Bf_o \leqslant d .$$

(6.17) EXERCISE. (a) Compute norco($A_c B_d$, f_o) in the Banach space setting.
(b) Show that f_o solves (6.14) if and only if

$$-J'(f_o) \in \mathrm{norco}(\, A_c B_d \,, f_o \,) .$$

This concludes the discussion of finitely many constraints. The situation becomes distinctly more troublesome when infinitely many constraints are allowed. In particular, the Euler equations may not hold and the solution may not be as smooth as one would hope or expect. The following is a standard example. Consider the one-dimensional minimization problem

(6.18)

$$\begin{aligned}
\text{minimize} \quad & \int_0^1 |f'(x)|^2 \, dx \\
\text{subject to} \quad & f \in W^{1,2}(0,1) , \\
& f(x) \geqslant 2 , \ \tfrac{3}{4} < x < 1 , \\
& f(x) \leqslant 0 , \ 0 < x < \tfrac{1}{4} .
\end{aligned}$$

Obviously, the solution is the function

(6.19)

$$f(x) = \begin{cases} 0 & , \quad \text{on } [\, 0 , \tfrac{1}{4} \,] , \\ 4x - 1 & , \quad \text{on } [\, \tfrac{1}{4} , \tfrac{3}{4} \,] , \\ 2 & , \quad \text{on } [\, \tfrac{3}{4} , 1 \,] . \end{cases}$$

However, the Euler equations will not hold: The solution is in $W^{1,2}(0,1)$, but not in $W^{2,2}(0,1)$. We do not consider this in any generality, but discuss some examples of interest in the next chapter.

(6.20) EXERCISE. Verify that (6.19) indeed gives the solution of (6.18).

EXERCISES: (6.6), (6.12), (6.17), (6.20).

7. Additional notes and comments

Ad §1: The treatment of the Gateaux differentiability of convex functions is standard. See, e.g., ROYDEN (1968) or HOLMES (1975). An easily accessible, short survey of various notions of differentiability is NASHED

(1966). For a comprehensive discourse, see NASHED (1971). For an extensive account of differentiability and convexity, PHELPS (1989) should be consulted. For the use of differentials in (parametric) estimation, see, e,g., FERNHOLZ (1983).

Ad §2:. The convexity condition (2.2) is very strong. It actually suffices that $\ell(x, u, p, q)$ is (strongly) convex in q, or that $\ell(x, u, p)$ is (strongly) convex in p. Moreover, the notion of convexity is often changed as well. See, e.g., TROUTMAN (1983) and DACOROGNA (1989) and references therein.

Ad §4:. The compactness arguments are but a summary of compactness arguments in functional analysis, but they suit our purposes. The books by HOLMES (1975) and DIESTEL (1984) are recommended.

11
Convexity in Action

1. Introduction

In this chapter, we study the existence of solutions to various convex minimization problems that we have run into in Chapter 5, as well as the strong convexity of the objective functions involved. In relevant cases, the Euler equations are derived. We also consider the existence and computation of optimal kernels for kernel density estimation, which, indeed, amounts to solving convex minimization problems. All of this material consists of applications of Chapter 10. Other applications in indirect estimation problems, already briefly discussed in Chapter 1, are treated in detail in Volume II.

In § 2, we consider the optimal kernels of § 4.7. Existence as well as some properties of the optimal kernels are derived, which serve as the basis for a computational scheme akin to Newton's method. In § 3, we study maximum penalized likelihood density estimation with the roughness penalization of the square root density. The existence is shown, as well as the Euler equations, which were needed for the comparison with kernel estimators. In § 4, maximum likelihood estimation with roughness penalization of the log-density is considered. The existence of solutions is shown by relating it to maximum likelihood estimation for an associated regular parametric exponential family of densities. Existence of log-concave maximum likelihood estimators is shown in § 5, whereas minimum L^1 distance estimators are considered in § 6.

2. Optimal kernels

In the chapter on kernel density estimation, the notion of kernels of order $2k$ was introduced. The determination of optimal $2k$-th order kernels was left as an exercise, but it is actually a problem in the calculus of variations, and to some extent, this was studied in the previous chapter. But, as is usual, some modifications must be made to make sense out of it. In § 4.7, we *verified* that the Epanechnikov kernel was the optimal (nonnegative) kernel, but it was unclear where the optimal kernel came from. Here, we *calculate* the optimal $2k$-th order kernels. The explicit construction of the Epanechnikov kernel is considered in an exercise.

The problem is/was as follows. Let $k \geqslant 1$ be an integer. We wish to determine a kernel $A(x)$ for which

$$(2.1) \qquad J_k(A) = \left\{ \int_{\mathbb{R}} x^{2k} \, |A(x)| \, dx \right\} \left\{ \int_{\mathbb{R}} |A(x)|^2 \, dx \right\}^{2k}$$

is as small as possible, and which satisfies

$$(2.2) \qquad \int_{\mathbb{R}} x^m \, A(x) \, dx = \begin{cases} 1 \; , & m = 0 \; , \\ 0 \; , & m = 1, 2, \cdots, 2k - 1 \; . \end{cases}$$

The solution or solutions, assuming they exist, are the $2k$-th order optimal kernels. These optimal kernels of order $2k$ were determined by ZHANG and FAN (2000), and we also refer to them as Zhang-Fan kernels.

Setting $A_h = h^{-1} A(h^{-1} x)$ for $h > 0$, one verifies that $J_k(A) = J_k(A_h)$. Thus, the feasible A may be scaled in such a way that $\int_{\mathbb{R}} x^{2k} \, | A(x) | \, dx = 1$, and the above problem translates to

$$\text{minimize} \qquad \int_{\mathbb{R}} |A(x)|^2 \, dx$$

$$(2.3) \qquad \text{subject to} \qquad \int_{\mathbb{R}} x^m \, A(x) \, dx = \delta_m \; , \quad 0 \leqslant m \leqslant 2k - 1 \; ,$$

$$\int_{\mathbb{R}} x^{2k} \, |A(x)| \, dx \leqslant 1 \; ,$$

where $\delta_0 = 1$, $\delta_m = 0$ for $1 \leqslant m \leqslant 2k - 1$. Now, (2.3) is a nice convex minimization problem, and it has a solution. Actually, we are jumping the gun a little: The last constraint in (2.3) should read as

$$(2.4) \qquad \int_{\mathbb{R}} x^{2k} \, |A(x)| \, dx = 1 \; ,$$

but by using the scaling argument, it is easy to see that if (2.3) has a solution, then the solution satisfies (2.4). Also, if (2.3) has a solution, it must be in $L^1(\mathbb{R})$, as the following exercise makes clear.

(2.5) EXERCISE. (a) Let A_o be a solution of (2.3). Show that

$$\int_{\mathbb{R}} x^{2k} \, |A_o(x)| \, dx = 1.$$

(b) Show that $A_o \in L^1(\mathbb{R})$, in particular, that

$$\| A_o \|_1 \leqslant 1 + \sqrt{2} \, \| A_o \|_2 \; .$$

[Hint: Prove the inequality, for measurable f,

$$\int_{\mathbb{R}} |f(x)| \, dx \leqslant \sqrt{2} \left\{ \int_{-1}^{1} |f(x)|^2 \, dx \right\}^{1/2} + \int_{|x| \geqslant 1} x^{2k} \, |f(x)| \, dx \; .]$$

The following theorem is a useful step in determining optimal kernels.

(2.6) THEOREM. *The solution of problem (2.3) exists and is unique.*

PROOF. Let B be a kernel of order $2k$. (In § 4.7, we have seen that kernels of order $2k$ exist.) Thus, the infimum of the objective function in the minimization problem (2.3) is finite. Let $\{A_n\}_n$ be a minimizing sequence. In particular, $\{A_n\}_n \subset \mathcal{A}$, where

$$(2.7) \qquad \mathcal{A} = \left\{ A \in L^1(\mathbb{R}) : \|A\|_2 \leqslant C , \ A \text{ is feasible for (2.3)} \right\} ,$$

where $C = \|B\|_2$. For later reference, we note that $\mathcal{A}$ is a closed convex subset of $L^2(\mathbb{R})$, and so it is weakly closed in the $L^2(\mathbb{R})$ topology. By Exercise (2.5)(b), it is also a bounded subset of $L^1(\mathbb{R})$. Now, the Dunford-Pettis Theorem (10.4.19) implies that the set $\mathcal{A}$ is a weakly compact subset of $L^1(\mathbb{R})$ (the relevant function is $V(t) = t^2$). So $\{A_n\}_n$ has a weakly convergent subsequence. Without loss of any generality, assume that $A_n \longrightarrow A_o$ weakly in $L^1(\mathbb{R})$. Since $\mathcal{A}$ is closed and convex, it is weakly closed, see Corollary (10.4.9), and so $A_o \in \mathcal{A}$. It follows that $\|A_o\|_2 \leqslant C$. But we may repeat the above with $C = \|A_n\|$, for any n. In particular, this shows that

$$\|A_o\|_2 \leqslant \|A_n\|_2 \quad \text{for all } n .$$

It follows that

$$\|A_o\|_2 = \liminf_{n \to \infty} \|A_n\|_2 ,$$

and so A_o is an optimal kernel.

The uniqueness is implied by the strict convexity of the objective function.   Q.e.d.

So optimal kernels exist. Can we actually compute them? The Lagrange Multiplier Theorem (10.6.16) does not quite apply to (2.3), since the function in the last constraint is not Gateaux differentiable, but since it is convex, it may be fudged by restricting the constraint set to those kernels A that have the same (fixed) sign as the optimal kernel A_o. Then, the last constraint is in fact linear, and so Gateaux differentiable. Consequently, we can apply the Lagrange Multiplier Theorem to obtain that the solution A satisfies

$$(2.8) \qquad A(x) = \sum_{m=0}^{2k-1} a_m \, x^m - \lambda \, x^{2k} \, \mathrm{sign}(\, A(x)\,) , \quad x \in \mathbb{R} ,$$

along with the constraints of (2.3). Also, since the last constraint in (2.3) is active (recall the scaling argument), then $\lambda > 0$. In general, the sign function is actually a multivalued function

$$
(2.9) \qquad \text{sign}(t) = \begin{cases} = +1 & , \quad t > 0 , \\ \in [-1, +1] , & t = 0 , \\ = -1 & , \quad t < 0 , \end{cases}
$$

but here sign$(A(x))$ is uniquely determined as follows. For known coefficients $a_0, a_1, \cdots, a_{2k-1}$, and $\lambda > 0$, let

$$
(2.10) \qquad P(x) = \sum_{m=0}^{2k-1} a_m x^m , \quad x \in \mathbb{R} .
$$

Then,

$$
(2.11) \qquad \text{sign}(A(x)) = \begin{cases} +1 & , \quad \text{if } P(x) > \lambda x^{2k} , \\ -1 & , \quad \text{if } P(x) < -\lambda x^{2k} , \\ \{\lambda x^{2k}\}^{-1} P(x) , & \text{otherwise} . \end{cases}
$$

In determining the optimal kernels, the following observations are useful. First, the support of A is bounded. This may be shown by way of contradiction. If the support of A was not bounded, then for $|x| \to \infty$, we infer from (2.8) that sign$(A(x)) = -\text{sign}(A(x))$, and this cannot happen on the support of A. So supp(A) is bounded: there exists a constant $T > 0$ such that

$$
(2.12) \qquad \text{supp}(A) \subset [-T, T] .
$$

The second observation looks innocent enough, but has a "strange" consequence. With P as in (2.10), the right-hand side of (2.8) may be written as $P(x) - \lambda x^{2k} \text{sign}(A(x))$, so it follows that for all x,

$$
(2.13) \qquad A(x)(A(x) - P(x)) = -\lambda x^{2k} |A(x)| .
$$

This implies that for each x, the value $A(x)$ lies between 0 and $P(x)$ (inclusive). In an unexpected way, this implies that the support of A is not connected (has gaps in it), as we now show. Let x_o be a point in supp(A), where A changes sign from positive to negative or vice versa (the graph of A "crosses" the x-axis at x_o, i.e., the graph cuts the x-axis in exactly one point, or jumps across the x-axis). It follows that P changes sign at x_o as well, whence A must be continuous at x_o. Thus, the right-hand side of (2.13) is continuous at x_o. However, since A does not change sign at $x = 0$ (A is even), then $x_o \neq 0$ and $x^{2k} \text{sign}(A(x))$ is not continuous at x_o: It follows from (2.8) that $\lambda = 0$. But we just argued that $\lambda > 0$, and we have a contradiction. Thus, the graph of A does not cross the x-axis anywhere. However, if A is a $2k$-th order kernel with $k \geqslant 2$, then A must have positive and negative values, so the conclusion is that the graph of A must have stretches of zero, or, more precisely, the support of A is not connected.

We now consider the computation of the optimal kernels. The first issue that needs clarification is the scaling (as in: location-scale) of the kernel.

Table 2.1. The optimal kernels of order $k = 4, 6, 8$, in the format (2.8). (Odd powers are absent.) The graphs are shown in Figure 2.1.

k	4	6	8
λ	3.115328086361	22.63313961910	413.9847783687
a_0	1.504310384541	2.524629983182	3.961675376084
a_2	−4.619638470902	−25.00127417349	−99.40959508772
a_4	-	45.10978380941	594.6960505276
a_6	-	-	−913.2329091846

What has been suggested so far is that we solve the system of equations

$$A(x) = \sum_{m=0}^{2k-1} a_m\, x^m - \lambda\, x^{2k} \operatorname{sign}(A(x)), \quad x \in \mathbb{R},$$

(2.14)
$$\int_{\mathbb{R}} x^\ell A(x)\, dx = \delta_\ell, \quad \ell = 0, 1, \cdots, 2k - 1,$$

$$\int_{\mathbb{R}} x^{2k}\, |A(x)|\, dx = 1,$$

the scale being provided by the last equation. It turns out to be simpler to just drop this scaling equation and to define the scale by setting (arbitrarily) $\lambda = 1$. That is, we consider the system

(2.15a)
$$B(x) = \sum_{m=0}^{2k-1} b_m\, x^m - x^{2k} \operatorname{sign}(B(x)), \quad x \in \mathbb{R},$$

(2.15b)
$$\int_{\mathbb{R}} x^\ell B(x)\, dx = \delta_\ell, \quad \ell = 0, 1, \cdots, 2k - 1.$$

(2.16) THEOREM. *The kernel A solves (2.14) if and only if there exists an $h > 0$ such that the kernel B defined by $B(x) = h^{-1}A(h^{-1}x)$ solves (2.15).*

(2.17) EXERCISE. Prove Theorem (2.16).

An iterative algorithm for solving (2.15) may now be developed. It is a simplified version of the ZHANG and FAN (2000) algorithm. A stylized version is displayed as Algorithm 2.1. The starting point is an initial guess for the coefficients $b_0, b_1, \cdots, b_{2k-1}$, which defines B^{old} by means of (2.15a), using (2.11) to determine $\operatorname{sign}(B^{\text{old}})$. That is, let

(2.18a)
$$P(x) = \sum_{i=1}^{2k-1} b_m\, x^m,$$

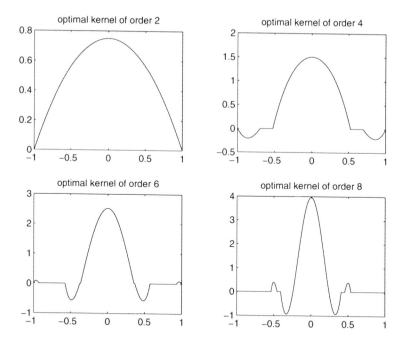

Figure 2.1. Optimal kernels of various orders, computed by algorithm (2.18), and then scaled to have supports $[-1, 1]$. The optimal kernel of order 8 has two tiny depressions near the endpoints. (Is this in fact true?)

and set

$$(2.18\text{b}) \qquad s(x) = \begin{cases} +1 & , \quad \text{if} \quad P(x) > x^{2k} \ , \\ -1 & , \quad \text{if} \quad P(x) < -x^{2k} \ , \\ \{x^{2k}\}^{-1} P(x) \ , & \text{otherwise} \ . \end{cases}$$

Now, define

$$(2.18\text{c}) \qquad B^{\text{old}}(x) = P(x) - x^{2k} s(x) \ .$$

Next, compute the support of B^{old}, that is, determine the sets

$$(2.18\text{d}) \qquad \begin{aligned} S^+ &= \{\, x \ : \ B^{\text{old}}(x) > 0 \,\} \ , \\ S^- &= \{\, x \ : \ B^{\text{old}}(x) < 0 \,\} \ , \end{aligned}$$

and approximate the system (2.15b) by computing the moment matrix $M \in \mathbb{R}^{2k \times 2k}$

$$(2.18\text{e}) \qquad M_{\ell,m} = \int_{S^+ \cup S^-} x^{\ell+m} \, dx \ , \quad \ell, m = 0, 1, \cdots, 2k-1 \ ,$$

Algorithm 2.1. Optimal Kernels

- Start with an initial guess for the coefficients $b_0, b_1, \cdots, b_{2k-1}$.

- Determine the sets

$$S^+ = \{\, x : \sum_{m=0}^{2k-1} b_m x^m - x^{2k} > 0 \,\} ,$$

$$S^- = \{\, x : \sum_{m=0}^{2k-1} b_m x^m + x^{2k} < 0 \,\} .$$

- Approximate the system (2.15b) by computing the moment matrix $M \in \mathbb{R}^{2k \times 2k}$

$$M_{\ell,m} = \int_{S^+ \cup S^-} x^{\ell+m} \, dx , \quad \ell, m = 0, 1, \cdots, 2k - 1 ,$$

and the moment right-hand sides

$$\mu_\ell = \int_{S^+} x^{2k+\ell} \, dx - \int_{S^-} x^{2k+\ell} \, dx , \quad \ell = 0, 1, \cdots, 2k - 1 .$$

- Solve the system of equations

$$M \, b^{\text{new}} = e_o + \mu$$

to obtain the new coefficients $b_0^{\text{new}}, b_1^{\text{new}}, \cdots, b_{2k-1}^{\text{new}}$.

Here, e_o is the first element of the standard basis for $\mathbb{R}^{2k}$.

- Repeat the above with the old coefficients replaced by the new, until the solution is accurate enough.

---

and the moment right-hand sides

$$(2.18f) \qquad \mu_\ell = \int_{S^+} x^{2k+\ell} \, dx - \int_{S^-} x^{2k+\ell} \, dx , \quad \ell = 0, 1, \cdots, 2k - 1 .$$

Then, solve the system of equations

$$(2.18g) \qquad\qquad M \, b^{\text{new}} = e_o + \mu$$

to obtain the new and improved coefficients b^{new}. Here, e_o is the first element of the standard basis for $\mathbb{R}^{2k}$. The above is the basic step of the iterative algorithm.

Note that in (2.18g), the matrix M and the moments μ depend on b^{old}. Displaying this functional dependence as $M(b^{\text{old}})$ and $\mu(b^{\text{old}})$, we may rewrite algorithm (2.18) as

$$(2.19) \qquad M(b^{\text{old}})\, b^{\text{new}} = e_o + \mu(b^{\text{old}}),$$

and it is easy to verify that the true b satisfies $M(b^{\text{true}})\, b^{\text{true}} = e_o + \mu(b^{\text{true}})$. This algorithm seems to work and (eventually) converges quadratically. In fact, if the analogy with the algorithm of ZHANG and FAN (2000) holds true, then it should be possible to prove the quadratic convergence. However, for $k \geqslant 4$, the initial rate of convergence is very slow, which means an algorithm for computing a good initial guess would be useful. It is possible to speed up things by using the fact that all kernels in question are even, so the dimension of the problem is only k rather than $2k$. In Table 2.1, we exhibit the coefficients of the optimal kernels of order 2, 4, 6, and 8. The graphs of these kernels are shown in Figure 2.1. As it should be, none of these kernels "cross" the x-axis. While hard to see visually (sic), the eighth-order kernel in fact vanishes on very short intervals around the points ± 0.2415, or thereabouts.

(2.20) EXERCISE. Implement Algorithm 2.1. If your favorite computing environment does not provide an easy way to compute zeros of polynomials, you should consider alternatives. Does your implementation of the algorithm work (well) for large values of k?

(2.21) EXERCISE. (a) Prove the quadratic convergence of Algorithm 2.1, starting from any choice of the b_ℓ (not all zero).
(b) Are there clever ways to choose the initial polynomial?

(2.22) EXERCISE. Compute the optimal second-order kernel (as opposed to verifying that the Epanechnikov kernel is indeed optimal).

EXERCISES: (2.5), (2.17), (2.19), (2.21), (2.22).

3. Direct nonparametric maximum roughness penalized likelihood density estimation

The flagship of maximum penalized likelihood estimation is the GOOD estimator of §5.2. For ease of presentation, we repeat its definition. With $X_1, X_2, \cdots, X_n$ iid random variables with common density f_o, this approach to estimating f_o translates to the problem

$$(3.1) \qquad \text{minimize} \quad L_n(u) \overset{\text{def}}{=} -\frac{2}{n} \sum_{i=1}^{n} \log u(X_i) + \| u \|^2 + h^2 \| u' \|^2$$

$$\text{subject to} \quad u \in W^{1,2}(\mathbb{R}),\ u \geqslant 0.$$

Here, and throughout this section, the $L^2(\mathbb{R})$ norm is denoted plainly by $\|\cdot\|$. The solution of (3.1) is denoted by u_{nh}, and the GOOD estimator for f_o is $f^{nh} = (u_{nh})^2$. This estimator was thoroughly analyzed in §5.2, except for the existence and the differential equation characterizing u_{nh} (Euler equations). Here, we set out to fill in the promised missing details, following largely DE MONTRICHER, TAPIA and THOMPSON (1975).

The first step toward establishing the existence of u_{nh} is the observation that if $L_n(u)$ is finite, then the minimum of $L_n(t\,u)$ as a function of $t > 0$ is achieved, and it follows that we may restrict the minimization in (3.1) further to the set of nonnegative u with

$$\|u\|^2 + h^2 \|u'\|^2 = 1 .$$

So everything happens on the closed convex subset of $W^{1,2}(\mathbb{R})$,

(3.2) $\mathcal{G}_h = \{ u \in W^{1,2}(\mathbb{R}) : u \geqslant 0 , \ \|u\|^2 + h^2 \|u'\|^2 \leqslant 1 \} .$

It should be noted that any $u \in \mathcal{G}_h$ is in fact Hölder continuous of order $\frac{1}{2}$, and so the point evaluations in (3.1) make sense.

(3.3) EXERCISE. (a) For $u \in W^{1,2}(\mathbb{R})$, show that for all $x, y \in \mathbb{R}$,

$$| u(x) - u(y) | \leqslant \| u' \| \, | x - y |^{1/2} .$$

(b) Show that $\mathcal{G}_h$ is equi-uniformly-continuous on the line.
(c) Show that there exists a constant c such that

$$\| u \|_\infty^3 \leqslant c \| u' \| \, \| u \|^2 , \quad u \in W^{1,2}(\mathbb{R}) .$$

The existence of solutions to (3.1) now follows from general compactness arguments. Specifically, for each fixed realization $X_1, X_2, \cdots, X_n$, the functional

$$J_n(u) = -\tfrac{2}{n} \sum_{i=1}^{n} \log u(X_i)$$

is continuous on the space of nonnegative continuous functions on the line, in the topology of uniform convergence on compact subsets of $\mathbb{R}$, and so also in the $W^{1,2}(\mathbb{R})$ topology. Since $J_n(u)$ is also convex, it is then weakly lower semicontinuous on $\mathcal{G}_h$. Now, $\mathcal{G}_h$ is a closed convex subset of a Hilbert space, thus, it is weakly sequentially compact, and so the question is whether a weakly lower semicontinuous function on a weakly compact set attains its minimum. The answer is yes, of course, since this development is just a rephrasing of Theorem (10.4.2). Moreover, the uniqueness of the solution follows by the strict (strong) convexity, and we have proven the following theorem.

(3.4) THEOREM. *The* GOOD *estimator exists and is unique, for every realization of* $X_1, X_2, \cdots, X_n$.

An alternative proof may be based on Theorem (10.3.5). This requires
the lower semicontinuity of L_n (as above), the strong convexity of L_n

$$(3.5) \quad L_n(u) - L_n(v) - \delta L_n(v;u-v) = \|u-v\|^2 + h^2 \|(u-v)'\|^2 ,$$

which we leave as an exercise, and the fact that L_n is bounded below. To
show that L_n is bounded below, observe that

$$-2 \int_{\mathbb{R}} \log u \, dF_n \geqslant 2 \int_{\mathbb{R}} (1-u) \, dF_n \geqslant 2 - 2\|u\|_\infty \geqslant 2 - c\|u'\| ,$$

where the last inequality comes from Exercise (3.3)(c). Thus,

$$L_n(u) \geqslant 3 - c\|u'\| + h^2 \|u'\|^2 \geqslant 3 - (c/2h)^2 > -\infty ,$$

where the last inequality comes from minimizing the quadratic polynomial,
and L_n is bounded below. Together with the next exercise, the premises
of Theorem (10.3.5) are fulfilled, and the minimum exists.

(3.6) EXERCISE. Show the strong convexity of L_n as described by (3.5).

Next, we show that u_{nh} satisfies the Euler equations for the problem
(3.1). The Euler equations should read as

$$(3.7) \quad -h^2 u'' + u = \frac{dF_n}{u} , \quad -\infty < x < \infty ,$$
$$u \longrightarrow 0 \quad , \quad x \longrightarrow \pm\infty .$$

A point of concern is that the right-hand side of the differential equations
is a sum of delta functions or point masses. This would imply that the
solution u is not in $W^{2,2}(\mathbb{R})$. The difficulty is thus the interpretation of
the differential equation. But we are lucky: Green's functions were just
about invented for the purpose of dealing with point sources, and for this
example, the Green's function is known. We recall from § 5.2 that for

$$(3.8) \quad \mathcal{B}_h(x) = (2h)^{-1} \exp(-h^{-1} |x|) ,$$

one has for fixed y,

$$(3.9) \quad -h^2 \frac{\partial^2}{\partial x^2} \mathcal{B}_h(x-y) + \mathcal{B}_h(x-y) = \delta(x-y) , \quad -\infty < x < \infty ,$$
$$\mathcal{B}_h(x-y) \longrightarrow 0 \quad , \quad x \longrightarrow \pm\infty .$$

Equivalently, for every $g \in L^2(\mathbb{R})$, or $L^1(\mathbb{R})$, the solution to

$$(3.10) \quad -h^2 \varphi'' + \varphi = g , \quad -\infty < x < \infty ,$$
$$\varphi \longrightarrow 0 , \quad x \longrightarrow \pm\infty ,$$

is given by $\varphi = \mathfrak{B}_h * g$ (convolution), and $\varphi \in W^{2,2}(\mathbb{R})$, or $W^{2,1}(\mathbb{R})$. Formally then, if u solves (3.7), then it satisfies the implicit equation

$$(3.11) \quad u(x) = \left[\mathfrak{B}_h * \frac{dF_n}{u}\right](x) = \frac{1}{n}\sum_{i=1}^{n}\frac{\mathfrak{B}_h(x - X_i)}{u(X_i)}, \quad -\infty < x < \infty.$$

The Green's function point of view shows that (3.11) and (3.7) are equivalent, but does not show either one holds. However, they do hold, as expressed in the following theorem.

(3.12) THEOREM. u_{nh} solves (3.1) if and only if it satisfies (3.11).

PROOF. The proof follows § 10.5, but we must deal with the point evaluations $u(X_i)$.

Let u be the solution of (3.1). Then, $u(X_i) > 0$ for all i. Define w by

$$(3.13) \qquad\qquad w = \mathfrak{B}_h * \{dF_n/u\}.$$

Since $w \in W^{1,2}(\mathbb{R})$ is strictly positive everywhere, then $L_n(w)$ is finite. If we can show that $u = w$, then u satisfies the Euler equations.

Now, the Gateaux variation $\delta L_n(u, w - u)$ may be computed as

$$(3.14) \quad \delta L_n(u, w - u) = -\frac{2}{n}\sum_{i=1}^{n}\left[u(X_i)\right]^{-1}\left\{w(X_i) - u(X_i)\right\} +$$

$$2\langle u, w - u\rangle + 2h^2\langle u', w' - u'\rangle.$$

With the reproducing kernel Hilbert space setting of Exercise (5.2.64), we have

$$\langle u, w - u\rangle + h^2\langle u', w' - u'\rangle = \langle u, w - u\rangle_{\mathfrak{H}}$$

and

$$\langle \mathfrak{B}_h(\cdot - y), w - u\rangle_{\mathfrak{H}} = w(y) - u(y), \quad y \in \mathbb{R}.$$

Then, working backwards,

$$\langle w, w - u\rangle_{\mathfrak{H}} = \langle \mathfrak{B}_h * (dF_n/u), w - u\rangle_{\mathfrak{H}}$$

$$= \frac{1}{n}\sum_{i=1}^{n}\left[u(X_i)\right]^{-1}\langle \mathfrak{B}_h(\cdot - X_i), w - u\rangle_{\mathfrak{H}}$$

$$(3.15) \qquad\qquad = \frac{1}{n}\sum_{i=1}^{n}\left[u(X_i)\right]^{-1}\left(w(X_i) - u(X_i)\right),$$

and consequently,

$$\delta L_n(u, w - u) = -2\|w - u\|_{\mathfrak{H}}^2.$$

Since u solves (3.1), this Gateaux variation must be nonnegative. It follows that the Gateaux variation vanishes, and thus, $u = w$. From (3.13), it follows that u satisfies the Euler equations (3.11). This is the "only if" part. The "if" part is left as an exercise.   Q.e.d.

(3.16) EXERCISE. (a) Verify (3.15) as is. Also verify it using integration by parts.
(b) Show the "if" part of the theorem.

The large sample asymptotic problem associated with (3.1) is

$$\text{(3.17)} \qquad \text{minimize} \quad L_\infty(u) \stackrel{\text{def}}{=} -2 \int_{\mathbb{R}} f_o(x) \, \log u(x) \, dx + \| u \|^2 + h^2 \| u' \|^2$$

$$\text{subject to} \quad u \in W^{1,2}(\mathbb{R}) \, , \; u \geqslant 0 \, ,$$

with the corresponding Euler equations

$$\text{(3.18)} \qquad -h^2 \, u'' + u = \frac{f_o}{u} \, , \qquad -\infty < x < \infty \, ,$$

$$u \longrightarrow 0 \quad , \qquad x \longrightarrow \pm \infty \, .$$

The existence, uniqueness, and necessity and sufficiency of the Euler equation in the solution of (3.17) follow as for the finite sample case, and they are left as an exercise.

(3.19) THEOREM. *The solution of problem (3.17) exists and is unique. Moreover, u solves (3.17) if and only if $u \in W^{2,2}(\mathbb{R})$ is positive everywhere and solves the Euler equations (3.18).*

(3.20) EXERCISE. Prove Theorem (3.19).

(3.21) EXERCISE. This problem deals with the constrained version of (3.1), viz. the original proposal of GOOD (1971),

$$\text{(3.22)} \qquad \text{minimize} \quad L_n(u) \stackrel{\text{def}}{=} -\frac{2}{n} \sum_{i=1}^{n} \log u(X_i) + \| u \|^2 + h^2 \| u' \|^2$$

$$\text{subject to} \quad u \in W^{1,2}(\mathbb{R}) \, , \; \| u \| = 1 \, , \; u \geqslant 0 \, .$$

(a) Show that $u = w^{nh}$ solves (3.22) if and only if there exists a Lagrange multiplier λ such that

$$\text{(3.23)} \qquad -h^2 \, u'' + \lambda \, u = \frac{dF_n}{u} \, , \qquad -\infty < x < \infty \, ,$$

$$u \longrightarrow 0 \quad , \qquad x \longrightarrow \pm \infty \, ,$$

$$\| u \| = 1 \qquad .$$

(b) Show that the solution w^{nh} of (3.23) exists and is unique.
(c) With the solution u_{nh} of (3.1) denoted as $u(\,\cdot\,; h)$, show that

$$w^{nh} = \lambda^{-1/2} \, u(\,\cdot\,; \lambda^{-1/2} h) \, ,$$

and that λ satisfies

$$\lambda = \| u(\,\cdot\,;\lambda^{-1/2}h) \|^2 .$$

(d) Show that

$$h^2 \| (w^{nh})' \|^2 = 1 - \| u(\,\cdot\,;\lambda^{-1/2}h) \|^2 .$$

(3.24) EXERCISE. Repeat this section for a special case of the second rough-ness penalization of GOOD and GASKINS (1971), viz. where $h^2 \| u' \|^2$ in (3.1) is replaced by

$$2\,h^2 \| u' \|^2 + h^4 \| u'' \|^2 .$$

[Hint: The Green's function is $\mathfrak{B}_h * \mathfrak{B}_h$, and is again nonnegative.]

(3.25) EXERCISE. Consider the least penalized squares problem

$$\text{minimize} \quad -\tfrac{2}{n} \sum_{i=1}^{n} f(X_i) + \| f \|^2 + h^2 \| f' \|^2$$

$$\text{subject to} \quad f \in W^{1,2}(\mathbb{R}) , \; f \geqslant 0 .$$

Show that the solution exists and is unique, derive the Euler equations, and show that the solution is given by

$$f(x) = \sum_{i=1}^{n} \mathfrak{B}_h(x - X_i) , \quad x \in \mathbb{R} .$$

EXERCISES: (3.3), (3.16), (3.19), (3.20), (3.21), (3.24), (3.25).

4. Existence of roughness penalized log-densities

In this section, we consider the existence of solutions to maximum pe-nalized likelihood estimation problems with roughness penalization of the log-density, as studied in § 5.3. Recall that the estimators are defined as solutions of the problem

(4.1) $\text{minimize} \quad -\displaystyle\int_{\mathbb{R}} \log f(x)\, dF_n(x) + \int_{\mathbb{R}} f(x)\, dx + h^2\, R(f) ,$

with the penalization functional

(4.2) $R(f) = \displaystyle\int_{\mathbb{R}} \left| \dfrac{d^3}{dx^3} \log f(x) \right|^2 dx .$

In this section, we prove the a.s. existence of solutions of (4.1)–(4.2) for $n \geqslant 2$. It is obvious that for $n = 1$, there is no solution. We also briefly discuss the generalization to higher order derivatives.

(4.3) EXERCISE. Why is it obvious that problem (4.1)–(4.2) has no solution for $n = 1$?

As in §5.3, we employ the transformation $u = \log f$, and consider the problem equivalent to (4.1)–(4.2)

$$
(4.4) \quad
\begin{aligned}
\text{minimize} \quad & L_{nh}(u) \overset{\text{def}}{=} -\int_{\mathbb{R}} u(x)\, dF_n(x) + \int_{\mathbb{R}} e^{u(x)}\, dx + h^2\, \| u^{(3)} \|^2 \\
\text{subject to} \quad & u \in C^2(\mathbb{R}), \; u^{(3)} \in L^2(\mathbb{R}).
\end{aligned}
$$

Here, again, $\| \cdot \|$ denotes the $L^2(\mathbb{R})$ norm. A very nice feature of this estimation problem is that the solution generates a pdf, if it exists.

(4.5) EXERCISE. (a) Suppose u_o solves (4.4). Show that $\exp(u_o)$ is a pdf. (b) Show that the problems (4.4) and (4.1)–(4.2) are equivalent, i.e., have the same solution(s) under the transformation $u = \log f$. (c) Assume that

$$
\int_{\mathbb{R}} \exp(\, u_o(x) + r\, x^2\,)\, dx < \infty \quad \text{for some } r > 0 \,.
$$

Show that

$$
\int_{\mathbb{R}} x^m \exp(\, u_o(x)\,)\, dx = \int_{\mathbb{R}} x^m\, dF_n(x)\,, \quad m = 0, 1, 2 \,.
$$

[Hint : Use perturbations of the form $f(x) = \exp(\, u(x) + a + b\, x + c\, x^2)$.] (d) Prove part (c) without the assumption. First show that regardless

$$
\int_{\mathbb{R}} x^2 \exp(\, u_o(x)\,)\, dx \leqslant \int_{\mathbb{R}} x^2\, dF_n(x)\,,
$$

and take it from there.

The first step in establishing the existence of solutions to (4.4) is to specify exactly the class of functions over which the minimization in (4.4) takes place. Since $u \in C^2 \cap W^{3,2}(\mathbb{R})$, Taylor's theorem with exact remainder gives

$$
(4.6) \quad u(x) = u(0) + x\, u'(0) + \tfrac{1}{2}\, x^2\, u''(0) + \tfrac{1}{2} \int_0^x (x - y)^2\, u^{(3)}(y)\, dy \,,
$$

which we write concisely as

$$
(4.7) \quad u(x) = p(u\,;x) + \big[\, \mathcal{V}\{\, u^{(3)}\,\}\,\big]\,(x) \,.
$$

We emphasize that $p(x) = p(u\,;x)$ is a quadratic polynomial, so the minimization in (4.4) takes place over the set

$$
(4.8) \quad \mathfrak{R} = \big\{\, p + \mathcal{V}q \,:\, q \in L^2(\mathbb{R})\,,\; p \text{ a quadratic polynomial}\,\big\} \,.
$$

Clearly, $\int_{\mathbb{R}} p(x)\, dF_n(x)$ is well defined. From Cauchy-Schwarz, it follows that for each x,

$$
(4.9) \quad |\, \mathcal{V}q(x)\,| \leqslant c\, |x|^{5/2}\, \| q \| \,,
$$

with $c = 1/(2\sqrt{5})$. This implies that $\int_{\mathbb{R}} [\mathcal{V}\{u^{(3)}\}](x)\, dF_n(x)$ is well defined also. The conclusion is that $L_{nh}(u)$ is finite for every $u \in \mathfrak{R}$.

(4.10) EXERCISE. Compute the Gateaux variation of L_{nh}.

The second step is to observe that L_{nh} is strongly convex in the following sense. For all $u, v \in \operatorname{domain}(L_{nh})$,

$$(4.11) \quad L_{nh}(u) - L_{nh}(v) - \delta L_{nh}(v; u - v) = \operatorname{KL}(e^v, e^u) + h^2 \,\| (u - v)^{(3)} \|^2 \,,$$

where $\operatorname{KL}(\varphi, \psi)$ is our old friend the Kullback-Leibler distance. This immediately implies the uniqueness of a solution of (4.4).

(4.12) EXERCISE. Use (4.11) to show that (4.4) has at most one solution.

The strong convexity (4.11) also implies the existence of a solution, provided we can show that L_{nh} is (almost surely) bounded from below.

(4.13) THEOREM. If L_{nh} is bounded away from $-\infty$, then the solution to (4.4) exists.

(4.14) EXERCISE. Prove the theorem, by applying Theorem (10.3.5).

The remainder of this sections deals with bounding L_{nh}. In particular, we prove the following result.

(4.15) LEMMA. Let $n \geqslant 2$. Then, L_{nh} is almost surely bounded from below.

PROOF. The starting point is (4.6)–(4.9), written in the form

$$(4.16) \qquad | u(x) - p(x) | \leqslant c\, |x|^{5/2} \,\| u^{(3)} \| \,,$$

with $c = 1/(2\sqrt{5})$. It is useful to define

$$(4.17) \qquad Q(u) \overset{\text{def}}{=} c\, \| u^{(3)} \| \,,$$

which we abbreviate to plain Q. Then, (4.16) may be rewritten as

$$p(x) - Q\, |x|^{5/2} \leqslant u(x) \leqslant p(x) + Q\, |x|^{5/2} \,, \quad x \in \mathbb{R} \,,$$

and it follows that

$$(4.18) \quad L_{nh}(u) \geqslant - \int_{\mathbb{R}} \{\, p(x) + Q\, |x|^{5/2} \,\}\, dF_n(x) +$$

$$\int_{\mathbb{R}} \exp\big(\, p(x) - Q\, |x|^{5/2} \,\big)\, dx + \lambda^2\, Q^2 \,,$$

with $\lambda = h/c$. We note that

$$(4.19) \qquad \int_{\mathbb{R}} p(x)\, dF_n(x) = \int_{\mathbb{R}} p(x)\, \phi(x)\, dx \,,$$

where ϕ is the normal density with mean $\overline{X}$, the sample mean, and variance

$$s^2 = \tfrac{1}{n} \sum_{i=1}^{n} |X_i - \overline{X}|^2 \ .$$

With the abbreviations

$$m = \int_{\mathbb{R}} |x|^{5/2} \, dF_n(x) \ , \quad \mu = \int_{\mathbb{R}} |x|^{5/2} \, \phi(x) \, dx \ , \quad r(x) = p(x) - Q \, |x|^{5/2} \ ,$$

the inequality (4.18) may be rewritten as

$$L_{nh}(u) \geqslant - \int_{\mathbb{R}} p(x) \, \phi(x) \, dx + \int_{\mathbb{R}} \exp(\, r(x) \,) \, dx - m \, Q + \lambda^2 \, Q^2$$

$$(4.20) \qquad \geqslant - \int_{\mathbb{R}} r(x) \, \phi(x) \, dx + \int_{\mathbb{R}} \exp(\, r(x) \,) \, dx + \mathrm{rem}_n(Q) \ ,$$

where

$$(4.21) \qquad\qquad \mathrm{rem}_n(Q) = \lambda^2 \, Q^2 - (m + \mu) \, Q \ .$$

By completing the square, one sees that

$$(4.22) \qquad\qquad \mathrm{rem}_n(Q) \geqslant -(\, 2\lambda \,)^{-2} \, (\, m + \mu \,)^2 > -\infty \ .$$

Thus, (4.20) may be rewritten as

$$L_{nh}(u) \geqslant - \int_{\mathbb{R}} r(x) \, \phi(x) \, dx + \int_{\mathbb{R}} \mathrm{e}^{r(x)} \, dx + \mathrm{rem}_n(Q)$$

$$\geqslant \int_{\mathbb{R}} \left\{ \phi(x) \, \log \frac{\phi(x)}{\mathrm{e}^{\, r(x)}} + \mathrm{e}^{r(x)} - \phi(x) \right\} dx \ +$$

$$+ \, \mathrm{rem}_n(Q) - \int_{\mathbb{R}} \phi(x) \, \log \phi(x) \, dx + 1$$

or

$$(4.23) \qquad\qquad L_{nh}(u) \geqslant \mathrm{KL}(\, \phi \,, \mathrm{e}^r \,) + c_n \ ,$$

where $c_n > -\infty$ almost surely. It follows that L_{nh} is a.s. bounded below for $n \geqslant 2$, and Lemma (4.15) is proven. Q.e.d.

Before considering the Euler equations, we add a few words about (4.1) with the penalization

$$(4.24) \qquad\qquad R(f) = \| \, \{\log f\}^{(m)} \, \|^2 \ , \quad m = 1, 2, \cdots \ .$$

The essential step in proving the existence of a solution to (4.1)–(4.24) is showing that L_{nh} is bounded from below. The proof of Lemma (4.15) for $m = 2$ depended on the construction of a density satisfying the moment conditions (4.19). Since $p(x)$ was quadratic, we were able to guess a solution. In the general case with the penalization functional given by (4.24), the polynomial $p(x)$ is of degree $\leqslant m - 1$, and it suffices to construct a density ψ with finite entropy so that for all polynomials p of degree not exceeding $m - 1$,

$$(4.25) \qquad \int_{\mathbb{R}} p(x) \, dF_n(x) = \int_{\mathbb{R}} p(x) \, \psi(x) \, dx \ .$$

One way (of presumably many) is to solve the *parametric* maximum likelihood problem for the exponential family

$$(4.26) \quad \mathcal{F}_m = \left\{ \exp(\pi(x) - |x|^m) \, : \, \pi \text{ polynomial of degree } \leqslant m - 1 \right\} .$$

That is, let $\psi(x) = \exp(\pi(x) - |x|^m)$, with π the solution to

$$(4.27) \qquad \begin{array}{ll} \text{minimize} & -\displaystyle\int_{\mathbb{R}} \pi(x) \, dF_n(x) + \int_{\mathbb{R}} \exp(\pi(x) - |x|^m) \, dx \\[2mm] \text{subject to} & \pi \text{ is a polynomial of degree } \leqslant m - 1 \ . \end{array}$$

This is as *regular* a problem as it gets : The second integral converges for all polynomials π of degree $\leqslant m - 1$. (This is the reason for subtracting $|x|^m$.) The existence of solutions to problems like this is discussed in BARNDORFF-NIELSEN (1978), with CRAIN (1976a) being very useful for the verification of the conditions. Then, (4.25) follows, similarly to Exercise (4.5)(c).

We note that SILVERMAN (1982) shows the existence on a bounded interval and proves for the penalization (4.24) (and more general ones actually) that the problem (4.1)–(4.24) has a solution if the parametric maximum likelihood problem

$$(4.28) \qquad \begin{array}{ll} \text{minimize} & -\displaystyle\int p(x) \, dF_n(x) + \int_{\mathbb{R}} e^{p(x)} \, dx \\[2mm] \text{subject to} & p \text{ is a polynomial of degree } \leqslant m - 1 \end{array}$$

has a solution. However, for $m \geqslant 4$, the exponential family in question is not *regular* [or *steep* in the parlance of BARNDORFF-NIELSEN (1978)], so it is not clear from the general theory that (4.28) has a solution. Of course, (4.1)–(4.24) does have a solution, since, in view of (4.27), the objective function is bounded below.

(4.29) EXERCISE. Work out the details for the penalizations (4.24).

(4.30) EXERCISE. Show that (4.28) has a solution for n large enough.

To finish this section, we consider the Euler equations for (4.1)–(4.2).

(4.31) THEOREM. *The function* $u \in \mathfrak{R}$ *solves* (4.4) *if and only if it satisfies*

$$\begin{array}{ll} -2 \, h^2 \, u^{(6)} + e^u = dF_n & \text{on } \mathbb{R} , \\[2mm] u^{(j)} \longrightarrow 0 & \text{at } \pm\infty , \ j = 3, 4, 5 \ . \end{array}$$

PROOF. Let u solve (4.4), and let v be defined by

(4.32) $$2\,h^2\,v^{(6)} = e^u - dF_n \;,$$

with the boundary conditions $v^{(j)} \longrightarrow 0$ at $\pm\infty$ for $j = 3, 4, 5$. Now, consider

$$\delta L_{nh}(u\,;v-u) = -\int_{\mathbb{R}} (v-u)\,dF_n + \int_{\mathbb{R}} e^u(v-u) + 2\,h^2 \int_{\mathbb{R}} u^{(3)}\,(v-u)^{(3)} \;,$$

which may be rewritten as

(4.33) $$\delta L_{nh}(u\,;v-u) = 2\,h^2 \int_{\mathbb{R}} (v-u)\,v^{(6)} + 2\,h^2 \int_{\mathbb{R}} u^{(3)}\,(v-u)^{(3)} \;.$$

Now, we have that

$$v(x) - u(x) = p(v-u\,;x) + \big[\, \mathcal{V}\,\{(v-u)^{(3)}\}\,\big]\,(x) \;,$$

and, in view of Exercise (4.5)(d), $\int_{\mathbb{R}} p(v-u\,;x)\,v^{(6)}(x)\,dx = 0$, whence

(4.34) $$\int_{\mathbb{R}} (v-u)v^{(6)} = \int_{\mathbb{R}} v^{(6)}(x)\,\big[\,\mathcal{V}\{(v-u)^{(3)}\}\,\big]\,(x)\,dx$$

$$= \int_{\mathbb{R}} \{(v-u)^{(3)}\}(x)\,\big[\,\mathcal{V}^* v^{(6)}\,\big]\,(x)\,dx \;.$$

Here, $\mathcal{V}^*$ is the adjoint of $\mathcal{V}$, and it is given by

(4.35) $$\big[\,\mathcal{V}^*\,v^{(6)}\,\big]\,(y) = \begin{cases} \frac{1}{2} \displaystyle\int_y^\infty (x-y)^2\,v^{(6)}(x)\,dx\;, & y > 0\;, \\[2mm] -\frac{1}{2} \displaystyle\int_{-\infty}^y (x-y)^2\,v^{(6)}(x)\,dx\;, & y < 0\;. \end{cases}$$

We now show that for any x,

(4.36) $$\big[\,\mathcal{V}^*\,v^{(6)}\,\big]\,(x) = -v^{(3)}(x) \;.$$

In view of (4.6), there exists a quadratic polynomial p such that for $x > 0$,

$$v^{(3)}(x) = p(x) + \frac{1}{2} \int_0^x (x-y)^2\,v^{(6)}(y)\,dy \;.$$

Consequently,

(4.37) $$v^{(3)}(x) = q(x) - \frac{1}{2} \int_x^\infty (x-y)^2\,v^{(6)}(y)\,dy \;,$$

in which

$$q(x) = p(x) + \frac{1}{2} \int_0^\infty (x-y)^2\,v^{(6)}(y)\,dy \;.$$

Note that q is also a quadratic polynomial. Since $v^{(3)} \longrightarrow 0$ at $+\infty$, and

$$\int_x^\infty (x-y)^2\,v^{(6)}(y)\,dy \longrightarrow 0 \quad (x \to \infty) \;,$$

may assume that $\{\,|\,m_n - m_o\,|\,\}_n$ is decreasing. Now, for $k = 2, 3, \cdots$, let

$$A_k = [\,m_o+1/k\,,\,m_o+k\,] \quad , \quad B_k = [\,m_o-k\,,\,m_o-1/k\,] \quad , \quad C_k = A_k \cup B_k \,.$$

Then, for all n large enough, we have $|\,m_n - m_o\,| < 1/(2k)$, and consequently, $f_n(x) \leqslant 1/(2k)$ on C_k. Now, we may appeal to Lemma (6.6) to conclude that $\{\,f_n\,\}_n$ has a subsequence, again denoted by $\{\,f_n\,\}_n$, and that there exists a measurable f_o such that

$$\int_{C_k} |\,f_n - f_o\,| \longrightarrow 0 \quad \text{for all } k \geqslant 2 \,.$$

It follows that for all $k \geqslant 2$,

$$\|\,f_n - f_o\,\|_1 \geqslant \int_{C_k} |\,f_n - \varphi\,| \longrightarrow \int_{C_k} |\,f_o - \varphi\,| \,,$$

and consequently,

(6.15)
$$\liminf_{n \to \infty} \|\,f_n - \varphi\,\|_1 \geqslant \|\,f_o - \varphi\,\|_1 \,,$$

and $f_o \in L^1(\mathbb{R})$. Also, f_o is decreasing on A_k, and increasing on B_k, for each k, whence f_o is unimodal on the line. Since $\{\,f_n\,\}_n$ was a minimizing sequence, then (6.15) says that f_o is a solution to (6.13).

If $\{\,m_n\,\}_n$ contains no bounded subsequences, then we may select a subsequence that tends to $+\infty$ or to $-\infty$. Without loss of generality, assume that $m_n \longrightarrow \infty$. Similar to the above, we may extract a subsequence of $\{\,f_n\,\}_n$, again denoted by $\{\,f_n\,\}_n$, such that on each interval $(-k, k)$, the subsequence converges in $L^1(-k, k)$. Thus, there exists a measurable f_o such that for all k,

$$\int_{-k}^{k} |\,f_n - f_o\,| \longrightarrow 0 \,,$$

and it follows that f_o is an increasing function. Now, for all k,

$$\liminf_{n \to \infty} \|\,f_n - \varphi\,\|_1 \geqslant \liminf_{n \to \infty} \int_{-k}^{k} |\,f_n - \varphi\,| = \int_{-k}^{k} |\,f_o - \varphi\,| \,,$$

and so $f_o \in L^1(\mathbb{R})$, and f_o is the solution. As a matter of fact, since f_o is also increasing, we must have $f_o = 0$.

All of this proves the following theorem.

(6.16) THEOREM. *The unimodal minimum L^1 distance problem (6.13) has a solution for every $\varphi \in L^1(\mathbb{R})$.*

The final problem to be considered is the log-concave minimum distance problem. Since the log-concave sub-pdfs are a closed subset of the unimodal sub-pdfs, the existence follows easily.

PROOF. Define $1\!\!1_h$ as $1\!\!1_h(x) = 1$ for $|x| \leqslant h^{1/2}$, and $= 0$ otherwise. Now, writing $\phi_h * \psi = (\phi_h - \phi_h 1\!\!1_h) * \psi + (\phi_h 1\!\!1_h) * \psi$ gives

$$\| \phi_h * \psi - \psi \|_2 \leqslant \| (\phi_h - \phi_h 1\!\!1_h) * \psi \|_2 + \| (\phi_h 1\!\!1_h) * \psi - \lambda \psi \|_2 + (1-\lambda) \| \psi \|_2 \, ,$$

where

$$\lambda = \int_{|y|<\sqrt{h}} \phi_h(y) \, dy = \int_{|y|<1/\sqrt{h}} \phi(y) \, dy \longrightarrow 1 \, , \quad h \to 0 \, .$$

For the first term, we have by Young's inequality (4.1),

$$\| (\phi_h - \phi_h 1\!\!1_h) * \psi \|_2 \leqslant \| \phi_h - \phi_h 1\!\!1_h \|_1 \| \psi \|_2 \longrightarrow 0$$

for $h \to 0$, since

$$\| \phi_h - \phi_h 1\!\!1_h \|_1 = \int_{|y|>\sqrt{h}} \phi_h(y) \, dy = \int_{|y|>1/\sqrt{h}} \phi(y) \, dy \longrightarrow 0 \, , \quad h \to 0 \, .$$

For the second term, we have

$$\| (\phi_h 1\!\!1_h) * \psi - \lambda \psi \|_2^2 =$$

$$= \int_{\mathbb{R}} \left\{ \int_{|y|<\sqrt{h}} \phi_h(y) \, (\psi(x-y) - \psi(x)) \, dy \right\}^2 dx$$

$$\leqslant \int_{\mathbb{R}} \left\{ \int_{|y|<\sqrt{h}} \phi_h(y) \, dy \right\} \left\{ \int_{|y|<\sqrt{h}} \phi_h(y) \, | \psi(x-y) - \psi(x) |^2 \, dy \right\} dx$$

$$\leqslant \lambda^2 \sup_{|y|<\sqrt{h}} \int_{\mathbb{R}} | \psi(x-y) - \psi(x) |^2 \, dx \, ,$$

where on the second line, we used Cauchy-Schwarz as in the proof of Young's inequality, and in the penultimate line, we interchanged the order of integration. Now, the last integral tends to 0, by the continuity of translation in $L^2(\mathbb{R})$. This concludes the proof.   Q.e.d.

We are now ready for Fourier transforms of L^2 functions. Let $\psi \in L^2(\mathbb{R})$, and let the function $1\!\!1_n$ be defined as $1\!\!1_n(x) = 1$ for $|x| \leqslant n$, and $= 0$ otherwise, and define ψ_n by

(4.3) $\psi_n = \phi_h * (\psi 1\!\!1_n) \, , \quad h \equiv 1/n \, ,$

still with ϕ the standard normal. The Approximation Theorem implies $\psi_n \longrightarrow \psi$ in $L^2(\mathbb{R})$. Moreover, then $\psi_n \in C_\downarrow^\infty(\mathbb{R})$ for all n.

(4.4) EXERCISE. For $\psi \in L^2(\mathbb{R})$, let ψ_n be defined by (4.3). Show that $\psi_n \in C_\downarrow^\infty(\mathbb{R})$ for all n.

Since $\psi_n \in C_\downarrow^\infty(\mathbb{R})$, for all n, their Fourier transforms exist and belong to $C_\downarrow^\infty(\mathbb{R})$. In particular, then Plancherel's formula (2.15) applies and gives

where

$$M(x, \omega) = \frac{d}{d\omega} \left\{ \frac{\omega^{\kappa-1}}{(x^{\kappa} + \omega^{\kappa})^2} \right\} .$$

One verifies that

$$M(x, \omega) = \frac{(\kappa - 1) \omega^{\kappa-2}}{(x^{\kappa} + \omega^{\kappa})^2} - \frac{2\kappa \omega^{2\kappa-2}}{x^{\kappa} + \omega^{\kappa}} \cdot \frac{1}{(x^{\kappa} + \omega^{\kappa})^2} ,$$

and we see that for $x > 1$, say, $M(x, \omega)$ is dominated in absolute value by

$$\frac{|\kappa - 1| \omega^{\kappa-2}}{x^{2\kappa}} + \frac{2\kappa \omega^{2\kappa-2}}{\omega^{\kappa} \cdot x^{2\kappa}} = (\kappa + 1) \omega^{\kappa-2} x^{-2\kappa} .$$

Also, it is easily verified that

$$\lim_{x \to \infty} x^{2\kappa} M(x, \omega) = (\kappa - 1) \omega^{\kappa-2} , \quad \text{for all } \omega > 0 .$$

Since $\omega^{\kappa-2}(\cos\omega - 1)$ is integrable over $(0, \infty)$, it follows again from the dominated convergence theorem that

$$\lim_{x \to \infty} \int_0^{\infty} (\cos\omega - 1) x^{2\kappa} M(x, \omega) \, d\omega = \int_0^{\infty} (\kappa - 1) \omega^{\kappa-2}(\cos\omega - 1) \, d\omega .$$

This proves the second part of the asymptotic behavior of g_κ, viz.

(5.6) $$g_\kappa(x) \sim \gamma_\kappa \, x^{-\kappa-1} , \quad x \to \infty ,$$

for a suitable constant γ_κ. We summarize the results.

(5.7) LEMMA. *For $\frac{1}{2} < \kappa \leqslant 1$, the functions g_κ are symmetric pdfs, and for suitable constants c_κ and γ_κ,*

$$g_\kappa(x) \sim c_\kappa \, |x|^{\kappa-1} , \quad |x| \longrightarrow 0 ,$$
$$g_\kappa(x) \sim \gamma_\kappa \, |x|^{-\kappa-1} , \quad |x| \longrightarrow \infty .$$

Lemma (5.7) actually also holds for $0 < \kappa \leqslant \frac{1}{2}$, but the difficulty is verifying the inversion formula (5.2).

The second example is somewhat similar to example (5.1). Let $\mathfrak{B}_{(\kappa)}$ be defined by means of its Fourier transform as

$$\widehat{\mathfrak{B}_{(\kappa)}}(\omega) = \{ 1 + (2\pi\omega)^2 \}^{-\kappa/2} , \quad \omega \in \mathbb{R} .$$

One verifies that $\kappa = 2$ gives the two-sided exponential distribution. In §§ 4.5 and 4.6, we denoted it as just $\mathfrak{B}$. For reasons to do with the roughness penalization of GOOD (1971), we are especially interested in the case $\kappa = 1$, and denote $\mathfrak{B}_{(1)}$ as $\mathfrak{C}$. So $\mathfrak{C}$ is defined via

(5.8) $$\widehat{\mathfrak{C}}(\omega) = \{ 1 + (2\pi\omega)^2 \}^{-1/2} , \quad \omega \in \mathbb{R} .$$

it follows that $q(x) = 0$ for all x. But then (4.37) says that (4.36) holds for $x > 0$. The same argument works for $x < 0$. Thus, (4.36) holds for all x.

Going back to (4.34), we see that

$$\int_{\mathbb{R}} (v - u)\, v^{(6)} = -\int_{\mathbb{R}} (v - u)^{(3)}\, v^{(3)} \; ,$$

and then going back to (4.33),

$$(4.38) \qquad \delta L_{nh}(u\,;v - u) = -2\, h^2\, \|\,(v - u)^{(3)}\,\|^2 \leqslant 0 \; .$$

On the other hand, since u is a solution to the problem (4.4), we have that $\delta L_{nh}(u\,;v - u) \geqslant 0$ and, thus, $v^{(3)} = u^{(3)}$. Consequently, u is as smooth as v, in particular, $u^{(6)} = v^{(6)}$, and from (4.32), then u satisfies the Euler equations.

The converse is left as an exercise.   Q.e.d.

(4.39) EXERCISE. Prove the "if" part of Theorem (4.31).

EXERCISES : (4.3), (4.5), (4.10), (4.12), (4.14), (4.29), (4.30), (4.39).

5. Existence of log-concave estimators

We next consider maximum smoothed likelihood estimation with constraints, in particular, the estimation of log-concave densities. As will be seen, this fits in well with the material of the previous two sections on log-densities. To prove the existence of log-concave maximum likelihood estimators, we again use the compactness arguments of Theorem (10.3.5). The existence of monotone, convex, and unimodal densities may be shown along similar lines, and are left as exercises.

Consider the problems (5.5.9) and (5.5.20) in the form

$$(5.1) \qquad \begin{array}{ll} \text{minimize} & -\int_{\mathbb{R}} \log f\, d\Psi + \int_{\mathbb{R}} f \\[2mm] \text{subject to} & f \text{ log-concave} \; , \end{array}$$

where either $\Psi = F_n$ or $\Psi = A_h * dF_n$, with F_n an empirical distribution fucntion. As in §§ 3 and 4, the transformation $u = \log f$ is useful and leads to the equivalent problem

$$(5.2) \qquad \begin{array}{ll} \text{minimize} & L(u) \stackrel{\text{def}}{=} -\int_{\mathbb{R}} u\, d\Psi + \int_{\mathbb{R}} e^u \\[2mm] \text{subject to} & u \in \mathcal{C} \; , \end{array}$$

where $\mathcal{C}$ is the set of concave functions on the line. This set is largish, but may be slimmed down to

$$(5.3) \qquad \mathcal{C} \stackrel{\text{def}}{=} \{\, u \text{ concave} : \int_{\mathbb{R}} e^u \leqslant 1 \,\} \; .$$

Note that the set $\{u = -\infty\}$ need not have measure 0.

(5.4) EXERCISE. Show that the problems (5.1) and (5.2)–(5.3) are equivalent, i.e., f is a solution of (5.1) iff $u = \log f$ is a solution of (5.2)–(5.3).

The reason for the transformation $u = \log f$ being useful here is that L is a strictly convex function and $\mathcal{C}$ is a convex set.

(5.5) EXERCISE. Show that $\mathcal{C}$ is a convex set and that L is a strictly convex function.

Also, the set $\mathcal{C}$ is closed. We formulate it precisely.

(5.6) LEMMA. If $\{u_n\}_n \subset \mathcal{C}$ and u_o satisfy $\|\exp(u_n) - \exp(u_o)\|_1 \longrightarrow 0$, then u_o differs at most on a set of measure 0 from an element $v_o \in \mathcal{C}$.

PROOF. It is obvious that $\int_{\mathbb{R}} \exp(u_o) \leqslant 1$. To show that $\exp(u_o)$ is essentially log-concave, it suffices to appeal to the characterization of log-concave functions of Lemma (5.5.4). Thus, let g be unimodal. Then, $\exp(u_n) * g$ is unimodal for every n. But the $L^1(\mathbb{R})$ limit of unimodal sub-pdfs is unimodal. So $\exp(u_o) * g$ is unimodal. It follows that $\exp(u_o)$ differs at most on a set of measure 0 from a log-concave sub-pdf.   Q.e.d.

Next, we note that $L(u)$ is strongly convex, in the sense that

(5.7) $L(u) - L(w) - \delta L(w\,;\,u - w) = \mathrm{KL}(e^w,\,e^u)\,.$

We leave its verification to the reader.

The final piece of the puzzle is to show that $L(u)$ is bounded from below. Here, we find it convenient to consider two cases.

Let $\Psi = A_h * F_n$ with A a log-concave density, and let $\psi = \Psi'$. Since ψ has exponential decay, it has finite entropy, and it follows that

(5.8) $L(u) = \mathrm{KL}(\psi,\,e^u) + 1 - \int_{\mathbb{R}} \psi \log \psi\,.$

Hence,

(5.9) $L(u) \geqslant 1 - \int_{\mathbb{R}} \psi \log \psi\,,$

and in this case, L is bounded from below.

Next, consider the case $\Psi = F_n$. For $n = 1$, there is no lower bound, but for $n \geqslant 2$ there is. First, let $n = 2$. Then, (5.2) may be written as

(5.10)
$$\text{minimize} \quad L(u\,;\,X_1, X_2) \stackrel{\text{def}}{=} -\tfrac{1}{2}\,u(X_1) - \tfrac{1}{2}\,u(X_2) + \int_{\mathbb{R}} e^u$$

$$\text{subject to} \quad u \in \mathcal{C}\,.$$

Supposing that $X_1 \neq X_2$, the solution of (5.10) is given by

$$
(5.11) \quad u(x) = \begin{cases} -\log|X_1 - X_2|, & x \text{ between } X_1 \text{ and } X_2 \text{ inclusive}, \\ -\infty, & \text{otherwise}. \end{cases}
$$

(5.12) EXERCISE. Prove that u as given by (5.11) solves (5.10).

Finally, writing L as $L(u) = \frac{1}{n} \sum_{i=1}^{n} L(u; X_i, X_{i+1})$, where $X_{n+1} = X_1$, it follows that for $n \geqslant 2$,

$$
L(u) \geqslant \frac{1}{n} \sum_{i=1}^{n} \min\{ L(u; X_i, X_{i+1}) : u \text{ concave}\} >_{as} -\infty.
$$

The almost sure character of this last statement comes from the fact that the X_i are a.s. distinct. So in this case too, L is bounded below.

We are done: By an appeal to Theorem (10.3.5), we have proven the desired existence and uniqueness result.

(5.13) THEOREM. *The solutions to the smoothed and plain maximum likelihood estimation problems with log-concavity constraints (5.5.9) and (5.5.20) exist and are unique.*

(5.14) EXERCISE. (a) Formulate and prove an analogous theorem for maximum likelihood estimators of monotone densities on $(0, \infty)$.
(b) Likewise for convex densities on $(0, \infty)$.
(c) Also for *completely monotone* densities on $(0, \infty)$. A function f is completely monotone if $f \in C^{\infty}(0, \infty)$ and for all $m \in \mathbb{N}$,

$$
(-1)^m f^{(m)}(x) \geqslant 0, \quad x \in (0, \infty).
$$

(d) Ditto for unimodal densities on $(-\infty, \infty)$. [Hint for (c): A peek at the minimum L^1 distance problem in the next section may be useful.]

EXERCISES: (5.4), (5.5), (5.12), (5.14).

6. Constrained minimum distance estimation

We now consider the existence of solutions to constrained minimum L^1 distance estimation problems

$$
(6.1) \quad \begin{aligned} &\text{minimize} \quad \| f - A_h * dF_n \|_1 \\ &\text{subject to} \quad f \geqslant 0, \ f \text{ satisfies shape constraints}. \end{aligned}
$$

The shape constraints intended are those of unimodality and log-concavity, but it is advantageous to explicitly consider monotonicity also. A case can be made for adding the pdf constraint on f to (6.1), but this is left as an exercise. The existence proofs are based on compactness arguments. Uniqueness is not considered. As a matter of fact, the solution of (6.1) is typically *not* unique. Throughout this section, $A_h * dF_n$ is denoted by φ.

We begin with monotone densities on $(0, \infty)$ and consider the problem

(6.2)
$$\text{minimize} \quad \| f - \varphi \|_1$$
$$\text{subject to} \quad f \in \mathcal{D} \, ,$$

where $\mathcal{D}$ denotes the set of decreasing sub-pdfs on $(0, \infty)$, that is,

(6.3) $\mathcal{D} \stackrel{\text{def}}{=} \{ f \in L^1(0, \infty) : f \geqslant 0 \, , \int_{\mathbb{R}} f(x) \, dx \leqslant 1 \, , \, f \text{ decreasing} \} \, .$

Note that the objective function in (6.1) is convex, but not strongly, nor strictly, convex.

(6.4) EXERCISE. Show this.

Thus, the existence of (6.3) must perforce be based on compactness arguments. However, $\mathcal{D}$ is not sequentially compact, for two reasons. Clearly, we can expect trouble at ∞, but $x = 0$ is a trouble spot also since the unit point mass at $x = 0$ is a limit point of elements in $\mathcal{D}$, considered as Borel measures restricted to, say, the interval $[0, 1]$. So it is a good idea to start with decreasing subdensities on a bounded interval $A = [a, b]$, with $0 < a < b < \infty$, that is,

(6.5) $$\mathcal{D}(A) = \{ f \, \mathbb{1}_A : f \in \mathcal{D} \} \, ,$$

where $\mathbb{1}_A$ is the indicator function of the set A. Note that $f(x) \leqslant 1/a$ for every $f \in \mathcal{D}([a, b])$. We then have the following lemma.

(6.6) LEMMA. *Let* $0 < a < b < \infty$. *Then,* $\mathcal{D}([a, b])$ *is a sequentially compact subset of* $L^1(0, \infty)$.

(6.7) EXERCISE. Prove it! [Hint: Theorem (10.4.18).]

Now, consider the problem (6.2). Let $\{ f_n \}_n \subset \mathcal{D}$ be a minimizing sequence, and define $A_k = [1/k, k]$. By a diagonal argument, one can extract a subsequence of $\{ f_n \}_n$, again denoted by $\{ f_n \}_n$, for which

(6.8) $\lim_{n \to \infty} \int_{A_k} f_n(x) \, dx$ exists for each integer $k > 1$.

Now, for each k, we have $\{ f_n \, \mathbb{1}_{A_k} \}_n \subset \mathcal{D}(A_k)$, and so it has a convergent subsequence in $L^1(0, \infty)$, with limit in $\mathcal{D}(A_k)$. Again, by a diagonal argument, we may extract a subsequence of $\{ f_n \}_n$, again denoted by $\{ f_n \}_n$, which converges to some measurable f_o in the sense that

$$(6.9) \qquad \int_{A_k} |f_n - f_o| \longrightarrow 0 \quad \text{for } n \to \infty , \quad \text{for every } k > 1 .$$

It follows that

$$\int_{A_k} f_o = \lim_{n \to \infty} \int_{A_k} f_n \leqslant 1 ,$$

so that $f_o \in L^1(0, \infty)$. Moreover, f_o is decreasing on $(0, \infty)$, so $f_o \in \mathcal{D}$. Finally, since $\{ f_n \}_n$ is a minimizing sequence,

$$\inf_{f \in \mathcal{D}} \| f - \varphi \|_1 = \lim_{n \to \infty} \| f_n - \varphi \|_1$$

$$\geqslant \lim_{n \to \infty} \int_{A_k} |f_n - \varphi| \geqslant \int_{A_k} |f_o - \varphi| .$$

This holds for all k, hence,

$$\| f_o - \varphi \|_1 \leqslant \inf_{f \in \mathcal{D}} \| f - \varphi \|_1 .$$

Since $f_o \in \mathcal{D}$, we have proven the following existence result.

(6.10) THEOREM. *The monotone minimum L^1 distance problem* (6.2) *has a solution for every $\varphi \in L^1(0, \infty)$.*

(6.11) EXERCISE. The above goes a long way toward showing existence for convexity constraints. Let $\mathcal{C}$ denote the set of convex sub-pdfs on $(0, \infty)$, and consider the problem

$$(6.12) \qquad \text{minimize} \quad \| f - \varphi \|_1 \quad \text{subject to} \quad f \in \mathcal{C} .$$

Show that (6.12) has a solution.

We now come to the case of unimodal density estimation. Let $\varphi \in L^1(\mathbb{R})$ be nonnegative, and consider the problem

$$(6.13) \qquad \begin{array}{c} \text{minimize} \quad \| f - \varphi \|_1 \\ \text{subject to} \quad f \text{ unimodal .} \end{array}$$

To prove the existence of a solution, consider a minimizing sequence $\{ f_n \}_n$, i.e.,

$$(6.14) \qquad \| f_n - \varphi \|_1 \longrightarrow \inf \{ \| f - \varphi \|_1 : f \text{ unimodal} \} .$$

For each n, let m_n be a mode of f_n. Now, either $\{ m_n \}_n$ contains a bounded subsequence or it does not.

If it does, then we may extract a convergent subsequence, which we denote again by $\{ m_n \}_n$. Thus, $m_n \longrightarrow m_o$ for some finite m_o. Also, we may assume that $\{\,|\,m_n - m_o\,|\,\}_n$ is decreasing. Now, for $k = 2, 3, \cdots$, let

$$A_k = [\,m_o + 1/k\,,\ m_o + k\,] \quad,\quad B_k = [\,m_o - k\,,\ m_o - 1/k\,] \quad,\quad C_k = A_k \cup B_k \,.$$

Then, for all n large enough, we have $|\,m_n - m_o\,| < 1/(2k)$, and consequently, $f_n(x) \leqslant 1/(2k)$ on C_k. Now, we may appeal to Lemma (6.6) to conclude that $\{\,f_n\,\}_n$ has a subsequence, again denoted by $\{\,f_n\,\}_n$, and that there exists a measurable f_o such that

$$\int_{C_k} |\,f_n - f_o\,| \longrightarrow 0 \quad \text{for all } k \geqslant 2 \,.$$

It follows that for all $k \geqslant 2$,

$$\|\,f_n - f_o\,\|_1 \geqslant \int_{C_k} |\,f_n - \varphi\,| \longrightarrow \int_{C_k} |\,f_o - \varphi\,| \,,$$

and consequently,

(6.15) $$\liminf_{n \to \infty} \|\,f_n - \varphi\,\|_1 \geqslant \|\,f_o - \varphi\,\|_1 \,,$$

and $f_o \in L^1(\mathbb{R})$. Also, f_o is decreasing on A_k, and increasing on B_k, for each k, whence f_o is unimodal on the line. Since $\{\,f_n\,\}_n$ was a minimizing sequence, then (6.15) says that f_o is a solution to (6.13).

If $\{\,m_n\,\}_n$ contains no bounded subsequences, then we may select a subsequence that tends to $+\infty$ or to $-\infty$. Without loss of generality, assume that $m_n \longrightarrow \infty$. Similar to the above, we may extract a subsequence of $\{\,f_n\,\}_n$, again denoted by $\{\,f_n\,\}_n$, such that on each interval $(-k, k)$, the subsequence converges in $L^1(-k, k)$. Thus, there exists a measurable f_o such that for all k,

$$\int_{-k}^{k} |\,f_n - f_o\,| \longrightarrow 0 \,,$$

and it follows that f_o is an increasing function. Now, for all k,

$$\liminf_{n \to \infty} \|\,f_n - \varphi\,\|_1 \geqslant \liminf_{n \to \infty} \int_{-k}^{k} |\,f_n - \varphi\,| = \int_{-k}^{k} |\,f_o - \varphi\,| \,,$$

and so $f_o \in L^1(\mathbb{R})$, and f_o is the solution. As a matter of fact, since f_o is also increasing, we must have $f_o = 0$.

All of this proves the following theorem.

(6.16) THEOREM. *The unimodal minimum L^1 distance problem (6.13) has a solution for every $\varphi \in L^1(\mathbb{R})$.*

The final problem to be considered is the log-concave minimum distance problem. Since the log-concave sub-pdfs are a closed subset of the unimodal sub-pdfs, the existence follows easily.

(6.17) THEOREM. *The log-concave minimum L^1 distance problem*

(6.18)
$$\text{minimize} \quad \| f - \varphi \|_1$$
$$\text{subject to} \quad f \text{ s a log-concave sub-pdf },$$

has a solution for every $\varphi \in L^1(\mathbb{R})$.

(6.19) EXERCISE. Prove the theorem.

(6.20) EXERCISE. We have not required that the solutions to our estimation problems be densities, and so the solutions might be sub-pdfs. Repeat this section with the pdf constraint added.

EXERCISES : (6.4), (6.7), (6.11), (6.19), (6.20).

Appendices

A1

Some Data Sets

1. Introduction

We present some data sets used in the text. The major source of data sets is of course the Web. Besides this, SILVERMAN (1986) contains many interesting data sets, and SHEATHER (1992) contains five more. One workhorse data set is the Old Faithful geyser data, see §2. Another workhorse data set is the Buffalo snow fall data presented in §3. A related data set containing the monthly and annual rain fall and hours of sunshine for Vancouver, may be found in GLICK (1978). NOAA maintains past weather data for all major cities in the United States. The rubber data are presented in §4. The abrasion loss part of the data is remarkably similar to the Buffalo snow fall data. Data from a cloud seeding experiment are reproduced in §5. Oil field drilling data are reported in §6.

2. Old Faithfull geyser data

This data set records the time in seconds between eruptions of the Old Faithful geyser in Yellowstone National Park. See WEISBERG (1985).

4.37	3.43	4.62	4.57	4.50	4.33	1.67	1.83	4.07
3.87	4.25	1.97	1.85	4.10	2.93	4.60	4.13	4.13
4.00	1.68	4.50	3.52	3.70	4.58	1.67	1.83	3.95
4.03	3.92	2.331	4.00	3.80	1.90	4.00	4.65	4.10
3.50	3.68	3.92	3.70	3.43	3.58	1.80	4.20	2.72
4.08	3.10	4.35	3.72	4.00	3.73	4.42	3.93	4.58
2.25	4.03	3.83	4.25	2.27	3.73	1.90	4.33	1.90
4.70	1.77	1.88	3.58	4.40	1.82	4.63	1.83	4.50
1.73	4.08	4.60	3.80	4.05	4.63	2.93	4.53	1.95
4.93	1.75	1.80	3.77	4.25	3.50	3.50	2.03	4.83
1.73	3.20	4.73	3.75	3.33	4.00	1.97	4.18	4.12
4.62	1.85	1.77	2.50	2.00	3.67	4.28	4.43	

3. The Buffalo snow fall data

The Buffalo Snow fall data record the annual snow fall, measured in inches, in Buffalo, New York, from 1910 to 1972. This data set was analyzed by CARMICHAEL (1976) and by PARZEN (1979).

yr	snow	yr	snow	yr	snow	yr	snow	yr	snow	yr	snow
10	126.4	21	53.5	32	49.6	43	85.5	54	89.9	65	70.9
11	82.4	22	39.8	33	54.7	44	58.0	55	84.8	66	98.3
12	78.1	23	63.6	34	71.8	45	120.7	56	105.2	67	55.5
13	51.1	24	46.7	35	49.1	46	110.5	57	113.7	68	66.1
14	90.9	25	72.9	36	103.9	47	65.4	58	124.7	69	78.4
15	76.2	26	79.6	37	51.6	48	39.9	59	114.5	70	120.5
16	104.5	27	83.6	38	82.4	49	40.1	60	115.6	71	97.0
17	87.4	28	80.7	39	83.6	50	88.7	61	102.4	72	110.0
18	110.5	29	60.3	40	77.8	51	71.4	62	101.4		
19	25.0	30	79.0	41	79.3	52	83.0	63	89.8		
20	69.3	31	74.4	42	89.6	53	55.9	64	71.5		

4. The rubber abbrasion data

The Rubber data give the abbrasion loss, hardness, and tensile strength of 30 specimen of rubber. S = specimen, H = hardness (in degrees Shore), and A = abbrasion loss (in g/hp hour). See DAVIES and GOLDSMITH (1974). (We treat the abbrasion data as if they were iid.)

S	H	A	S	H	A	S	H	A
1	372	45	11	164	64	21	219	71
2	206	55	12	113	68	22	186	80
3	175	61	13	82	79	23	155	82
4	154	66	14	32	81	24	114	89
5	136	71	15	228	56	25	341	51
6	112	71	16	196	68	26	340	59
7	55	81	17	128	75	27	283	65
8	45	86	18	97	83	28	267	74
9	221	53	19	64	88	29	215	81
10	166	60	20	249	59	30	148	86

5. Cloud seeding data

The Cloud Seeding data set deals with a weather control experiment to test whether seeding of clouds with silver oxide has an effect on rainfall. The rainfall from seeded and unseeded (control) clouds is given in acre-feet. It was estimated using radar. See SIMPSON, OLSEN and EDEN (1975). Note that this is the only data set in this appendix with truly iid data.

Seeded Clouds				Unseeded Clouds			
2745.6	334.1	198.6	32.7	1202.6	147.8	36.6	17.3
1697.8	302.8	129.6	31.4	830.1	95.0	29.0	11.5
1656.0	274.7	119.0	17.5	372.4	87.0	28.6	4.9
978.0	274.7	118.3	7.7	345.5	81.2	26.3	4.9
703.4	255.0	115.3	4.1	321.2	68.5	26.1	1.0
489.1	242.5	92.4		244.3	47.3	24.4	
430.0	200.7	40.6		163.0	41.1	21.7	

6. Texo oil field data

This data set consists of the recorded sizes of oil fields in the Frio strand plane in the central coast of Texas, discovered after the first 318 drillings. In all, there were 695 drillings. See SCHUENEMEYER and DREW (1991). NAIR and WANG (1989) have a similar, but smaller, data set.

nr.	size	nr.	size	nr.	size	nr.	size	nr.	size	nr.	size
318	0.02	319	0.04	320	0.14	321	0.23	322	0.42	323	0.86
324	0.48	325	1.04	326	0.09	327	0.04	328	4.06	329	0.37
330	11.45	331	2.15	332	2.70	333	0.26	334	2.40	335	0.11
336	0.03	337	0.44	338	2.19	339	1.49	340	0.67	341	11.20
342	0.18	343	0.12	344	0.10	345	0.16	346	0.45	347	8.37
348	0.39	349	7.38	350	22.03	351	2.27	352	0.12	353	0.04
354	2.06	355	0.06	356	2.88	357	0.48	358	0.16	359	0.05
360	1.30	361	0.05	362	0.09	363	1.94	364	0.08	365	1.15
366	0.03	367	0.28	368	0.07	369	1.13	370	14.83	371	15.07
372	1.30	373	6.67	374	1.67	375	1.30	376	0.63	377	0.19
378	0.09	379	0.03	380	1.66	381	0.74	382	0.87	383	0.03
384	1.83	385	1.08	386	5.34	387	0.16	388	15.77	389	0.25
390	1.69	391	0.32	392	0.33	393	1.62	394	17.63	395	2.75
396	0.05	397	0.10	398	0.65	399	0.03	400	0.23	401	1.88
402	0.10	403	0.56	404	0.10	405	0.67	406	0.13	407	0.08
408	0.25	409	0.21	410	0.17	411	1.12	412	0.09	413	0.04
414	0.80	415	3.93	416	0.24	417	3.17	418	0.32	419	0.03
420	0.12	421	15.81	422	0.10	423	2.24	424	1.87	425	0.70
426	0.34	427	34.92	428	0.04	429	371.77	430	0.26	431	1.72
432	0.03	433	0.53	434	0.04	435	28.17	436	1.00	437	0.57
438	0.08	439	2.70	440	3.84	441	1.36	442	0.07	443	0.08
444	5.55	445	8.06	446	9.36	447	0.28	448	0.76	449	11.52

nr.	size	nr.	size	nr.	size	nr.	size	nr.	size	nr.	size
450	0.15	451	0.44	452	0.44	453	0.13	454	0.18	455	1.00
456	1.52	457	0.17	458	0.64	459	0.27	460	7.27	461	0.05
462	0.13	463	0.72	464	0.72	465	0.03	466	0.06	467	0.03
468	6.54	469	0.64	470	0.06	471	4.83	472	0.36	473	3.31
474	7.20	475	0.24	476	1.45	477	0.04	478	5.65	479	2.62
480	0.04	481	0.10	482	0.11	483	5.09	484	0.04	485	0.07
486	0.06	487	2.31	488	0.62	489	0.04	490	0.25	491	0.63
492	0.40	493	0.09	494	0.90	495	0.78	496	0.07	497	13.92
498	0.03	499	4.66	500	0.02	501	0.90	502	0.25	503	1.01
504	0.10	505	0.26	506	3.00	507	0.33	508	1.13	509	5.77
510	2.76	511	0.05	512	1.05	513	1.04	514	0.05	515	4.15
516	0.08	517	0.03	518	0.07	519	0.10	520	0.50	521	0.09
522	0.11	523	4.15	524	3.90	525	0.45	526	0.68	527	1.37
528	0.05	529	0.33	530	2.35	531	0.13	532	0.05	533	0.72
534	1.03	535	0.09	536	0.09	537	1.26	538	0.69	539	0.90
540	0.04	541	0.11	542	0.08	543	1.00	544	0.06	545	0.09
546	0.50	547	12.50	548	0.14	549	0.55	550	0.34	551	0.05
552	0.28	553	0.02	554	1.04	555	0.62	556	0.07	557	5.23
558	0.30	559	0.19	560	0.25	561	0.45	562	0.04	563	0.35
564	0.52	565	0.03	566	0.08	567	0.10	568	0.03	569	2.57
570	0.15	571	0.21	572	0.83	573	0.04	574	0.09	575	7.34
576	1.37	577	0.62	578	0.12	579	0.59	580	4.14	581	0.43
582	0.10	583	0.04	584	0.05	585	0.18	586	38.93	587	0.06
588	0.15	589	1.07	590	0.15	591	0.05	592	0.24	593	10.93
594	0.05	595	0.03	596	4.30	597	0.06	598	0.25	599	2.65
600	0.11	601	5.92	602	0.04	603	0.17	604	0.07	605	0.04
606	2.38	607	0.38	608	1.37	609	0.10	610	0.10	611	0.04
612	0.10	613	1.20	614	0.08	615	0.17	616	3.48	617	1.36
618	0.04	619	1.27	620	0.68	621	0.11	622	1.64	623	0.26
624	0.05	625	0.69	626	0.45	627	0.37	628	0.13	629	0.75
630	0.34	631	0.14	632	0.58	633	0.17	634	0.27	635	1.40
636	0.21	637	0.69	638	0.11	639	0.36	640	0.95	641	3.40
642	0.86	643	0.06	644	35.92	645	0.15	646	0.27	647	0.06
648	0.37	649	0.03	650	0.23	651	0.21	652	0.28	653	0.41
654	0.04	655	0.33	656	1.42	657	0.05	658	0.25	659	0.04
660	1.67	661	0.27	662	0.03	663	0.33	664	0.50	665	0.08
666	0.13	667	0.33	668	0.46	669	2.53	670	0.15	671	0.07
672	0.30	673	0.22	674	0.33	675	0.07	676	0.50	677	4.19
678	0.25	679	0.10	680	0.06	681	0.27	682	0.53	683	0.16
684	0.09	685	0.58	686	8.00	687	0.20	688	0.05	689	0.33
690	0.25	691	0.42	692	0.39	693	0.33	694	0.08	695	0.38

A2

The Fourier Transform

1. Introduction

In this appendix, we discuss the Fourier transform to prepare for its use in fractional integration by parts in Chapter 4 and for the study of Fourier deconvolution methods in Volume II. The Fourier transform of an integrable function f on the line is defined as

$$(1.1) \qquad \widehat{f}(\omega) = \int_{\mathbb{R}} f(x)\,e^{-2\pi i \omega x}\,dx\ , \quad \omega \in \mathbb{R}\ ,$$

but this definition needs to be extended, most notably, to square integrable functions. We also need to establish the inversion formula, as well as the interaction of the Fourier transform with scaling, convolutions, and derivatives. Note that scaling and convolution are basic operations in statistics and probability, cf. kernel density estimation and the classical proof of the Central Limit Theorem.

We begin with the Fourier transform of infinitely smooth functions and extend the results to L^1 and L^2 functions in §§ 3 and 4. In § 5, we consider some examples that were used in the fractional integration by parts of § 4.3. In § 6, we briefly discuss the Wiener-Lévy theorem, which is useful in the Fourier deconvolution method, see Volume II.

Our treatment of Fourier transform is quite standard and follows (largely) parts of DYM and McKEAN (1972), but it avoids complex integration. It is assumed that the reader is familiar with the basic theorems of Lebesgue integration (dominated convergence, Fubini, Fatou), see ROYDEN (1968) or WHEEDEN and ZYGMUND (1977).

2. Smooth functions

We begin with the easiest case, that of infinitely smooth functions that decay faster than any power of x at infinity. In particular, we define $C_{\downarrow}^{\infty}(\mathbb{R})$

to be the set of all functions ψ on $\mathbb{R}$ that satisfy

(2.1) ψ is infinitely many times continuously differentiable, and

(2.2) $\lim\limits_{|x|\to\infty} |x|^m \psi^{(k)}(x) = 0$, $m = 1, 2, \cdots$, $k = 0, 1, \cdots$,

where $\psi^{(k)}$ denotes the k-th derivative of ψ. The last property is equivalent to $|x|^m\psi^{(k)}(x) = \mathcal{O}(|x|^{-2})$, $|x| \to \infty$, for all m and k. This clearly implies that $|x|^m\psi^{(k)}(x)$ is bounded and integrable.

Since a function $\psi \in C_{\downarrow}^{\infty}(\mathbb{R})$ is certainly integrable, its Fourier transform is defined as

(2.3) $$\widehat{\psi}(\omega) = \int_{\mathbb{R}} \psi(x)\, e^{-2\pi i\omega x}\, dx , \quad \omega \in \mathbb{R} .$$

There are alternative definitions, which in our notation correspond to $\widehat{\psi}(\omega/2\pi)$ or $(2\pi)^{1/2}\,\widehat{\psi}(\omega/2\pi)$, but they are all "equivalent".

It is obvious that $\widehat{\psi}$ is bounded and continuous. but much more is true. First, there is the nearly trivial inequality

(2.4) $$\|\widehat{\psi}\|_{\infty} \leqslant \|\psi\|_{1} ,$$

which the reader should verify. Next, we have:

(2.5) THEOREM. *If $\psi \in C_{\downarrow}^{\infty}(\mathbb{R})$, then $\widehat{\psi} \in C_{\downarrow}^{\infty}(\mathbb{R})$.*

PROOF. The proof is broken up into a number of parts.

(2.6) LEMMA. *If $\psi \in C_{\downarrow}^{\infty}(\mathbb{R})$, then (in a somewhat loose notation)*

$$\{\,\widehat{\psi}\,\}^{(k)} = (\,\{-2\pi ix\}^k\,\psi)^{\wedge} ,$$

for $k = 1, 2, \cdots$.

PROOF. Begin with $k = 1$. Note that

$$\{\,\widehat{\psi}\,\}'(\omega) = \lim_{h\to 0} h^{-1}\{\,\widehat{\psi}(\omega + h) - \widehat{\psi}(\omega)\,\} ,$$

provided the limit exists. Now, observe that

$$h^{-1}\{\,\widehat{\psi}(\omega + h) - \widehat{\psi}(\omega)\,\} = \int_{\mathbb{R}} \frac{e^{-2\pi ihx} - 1}{h}\, \psi(x)\, e^{-2\pi i\omega x}\, dx ,$$

and that $h^{-1}|e^{-2\pi ihx} - 1| \leqslant 2\pi|x|$. Since $|x|\psi(x)$, $x \in \mathbb{R}$, is integrable, the dominated convergence theorem then gives (2.6). This proves the lemma for $k = 1$. By induction, the cases $k = 2, 3, \cdots$ follow.   Q.e.d.

The above shows that $\widehat{\psi}$ is infinitely many times (continuously) differentiable. The decay properties of $\widehat{\psi}^{(k)}$ go by induction as well. We have:

(2.7) DIFFERENTIATION THEOREM. *If* $\psi \in C_\downarrow^\infty(\mathbb{R})$, *then for* $m = 1, 2, \cdots$,

$$\{\psi^{(m)}\}^\wedge(\omega) = \{-2\pi i \omega\}^m \,\widehat{\psi}(\omega) , \quad \omega \in \mathbb{R} .$$

PROOF. For $m = 1$, this follows by integration by parts, and for the remaining m by induction.   Q.e.d.

Now, to the actual proof of Theorem (2.5). Let $\psi \in C_\downarrow^\infty(\mathbb{R})$. By Lemma (2.6), then $\widehat{\psi}$ is infinitely many times continuously differentiable. Also, since $(\psi^{(m)})^\wedge$ is bounded for each m, it follows that

$$\widehat{\psi}(\omega) = \mathcal{O}(|\omega|^{-m}) \quad \text{for } |\omega| \to \infty .$$

Now, fix $k \geqslant 1$, and apply this last result to the function φ, defined as $\varphi(x) = \{-2\pi i x\}^k \psi(x)$. Obviously, $\varphi \in C_\downarrow^\infty(\mathbb{R})$, and $\widehat{\varphi}(\omega)$ decays faster than any power of ω. But $\widehat{\varphi} = \{\widehat{\psi}\}^{(k)}$. Thus, $\widehat{\psi}$ satisfies (2.2) for (all values of) k, and (2.5) is proved.   Q.e.d.

A useful and important example is the Fourier transform of the normal density $\phi(x) = (2\pi)^{-1/2} \exp(-\frac{1}{2} x^2)$, which is

(2.8) $\widehat{\phi}(\omega) = \exp(-2(\pi\omega)^2) , \quad \omega \in \mathbb{R} .$

(2.9) EXERCISE. Verify (2.8), as follows. Let

$$I(\omega) = \int_{\mathbb{R}} \exp(-\tfrac{1}{2}x^2 - 2\pi i \omega x) \, dx , \quad \omega \in \mathbb{R} .$$

(a) Show that $I(\omega)$ is differentiable, with

$$I'(\omega) = \int_{\mathbb{R}} (-2\pi i x) \, \exp(-\tfrac{1}{2}x^2 - 2\pi i \omega x) \, dx,$$

and that $I'(\omega) + 4\pi^2\omega\, I(\omega) = 0$.
(b) Show that $I(0) = (2\pi)^{1/2}$.
(c) Solve the initial value problem for the differential equation.

The Fourier transform interacts nicely with scaling and convolutions.

(2.10) SCALING LEMMA. *If* $\psi \in C_\downarrow^\infty(\mathbb{R})$, *and* $\psi_h(x) = h^{-1} \psi(h^{-1} x)$, $x \in \mathbb{R}$, *then*

$$(\psi_h)^\wedge(\omega) = \widehat{\psi}(h\omega) , \quad \omega \in \mathbb{R} .$$

The Fourier transform owes its existence to convolutions. Recall that the convolution of $\varphi, \psi \in C_\downarrow^\infty(\mathbb{R})$ is defined as

(2.11) $\varphi * \psi(x) = \int_{\mathbb{R}} \varphi(x - y)\, \psi(y) \, dy , \quad x \in \mathbb{R} .$

(2.12) THE CONVOLUTION THEOREM. *If φ and ψ belong to $C_{\downarrow}^{\infty}(\mathbb{R})$, then $\varphi * \psi \in C_{\downarrow}^{\infty}(\mathbb{R})$ and*

$$(\varphi * \psi)^{\wedge}(\omega) = \widehat{\varphi}(\omega)\,\widehat{\psi}(\omega)\,, \quad \omega \in \mathbb{R}\,.$$

This is proved by changing the order of integration.

We next come to the inverse Fourier transform. The inverse Fourier transform of $\psi \in C_{\downarrow}^{\infty}(\mathbb{R})$ is defined as

(2.13)
$$\overset{\vee}{\psi}(x) = \int_{\mathbb{R}} \psi(\omega)\,e^{2\pi i \omega x}\,d\omega\,, \quad x \in \mathbb{R}\,.$$

Of course, for the name inverse Fourier transform to be justified, the following theorem must be proven.

(2.14) THEOREM. *For every $\psi \in C_{\downarrow}^{\infty}(\mathbb{R})$,*

$$(\widehat{\psi})^{\vee}(x) = \psi(x)\,, \quad x \in \mathbb{R}\,.$$

PROOF. The inversion formula is easily verified for the function ϕ of (2.8). So consider an arbitrary $\psi \in C_{\downarrow}^{\infty}(\mathbb{R})$. From Theorem (4.2.9), we know the approximation property $\phi_h * \psi - \psi \to 0$ in L^1 and L^{∞}, and from the convolution theorem that $(\phi_h * \psi)^{\wedge}(\omega) = \widehat{\phi}(h\omega)\,\widehat{\psi}(\omega)$ for all ω. Now, write

$$(\widehat{\psi})^{\vee} = \{(\phi_h * \psi)^{\wedge}\}^{\vee} - \{(\phi_h * \psi)^{\wedge} - \widehat{\psi}\}^{\vee}\,.$$

For the first term, we have

$$\{(\phi_h * \psi)^{\wedge}\}^{\vee}(x) = \int_{\mathbb{R}} e^{2\pi i \omega x}\,\widehat{\phi}(h\omega)\,\widehat{\psi}(\omega)\,d\omega$$

$$= \int_{\mathbb{R}} e^{2\pi i \omega x} \left\{ \int_{\mathbb{R}} \psi(y)\,e^{-2\pi i \omega y}\,dy \right\} \widehat{\phi}(h\omega)\,d\omega$$

$$= \int_{\mathbb{R}} \psi(y) \left\{ \int_{\mathbb{R}} e^{2\pi i \omega(x-y)}\,\widehat{\phi}(h\omega)\,d\omega \right\} dy\,,$$

the last equality by an appeal to Fubini's theorem. The appeal is justified since $\psi(y)\widehat{\phi}(h\omega)$ is an integrable function of $(y, \omega) \in \mathbb{R} \times \mathbb{R}$. But the last integral equals

$$\int_{\mathbb{R}} \psi(y)\,\phi_h(x - y)\,dy = \phi_h * \psi(x)\,,$$

and hence, $\{(\phi_h * \psi)^{\wedge}\}^{\vee} = \phi_h * \psi$.

For the second term, we note that

$$(\phi_h * \psi)^{\wedge}(\omega) - \psi^{\wedge}(\omega) = \left(1 - \exp(-2(\pi h \omega)^2)\right) \psi^{\wedge}(\omega),$$

so that

$$\| \{ (\phi_h * \psi)^\wedge - \psi^\wedge \}^\vee \|_\infty \leqslant \int_{\mathbb{R}} | 1 - \exp(-2(\pi h \omega)^2 | \, | \widehat{\psi}(\omega) | \, d\omega$$

$$\leqslant 2 \pi^2 h^2 \int_{\mathbb{R}} \omega^2 \, | \widehat{\psi}(\omega) | \, d\omega \ .$$

Since the integral converges, by Theorem (2.5), then

$$\| \{ (\phi_h * \psi)^\wedge - \psi^\wedge \}^\vee \|_\infty \longrightarrow 0 \ , \quad h \to 0 \ .$$

Thus, for $h \longrightarrow 0$,

$$\| (\widehat{\psi})^\vee - \psi \|_\infty \leqslant \| \{ (\psi - \phi_h * \psi)^\wedge \}^\vee \|_\infty + \| \{ (\phi_h * \psi)^\wedge - \widehat{\psi} \}^\vee \|_\infty$$
$$\longrightarrow 0 \ . \qquad\qquad\qquad \text{Q.e.d.}$$

To conclude the discussion of the Fourier transform on $C_\downarrow^\infty(\mathbb{R})$, we note the following. If $\varphi, \psi \in C_\downarrow^\infty(\mathbb{R})$, then by Fubini's theorem,

$$\int_{\mathbb{R}} \varphi(x) \overset{\vee}{\psi}(x) \, dx = \int_{\mathbb{R}} \overset{\vee}{\varphi}(\omega) \, \psi(\omega) \, d\omega \ .$$

So then

$$\int_{\mathbb{R}} \overline{\varphi(x)} \, \psi(x) \, dx = \int_{\mathbb{R}} \overline{\varphi(x)} \, (\widehat{\psi})^\vee(x) \, dx = \int_{\mathbb{R}} (\overline{\varphi})^\vee(\omega) \, \widehat{\psi}(\omega) \, d\omega \ .$$

But $(\overline{\varphi})^\vee(\omega) = \overline{\widehat{\varphi}(\omega)}$, so

$$\int_{\mathbb{R}} \overline{\varphi(x)} \, \psi(x) \, dx = \int_{\mathbb{R}} \overline{\widehat{\varphi}(\omega)} \, \widehat{\psi}(\omega) \, d\omega \ .$$

In particular, by taking $\varphi = \psi$, we have:

(2.15) PLANCHEREL'S FORMULA. For all $\psi \in C_\downarrow^\infty(\mathbb{R})$, we have

$$\| \widehat{\varphi} \|_2 = \| \varphi \|_2 \ .$$

3. Integrable functions

The definition (2.3) of the Fourier transform extends to integrable functions, as do some of its important properties.

If $\psi \in L^1(\mathbb{R})$, then its Fourier transform is still defined as

(3.1) $$\widehat{\psi}(\omega) = \int_{\mathbb{R}} \psi(x) \, e^{-2\pi i \omega x} \, dx \ , \quad \omega \in \mathbb{R} \ ,$$

and we have again that

(3.2) $$\| \widehat{\psi} \|_\infty \leqslant \| \psi \|_1 \ ,$$

Further, we note that $\widehat{\psi}$ is uniformly continuous on the line, since

$$|\widehat{\psi}(\omega + h) - \widehat{\psi}(\omega)| \leqslant \int_{\mathbb{R}} |\psi(x)||e^{-2\pi ihx} - 1| \, dx \, ,$$

and the right-hand side does not depend on ω. Moreover, it tends to 0 by the dominated convergence theorem.

The convolution of two functions $\varphi, \psi \in L^1(\mathbb{R})$ is still defined by (2.11), and we have:

(3.3) SCALING LEMMA. *If $\psi \in L^1(\mathbb{R})$, and $\psi_h(x) = h^{-1}\psi(h^{-1}x)$, $x \in \mathbb{R}$, for $h > 0$, then*

$$(\psi_h)^\wedge(\omega) = \widehat{\psi}(h\omega) \, , \quad \omega \in \mathbb{R} \, .$$

(3.4) CONVOLUTION THEOREM. *If $\varphi, \psi \in L^1(\mathbb{R})$, then $\varphi * \psi \in L^1(\mathbb{R})$ with*

$$\| \varphi * \psi \|_1 \leqslant \| \varphi \|_1 \| \psi \|_1$$

and

$$(\varphi * \psi)^\wedge(\omega) = \widehat{\varphi}(\omega) \, \widehat{\psi}(\omega) \, , \quad \omega \in \mathbb{R} \, .$$

(3.5) DIFFERENTIATION THEOREM. *If $\psi \in W^{1,1}(\mathbb{R})$, then*

$$(\psi')^\wedge(\omega) = 2\pi i\omega \, \widehat{\psi}(\omega) \, , \quad \omega \in \mathbb{R} \, .$$

There is also a need to consider distribution functions. If Ψ is bounded, increasing, and continuous from the right, then we define its norm as

$$(3.6) \qquad \| \Psi \|_{BV} = \int_{\mathbb{R}} d\Psi(x) \, ,$$

where the integral is in the sense of Lebesgue-Stieltjes. The denomination BV stands for bounded variation. It is obviously related to the total variation, see (1.3.14). Note that if Ψ is absolutely continuous with derivative ψ, then $\| \Psi \|_{BV} = \| \psi \|_1$.

The Fourier transform of a distribution function Ψ is defined as

$$(3.7) \qquad (d\Psi)^\wedge(\omega) = \int_{\mathbb{R}} e^{-2\pi i\omega x} \, d\Psi(x) \, , \quad \omega \in \mathbb{R} \, .$$

We have again the inequality

$$(3.8) \qquad \| (d\Psi)^\wedge \|_\infty \leqslant \| \Psi \|_{BV} \, .$$

If $\varphi \in L^1(\mathbb{R})$ and Ψ is a distribution function, then the convolution is defined as

$$(3.9) \qquad \varphi * d\Psi(x) = \int_{\mathbb{R}} \varphi(x - y) \, d\Psi(y) \, , \quad \text{a.e. } x \in \mathbb{R} \, ,$$

and the convolution theorem takes the form:

(3.10) CONVOLUTION THEOREM. *If $\varphi \in L^1(\mathbb{R})$ and Ψ is a distribution function, then $\varphi * d\Psi \in L^1(\mathbb{R})$, with*

$$\| \varphi * d\Psi \|_1 \leqslant \| \varphi \|_1 \| \Psi \|_{BV}$$

and

$$\{ \varphi * d\Psi \}^\wedge(\omega) = \widehat{\varphi}(\omega) \{ d\Psi \}^\wedge(\omega) , \quad \omega \in \mathbb{R} .$$

There is obvious difficulty in introducing the inverse Fourier transform, since the Fourier transform of an integrable function need not be integrable. But we quote, without proof,

(3.11) UNIQUENESS THEOREM. *If $\psi \in L^1(\mathbb{R})$ satisfies $\widehat{\psi} = 0$ a.e., then $\psi = 0$ almost everywhere.*

4. Square integrable functions

Due to the Plancherel formula (2.15), the Fourier transform is much nicer on $L^2(\mathbb{R})$ than on $L^1(\mathbb{R})$. Initially, this does not appear to be the case, since the definition (1.1) of the Fourier transform as an integral does not make sense on $L^2(\mathbb{R})$, at least not as is. However, properly interpreted, everything goes through.

We recall the following useful tool.

(4.1) YOUNG'S INEQUALITY. *If $\varphi \in L^1(\mathbb{R})$ and $\psi \in L^2(\mathbb{R})$, then*

$$\varphi * \psi \in L^2(\mathbb{R}) \quad \text{and} \quad \| \varphi * \psi \|_2 \leqslant \| \varphi \|_1 \| \psi \|_2 .$$

PROOF. This follows from a judicious use of Cauchy-Schwarz:

$$\left\{ \int_{\mathbb{R}} \varphi(x-y) \, \psi(y) \, dy \right\}^2 \leqslant \left\{ \int_{\mathbb{R}} |\varphi(x-y)| \, dy \right\} \left\{ \int_{\mathbb{R}} |\varphi(x-y)| \, | \psi(y) |^2 \, dy \right\} .$$

Note that the first integral on the right is just $\| \varphi \|_1$. Thus,

$$\| \varphi * \psi \|_2^2 \leqslant \| \varphi \|_1 \| \varphi * (\psi^2) \|_1 ,$$

and the Convolution Theorem (3.4) does the trick.   Q.e.d.

It is unfortunate that $L^2(\mathbb{R})$ functions need not be integrable, but we can approximate them by functions that are.

(4.2) APPROXIMATION THEOREM. *If ϕ is the standard normal density, then for all $\psi \in L^2(\mathbb{R})$,*

$$\| \phi_h * \psi - \psi \|_2 \longrightarrow 0 , \quad h \to 0 .$$

PROOF. Define $\mathbb{1}_h$ as $\mathbb{1}_h(x) = 1$ for $|x| \leqslant h^{1/2}$, and $= 0$ otherwise. Now, writing $\phi_h * \psi = (\phi_h - \phi_h \mathbb{1}_h) * \psi + (\phi_h \mathbb{1}_h) * \psi$ gives

$$\| \phi_h * \psi - \psi \|_2 \leqslant \| (\phi_h - \phi_h \mathbb{1}_h) * \psi \|_2 + \| (\phi_h \mathbb{1}_h) * \psi - \lambda \psi \|_2 + (1 - \lambda) \| \psi \|_2 \,,$$

where

$$\lambda = \int_{|y| < \sqrt{h}} \phi_h(y) \, dy = \int_{|y| < 1/\sqrt{h}} \phi(y) \, dy \longrightarrow 1 \,, \quad h \to 0 \,.$$

For the first term, we have by Young's inequality (4.1),

$$\| (\phi_h - \phi_h \mathbb{1}_h) * \psi \|_2 \leqslant \| \phi_h - \phi_h \mathbb{1}_h \|_1 \| \psi \|_2 \longrightarrow 0$$

for $h \to 0$, since

$$\| \phi_h - \phi_h \mathbb{1}_h \|_1 = \int_{|y| > \sqrt{h}} \phi_h(y) \, dy = \int_{|y| > 1/\sqrt{h}} \phi(y) \, dy \longrightarrow 0 \,, \quad h \to 0 \,.$$

For the second term, we have

$$\| (\phi_h \mathbb{1}_h) * \psi - \lambda \psi \|_2^2 =$$

$$= \int_{\mathbb{R}} \left\{ \int_{|y| < \sqrt{h}} \phi_h(y) \, (\psi(x - y) - \psi(x)) \, dy \right\}^2 dx$$

$$\leqslant \int_{\mathbb{R}} \left\{ \int_{|y| < \sqrt{h}} \phi_h(y) \, dy \right\} \left\{ \int_{|y| < \sqrt{h}} \phi_h(y) \, | \psi(x - y) - \psi(x) |^2 \, dy \right\} dx$$

$$\leqslant \lambda^2 \sup_{|y| < \sqrt{h}} \int_{\mathbb{R}} | \psi(x - y) - \psi(x) |^2 \, dx \,,$$

where on the second line, we used Cauchy-Schwarz as in the proof of Young's inequality, and in the penultimate line, we interchanged the order of integration. Now, the last integral tends to 0, by the continuity of translation in $L^2(\mathbb{R})$. This concludes the proof.   Q.e.d.

We are now ready for Fourier transforms of L^2 functions. Let $\psi \in L^2(\mathbb{R})$, and let the function $\mathbb{1}_n$ be defined as $\mathbb{1}_n(x) = 1$ for $|x| \leqslant n$, and $= 0$ otherwise, and define ψ_n by

(4.3) $\psi_n = \phi_h * (\psi \mathbb{1}_n) \,, \quad h \equiv 1/n \,,$

still with ϕ the standard normal. The Approximation Theorem implies $\psi_n \longrightarrow \psi$ in $L^2(\mathbb{R})$. Moreover, then $\psi_n \in C^\infty_\downarrow(\mathbb{R})$ for all n.

(4.4) EXERCISE. For $\psi \in L^2(\mathbb{R})$, let ψ_n be defined by (4.3). Show that $\psi_n \in C^\infty_\downarrow(\mathbb{R})$ for all n.

Since $\psi_n \in C^\infty_\downarrow(\mathbb{R})$, for all n, their Fourier transforms exist and belong to $C^\infty_\downarrow(\mathbb{R})$. In particular, then Plancherel's formula (2.15) applies and gives for all n, m,

(4.5) $$\| \widehat{\psi}_n - \widehat{\psi}_m \|_2 = \| \psi_n - \psi_m \|_2 \longrightarrow 0 \ ,$$

so that $\{\widehat{\psi}_n\}_n$ is a Cauchy sequence in $L^2(\mathbb{R})$. Its limit thus exists, and we define the Fourier transform of ψ as this limit:

(4.6) $$\widehat{\psi} = \lim_{n \to \infty} \widehat{\psi}_n \quad \text{in } L^?(\mathbb{R}) \ .$$

Note that we may not conclude that $\widehat{\psi}_n \longrightarrow \widehat{\psi}$ almost everywhere, but it will hold along a subsequence. The definition (4.6) is somewhat involved, because

$$\widehat{\psi}_n(\omega) = \widehat{\phi}(\omega/n) \int_{|x|<n} \psi(x) \, e^{-2\pi i \omega x} \, dx \ .$$

However, since $\psi \mathbb{1}_n \longrightarrow \psi$ in $L^2(\mathbb{R})$, then again Plancherel's formula (2.15) implies that $\| \{\psi \mathbb{1}_n\}^\wedge - \widehat{\psi} \|_2 = \| \psi \mathbb{1}_n - \psi \|_2 \longrightarrow 0$, so that

(4.7) $$\widehat{\psi}(\omega) = \lim_{n \to \infty} \int_{|x|<n} \psi(x) \, e^{-2\pi i \omega x} \, dx \quad \text{in the } L^2(\mathbb{R}) \text{ sense} \ .$$

HISTORICAL/HYSTERICAL COMMENT. WIENER (1930) invented the clever notation $\widehat{\psi} = \text{l.i.m.}_{n \to \infty} \widehat{\psi}$, for (4.7), where l.i.m. stands for "limit in mean", but it never caught on. Had he been a present day statistician he would have chosen LIM, and it would have been a smashing success.

With the Fourier transform defined for square integrable functions, we can now state some of its standard properties.

(4.8) PLANCHEREL'S FORMULA. *For all $\psi \in L^2(\mathbb{R})$, the Fourier transform $\widehat{\psi}$ lies in $L^2(\mathbb{R})$, and $\| \widehat{\psi} \|_2 = \| \psi \|_2$ for all $\psi \in L^2(\mathbb{R})$.*

The Plancherel formula shows that the Fourier transform is an *isometry* on $L^2(\mathbb{R})$. So the Fourier transform is really nice on $L^2(\mathbb{R})$, as is *mutatis mutandis* the inverse Fourier transform.

(4.9) INVERSION FORMULA. *If $\psi \in L^2(\mathbb{R})$, then $\widehat{\psi} \in L^2(\mathbb{R})$ and*

$$(\widehat{\psi})^\vee = \psi \quad \text{in } L^2(\mathbb{R}) \ .$$

(4.10) DIFFERENTIATION THEOREM. *If $\psi \in W^{1,2}(\mathbb{R})$, then*

$$(\psi')^\wedge(\omega) = 2\pi i \omega \, \widehat{\psi}(\omega) \ , \quad \text{a.e. } \omega \in \mathbb{R} \ .$$

(4.11) CONVOLUTION THEOREM. *If $\varphi \in L^1(\mathbb{R})$ and $\psi \in L^2(\mathbb{R})$, then*

$$\varphi * \psi \in L^2(\mathbb{R}) \quad \text{and} \quad (\varphi * \psi)^\wedge = \widehat{\varphi} \, \widehat{\psi} \ .$$

We finish this section with a brief discussion of Sobolev spaces. The Fourier transform is a very handy tool to introduce Sobolev spaces of fractional order, without having to talk about derivatives of fractional order. Let $s > 0$ be real, and define

(4.12) $W^{s,2}(\mathbb{R}) = \{ \varphi \in L^2(\mathbb{R}) : (1 + \omega^2)^{s/2} \, \widehat{\varphi}(\omega) \text{ is square integrable} \}$.

With the inner product

$$\langle \varphi, \psi \rangle_{W^{s,2}(\mathbb{R})} = \int_{\mathbb{R}} (1 + \omega^2)^s \, \widehat{\varphi}(\omega) \, \overline{\widehat{\psi}(\omega)} \, d\omega \, ,$$

and induced norm, $W^{s,2}(\mathbb{R})$ is a Hilbert space. Note that there are equivalent norms, e.g., with $(1 + \omega^2)^s$ replaced by $1 + |\omega|^{2s}$.

(4.14) INTERPOLATION INEQUALITY. *If* $0 \leqslant r < s < t$, *and* $\psi \in W^{t,2}(\mathbb{R})$, *then with* $\theta = (t - s)/(t - r)$,

$$\| \psi \|_{W^{s,2}(\mathbb{R})} \leqslant \| \psi \|_{W^{r,2}(\mathbb{R})}^{\theta} \, \| \psi \|_{W^{t,2}(\mathbb{R})}^{1-\theta} \, .$$

This follows from Hölder's inequality.

(4.15) EXERCISE. If s is a positive integer, show that the space $W^{s,2}(\mathbb{R})$ defined above coincides with the previous definition of (4.1.45).

5. Some examples

We consider three examples of interest. The first one arose in fractional integration by parts, see § 4.3. Let $\frac{1}{2} < \kappa \leqslant 1$, and set

(5.1) $\widehat{g}_{\kappa}(\omega) = \{ 1 + | 2\pi i \omega \, x |^{\kappa} \}^{-1} \, , \quad \omega \in \mathbb{R} \, .$

We show that the g_{κ} are pdfs that are bounded except near 0, and determine their asymptotic behavior near 0 and $\pm \infty$.

Since $\widehat{g}_{\kappa} \in L^2(\mathbb{R})$ for $\kappa > \frac{1}{2}$, it is indeed the Fourier transform of a function $g_{\kappa} \in L^2(\mathbb{R})$. We may then use the inverse Fourier transform to retrieve g_{κ}

$$g_{\kappa}(x) = \int_{\mathbb{R}} \{ 1 + | 2 \pi \omega |^{\kappa} \}^{-1} e^{2\pi i \omega x} \, d\omega$$

(5.2) $= 2 \int_0^{\infty} \{ 1 + | 2 \pi \omega |^{\kappa} \}^{-1} \cos 2\pi \omega x \, d\omega \, , \quad \text{a.e. } x \in \mathbb{R} \, .$

Note that integration by parts shows that the integral in (5.2) converges as an improper Riemann integral, so there is no (explicit) need for the definition (4.7).

(5.3) LEMMA. g_{κ} *is a symmetric pdf on the line for* $\frac{1}{2} < \kappa \leqslant 1$.

PROOF. The symmetry is obvious. The remainder of the proof boils down to showing that g_κ is nonnegative. Once this has been established, then $\widehat{g}_\kappa(0) = 1$ implies that g_κ is a pdf.

To show the nonnegativity, let $x > 0$, and write (5.2) as

$$
g_\kappa(x) - 2 \sum_{n=0}^{\infty} \int_{n/x}^{(n+1)/x} \{ 1 + |2\pi w|^\kappa \}^{-1} \cos(2\pi w x)\, dw
$$

$$
= \frac{1}{\pi x} \sum_{n=0}^{\infty} \int_0^{2\pi} \{ 1 + |(w+n)/x|^\kappa \}^{-1} \cos(w)\, dw \ .
$$

The nonnegativity of each of these integrals then follows from the following well-known simple lemma.   Q.e.d.

(5.4) LEMMA. *Suppose ψ has the following properties.*
(a) *ψ is bounded on $[\, 0\, , 2\pi\,]$.*
(b) *ψ is twice continuously differentiable on the half-open interval $(\, 0\, , 2\pi\,]$.*
(c) *$\psi^{(m)}(x) = o(\, x^{-m}\,)$ as $x \to 0$, for $m = 1, 2$.*
If $\psi^{(2)} \geqslant 0$ on $(\, 0\, , 2\pi\,)$, then

$$
\int_0^{2\pi} \psi(w)\, \cos(w)\, dw \geqslant 0 \ .
$$

PROOF. The proof follows after integrating by parts twice. The conditions on ψ are sufficient to make the boundary terms vanish. The result is that

$$
\int_0^{2\pi} \psi(w)\, \cos(w)\, dw = \int_0^{2\pi} \psi''(w)\, \{\, 1 - \cos(w)\, \}\, dw \geqslant 0 \ . \qquad \text{Q.e.d.}
$$

Next, we consider the asymptotic behavior of $g_\kappa(x)$. From the representation (5.2), we have for $x > 0$,

$$
g_\kappa(x) = \frac{x^{\kappa-1}}{\pi} \int_0^{\infty} (\, x^\kappa + w^\kappa\,)^{-1} \cos w\, dw \ ,
$$

and by integration by parts, then

$$
g_\kappa(x) = \frac{\kappa\, x^{\kappa-1}}{\pi} \int_0^{\infty} \frac{w^{\kappa-1} \sin w}{(\, x^\kappa + w^\kappa\,)^2}\, dw \ .
$$

The integrand is bounded by $|\sin w|/w^{\kappa+1}$, which is integrable on $(0, \infty)$ if $\frac{1}{2} < \kappa < 1$. So, by the dominated convergence theorem,

$$
(5.5) \qquad x^{-\kappa+1}\, g_\kappa(x) \longrightarrow c_\kappa \overset{\text{def}}{=} \frac{\kappa}{\pi} \int_0^{\infty} w^{-\kappa-1} \sin w\, dw \ , \qquad x \to 0 \ .
$$

What happens for $x \to \infty$? One more integration by parts yields

$$
g_\kappa(x) = \frac{\kappa\, x^{\kappa-1}}{\pi} \int_0^{\infty} (\cos w - 1)\, M(x, w)\, dw \ ,
$$

where

$$M(x,\omega) = \frac{d}{d\omega} \left\{ \frac{\omega^{\kappa-1}}{(x^{\kappa} + \omega^{\kappa})^2} \right\} .$$

One verifies that

$$M(x,\omega) = \frac{(\kappa-1)\,\omega^{\kappa-2}}{(x^{\kappa} + \omega^{\kappa})^2} - \frac{2\,\kappa\,\omega^{2\kappa-2}}{x^{\kappa} + \omega^{\kappa}} \cdot \frac{1}{(x^{\kappa} + \omega^{\kappa})^2} ,$$

and we see that for $x > 1$, say, $M(x,\omega)$ is dominated in absolute value by

$$\frac{|\kappa-1|\,\omega^{\kappa-2}}{x^{2\kappa}} + \frac{2\kappa\,\omega^{2\kappa-2}}{\omega^{\kappa} \cdot x^{2\kappa}} = (\kappa+1)\,\omega^{\kappa-2}x^{-2\kappa} .$$

Also, it is easily verified that

$$\lim_{x\to\infty} x^{2\kappa}\, M(x,\omega) = (\kappa-1)\,\omega^{\kappa-2} , \quad \text{for all } \omega > 0 .$$

Since $\omega^{\kappa-2}(\cos\omega - 1)$ is integrable over $(0,\infty)$, it follows again from the dominated convergence theorem that

$$\lim_{x\to\infty} \int_0^{\infty} (\cos\omega - 1)\, x^{2\kappa}\, M(x,\omega)\, d\omega = \int_0^{\infty} (\kappa-1)\,\omega^{\kappa-2}(\cos\omega - 1)\, d\omega .$$

This proves the second part of the asymptotic behavior of g_{κ}, viz.

(5.6) $$g_{\kappa}(x) \sim \gamma_{\kappa}\, x^{-\kappa-1} , \quad x \to \infty ,$$

for a suitable constant γ_{κ}. We summarize the results.

(5.7) LEMMA. *For $\frac{1}{2} < \kappa \leqslant 1$, the functions g_{κ} are symmetric pdfs, and for suitable constants c_{κ} and γ_{κ},*

$$g_{\kappa}(x) \sim c_{\kappa}\,|x|^{\kappa-1} , \quad |x| \longrightarrow 0 ,$$
$$g_{\kappa}(x) \sim \gamma_{\kappa}\,|x|^{-\kappa-1} , \quad |x| \longrightarrow \infty .$$

Lemma (5.7) actually also holds for $0 < \kappa \leqslant \frac{1}{2}$, but the difficulty is verifying the inversion formula (5.2).

The second example is somewhat similar to example (5.1). Let $\mathfrak{B}_{(\kappa)}$ be defined by means of its Fourier transform as

$$\widehat{\mathfrak{B}_{(\kappa)}}(\omega) = \{ 1 + (2\pi\omega)^2 \}^{-\kappa/2} , \quad \omega \in \mathbb{R} .$$

One verifies that $\kappa = 2$ gives the two-sided exponential distribution. In §§ 4.5 and 4.6, we denoted it as just $\mathfrak{B}$. For reasons to do with the roughness penalization of GOOD (1971), we are especially interested in the case $\kappa = 1$, and denote $\mathfrak{B}_{(1)}$ as $\mathfrak{C}$. So $\mathfrak{C}$ is defined via

(5.8) $$\widehat{\mathfrak{C}}(\omega) = \{ 1 + (2\pi\omega)^2 \}^{-1/2} , \quad \omega \in \mathbb{R} .$$

So, formally, if $\psi \in L^2(\mathbb{R})$, and φ solves the boundary value problem

(5.9)
$$-\varphi'' + \varphi = \psi \qquad \text{on } \mathbb{R} ,$$
$$\varphi \longrightarrow 0 \qquad \text{at } \pm\infty ,$$

then

(5.10)
$$\| \varphi \|_2^2 + \| \varphi' \|_2^2 = \| \mathfrak{C}\psi \|_2^2 .$$

(5.11) EXERCISE. Verify (5.10).

As in the previous example, we show that $\mathfrak{C}$ is a symmetric pdf, determine its (exact) asymptotic behavior near 0, and show that it has exponential decay at ∞. Note that apart from a constant factor depending on κ, the function $|x|^{-(\kappa-1)/2} \mathfrak{B}_{(\kappa)}(x)$ is equal to the modified Bessel function of the second kind $K_\nu(x)$, where $\nu = \frac{1}{2}(\kappa - 1)$, and the Fourier inversion formula is known as Basset's formula, see WATSON (1944), Chapter VI, § 6.16.

First, let us consider the asymptotic behavior near 0. The symmetry about 0 of $\mathfrak{C}$ is obvious, so we only consider $\mathfrak{C}(x)$ for $x > 0$. Manipulations that resemble those of Example (5.1) show that

$$\mathfrak{C}(x) = \frac{1}{\pi} \int_0^\infty \frac{\cos\omega}{\sqrt{x^2 + \omega^2}} \, d\omega .$$

The improper integral is troublesome both at $\omega \to 0$ and at $\omega \to \infty$. So we split it. For the first part, integration by parts gives

$$\int_{\pi/2}^\infty \frac{\cos\omega}{\sqrt{x^2 + \omega^2}} \, d\omega = \int_{\pi/2}^\infty \frac{\omega \sin\omega}{(x^2 + \omega^2)^{3/2}} \, d\omega .$$

Now, for all $x > 0$, the integrand is dominated in absolute value by $|\omega|^{-2}$, which is integrable on $(\pi/2, \infty)$. So by dominated convergence,

$$\int_{\pi/2}^\infty \frac{\cos\omega}{\sqrt{x^2 + \omega^2}} \, d\omega \longrightarrow \int_{\pi/2}^\infty \frac{\sin\omega}{\omega^2} \, d\omega , \qquad x \to 0 .$$

Thus, this part converges to a constant. For the integral from 0 to $\pi/2$, we again do integration by parts, using the fact that the derivative of $\log(\omega + \sqrt{x^2 + \omega^2})$ is $1/\sqrt{x^2 + \omega^2}$. This gives

$$\int_0^{\pi/2} \frac{\cos\omega}{\sqrt{x^2 + \omega^2}} \, d\omega = -\log x + \int_0^{\pi/2} \sin\omega \, \log(\omega + \sqrt{x^2 + \omega^2}) \, d\omega .$$

Again, a dominated convergence argument shows that the last integral converges to a constant as $x \to 0$. Putting the integral back together again shows that

(5.12)
$$\mathfrak{C}(x) = -\frac{1}{\pi} \log|x| + \text{Const} + o(1) , \qquad x \to 0 .$$

The case $x \to \infty$ is difficult if one wants the exact asymptotic behavior, but showing that it decays faster than any power of x is just integration by parts as many times as needed. Starting with the Fourier inversion formula, integrating by parts n times gives

$$\mathfrak{C}(x) = \frac{(-1)^n}{(2\pi i x)^n} \int_{\mathbb{R}} e^{2\pi i \omega x} \frac{d^n}{d\omega^n} \frac{1}{\sqrt{1+\omega^2}} \, d\omega \ ,$$

and it is not too hard to show that

$$\left| \frac{d^n}{d\omega^n} \frac{1}{\sqrt{1+\omega^2}} \right|$$

is integrable over $\mathbb{R}$. This shows that

(5.13) $\mathfrak{C}(x) = \mathcal{O}\left(|x|^{-n} \right) , \quad |x| \longrightarrow \infty , \quad$ for all $n > 0 .$

But see (5.15) below!

To show that $\mathfrak{C}$ is a pdf is not as easy. Again, it suffices to demonstrate that $\mathfrak{C}(x) > 0$ for all x, since $\mathfrak{C}^{\wedge}(0) = 1$. Showing the positivity requires a "better" representation for $\mathfrak{C}$. It starts with the following elementary integral:

$$\frac{1}{\sqrt{1+(2\pi\omega)^2}} = \frac{2}{\pi} \int_0^\infty \frac{dt}{1+(2\pi\omega)^2 + t^2} \ ,$$

the integrand of which may be rewritten as

$$(1+t^2)^{-1} \left[1 + \frac{(2\pi\omega)^2}{1+t^2} \right]^{-1} .$$

For fixed $t > 0$, the last factor is the Fourier transform of a scaled version of the two-sided exponential, viz. $\psi_t(x) = \frac{1}{2}\sqrt{1+t^2} \exp(-|x|\sqrt{1+t^2})$, see the Scaling Lemma (3.3). Obviously, ψ_t is a nonnegative function. Now, consider

$$\mathfrak{C}_T = \frac{2}{\pi} \int_0^T (1+t^2)^{-1} \psi_t(x) \, dt \ , \quad x \in \mathbb{R} .$$

Note that $\mathfrak{C}_T(x)$, is increasing in T for each x, and that its Fourier transform is what it should be, viz.

$$(\mathfrak{C}_T)^{\wedge}(\omega) = \frac{2}{\pi} \int_0^T (1+\omega^2 + t^2)^{-1} \, dt \ ,$$

and it follows after taking limits in $L^2(\mathbb{R})$ that

(5.14) $\mathfrak{C}(x) = \dfrac{1}{\pi} \displaystyle\int_0^\infty \dfrac{\exp(-|x|\sqrt{1+t^2})}{\sqrt{1+t^2}} \, dt \ , \quad x \in \mathbb{R} .$

The positivity of $\mathfrak{C}(x)$ is now obvious, and so $\mathfrak{C}$ is a pdf. As an added reward for clean living, it is now easy to show that

(5.15) $e^{|x|} \mathfrak{C}(x) \longrightarrow 0 , \quad |x| \to \infty ,$

as follows. In (5.14), multiply both sides by $e^{|x|}$. Note that for each $x \neq 0$, the modified integrand

$$\frac{\exp(-|x| \{ \sqrt{1+t^2} - 1 \})}{\sqrt{1+t^2}}$$

belongs to $L^1(0, \infty)$, is positive, and pointwise decreases to 0 as $|x| \to \infty$. Thus, (5.15) follows by the Monotone Convergence Theorem. The exact asymptotic behavior is

$$\mathfrak{C}(x) \sim \text{const} \cdot |x|^{-1/2} e^{-|x|} , \quad |x| \to \infty ,$$

which may be found in WATSON (1944), but never mind. We summarize the results for this example.

(5.16) LEMMA. *The function $\mathfrak{C}$ defined by (5.8) is a symmetric pdf and satisfies*

$$\mathfrak{C}(x) \sim -\tfrac{1}{\pi} \log |x| + \text{Const} + o(1) , \quad |x| \to 0 ,$$
$$\mathfrak{C}(x) = o\big(\exp(-|x|)\big) , \quad |x| \to \infty .$$

A last word concerning $\mathfrak{C}$ being a pdf. TEICHROW (1957) observes that for $0 < \alpha < 1$,

$$\{ 1 + (2\pi\omega)^2 \}^{-\alpha} , \quad -\infty < \omega < \infty$$

is the Fourier transform of the density of a random variable whose distribution, conditioned on the variance σ^2, is a Gaussian with mean 0 and variance σ^2, where σ^2 is itself a random variable with a Gamma distribution. The source of this reference is DEVROYE (1986).

Before considering the last example, we make a small detour via the Hilbert transform. The Hilbert transform is informally defined via Fourier transforms as

$$(5.17) \qquad (\mathcal{H} \psi)^{\wedge}(\omega) = -i \, \text{sign}(\omega) \, \widehat{\psi}(\omega) , \quad \omega \in \mathbb{R} .$$

We discuss the L^2 setting and, briefly, the situation on L^1.

It is clear that $\mathcal{H} : L^2(\mathbb{R}) \longrightarrow L^2(\mathbb{R})$ is a bounded linear operator, and that its inverse is $-\mathcal{H}$. Moreover,

$$(5.18) \qquad \| \mathcal{H} \psi \|_2 = \| \psi \|_2 \quad \text{for all } \psi \in L^2(\mathbb{R}) .$$

With some difficulty, the Hilbert transform may be represented as an integral. It turns out to be easier to start at the end and to work our way back.

(5.19) THEOREM. *Let $\psi \in L^2(\mathbb{R})$. Then,*

$$\mathcal{H} \psi(x) = \lim_{n \to \infty} -\frac{1}{\pi} \int_{|x-y|>1/n} \frac{\psi(y)}{x-y} \, dy$$

defines a.e. a function in $L^2(\mathbb{R})$ whose Fourier transform equals

$$(\mathcal{H}\,\psi)^\wedge(\omega) = -i\,\text{sign}(\omega)\,\widehat{\psi}(\omega)\,, \quad \omega \in \mathbb{R}\,.$$

$\mathcal{H}\,\psi$ *is called the Hilbert transform of* ψ.

PROOF. For $n,\, m \in \mathbb{N}$, let

$$g_{n,m}(x) = -(\pi x)^{-1}\,\mathbb{1}(\tfrac{1}{n} < |x| < m)\,, \quad x \in \mathbb{R}\,.$$

Then, $g_{n,m} \in L^1(\mathbb{R})$, and so the Convolution Theorem (4.11) applies to $g_{n,m} * \psi$ for $\psi \in L^2(\mathbb{R})$. Consequently,

$$(\,g_{n,m} * \psi)^\wedge(\omega) = c_{n,m}(\omega)\,\widehat{\psi}(\omega)\,, \quad \omega \in \mathbb{R}\,,$$

with

$$c_{n,m}(\omega) = \frac{2}{\pi i}\int_{1/n}^{m}\frac{\sin(2\pi i\omega x)}{x}\,dx\,.$$

One verifies that

$$\lim_{\substack{n\to\infty \\ m\to\infty}} c_{n,m}(\omega) = -i\,\text{sign}(\omega) \quad \text{for all } \omega \in \mathbb{R}\,,$$

and so $\|\,g_{n,m} * \psi - \mathcal{H}\,\psi\,\|_2 \longrightarrow 0$. Now, for a.e. $x \in \mathbb{R}$ and uniformly in n,

$$\lim_{m\to\infty} g_{n,m} * \psi(x) = h_n * \psi(x)\,,$$

where $h_n(x) = -(\pi x)^{-1}\,\mathbb{1}(|x| > 1/n)$. Thus, by Fatou's Lemma, for all n,

$$\|\,h_n * \psi - \mathcal{H}\,\psi\,\|_2 \leqslant \liminf_{m\to\infty} \|\,g_{n,m} * \psi - \mathcal{H}\,\psi\,\|_2\,,$$

and the theorem follows.   Q.e.d.

The Hilbert transform is not so nice on $L^1(\mathbb{R})$. It may be proven that for $\psi \in L^1(\mathbb{R})$,

$$(5.20) \qquad \mathcal{H}\,\psi(x) = \lim_{n\to\infty} -\frac{1}{\pi}\int_{|x-y|>1/n}\frac{\psi(y)}{x-y}\,dy$$

again, almost everywhere defines a measurable function. However, $\mathcal{H}\,\psi$ need not be in $L^1(\mathbb{R})$, as the following example shows, see, e.g., TITCH-MARSH (1975). Let $\psi(x) = (x\,(\log x)^2\,)^{-1}$ for $0 < x < \tfrac{1}{2}$, and $= 0$ otherwise. Obviously, $\psi \in L^1(\mathbb{R})$. Now, for $x < 0$,

$$-\frac{1}{\pi}\lim_{n\to\infty}\int_{|x-y|>1/n}\frac{\psi(y)}{x-y}\,dy = \frac{1}{\pi}\int_0^{\frac{1}{2}}\frac{dy}{(y-x)\,y\,(\log y)^2}\,.$$

For $-\tfrac{1}{2} < x < 0$, the last integrand dominates

$$\frac{1}{\pi}\int_0^{-x}\frac{dy}{(y-x)\,y\,(\log y)^2}\,,$$

which in turn dominates

$$\frac{1}{2\pi x} \int_0^{-x} \frac{dy}{y \, (\log y)^2} = \frac{1}{2\pi |\, x \, \log(-x)\,|} \; .$$

Consequently, $\mathcal{H}\,\psi \notin L^1(\mathbb{R})$. We have proven:

(5.21) THEOREM. $\mathcal{H} : L^1(\mathbb{R}) \longrightarrow L^1(\mathbb{R})$ is an unbounded operator.

The last example is peanuts now. Let $0 < \kappa < 1$. We want to make sense of the operator R_κ informally defined via Fourier transforms as

$$(5.22) \qquad\qquad (R_\kappa \, \psi)^{\wedge}(\omega) = |\, 2\pi\omega\,|^\kappa \, \widehat{\psi}(\omega) \,, \quad \omega \in \mathbb{R} \, .$$

The symbol R stands for M. RIESZ, who considered the case $-1 < \kappa < 0$.

The L^2 setting is fairly straightforward, especially with the use of the Sobolev spaces of fractional order of (4.12). However, the interest is in the L^1 setting. Here, it is easier to first do some rough calculations and then to verify that they are correct. To begin, we may write

$$|\, 2\pi\omega\,|^\kappa \widehat{\psi}(\omega) = \big\{ -i\,\mathrm{sign}(\omega)\,|\, 2\pi\omega\,|^{\kappa-1} \big\} \big\{ 2\pi i\omega \, \widehat{\psi}(\omega) \big\} \,, \quad \omega \in \mathbb{R} \, .$$

The last expression in curly brackets equals $(\psi')^{\wedge}(\omega)$ if $\psi \in W^{1,1}(\mathbb{R})$. This is fair and square. Now, keeping in mind the Convolution Theorem (3.4), we hope that the expression in the first pair of curly brackets is the Fourier transform of something. Performing the inverse Fourier transform accordingly gives the expression

$$-\int_{\mathbb{R}} \big\{ i\,\mathrm{sign}(\omega)\,|\, 2\pi\omega\,|^{\kappa-1} \big\}\, e^{2\pi i\omega x}\, d\omega$$

$$= \frac{\mathrm{sign}(x)}{\pi\,|\, x\,|^\kappa} \int_0^\infty \varpi^{\kappa-1} \, \sin(\varpi)\, d\varpi$$

$$= c_\kappa \, \mathrm{sign}(x)\,|\, x\,|^{-\kappa} \,, \quad \text{for all } x \neq 0 \,,$$

with c_κ an appropriate constant, viz. $c_\kappa = \pi^{-1}\,\sin(\pi\kappa/2)\,\Gamma(\kappa)$, with Γ the Gamma function. So, formally ($\equiv$ informally), we are lead to the following result.

(5.23) THEOREM. Let $0 < \kappa < 1$. If $\psi \in W^{1,1}(\mathbb{R})$, then its Riesz transform $R_\kappa\psi$, defined as

$$R_\kappa\psi(x) = c_\kappa \int_{\mathbb{R}} \mathrm{sign}(y)\,|\, y\,|^{-\kappa}\, \psi'(x - y)\, dy \,, \quad \text{a.e. } x \in \mathbb{R} \,,$$

is an element of $L^1(\mathbb{R})$, and its Fourier transform satisfies

$$(R_\kappa\psi)^{\wedge}(\omega) = |\, 2\pi\omega\,|^\kappa \, \widehat{\psi}(\omega) \,, \quad \text{a.e. } \omega \in \mathbb{R} \, .$$

PROOF. We first show that $R_\kappa \psi \in L^1(\mathbb{R})$. Note that $R_\kappa \psi$ may be written as

$$R_\kappa \psi(x) = c_\kappa \int_0^\infty |y|^{-\kappa} \{ \psi'(x-y) - \psi'(x+y) \} \, dy , \quad \text{a.e. } x \in \mathbb{R} .$$

Integration by parts shows that this equals

$$(5.24) \quad -c_\kappa |y|^{-\kappa} \{ \psi(x+y) - 2\psi(x) + \psi(x-y) \} \Big|_0^\infty$$

$$+ \kappa c_\kappa \int_0^\infty |y|^{-\kappa-1} \{ \psi(x+y) - 2\psi(x) + \psi(x-y) \} \, dy .$$

To handle the boundary terms, we note the following factoid, which is the Fundamental Theorem of Calculus, with a change of variables.

(5.25) LEMMA. *If $\psi \in W^{1,1}(\mathbb{R})$, then for a.e. $x, y \in \mathbb{R}$,*

$$\psi(x+y) - \psi(x) = y \int_0^1 \psi'(x+yt) \, dt .$$

(5.26) EXERCISE. Prove the lemma.

Observing that $\psi' \in L^1(\mathbb{R})$ implies

$$\int_0^1 \psi'(x+yt) \, dt \longrightarrow \psi'(x) \quad \text{as } y \to 0 \quad \text{for a.e. } x \in \mathbb{R} ,$$

Lemma (5.25) shows that in (5.24) the boundary term at $y = 0$ vanishes. The boundary term at $y = \infty$ vanishes also, by the boundedness of ψ on the line. So we have the representation

$$(5.27) \qquad R_\kappa \psi(x) = \kappa c_\kappa \int_0^\infty \frac{\psi(x+y) - 2\psi(x) + \psi(x-y)}{|y|^{\kappa+1}} \, dy .$$

It then follows immediately that

$$(5.28) \qquad \| R_\kappa \psi \|_1 \leqslant 2 \kappa c_\kappa \| \psi \|_{V^{\kappa,1}(\mathbb{R})} ,$$

where

$$(5.29) \qquad \| \psi \|_{V^{\kappa,1}(\mathbb{R})} = \int_{\mathbb{R} \times \mathbb{R}} \frac{| \psi(x+y) - \psi(x) |}{|y|^{\kappa+1}} \, dx \, dy .$$

This is the norm of the Sobolev space of fractional order $V^{\kappa,1}(\mathbb{R})$, see ADAMS (1975). Below, in Remark (5.31), we explain the reason for the notation $V^{\kappa,1}(\mathbb{R})$ rather than $W^{\kappa,1}(\mathbb{R})$. In Lemma (5.30), we demonstrate that $W^{1,1}(\mathbb{R}) \subset V^{\kappa,1}(\mathbb{R})$, and so $R_\kappa \psi \in L^1(\mathbb{R})$ whenever $\psi \in W^{1,1}(\mathbb{R})$.

Since $R_\kappa \psi \in L^1(\mathbb{R})$, its Fourier transform is defined pointwise in the usual way, see (3.1). Using the representation (5.27) for $R_\kappa \psi$, we may

References

Abou-Jaoudé, S. (1977), *La convergence L_1 et L_∞ de certains estimateurs d'une densité de probabilité*, Thèse de Doctorat d'État, Université Paris VI, Paris.

Abramson, I.S. (1982), *On bandwidth variation in kernel estimates–a square root law*, Ann. Statist. 10, 1217–1223.

Achieser, N.I. (1956), *Theory of approximation*, Frederic Ungar, New York (Reprinted: Dover, New York, 1993).

Acusta, A.P. (1998), *Nonparametric density estimation with randomly censored data*, Ph.D. Disseration, Univeristy of Delaware.

Adams, R.A. (1975), *Sobolev spaces*, Academic Press, New York.

Ahlberg, J.H., Nilson, E.N., Walsh, J.L. (1967), *The theory of splines and their applications*, Academic Press, New York.

Ahmad, I.A., Lin, P.E. (1976), *A nonparametric estimation of the entropy for absolutely continuous distributions*, IEEE Trans. Information Theory 22, 372–375.

Akaike, H. (1954), *An approximation to the density function*, Ann. Inst. Statist. Math. 6, 127–132.

Akhiezer, N.I., Glazman, I.M. (1981), *Theory of linear operators in Hilbert space*, Pitman, Boston (Reprinted: Dover, New York, 1996).

Arcangeli, R. (1966), *Pseudo-solution de l'equation $Ax = y$*, C.R. Acad. Sci. Paris Sér. A-B 263, A282–A285.

Aronszajn, N. (1950), *Theory of reproducing kernels*, Trans. Amer. Math. Soc. 68, 337–404.

Barlow, R.E., Bartholomew, D.J., Bremner, J.M., Brunk, H.D. (1972), *Statistical inference under order restrictions*, John Wiley and Sons, New York.

Barndorff-Nielsen, O. (1978), *Information and exponential families in statistical theory*, John Wiley and Sons, New York.

Barrodale, I., Roberts, F.D.K. (1978), *An efficient algorithm for discrete ℓ_1 linear approximation with linear constraints*, SIAM J. Numer. Anal. 15, 603–611.

Barron, A.R., Birgé, L., Massart, P. (1999), *Risk bounds for model selection via penalization*, Probab. Theory Related Fields 113, 301–413.

Barron, A.R., Sheu, C.-H., (1991), *Approximation of density functions by sequences of exponential families*, Ann. Statist. 19, 1347–1369.

Bartlett, M.S. (1963), *Statistical estimation of density functions*, Sankhya A 25, 245–254.

Beirlant, J., Dudewicz, E., Györfi, L., van der Meulen, E.C. (1997), *Nonparametric entropy estimation: an overview*, Int. J. Math. Stat. Sci. 6, 17–39.

Beirlant, J., Mason, D.M. (1995), *On the asymptotic normality of L_p norms of empirical functionals*, Math. Meth. Statist. 4, 1–19.

Beran, R. (1977), *Minimum Hellinger distance estimates for parametric models*, Ann. Statist. 5, 445–463.

Berkson, J. (1980), *Minimum chi-square, not maximum likelihood (with discussion)*, Ann. Statist. 8, 457–487.

Berlinet, A. (1993), *Hierarchies of higher order kernels*, Probab. Theory Related Fields 94, 489–504.

Berlinet, A., Devroye, L. (1994), *A comparison of kernel density estimates*, Publ. Inst. Stat. Univ. Paris 38, 3–59.

Bickel, P.J., Fan, J. (1996), *Some problems on the estimation of unimodal densities*, Statist. Sinica 6, 23–45.

Bickel, P.J., Rosenblatt, M. (1973), *On some global measures of the deviations of density function estimates*, Ann. Statist. 1, 1071–1095.

Billingsley, P. (1968), *Convergence of probability measures*, John Wiley and Sons, New York.

Birgé, L. (1987a), *Estimating a density under restrictions: nonasymptotic minimax risk*, Ann. Statist. 15, 995–1012.

Birgé, L. (1987b), *On the risk of histograms for estimating decreasing densities*, Ann. Statist. 15, 1113–1022.

Birgé, L. (1989), *The Grenander estimator: a nonasymptotic approach*, Ann. Statist. 17, 1532–1549.

Birgé, L. (1997), *Estimation of unimodal densities without smoothness assumptions*, Ann. Statist. 25, 970–981.

Boggs, P.T., Tolle, J.W. (1995), *Sequential quadratic programming*, Acta Numerica 1995, Cambridge University Press, Cambridge, pp. 1–51.

Bowman, A.W. (1984), *An alternative method of cross validation for the smoothing of density estimates*, Biometrika 71, 353–360.

Breiman, L., Meisel, W., Purcell, E. (1977), *Variable kernel estimates of multivariate densities*, Technometrics 19, 135–144.

Brunk, H.D., (1965), *Conditional expectation given a σ lattice and applications*, Ann. Math. Statist. 36, 1339–1350.

Cao, R., Cuevas, A., Fraiman, R. (1995), *Minimum distance density-based estimation*, Comput. Statist. Data Anal. 20, 611–631.

Cao, R., Cuevas, A., González-Manteiga, W. (1994), *A comparative study of several smoothing methods in density estimation*, Comput. Statist. Data Anal. 17, 153–176.

Cao, R., Devroye, L. (1995), *The consistency of a smoothed minimum distance estimate*, Scand. J. Statist. 23, 405–418.

Carmichael, J.P. (1976), *The autoregressive method*, Ph.D. Dissertation, State University of New York at Buffalo, Buffalo, New York.

Censor, Y., Zenios, S. (1997), *Parallel optimization algorithms*, Oxford University Press, Oxford.

Cheng, R.C.H., Amin, N.A.K. (1983), *Estimating parameters in continuous univariate distributions with a shifted origin*, J. R. Statist. Soc. B 45, 294–403.

Chung, K.L. (1949), *An estimate concerning the Kolmogorov limit distribution*, Trans. Amer. Math. Soc. 67, 36–50.

Collins, J.C. (1997), *Functional estimation: the asymptotic regression approach*, Ph.D. Dissertation, University of Delaware.

Courant, R., Hilbert, D. (1953), *Methods of mathematical physics. Volume I*, Wiley-Interscience, New York.

Cox, D.D., O'Sullivan, F. (1990), *Asymptotic analysis of penalized likelihood and related estimators*, Ann. Statist. 18, 1676–1695.

Crain, B.R. (1976a), *Exponential models, maximum likelihood estimation, and the Haar condition*, J. Amer. Statist. Assoc. 71, 737–740.

Crain, B.R. (1976b), *More on estimation of distributions using orthogonal expansions*, J. Amer. Statist. Assoc. 71, 741–745.

Cramér, H. (1946), *Mathematical methods of statistics*, Princeton University Press, Princeton.

Dacorogna, B. (1989), *Direct methods in the calculus of variations*, Springer-Verlag, New York.

Davies, O.L., Goldsmith, P.L. (1974), *Statiscal methods in research and production, Fourth edition*, Longman, New York.

de Acosta, A. (1981), *Inequalities for B-valued random vectors with applications to the strong law of large numbers*, Ann. Probab. 9, 157–161.

Deheuvels, P. (1977), *Estimation nonparamétric de la densité par histogrammes generalisés*, Rev. Statist. Appl. 25, 5–42.

Deheuvels, P. (2000a), *Uniform limit laws for kernel density estimators on possibly unbounded intervals*, Recent advances in reliability theory: methodology, practice and inference (N. Limnios and M. Nikulin, eds.), Birkhäuser, Boston, pp. 477–492.

Deheuvels, P. (2000b), *Limit laws for kernel density estimators for kernels with unbounded supports*, Asymptotics in statistics and probability. Papers in Honor of George Gregory Roussas (M.L. Puri, ed.), VSP International Science Publishers, Zeist.

Deheuvels, P., Hominal, P. (1980), *Estimation automatique de la densité*, Rev. Statist. Appl. 28, 25–55.

Deheuvels, P., Mason, D.M. (1992), *Functional laws of the iterated logarithm for the increments of empirical and quantile processes*, Ann. Probab. 20, 1246–1287.

de Montricher, G.F., Tapia, R.A., Thompson, J.R. (1975), *Nonparametric maximum likelihood estimation of probability densities by penalty function methods*, Ann. Statist. 3, 1329–1348.

Dempster, A.P., Laird, N.M., Rubin, D.B. (1977), *Maximum likelihood from incomplete data via the EM algorithm (with discussion)*, J. R. Statist. Soc. B 39, 1–38.

Devroye, L. (1983), *The quivalence of weak, strong, and complete convergence in L_1 for kernel density estimates*, Ann. Statist. 11, 896–904.

Devroye, L. (1986), *Non-uniform random variate generation*, Springer-Verlag, New York.

Devroye, L. (1987), *A course in density estimation*, Birkäuser, Boston.

Devroye, L. (1988), *The kernel estimate is relatively stable*, Probab. Theory Rel. Fields 77, 521-536.

Devroye, L. (1989), *The double kernel method in density estimation*, Ann. Inst. H. Poincaré Probab. Statist. 25, 533–580.

Devroye, L. (1991), *Exponential inequalities in nonparametric estimation*, Nonparametric functional estimation and related topics (G. Roussas, ed.), Kluwer, Dordrecht, pp. 31–44.

Devroye, L. (1997), *Universal smoothing factor selection in density estimation: theory and practice*, Test 6, 223–320.

Devroye, L., Györfi, L. (1985), *Density estimation: the L_1-view*, John Wiley and Sons, New York.

Devroye, L., Lugosi, G. (1996), *A universally acceptable smoothing factor for kernel density estimates*, Ann. Statist. 24, 2499–2512.

Devroye, L., Lugosi, G. (1997), *Nonasymptotic universal smoothing factors, kernel complexity and Yatracos classes*, Ann. Statist. 25, 2626–2637.

Devroye, L., Lugosi, G. (2000), *Variable kernel estimates: on the impossibility of tuning the parameters*, High dimensional probability, II (E. Giné, D.M. Mason, J.A. Wellner, eds.), Birkhäuser, Boston, pp. 405–424.

Devroye, L., Lugosi, G., Udina, K. (1998), *Inequalities for a new data-based method for selecting nonparametric density estimates*, Econometrics Abstracts.

Dey, A.K., Ruymgaart, F.H., Mair, B.A. (1996), *Cross-validation for parameter selection in inverse estimation problems*, Scand. J. Statist. 23, 609–620.

Dharmadhikari, S., Joag-dev, K. (1988), *Unimodality, convexity, and applications*, Academic Press, New York.

Diestel, J. (1984), *Sequences and series in Banach spaces*, Springer-Verlag, New York.

Dumonceaux, R., Antle, C.E. (1973), *Discrimination between the lognormal and the Weibull distributions*, Technometrics 15, 923–926.

Dym, H., McKean, H.P. (1972), *Fourier series and integrals*, Academic Press, New York.

Eastham, J.F., LaRiccia, V.N., Schuenemeyer, J.H. (1987), *Small sample properties of maximum likelihood estimators for an alternative parameterization of the three-parameter lognormal distribution*, Comm. Statist. Simul. 16, 871–884.

Eddy, W.F. (1980), *Optimal kernel estimators of the mode*, Ann. Statist. 8, 870–882.

Eggermont, P.P.B., LaRiccia, V.N. (1996), *A simple and effective bandwidth selector for kernel density estimation*, Scand. J. Statist. 23, 285–301.

Eggermont, P.P.B., LaRiccia, V.N. (1997), *Nonlinearly smoothed EM density estimation with automated smoothing parameter selection for nonparametric deconvolution problems*, J. Amer. Statist. Assoc. 92, 1451–1458.

Eggermont, P.P.B., LaRiccia, V.N. (1999a), *Best asymptotical normality of the kernel density entropy estimator for smooth densities*, IEEE Trans. Information Theory 45, 1321–1326.

Eggermont, P.P.B., LaRiccia, V.N. (1999b), *Optimal convergence rates for Good's nonparametric maximum likelihood density estimator*, Ann. Statist. 28, 1600–1615.

Eggermont, P.P.B., LaRiccia, V.N. (2000), *Maximum smoothed likelihood estimation of monotone and unimodal densities*, Ann. Statist. 29, 922–947.

Epanechnikov, V.A. (1969), *Nonparametric estimates of a multivariate probability density*, Theory Probab. Appl. 14, 153–158.

Eubank, R.L. (1999), *Spline smoothing and nonparametric regression, Second edition*, Marcel Dekker, New York.

Eubank, R.L., LaRiccia, V.N., Schuenemeyer, J.H. (1995), *Component type tests with estimated parameters*, Prob. Math. Statist. 15, 275–289.

Fernholz, L.T. (1983), *Von Mises calculus for statistical functionals*, Lecture Notes in Statistics 19, Springer-Verlag, New York.

Fiacco, A.V., McCormick, G.P. (1968), *Nonlinear programming; sequential unconstrained minimization techniques*, John Wiley and Sons, New York (Reprinted: SIAM, Philadelphia, 1990).

Fisher, R.A. (1922), *On the mathematical foundations of theoretical statistics*, Philos. Trans. Royal. Soc. London A 222, 309–368.

Fougères, A.-L. (1997), *Estimation de densités unimodales*, Canad. J. Statist. 25, 375–387.

Gajek, L. (1987), *Estimation of a density function and its derivatives by the minimum distance method*, Sci. Bull. Łódź Tech. Univer. 533, Thesis 103 (Polish).

Geyer, C.J. (1994), *On the asymptotics of constrained M-estimators*, Ann. Statist. 22, 1993–2010.

Giné, E., Guillou, A. (2000), *Rates of strong uniform consistency for multivariate kernel density estimators*, Tech. Report 00–39 (October 2000), Department of Statistics, University of Connecticut.

Giné, E., Mason, D.M, Zaitsev, A. (2001), *The L_1-norm density estimator process*, Manuscript (January, 2001).

Glick, N. (1978), *Breaking records and breaking boards*, Amer. Math. Monthly 85, 2–26.

Goldenshluger, A. (2000), *On pointwise adaptive nonparametric deconvolution*, Manuscript, Department of Statistics, University of Haifa.

Good, I.J. (1971), *A nonparametric roughness penalty for probability densities*, Nature 229, 29–30.

Good, I.J., Gaskins, R.A. (1971), *Nonparametric roughness penalties for probability densities*, Biometrika 58, 255-277.

Gorenflo, R., Vessella, S. (1991), *Abel integral equations*, Lecture Notes in Mathematics 1461, Springer-Verlag, New York.

Gosh, K., Rao Jammalamadaka, S. (2001), *A general estimation method using spacings*, J. Statist. Plann. Inference 93, 71–82.

Grenander, U. (1956), *On the theory of mortality measurements. Part II*, Skand. Akt. 39, 125–153.

Grenander, U. (1981), *Abstract inference*, John Wiley and Sons, New York.

Groeneboom, P. (1985), *Estimating a monotone density*, Proc. Berkeley Conf. in honor of Jerzy Neyman and Jack Kiefer, Volume II (L.M. Le Cam, R.A. Olshen, eds.), Wadsworth, Belmont, CA, pp. 539–555.

Groeneboom, P., Hooghiemstra, G., Lopuhaä, H.P. (1999), *Asymptotic normality of the L_1 error of the Grenander estimator*, Ann. Statist. 28, 1316–1347.

Groeneboom, P., Jongbloed, G., Wellner, J.A. (2000), *Estimation of a convex function: characterizations and asymptotic theory*, Manuscript.

Groeneboom, P., Wellner, J.A. (1992), *Information bounds and nonparametric maximum likelihood estimation*, Birkhäuser, Basel.

Grund, B., Hall, P. (1995), *On the minimisation of L^p error in mode estimation*, Ann. Statist. 23, 2264–2284.

Gu, C. (1993), *Smoothing spline density estimation: a dimensionless automatic algorithm*, J. Amer. Statist. Assoc. 88, 495–503.

Gu, C., Qiu, C. (1993), *Smoothing spline density estimation: theory*, Ann. Statist. 21, 217–234.

Györfi, L., van der Meulen, E.C. (1987), *Density-free convergence properties of various estimators of the entropy*, Comput. Statist. Data Anal. 5, 425–436.

Györfi, L., van der Meulen, E.C. (1990), *An entropy estimate based on a kernel density estimation*, Colloq. Math. Soc. J. Bolyai, 57: Limit Theorems in Probability and Statistics (I. Berkes, E. Csáki, P. Révész, eds.), North-Holland, Amsterdam.

Hall, P. (1983), *Large sample optimality of least squares cross validation in density estimation*, Ann. Statist. 11, 1156–1174.

Hall, P., Johnstone, I. (1992), *Empirical functionals and efficient smoothing parameter selection (with discussion)*, J. R. Statist. Soc. B 54, 475–530.

Hall, P., Marron, J.S., Park, B.U. (1992), *Smoothed cross validation*, Probab. Theory Related Fields 92, 1–20.

Hall, P., Morton, S.C. (1993), *On the estimation of entropy*, Ann. Inst. Statist. Math. 45, 69–88.

Hall, P., Wand, M.P. (1988), *Minimizing L_1 distance in nonparametric density estimation*, J. Multivariate Anal. 26, 59–88.

Hall, P., Wehrly, T.E. (1991), *A geometrical method for removing edge effects from kernel-type nonparametric regression estimators*, J. Amer. Statist. Assoc. 86, 665–672.

Hardy, G.H., Littlewood, G.E. (1928), *Some properties of fractional integrals*, Math. Zeitschrift. 27, 565-606.

Hardy, G.H., Littlewood, G.E., Polya, G. (1951), *Inequalities*, Cambridge University Press, Cambridge.

Hartley, H.O. (1958), *Maximum likelihood estimation from incomplete data*, Biometrics 14, 174–194.

Hartley, H.O., Hocking, R.R. (1971), *The analysis of incomplete data*, Biometrics 27, 783–808.

Healy, M., Westmacott, M. (1956), *Missing values in experiments analyzed on automatic computers*, Appl. Statist. 5, 203–206.

Hille, E. (1972), *Methods in classical and functional analysis*, Addison-Wesley, Reading.

Hiriart-Urruty, J.-B., Lemaréchal, C. (1993), *Convex analysis and minimization algorithms, 2 Volumes*, Springer-Verlag, New York.

Hirose, H. (1995), *Maximum likelihood parameter estimation in the three parameter Gamma distribution*, Comput. Statist. Data Anal. 20, 343–354.

Hodges, J.L., Lehmann, E.L. (1956), *The efficiency of some nonparametric competitors of the t-test*, Ann. Math. Statist. 27, 324–335.

Hoeffding W. (1963), *Probability inequalities for sums of bounded random variables*, J. Amer. Statist. Assoc. 58, 13–30.

Hoerl, A.E., Kennard R.W. (1970), *Ridge regression: biased estimation for non-orthogonal problems*, Technometrics 12, 55–67.

Holmes, R.B. (1975), *Geometric functional analysis and its applications*, Springer-Verlag, New York.

Huber, P. (1967), *The behavior of maximum likelihood estimates under nonstandard conditions*, Proc. Fifth Berkeley Symp. Math. Statist. Probab. 1, University of California Press, Berkeley, 221–233.

Huber, P. (1977), *Robust statistical procedures*, SIAM, Philadelphia.

Huber, P. (1981), *Robust statistics*, John Wiley and Sons, New York.

Ibragimov, I.A. (1956), *On the composition of unimodal distributions*, Theory Probab. Appl. 1, 255–260.

Ibragimov, I.A., Has'minskii, R.Z. (1981), *Statistical estimation, asymptotic theory*, Springer-Verlag, New York.

Ibragimov, I.A., Has'minskii, R.Z. (1982), *An estimate of the density of a distribution belonging to a class of entire functions*, Theory Probab. Appl. 27, 551–562.

Joe, H. (1989), *Estimation of entropy and other functionals of a multivariate density*, Ann. Inst. Statist. Math. 41, 683–697.

Johnson, N.L., Kotz, S., Balakrishnan, N. (1994), *Continuous univariate distributions, 2 Volumes, Second edition*, John Wiley and Sons, New York.

Jones, M.C. (1993), *Simple boundary correction for kernel density estimation*, Statist. Comput. 3, 135–146.

Kaluszka, M. (1998), *On the Devroye-Györfi methods of correcting density estimators*, Statist. Probab. Lett. 37, 249–257.

Katznelson, Y. (1968), *An introduction to harmonic analysis*, John Wiley and Sons, New York (Reprinted: Dover, New York, 1976).

Kemperman, J.H.B. (1967), *On the optimum rate of transmitting information*, Springer Lecture Notes in Mathematics 23, 126–169.

Klonias, V.K. (1982), *Consistency of two nonparametric maximum penalized likelihood estimators of the probability density function*, Ann. Statist. 10, 811–824.

Klonias, V.K. (1984), *On a class of nonparametric density and regression estimators*, Ann. Statist. 12, 1263–1284.

Klonias, V.K. (1991), *On the influence function of maximum penalized likelihood density estimators*, Nonparametric functional estimation and related topics (G. Roussas, ed.), Kluwer, Dordrecht, pp. 125–131.

Kooperberg, C., Stone, C.J. (1991), *A study of log-spline density estimation*, Comput. Statist. Data Anal. 12, 327–347.

Kruskal, J.B. (1964), *Multidimensional scaling by optimizing goodness of fit to a nonmetric hypothesis*, Psychometrika 29, 1–27.

Kübler, H. (1979), *On the fitting of the three-parameter distributions - lognormal, Gamma, and Weibull*, Statist. Hefte 20, 68–125.

Landsman, Z. (2000), *On the minimum of the Fisher information about the scale parameter and the singular Sturm-Liouville problem*, J. Statist. Plann. Inference 88, 29–35.

Lawson, C.L., Hanson, R.J. (1995), *Solving least squares problems*, Prentice-Hall, New York (Reprinted: SIAM, Philadelphia, 1995).

Le Cam, L. (1970), *On the assumptions used to prove asymptotic normality of maximum likelihood estimates*, Ann. Math. Statist. 41, 802–828.

Lehmann, E.L. (1983), *Theory of point estimation*, John Wiley and Sons, New York.

Lekkerkerker, C.G. (1953), *A property of log-concave functions*, Indag. Math. 15, 505–521.

Levit, B.Ya. (1978), *Asymptotically efficient estimation of nonlinear functionals*, Problems Inform. Transmission 14, 204–209.

Li, W., Swettits, J.J. (1998), *The linear ℓ_1 estimator and the Huber M-estimator*, SIAM J. Optimiz. 8, 457–475.

Lieb, E.H., Loss, M. (1996), *Analysis*, AMS, Providence.

Lockhart, R.A., Stephens, M.A. (1994), *Estimation and tests of fit for the three-parameter Weibull distribution*, J. R. Statist. Soc. B 56, 491–500.

Luenberger, D.G. (1968), *Optimization by vector psace methods*, John Wiley and Sons, New York.

Madsen, K., Nielsen, H.B. (1993), *A finite smoothing algorithm for linear ℓ_1 estimation*, SIAM J. Optimiz. 3, 223–235.

Mammen, E. (1991), *Estimating a smooth monotone regression function*, Ann. Statist. 19, 724–740.

Mangasarian, O.L. (1969), *Nonlinear programming*, McGraw-Hill, New York (Reprinted: SIAM, Philadelphia, 1994).

Marron, J.S., Wand, M.P. (1989), *Exact mean integrated squared error*, J. Amer. Statist. Assoc. 20, 712–736.

Marshall, A.W. (1970), *Discussion on Barlow and van Zwet's paper*, Nonparametric techniques in statistical inference (M.L. Puri, ed.), Cambridge University Press, Cambridge, pp. 175–176.

Mason, D.M. (2000), *Private communication.*

Maz'ja, V.G. (1985), *Sobolev spaces*, Springer-Verlag, Berlin.

McDiarmid, C. (1989), *On the method of bounded differences*, Surveys in combinatorics 1989, Cambridge University Press, Cambridge, pp. 148–188.

McLachlan, G.J., Basford, N. (1988), *Mixture models*, Marcel Dekker, New York.

Milman, V., Schechtman, G. (1986), *Asymptotic theory of finite dimensional normed spaces*, Lecture Notes in Mathematics 1200, Springer-Verlag, New York.

Mokkadem, A. (1989), *Estimation of entropy and information of absolutely continuous random variables*, IEEE Trans. Information Theory 35, 193–196.

Morozov, V.A. (1966), *On the solution of functional equations by the method of regularization*, Soviet Math. Dokl. 7, 414–417.

Müller, H.-G. (1993), *On the boundary kernel method for nonparametric curve estimation near endpoints*, Scand. J. Statist. 20, 313–328.

Munroe, A.H., Wixley, R.A.J. (1970), *Estimators based on order statistics of small samples from a three-parameter log-normal distribution*, J. Amer. Statist. Assoc. 65, 212–225.

Nair, V.N., Wang, P.C.C. (1989), *Maximum likelihood estimation under a successive sampling discovery model*, Technometrics 31, 423–436.

Nashed, M.Z. (1966), *Some remarks on variations and differentials*, Amer. Math. Monthly 73, 63–76.

Nashed, M.Z. (1971), *Differentiability and related properties of nonlinear operators*, Nonlinear functional analysis and applications (L.B. Rall, ed.), Academic Press, New York, pp. 103–309.

Neveu, J. (1975), *Discrte parameter martingales*, North-Holland, Amsterdam.

O'Sullivan, F. (1988), *Fast computation of fully automated log-density and log-hazard estimators*, SIAM J. Sci. Statist. Comput. 9, 363–379.

Park, B.U., Turlach, B.A. (1992), *Practical performance of several data driven bandwidth selectors (with discussion)*, Comput. Statist. 7, 251–270.

Parzen, E. (1962), *On estimation of a probability density function and mode*, Ann. Math. Statist. 33, 1065–1076.

Parzen, E. (1967), *Time series analysis papers*, Holden-Day, San Francisco.

Parzen, E. (1979), *Nonparametric statistical data modeling (with discussion)*, J. Amer. Statist. Assoc. 74, 105–131.

Phelps, R.R. (1989), *Convex functions, monotone operators and differentiability*, Springer Lecture Notes in Mathematics 1364, Springer-Verlag, New York.

Pike, M.C. (1966), *A method of analysis of a certain class of experiments in carcinogenesis*, Biometrics 22, 142–161.

Pinelis, I.F. (1990), *Inequalities for sums of independent random vectors and their application to estimating a density*, Theory Probab. Appl. 35, 605–607.

Pinelis, I.F. (1994), *On a majorization inequality for sums of independent random variables*, Probab. Statist. Lett. 19, 97–99.

Prakasa Rao, B.L.S (1983), *Nonpurumetric functional estimation*, Academic Press, New York.

Prenter, P.M. (1975), *Splines and variational methods*, Wiley-Interscience, New York.

Pyke, R. (1965), *Spacings*, J. R. Statist. Soc. B 27, 395–449.

Rachev, S.T. (1991), *Probability metrics and the stability of stochastic models*, John Wiley and Sons, New York.

Ranneby, B. (1984), *The maximum spacings method. An estimation method related to the maximum likelihood method*, Scand. J. Statist. 11, 93–112.

Rao, C.R. (1973), *Linear statistical inference and its applications*, John Wiley and Sons, New York.

Rao, C.R., Mitra, S.K. (1971), *Generalized inverse of matrices and its applications*, John Wiley and Sons, New York.

Redner, R.A., Walker, H.F. (1984), *Mixture densities, maximum likelihood and the EM algorithm*, SIAM Rev. 26, 195–239.

Reinsch, Ch. (1967), *Smoothing by spline functions*, Numer. Math. 10, 177–183.

Rieder, H. (1994), *Robust asymptotic statistics*, Springer-Verlag, New York.

Riesz, F., Sz-Nagy, B. (1955), *Functional analysis*, Ungar, New York (Reprinted: Dover, New York, 1996).

Robbins, H. (1956), *An empirical Bayes approach to statistics*, Proc. Third Berkeley Symp. Math. Statist. Probab. 1, University of California Press, Berkeley, pp. 157–164.

Robertson, T., Wright, F.T. (1974), *A norm reducing property for isotonized Cauchy mean value functions*, Ann. Statist. 2, 1302–1307.

Rockafellar, R.T. (1970), *Convex analysis*, Princeton Univeristy Press, Princeton.

Rockette, H., Antle, C., Klimko, L.A. (1974), *Maximum likelihood estimation with the Weibull model*, J. Amer. Statist. Assoc. 69, 246–249.

Roeder, K. (1992), *Semiparametric estimation of normal mixture densities*, Ann. Statist. 21, 929–943.

Rosenblatt, M. (1956), *Remarks on some nonparametric estimates of a density function*, Ann. Math. Statist. 27, 832–835.

Rousseeuw, P.J., Leroy, A.M. (1987), *Robust regression and outlier detection*, John Wiley and Sons, New York.

Royden, H.L. (1968), *Real analysis*, MacMillan, New York.

Rudemo, M. (1982), *Empirical choice of histograms and kernel density estimators*, Scand. J. Statist. 9, 65–78.

Ruud, P.A. (1991), *Extensions of estimation methods using the EM algorithm*, J. Econometrics 49, 305–341.

Sansone, G. (1991), *Orthogoanl functions*, John Wiley and Sons, New York (Reprinted: Dover, New York, 1991).

Schuenemeyer, J.H., Drew, J. (1991), *A forecast of undiscovered oil and gas in the Frio strand plain trend: the unfolding of a very large exploration play*, Amer. Assoc. Petrol. Geol. Bull. 75, 1107–1115.

Scott, D.W., Terrell, G.R. (1987), *Biased and unbiased cross validation in density estimation*, J. Amer. Statist. Assoc. 82, 1131–1146.

Seber, G.A.F. (1977), *Linear regression analysis*, John Wiley and Sons, New York.

Shapiro, H.S. (1969), *Smoothing and approximation of functions*, Van Nostrand Reinhold, New York.

Sheather, S.J. (1992), *The performance of six popular bandwidth selection methods on some real data sets*, Comput. Statist. 7, 225–250.

Sheather, S.J., Jones, M.C. (1991), *A reliable data-based bandwidth selection method for kernel density estimation*, J. R. Statist. Soc. B 53, 683–690.

Shepp, L.A., Vardi, Y. (1982), *Maximum likelihood reconstruction in emission tomography*, IEEE Trans. Medical Imaging 1, 113–122.

Shergin, V.V. (1979), *On the rate of convergence in the central limit theorem for m-dependent random vectors*, Theory Probab. Appl. 24, 782–796.

Shorack, G.R., Wellner, J.A. (1986), *Empirical processes with applications to statistics*, John Wiley and Sons, New York.

Silverman, B.W. (1978), *Weak and strong uniform consistency of the kernel estimate of a density and its derivatives*, Ann. Statist. 6, 177-184.

Silverman, B.W. (1982), *On the estimation of a probability density function by the maximum penalized likelihood method*, Ann. Statist. 10, 795–810.

Silverman, B.W. (1986), *Density estimation for statistics and data analysis*, Chapman and Hall, London.

Silverman, B.W., Jones, M.C., Wilson, J.D., Nychka, D.W. (1990), *A smoothed EM algorithm approach to indirect estimation problems, with particular reference to stereology and emission tomography (with discussion)*, J. R. Statist. Soc. B 52, 271–324.

Simpson, J., Olsen, A., Eden, J.C. (1975), *A Bayesian analysis of a multiplicative treatment effect in weather modification*, Technometrics 17, 161–166.

Sobczyk, K., Spencer, B.F. (1992), *Random fatigue*, Academic Press, New York.

Stefanski, L.A., Carroll, R.J. (1990), *Deconvoluting kernel density estimators*, Statistics 21, 169–184.

Stone, C.J. (1984), *An asymptotically optimal window selection rule for kernel density estimates*, Ann. Statist. 12, 1285–1297.

Stone, C.J. (1990), *Large-sample inference for log-spline models*, Ann. Statist. 18, 717–741.

Stone, C.J., Koo, C.-Y. (1986), *Log-spline density estimation. Function estimates*, Contemp. Math. 59, AMS, Providence, pp. 1–15.

Stuetzle, W., Mittal, Y. (1979), *Some comments on the asymptotic behavior of robust smoothers*, Proc. Heidelberg Workshop (T. Gasser, M. Rosenblatt, eds.), Springer-Verlag, Heidelberg, pp. 191–195.

Sweeting, T.J. (1977), *Speed of convergence in the multidimensional central limit theorem*, Ann. Probab. 5, 28–41.

Teichrow, D. (1957), *The mixture of normal distributions with different variances*, Ann. Math. Statist. 28, 510–511.

Thompson, J.R., Tapia, G.F. (1990), *Nonparametric function estimation, modeling, and simulation*, SIAM, Philadelphia.

Titchmarsh, E.C. (1975), *Introduction to the theory of Fourier integrals*, Clarendon Press, Oxford.

Titterington, D.M., Smith, A.F.M., Makov, U.E. (1985), *Statistical analysis of finite mixture distributions*, John Wiley and Sons, New York.

Troutman, J.L. (1983), *Variational calculus with elementary convexity*, Springer-Verlag, New York.

Tsybakov, A.B., van der Meulen, E.C. (1996), *Root-n consistent estimators of entropy for densities with unbounded support*, Scand. J. Statist. 23, 75–83

van de Geer, S. (1993), *Hellinger-consistency of certain nonparametric maximum likelihood estimators*, Ann. Statist. 21, 14–44.

van de Geer, S. (2000), *Empirical processes in M-estimation*, Cambridge University Press, Cambridge.

van Es, B. (1992), *Estimating functionals related to a density by a class of statistics based on spacings*, Scand. J. Statist. 19, 61–72.

Wahba, G. (1981), *Data-based optimal smoothing of orthogonal series density estimates*, Ann. Statist. 9, 146–156.

Wahba, G. (1990), *Spline models for observational data*, SIAM, Philadelphia.

Wald, A. (1949), *Note on the consistency of the maximum likelihood estimate*, Ann. Math. Statist. 20, 591–601.

Walter, G., Blum, J. (1984), *A simple solution to a nonparametric maximum likelihood estimation problem*, Ann. Statist. 12, 372–379.

Wand, M.P., Jones, M.C. (1995), *Kernel smoothing*, Chapman and Hall, London.

Wang, Y. (1995), *The L_1 theory of estimation of monotone and unimodal densities*, J. Nonparam. Statist. 4, 249–261.

Watson, G.N. (1944), *Theory of Bessel functions*, Cambridge University Press, Cambridge.

Watson, G.S., Leadbetter, M.R. (1963), *On the estimation of the probability density*, Ann. Math. Statist. 34, 480–491.

Weyl, H. (1917), *Bemerkungen zum Begriff des Differentialquotienten gebrochener Ordnung*, Vierteljahreszeitschrift d. Naturf. Gesellsch. Zürich 62, 296–302.

Wegman, E.J. (1970), *Maximum likelihood estimation of a unimodal density, II*, Ann. Math. Statist. 6, 2169–2174.

Weisberg, S. (1985), *Applied linear regression, Second edition*, John Wiley and Sons, New York.

Wheeden, R.L., Zygmund, A. (1977), *Measure and integral*, Marcel Dekker, New York.

Whittle, P. (1958), *On the smoothing of a probability density function*, J. R. Statist. Soc. B 20, 334–343.

Wiener, N. (1930), *Generalized harmonic analysis*, Acta Math. 55, 117–258.

Williams, D. (1991), *Probability with martingales*, Cambridge University Press, Cambridge.

Wolfowitz, J. (1949), *On Wald's proof of the consistency of the maximum likelihood estimate*, Ann. Math. Statist. 20, 601–602.

Woodroofe, M. (1970), *On choosing delta sequences*, Ann. Math. Statist. 41, 1665–1671.

Woodroofe, M., Sun, J. (1993), *A penalized maximum likelihood estimate of $f(0+)$ when f is non-increasing*, Statist. Sinica 3, 501–515.

Wu, C.F.J. (1983), *On the convergence properties of the EM algorithm*, Ann. Statist. 11, 95–103.

Yatracos, Y.G. (1985), *Rates of convergence of minimum distance estimators and Kolmogorov's entropy*, Ann. Statist. 13, 768–774.

Yatracos, Y.G. (1998), *A small sample optimality property of the M.L.E.*, Sankhyā Ser. A 60, 90–101.

Zhang, J., Fan, J. (2000), *Minimax kernels for nonparametric curve estimation*, J. Nonparam. Statist. 12, 417–445.

Author Index

Subject Index

Springer Series in Statistics *(continued from p. ii)*